Lecture Notes in Computer Science 12632

Services Science

Subline of Lectures Notes in Computer Science

More information about this subseries at http://www.springer.com/series/7408

Hakim Hacid · Fatma Outay ·
Hye-young Paik · Amira Alloum ·
Marinella Petrocchi · Mohamed Reda Bouadjenek ·
Amin Beheshti · Xumin Liu ·
Abderrahmane Maaradji (Eds.)

Service-Oriented Computing – ICSOC 2020 Workshops

AIOps, CFTIC, STRAPS, AI-PA, AI-IOTS, and Satellite Events
Dubai, United Arab Emirates, December 14–17, 2020
Proceedings

 Springer

Editors
Hakim Hacid (iD)
Zayed University
Dubai, United Arab Emirates

Fatma Outay (iD)
Zayed University
Dubai, United Arab Emirates

Hye-young Paik (iD)
University of New South Wales
Sydney, NSW, Australia

Amira Alloum
Huawei Paris Research Center
Paris, France

Marinella Petrocchi (iD)
National Research Council C.N.R.
Pisa, Italy

Mohamed Reda Bouadjenek (iD)
Deakin University
Waurn Ponds, VIC, Australia

Amin Beheshti (iD)
Macquarie University
Sydney, NSW, Australia

Xumin Liu
Rochester Institute of Technology
Rochester, NY, USA

Abderrahmane Maaradji (iD)
Université de Paris
Paris, France

ISSN 0302-9743   ISSN 1611-3349   (electronic)
Lecture Notes in Computer Science
ISBN 978-3-030-76351-0   ISBN 978-3-030-76352-7   (eBook)
https://doi.org/10.1007/978-3-030-76352-7

LNCS Sublibrary: SL2 – Programming and Software Engineering

This Springer imprint is published by the registered company Springer Nature Switzerland AG
The registered company address is: Gewerbestrasse 11, 6330 Cham, Switzerland

Preface

This volume presents the proceedings of the scientific satellite events that were held in conjunction with the 18th International Conference on Service-Oriented Computing (ICSOC 2020), held virtually during December 14–17, 2020. The satellite events provide an engaging space for specialist groups to meet, generating focused discussions on specific sub-areas within service-oriented computing, which contributes to ICSOC community-building. These events significantly helped enrich the main conference by both expanding the scope of research topics and attracting participants from a wider community.

As is customary for ICSOC, this year, these satellite events were organized around three main tracks, comprising a workshop track, a PhD symposium track, and a demonstration track.

The ICSOC 2020 workshop track consisted of the following five workshops covering a wide range of topics that fall into the general area of service computing.

- International Workshop on Artificial Intelligence for IT Operations (AIOps 2020)
- International Workshop on Cyber Forensics and Threat Investigations Challenges in Emerging Infrastructures (CFTIC 2020)
- 2nd Workshop on Smart Data Integration and Processing (STRAPS 2020)
- International Workshop on AI-enabled Process Automation (AI-PA 2020)
- International Workshop on Artificial Intelligence in the IoT Security Services (AI-IOTS 2020)

This year in the workshop track, the theme of artificial intelligence and its applications in service computing was particularly noticeable. All papers accepted to the workshops were peer reviewed and overall acceptance rates for the workshops were less than 50%. The workshops were held on December 14, 2020, and included keynote talks from prominent speakers from industry and academia.

The PhD symposium is an international forum for PhD students to present, share, and discuss their research in a constructive and critical atmosphere. It also provides students with fruitful feedback and advice on their research approach and thesis. The PhD symposium track was held over a half-day session and included nine accepted papers. This year, and due to COVID-19, the conference supported all PhD students and their participation was fully free of charge.

The demonstration track offers an exciting and highly interactive way to show research prototypes/work in service-oriented computing and related areas. The demonstration track was held over a two-hour session for the presentations and then dedicated sessions for real-time demonstrations, running in parallel. Four demonstrations were accepted and presented during the conference.

We would like to thank the workshops, PhD symposium, and demonstration track authors, as well as the Organizing Committees, who together contributed to these important events of the conference. We hope that these proceedings will serve as a valuable reference for researchers and practitioners working in the service-oriented computing domain and its emerging applications.

Hakim Hacid
Fatma Outay
Hye-young Paik
Amira Alloum
Marinella Petrocchi
Mohamed Reda Bouadjenek
Amin Beheshti
Xumin Liu
Abderrahmane Maaradji

Organization

Workshop Chairs

Fatma Outay — Zayed University, UAE
Hye-young Paik — University of New South Wales, Australia
Amira Alloum — Huawei, France

Demonstration Chairs

Amin Beheshti — Macquarie University, Australia
Xumin Liu — Rochester Institute of Technology, USA
Abderrahmane Maaradji — Université de Paris, France

PhD Symposium Chairs

Marinella Petrocchi — Institute of Informatics and Telematics, Italy
Mohamad Badra — Zayed University, UAE

Finance Chair

Bernd J. Krämer — Fern University, Germany

Publication Chair

Hakim Hacid — Zayed University, UAE

Publicity Chairs

Noura Faci — Claude Bernard Lyon 1 University, France
Guilherme Horta Travassos — Federal University of Rio de Janeiro, Brazil
Hai Dong — RMIT, Australia

Web Chairs

Emir Ugljanin — State University of Novi Pazar, Serbia
Emerson Bautista — Zayed University, UAE

International Workshop on Artificial Intelligence for IT Operations (AIOps)

Organizers

Odej Kao   TU Berlin, Germany
Jorge Cardoso   Huawei, Germany

International Workshop on Cyber Forensics and Threat Investigations Challenges in Emerging Infrastructures (CFTIC 2020)

Organizers

John William Walker   Nottingham Trent University, UK
Ahmed Elmesiry   University of South Wales, UK

2nd Workshop on Smart Data Integration and Processing (STRAPS 2020)

Organizers

Genoveva Vargas-Solar   CNRS, France
Nadia Bennani   INSA Lyon, France
Chirine Ghedira Guegan   University Lyon 3, France

International Workshop on AI-enabled Process Automation (AI-PA 2020)

Organizers

Amin Beheshti   Macquarie University, Australia
Boualem Benatallah   University of New South Wales, Australia
Ladjel Bellatreche   Poitiers University, France
Francois Charoy   Inria-Loriria, France
Hamid Motahari   EY, USA
Mohamed Adel Serhani   UAE University, UAE
Li Qing   Hong Kong Polytechnic University, Hong-Kong

International Workshop on Artificial Intelligence in the IoT Security Services (AI-IOTS 2020)

Organizers

S. Selvakumar   Indian Institute of Information Technology Una, India
R. Kanchana   SSN College of Engineering, India

Contents

Artificial Intelligence for IT Operations (AIOPS 2020)

AI-Enabled Process Automation (AI-PA 2020)

Artificial Intelligence in the IoT Security Services (AI-IOTS 2020)

**Cyber Forensics and Threat Investigations Challenges in Emerging
Infrastructures (CFTIC 2020)**

Ph.D Symposium

Staking Assets Management on Blockchains: Vision and Roadmap

Stefan Driessen(✉)

Jheronimus Academy of Data Science, Tilburg University,
Sint Janssingel 92, 5211 DA 's-Hertogenbosch, The Netherlands
s.w.driessen@jads.nl

Abstract. This paper introduces and explores the vision wherefore stakeholders and the process of staking —that is, the idea of guaranteeing the quality of a process by risking valuable assets on their correct execution— may run both on and off a blockchain while in the context of cloud-enabled services and processes. The emerging trend behind blockchain-oriented computing and the reliance on stakeholders therein make distilling and evaluating this vision a priority to deliver high-quality, sustainable services of the future. We identify key defining concepts of stakeholders and the staking process, using three very different staking scenarios as a base. Subsequently, we analyze the key challenges that these stakeholders face and propose the development of a framework that can help overcome these challenges. Finally, we give a road-map to steer systematic research stemming from the proposed vision, leveraging design science along with short-cyclic experimentation.

Keywords: Blockchain · Staking · Service monitoring · Cloud

1 Introduction

The continued rise in popularity of blockchain technology has sparked new solutions for a wide variety of service processes, such as smart contracts, that place a strong emphasis on trusted computing and transparency [1].

The trust or *trustlessness* in most of these solutions is realized by the provision of proofs, which are publicly shared on the blockchain. These proofs manifest themselves in many different forms and shapes, such as Proof-of-Work (PoW), Proof-of-Stake (PoS) or Proof-of-Authority (PoA) and it is a hot topic of debate to what extent these proofs actually realize a trusted or trustless environment, see, e.g. [5]. An emerging key actor in blockchain services is the *staker*; an actor who proves that they are invested in the quality/correctness of the service, its environment and its execution in an attempt to add trustworthiness to it. Usually, this is done by *locking* an asset (such as a cryptocurrency) on the blockchain, which will either lose value or fail to generate revenue if the service fails to live up to the promised/expected quality.

Supervisors: W.-J. Willem-Jan Heuvel (w.j.a.m.v.d.heuvel@jads.nl) and Damian Tamburri (d.a.tamburri@tue.nl).

H. Hacid et al. (Eds.): ICSOC 2020 Workshops, LNCS 12632, pp. 3–9, 2021.
https://doi.org/10.1007/978-3-030-76352-7_1

Stakers often come into play when services run partly on and partly off the blockchain; in a common scenario, a transaction may be comprised of *agreements* (including, for example, classically-defined business-level agreements and application-level agreements) that may be stored on the blockchain. The actual execution of the process, however, (e.g., delivery of a digital asset), will take place off-chain. The staker pertains to that actor in such scenarios that improves trustworthiness in the sense that it checks, verifies, or otherwise *witnesses* that, what has been promised initially, is (likely to be) actually delivered. For example, the staker can help to maintain a sufficient level of network security by replicating network assets. Roles that such a staker may play include, but are not restricted to, logging, monitoring, metering, provisioning, and compliance assurance.

Our vision in this sense is, therefore, to investigate those scenarios in which stakers attempt to add trustworthiness to a blockchain environment and to support the staker's endeavor to do so. This proposition has, so far, been neglected at best and deserves further attention.

This paper sets out to define in an abstract manner the concept of a *staker* in a staking scenario; we conclude the definition by defining the main challenges behind such a scenario. Subsequently, the paper proposes the development of a novel staking methodology and associated toolkit to support stakers in decision-making scenarios where actual staking can take place in a controllable and repeatable manner. The paper furthermore explores this approach against three prototypical blockchain-oriented orchestration service scenarios and plots a road-map for future work.

2   Background

This section discusses 3 prototypical application staking scenarios and generalizes from them into several fundamental key characteristics to underpin a generic definition of a staker as well as a rudimentary staking framework.

1. Proof-of-Stake Consensus Protocol
The most popular alternative to the Proof-of-Work (PoW) consensus protocol for blockchains is Proof-of-Stake (PoS) [4]. The PoS protocol (semi-)randomly assigns a staker, who has locked some cryptocurrency, the authority to create a new block, and update the blockchain and rewards them if the updated chain achieves consensus (i.e., is approved by other stakers) [6]. Actors sending transactions on the blockchain benefit from stakers who ensure that the blockchain is in a trustworthy state.

2. Staking in Goods and Services
Distributed marketplaces for (digital) goods and services use blockchain and smart contracts for impartial enforcement of purchasing and are becoming more and more popular[1]. In order to guarantee the quality of an (off-chain) product

---

[1] See for example Ocean: https://oceanprotocol.com/, OpenBazaar: https://www.openbazaar.org/, CanYa: https://canya.io/, BitBay: https://bitbay.market/decentralized-marketplace/.

being offered through such an (on-chain) decentralized marketplace, the marketplace can ask the stakers to stake some assets on high-quality products and reward stakers that stake in popular products. Both the seller and the buyer on the market can profit from the independent quality assurance provided by the staker in this scenario.

3. Staking in Service Monitoring
Recent initiatives have proposed moving the management of Service Level Agreements (SLAs) to smart contracts on a blockchain [3,7]. Checking that the (off-chain) service lives up to the Quality-of-Service (QoS) and non-functional requirements captured in the (on-chain) smart contract representation of the SLA depends on the correct monitoring of the service, which necessarily happens off-chain. This monitoring can be done by third parties who stake on their ability to provide independent, high-quality monitoring. These stakers take up the role of an oracle and are rewarded by the smart contract if they manage to achieve consensus on their measurements.

Based on the 3 cases introduced above, we generalize here four fundamental characteristics of the staker actor:

1. The staker provides supportive services with respect to trustworthiness to a process by *measuring/testing* or *guaranteeing* some of its quality aspects.
2. This process often runs, at least partially, off-chain and the staker is generally not the main actor in the process.
3. The staker demonstrates that they are invested in the quality of the process by backing up the accuracy of their measurements/tests/guarantee with some staked asset(s) on the blockchain (e.g., cryptocurrency).
4. The staker follows a protocol that allows it to check and be checked by other stakers.
5. The staker is rewarded by the process, depending on the value of their contribution and the size of their stake.

3   Problem Definition

As argued in Sect. 2, there is increasing recognition of the usefulness of staking as a means to leverage trustworthiness to a decentralized process. Unfortunately, the perspective of the staker in these processes has thus far been rather rudimentary and overly simplistic in nature. Typically, existing methods and tools merely assume the staker's decisions to be taken without considering how exactly this decision process takes place.

However, staking is indeed an exceedingly challenging and critical endeavor since it concerns guaranteeing the quality of processes that may reside outside the staker's sphere of control. We claim that the only type of staking already happening in practice is staking on fully transparent processes such as the PoS in Scenario 1. The challenging nature of such an actor, who has little to no control over the process, has already been recognized in the past, for example, in the context of Service-Oriented-Architecture (SOA) testing by Canfora and

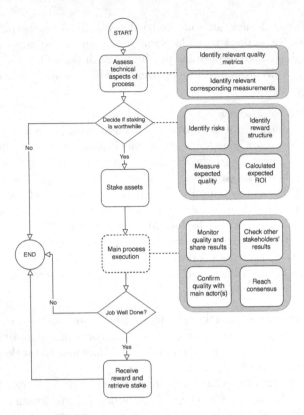

Fig. 1. The process of staking and the different steps involved in the necessary decisions of stakers. The main process execution is outside of the stakers control and is therefore not shown with a solid line around the box.

Di Penta [2]. We illustrate the stakers perspective using the generalized process of a staking process depicted in Fig. 1.

When selecting a process to stake on, the staker should start by assessing the technical aspects of the main process that affect its (promised) quality and decide how and when these should be measured. These can be either the quality of a good, such as in Scenario 2, the relevant quality metrics of the process, such as in Scenario 1, or both, such as in Scenario 3. This decision is ultimately based on the staker's expected Return-on-Investment (ROI), which in turn depends on the perceived (expected) quality, the perceived risk, and the expected reward that results from the staking process. If the staker decides to stake, they lock some of their assets on the blockchain, at which point the main process can commence. The role of the staker during the main process depends entirely on the nature of said process; it can be that the staker is not required to do anything, as in Scenario 2, or that the staker has to monitor and guarantee the quality during the main process and, achieve consensus with other stakers, such as in Scenarios

1 and 3. Either way, once the process is over, the staker gets rewarded, usually in proportion to the usefulness of their contribution and the size of their stake.

A crucial step in the flowchart in Fig. 1 is the decision of whether staking is worthwhile. The staker wants to maximize their expected ROI. This requires that the staker balances the promised reward against the risk of unintended behavior in the process, which can violate the quality that the staker is guaranteeing. In order to assess this risk, the staker has to have a good understanding of the process it is staking in from the perspective of the main actor(s), who control the process, *as well* as the staking and/or consensus protocol. By staking, the staker effectively guarantees the quality of both the main process, which they deem likely enough to be used, *and* the (un)likelihood that the process will contain unintended behavior. We have found a surprising lack of literature, theory, and tools that address this challenge and propose our own ideas on this in the next section.

The systems architect should take into account that the staker's objective (maximizing ROI) does not necessarily align well with the objectives of the other actors in the system. This, in turn, means that the service that the supportive services mentioned in Sect. 2 staker is another problem that arises from the lack of consideration for the staker's perspective. We have found that existing literature on staking in blockchain-based scenarios consistently fails to address this potential misalignment between the staker and the other actors in the system.

4   Proposed Solution

As a way to address the challenges introduced in Sect. 3, we propose the introduction of a staking framework, which can assist both aspiring stakers as well as systems architects in the process of staking and the decisions associated with this process. This framework is bound to contain at least the following supportive elements:

1. A "cookbook" which will contain standard patterns, best practices in staking scenarios and, how to deal with them;
2. A number of techniques that can be leveraged for assessing the process, the staking protocol, the alignment between the staker and, the other actors;
3. A tool suite which can be used to streamline the staking process outlined in Fig. 1 by automating (part of the) the required risk assessment and the connected orchestration machinery;

We illustrate the usefulness and expected impact of our proposed framework by describing a hypothetical staker who participates in scenario 3 following the flowchart in Fig. 1.

When designing a blockchain-based solution that involves staking, a systems architect considers the framework's standard patterns and best practices to help them select the best option for their system. Additionally, they consider whether the objectives of the staker and the other actors line up well. Before selecting a service, the staker uses our proposed framework to identify relevant patterns and best practices regarding quality metrics (e.g., QoS & non-functionals). They leverage our assessment techniques to make sure they understand which measurements allow them to monitor these quality metrics effectively. Using this knowledge and our proposed tool suite, the staker identifies scenarios where they could fail to achieve consensus (and thus not get rewarded for staking). Finally, after having confirmed their understanding of both the process they are staking in *and* the process of staking itself can they select processes that have acceptable expected ROIs.

Once the decision is made, the staker stakes their assets and participates in the monitoring process and, if consensus is achieved, it gets rewarded by the smart contract that enforces the SLA. For a more in-depth explanation of this scenario, we invite the reader to check Uriarte et al. [7].

5   Roadmap and Contributions

This paper has outlined the contours of a methodological framework for effectively supporting stakers, -and the process of staking- in a blockchain environment. The results of this paper are core research results in nature. More research is required in several main directions.

Firstly we intend to conduct one or more case studies, possibly followed by a survey. This will help us map more clearly how the proposed challenges are experienced first-hand by the actors in blockchain-based staking scenarios. It is likely this will result into a call for research based on the challenges we identify.

Secondly, we intend to apply action research to solve the problems arising from the challenges identified in this paper. This will provide a proof-of-concept for future endeavors, both in academics and in business, that aim to solve these challenges.

Finally, we hope to generalize a framework, supported by theory, that can be applied to the set of problems described in this paper and addressed in the previous step. This will both open up the door for further theory development regarding staking in blockchain scenarios, as well as directly benefit both stakers and systems architects.

References

1. Butijn, B.J., Tamburri, D.A., Heuvel, W.J.V.D.: Blockchains: a systematic multivocal literature review. arXiv preprint arXiv:1911.11770 (2019)
2. Canfora, G., Di Penta, M.: Service-oriented architectures testing: a survey. In: De Lucia, A., Ferrucci, F. (eds.) ISSSE 2006-2008. LNCS, vol. 5413, pp. 78–105. Springer, Heidelberg (2009). https://doi.org/10.1007/978-3-540-95888-8_4
3. Ocean Protocol Foundation and BigChainDB GmbH and Newton Circus: Ocean protocol: a decentralized substrate for AI data & services technical whitepaper (2019). https://oceanprotocol.com/tech-whitepaper.pdf
4. Irresberger, F., John, K., Saleh, F.: The public blockchain ecosystem: an empirical analysis. Available at SSRN (2020)

5. Mik, E.: Smart contracts: terminology, technical limitations and real world complexity. Law Innov. Technol. **9**(2), 269–300 (2017)
6. Saleh, F.: Blockchain without waste: proof-of-stake. Available at SSRN **3183935** (2020)
7. Uriarte, R.B., Zhou, H., Kritikos, K., Shi, Z., Zhao, Z., De Nicola, R.: Distributed service-level agreement management with smart contracts and blockchain. Concurrency Comput. Pract. Experience **1–17**(March) (2020). https://doi.org/10.1002/cpe.5800

Hybrid Context-Aware Method for Quality Assessment of Data Streams

Mostafa Mirzaie(✉) (iD)

Ferdowsi University of Mashhad (FUM), Mashhad, Iran
mostafa.mirzaie@mail.um.ac.ir

Abstract. Data quality is one of the most important issues that if not taken into consideration appropriately, results in the low reliability of the knowledge extracted through big data analytics. Furthermore, the challenges with data quality management are even greater with streaming data. Most of the methods introduced in the literature for processing streaming data do not use contextual information for the purpose of addressing data quality issues, however, it is possible to improve the performance of these methods by considering the contextual information, especially those obtained from the external resources. Based on this point of view, our main objective in this thesis is to propose a hybrid multivariate context-aware approach for data quality assessment in streaming environments, such as smart city applications.

Keywords: Data quality assessment · Context awareness · Streaming data

1 Problem Statement and Contributions

According to Statista website[1] report, the total amount of data created in the world reaches 175 zettabytes by the year 2025. This data is generated from various sources including sensors in a smart city platform [1]. However, an important point is that the observations of a sensor might be of insufficient quality due to various constraints, like environmental conditions or hardware malfunctioning [2]. The poor quality data may result in wrong business decisions being made by organizations [3]. Therefore, it is important to find data quality issues and clean poor data before using for any knowledge extraction or decision making. Some researchers use context-aware methods that quality of data is determined not only through analysis of the local application-specific information, but also using information from a global context, which in turn enhances the performance of big data quality management [4]. Based on our recent

---

[1] https://www.statista.com/statistics/871513/worldwide-data-created/

Supervised by Behshid Behkamal, Samad Paydar (Ferdowsi University of Mashhad (FUM), Mashhad, Iran) and Mohammad Allahbakhsh (University of Zabol, Zabol, Iran).

© Springer Nature Switzerland AG 2021
H. Hacid et al. (Eds.): ICSOC 2020 Workshops, LNCS 12632, pp. 10–16, 2021.
https://doi.org/10.1007/978-3-030-76352-7_2

systematic literature review on big data quality[2], we have observed that although a number of techniques have been proposed in the literature to improve data quality in the big data field, only a few consider contextual information in the process of data quality assessment. Since no context model for big data quality assessment has been proposed, in another study[3] we reviewed context-aware studies to provide a context model for big data quality, according to which we found that in all studies, only internal contexts (available in the subject data set) are used and none of them has considered the external context (available from other data sources). In addition, none of the context-aware techniques have used the stored data to increase the accuracy of quality assessment. In what follows, some of the challenges are mentioned:

- **Variety of arrival rate:** Data values arrives at the different rate, so the evaluation algorithm should provide a mechanism for processing existing data before the arrival of the new incoming data.
- **Infinite:** In streaming data, data is continuously being received, and the evaluation process must be done online and without interruption of the main retrieval process.
- **Volatility:** In data stream, volatility is a significant challenge that data expires after a while and lose credibility, so data processing should be done before it expires.
- **Heterogeneous sources:** Data may be received from different sources, in which case it is necessary to integrate and extract the correlation of these data in order to obtain the appropriate context information.

Based on the discussion above, in this early stage proposal, we intend to present a novel hybrid context-aware method using environmental information to assess the quality of data stream. The novel contributions of our work can be summarized as follows:

- We use external context (related information extracted from other sources), in order to improve data stream quality assessment performance.
- We benefit historical data that enables tracking of data values over time which gives key insights, in order to increase detection precision.
- We propose a grid-based clustering to decrease execution time.

2 State of the Art

In this section, we discuss and compare the context-aware quality assessment studies in streaming data, based on our systematic literature review. Studies have been compared based on several criteria, including level of management, type of contextual information, processing type, variable quantity, and technique used. From the point of view of level of management, there are two approaches: sensor-level, in which all quality controls are performed by the sensors without any interference by users, and user-level, in which pre-processing phase is performed by the user after receiving data from the sensors. From another point of view, we have classified these methods into two groups based on the type

---

[2] State of the Art on the Quality of Big Data: A Systematic Literature Review and Classification Framework.

[3] Contextualization of Big Data Quality: A framework for comparison.

regression. It is used when we want to predict the value of a variable based on the value of two or more other variables. This process continues until the data distribution model is obtained in all clusters. After this process, these models are used in the online quality evaluation step. In this proposal, we do not take concept drift, i.e. changes in the data distribution over time, into consideration. Consequently, it is not needed to update model during evaluation time and this will be considered as future work.

3.2 Assessing Quality of Streaming Data (Online)

After creating a distribution model for each cluster, by observing new data streams, the online quality assessment process begins, which has several steps including assigning data streams to the desired cluster, recording contextual information in the corresponding log, obtaining the distribution function and the predicted value, detecting and cleaning poor data quality.

Assign Data Streams to the Desired Cluster: At this stage, if the number of clusters is large, the allocation of data streams to the corresponding cluster will be time-consuming, and on the other hand, if the number of clusters is small, the accuracy of the distribution function will be reduced. As explained before, the proposed method will be able to make a trade-off by providing a grid-based clustering algorithm. In each of the first depth clusters, there is metadata for faster search and allocation of data to the desired cluster. Therefore, using these metadata, the data streams and their contextual information are quickly allocated to the desired cluster.

Record Contextual Information in the Corresponding Log: When each of the given data values is assigned to the desired cluster, the final log for that time interval must be updated with the new data values. This will continue until the data value, which should be evaluated in terms of quality, is seen, the log record is closed.

Obtain the Distribution Function and the Prediction Value: After closing the data log, the distribution function obtained from the first step (offline processing) is calculated with the values of the last log record and the output of the function is considered as the prediction value.

Detect Poor Quality Data and Clean It: After calculating the function and the prediction value, this value is compared with the data value. The threshold value is specified by the expert, which determines the difference between the predicted value and the data value. If the data value is within the predicted value range, the data is normal, otherwise the data is of poor quality and must be improved and replaced with predicted value.

4 Evaluation Plan

To support our claim, we will theoretically and empirically evaluate our proposed approach. In theoretical evaluation, we will compare our method with previous methods presented in Table 1 in terms of both accuracy and performance. For empirical evaluation, we develop an automated tool to measure the quality of data values of input datasets using both historical and contextual data. The details of experimental datasets are presented in Table 2. These datasets are available at Chicago city data portal[4] and all are related to past three years. A common feature in all datasets is geographic location, so we are able to find all contextual data and filter out low-quality data based on location.

Table 2. The details of the datasets used in our experiments

Datasets	No. of records	No. of features	Contextual features
Chicago traffic tracker	119M	22	Main dataset enriched with location
Chicago traffic crashes	417K	49	Date, time, location
Roadway construction events	17.5K	8	Start and end time, location
Public health department events	494	13	Time, location
Chicago weather	API	10	Weather type, visibility, wind speed

5 Conclusions

The goal of this research is to propose a hybrid multivariate context-aware data quality assessment method for data streams. Although many methods have been proposed to improve the quality of streaming data, none of them have used external contextual information. In this thesis, we are going to use external contextual information as well as historical data in order to improve the performance of the method. The proposed method will be evaluated on real datasets.

References

1. Perez-Castillo, R., et al.: DAQUA-MASS: an ISO 8000–61 based data quality management methodology for sensor data. Sensors **18**(9), 3105 (2018)
2. Bu, Y., Chen, L., Fu, A.W.-C., Liu, D.: Efficient anomaly monitoring over moving object trajectory streams. In: Proceedings of the 15th ACM SIGKDD International Conference on Knowledge Discovery and Data Mining - KDD 2009, p. 159 (2009)
3. Sidi, F., Panahy, P.H.S., Affendey, L.S., Jabar, M.A., Ibrahim, H., Mustapha, A.: Data quality: a survey of data quality dimensions. In: Proceedings of 2012 International Conference on Information Retrieval and Knowledge Management CAMP 2012, pp. 300–304, June 2014
4. Ardagna, D., Cappiello, C., Samá, W., Vitali, M.: Context-aware data quality assessment for big data. Future Gener. Comput. Syst. **89**, 548–562 (2018)

---

[4] https://data.cityofchicago.org/.

5. Anusha, A., Rao, I.S., Student, M.T.: A study on outlier detection for temporal data. Int. J. Eng. Sci. Comput. **8**(3), 16354–16356 (2018)
6. Chen, L., Gao, S., Cao, X.: Research on real-time outlier detection over big data streams. Int. J. Comput. Appl. **42**(8), 1–9 (2017)
7. Zhang, Y., Hamm, N.A.S., Meratnia, N., Stein, A., van de Voort, M., Havinga, P.J.M.: Statistics-based outlier detection for wireless sensor networks. Int. J. Geor. Inf. Sci. **26**(8), 1373–1392 (2012)
8. Iyer, V.: Ensemble Stream Model for Data-Cleaning in Sensor Networks (2013)
9. Zhang, Y., Meratnia, N., Havinga, P.J.M.: Distributed online outlier detection in wireless sensor networks using ellipsoidal support vector machine. Ad Hoc Netw. **11**(3), 1062–1074 (2013)
10. Zhang, Y., Szabo, C., Sheng, Q.: Cleaning environmental sensing data streams based on individual sensor reliability. In: Benatallah, B., Bestavros, A., Manolopoulos, Y., Vakali, A., Zhang, Y. (eds.) WISE 2014. LNCS, vol. 8787, pp. 405–414. Springer, Cham (2014). https://doi.org/10.1007/978-3-319-11746-1_29
11. Hayes, M.A., Capretz, M.A.M.: Contextual anomaly detection framework for big sensor data. J. Big Data **2**(1), 1–22 (2015). https://doi.org/10.1186/s40537-014-0011-y
12. Rassam, M.A., Maarof, M.A., Zainal, A.: A distributed anomaly detection model for wireless sensor networks based on the one-class principal component classifier. Int. J. Sens. Netw. **27**(3), 200 (2018)

Container-Based Network Architecture for Mobility, Energy and Security Management as a Service in IoT Environments

Zahid Iqbal[✉][iD]

China University of Petroleum, Qingdao, China
lb1901002@s.upc.edu.cn

Abstract. Internet of Things is being used in every field of life. The increased IoT devices are building heterogeneous networks which result in introducing some serious challenges. The presence of mobile nodes makes the network unstable, the protocols used in these network do not offer the required security level and network growth results in more energy consumption. On the other hand, the Container-Based solutions are getting huge attention because they are lightweight then Virtual Machines (VMs). They are reusable, flexible and offer dynamic allocation of resources. In this article, we have proposed a Container-Based architecture which offers Mobility, Energy and Security Management as a Service (MESMaaS). Our main objective is to implement MESMaaS at the core of the network to address Network issues and to achieve improved network performance in terms of network life time, network stability, re-transmission of data, signaling cost, packet loss, data & network security and other communication issues.

Keywords: Internet of Things (IoT) · Dockers · Virtual Machines (VMs) · Container · Virtualization

1 Introduction

Container-Based solutions are widely used and adopted solutions being lightweight virtualization instances as compared to virtual machines (VMs), further they offer many advantages such as performance optimization, resource utilization, dynamic allocation of resources and agile environments [8]. Introduction of the Internet of Things (IoT) is one of the remarkable advancement of the last decade which resulted in the development of various communication protocols. The IoT environments provide us an opportunity to convert a secluded device into a smart object. These object have communicational & computational ability and can act independently without any human assistance [10]. IoT covers a huge range of industries and other environments formed from few constraint devices to massive cross-platform networks [13]. But the main problem with IoT environments is that the devices used by IoT are constraint devices and the protocols

© Springer Nature Switzerland AG 2021
H. Hacid et al. (Eds.): ICSOC 2020 Workshops, LNCS 12632, pp. 17–24, 2021.
https://doi.org/10.1007/978-3-030-76352-7_3

used in these networks do not offer the required support for mobility, energy efficiency and security as offered by the traditional communication protocols [1, 5]. Considering the network and user requirements, the provision of services such as mobility, energy and security management are matters which still need to be addressed. Most of the applications of the Internet of Thing today use mobility as a way for sharing and processing of information. It also facilitates the integration of Low power wireless networks with other Internet Protocol (IP)-based networks. But the presence of mobile nodes in the network makes the network unstable which result in increased signaling cost and increased re-transmission of data. Therefore, the support for mobility in an IoT environment is paramount to make network stable. Further, the energy efficiency increases the lifetime of a network and security ensures the reliability of the network.

Recently, container based virtualization has already been introduced to IoT devices which is mainly focusing on resource utilization such as single device providing multiple sensing requirements. The intended research will provide a Container-Based network architecture which operates at the core of the network and offers the Mobility, Energy and Security Management as a Service (MES-MaaS) for the IoT networks. Our basic idea is to address issues created at the network level due to network scalability in terms of energy consumption, network instability due to presence of mobile node and lake of security due to constraint communication protocols. By addressing these challenges at the core of network, we can achieve quality of service by eliminating performance anomalies in constraint networks. This article will invite the attention of researchers towards container based network optimization and direction for the future research. The research article is organized as follows:

The Sect. 2 will provide the research work already conducted in this field and benefits offered by the container-based services. In Sect. 3, we will discuss a brief overview regarding challenges introduced by the mobility, energy & security in IoT environments. The Sect. 4 will provide the detailed architecture and its implementation. The Sect. 5 will conclude the research and we will discuss about the future direction.

1.1 Container-Based Virtualization

Resource utilization through virtualization is one of the most important concepts in cloud computing as it involves the process of creating virtual instance of an object instead of creating it physically. Now it is not limited to creating a virtual machine, hardware platform, utilizing computer network resources, storage server or an operating system. The virtualization provides an intermediate software layer on the top of a system to provide abstraction of a system to utilize resources at optimum level [11]. Several Virtualization techniques are being used now a day. Mainly they can be categories as hypervisor-based virtualization and container-based virtualization. The Fig. 1 shows the comparison between the two technologies.

Containers are comparatively more resource efficient as they eliminate the execution of guest operating system and more time efficient as they avoid boot-

Hypervisor-Based Applications   **Container-Based Applications**

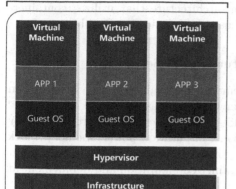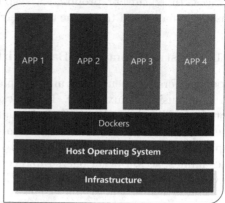

Fig. 1. Comparison between hypervisor-based and container-based virtualization

ing and shutting down an operating system [9]. This newly adopted virtualization technique can provide many beneficial services for heterogeneous IoT environments. It enables to develop dynamically on demand service provisioning according to device capabilities and user requirements. Further, this allows the reconfiguration of operational behavior of the deployed nodes.

2 Related Work

The base of container based technology was established as chroot command back in 1979. With passage of time and evaluation, eventually it came as virtualization mechanisms like Open VZ, Linux Vserver and Container Engine (LXC). For the past few years the software oriented solutions were being considered as the most promising trends to overcome the challenges raised by the IoT smart environments. Majority of studies focused on improving the inter-connectivity in these smart environments through implementing software defined networking [2]. The 5G is turned out as the future of the technology providing anything as a service. To enable anything as a service, there is dire need to develop the virtualized ecosystem to support new technologies [14]. In [12] the concept for electing the coordinator among the mobile node was proposed to manage between node resources and application requirements. The authors in [3,11] proposed container based virtualization implementation to IoT devices to solve various management issues. Tools like Kubernetes and Dockers Swarm provide us opportunity to automate container operations. On the other hand different IoT platforms [4] provide us different APIs with different functionality to develop applications for a specific platform. Containers enable us to run multiple platforms on same IoT device simultaneously to run a specific application. Research studies [6,11] were conducted to compare the performance matrix of native IoT environments with container based environments where containers were deployed to the IoT devices. In [7] author proposed a system architecture that utilizes container-based virtualization technique on widely used IoT device i.e. Raspberry pi3.

2.1   Existing Container-Based Virtualization Solutions

Docker offers an open platform for implementing container-based virtualization on Linux using Linux kernel. The Docker Container as compared to VM does not demand independent OS. It is build on the functionality provisioned by the kernel to achieve resource isolation i.e. computation power, memory, block input/output and network resources. To isolate application from the host OS, it uses namespaces. Docker offers rapid container configuration and deployment at any place such as cloud within local VMs or constraint devices used for IoT.

3   Mobility, Energy Optimization and Security Challenges in IoT

Dealing and designing of sensor network is always considered as a challenge due to many factors i.e. network scalability, installation, hardware cost, operating environment, mobility management, network topology, topological changes, constraint devices, fault detection and tolerance etc. The Table 1 below describes the mobility, energy optimization and security challenges in constraint networks.

Table 1. Mobility, energy & security management challenges

Challenges			
QoS & scalability	**Resource management**	**Security**	**Topology control**
Data loss rate	Bandwidth	Authentication	Network connectivity
Continuous connectivity	Reduce signaling cost	Authorization	Coverage area
Fast movements detection	Power consumption	Integrity	**Routing protocol**
Reduce handover delay	Duty cycle	Confidentiality	Shortest path
End-to-end delay	Reduce signaling cost	Encryption	Alternate path
Route optimization	Multi-hop		Path stability
Avoid triangle routing	Network-based communication	◂	Routing metrics

4   Proposed Container-Based Network Architecture

In this section, we describe our proposed Container-Based Network Architecture using diagram shown in Fig. 2. The Fig. 3 below describe the process of container scheduling, selection and migration. The proposed architecture consists of five layers as shown in Fig. 2.

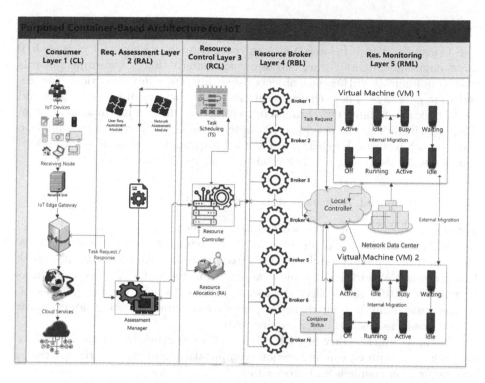

Fig. 2. Proposed container-based network architecture for mobility, energy & security management as a service in IoT environments

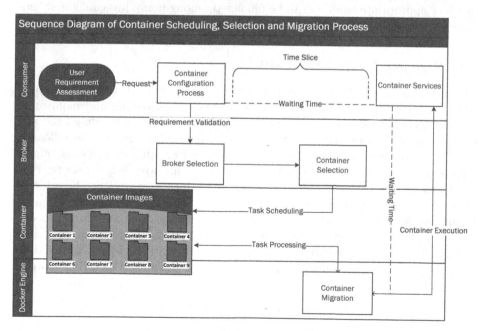

Fig. 3. Sequence diagram for container scheduling, selection & migration process.

4.1 Consumer Layer (CL)

The Consumer Layer belongs to resource consumers such as IT clients, IoT smart environments, and other end users using smart things. Consumers at this layer communicate with smart objects using different communication channels and edge gateways. These communications enable heterogeneous networks and large amount of data. Our proposed architecture deals with management of challenges introduced by heterogeneity in terms of mobility, energy consumption and network security. Further, it is intended to provide quality of service (QoS) guarantees by increasing network efficiency and its life time.

4.2 Requirement Assessment Layer (RAL)

The second layer of our proposed architecture deals with assessment of overall user and network requirements. Our proposed architecture divides these requirements in three types of modules i.e. User Requirement Collector (URC), Network Requirement Collector (NRC) and Configuration Agent (CCA.) The main functionality of (URC) is to collect the user requirements such as sensing needs i.e. temperature, humidity, air, light etc., number of communicating nodes, static nodes, mobile nodes and other requirements.

The main functionality of (NRC) is to collect network requirements such as network size, type of communication, computation and storage requirements, communication channels and protocol etc.

The main functionality of (CCA) is to make a container configuration file according to a preset configuration value. Later, this configuration will be used to select appropriate container to be offloaded among many pre-configured containers for mobility, energy and security management.

4.3 Resource Control Layer (RCL)

The (RCL) is very important layer of our proposed architecture as it is responsible to control and manage overall task scheduling, offloading and monitoring process. This layer divides tasks into two modules i.e. task scheduling (TS) and resource management (RM). (TS) module schedules all task to be offloaded and (RM) continuously monitors all the available, running and idle containers images in coordination with resource broker layer. This layer helps Resource Broker Layer (RBL) to select, combine, rank and allocate/reserve combination of resources in the form of containers. This layer also provides mechanism for monitoring resource current status and container internal/external migration process. Docker provides efficient migration of container images (container-to-container – internal) or (container-to-VMs – external). Further, it provides real time data collection and resource isolation by sharing Kernel Instance, Operating System, and Network Connections.

4.4 Resource Broker Layer (RBL)

The Resource Broker acts as the back bone of our proposed architecture. This layer works in coordination with Resource Control Layer and Resource Monitoring Layer. Primarily this layer is used to rank, reserve and assign combination of resources to containers. This layer also used to keep the information of available VMs, containers and requests for resource provisioning.

4.5 Resource Monitoring Layer (RML)

The (RML) is responsible to monitor all running, consumed and available resources. This layer consists of Clusters and Network Data Centers from where resources are provisioned. This layer helps resource broker layer and resource controller layer for task scheduling by giving the information about the current status of containers.

5 Conclusion

Growing IoT environments result in increase in number of connected devices and heterogeneous networks. This exponential growth is introducing various communication, integration and security challenges. Containers being lightweight solution provide mechanism for efficient utilization of resources and new provisions. By adopting new technologies like Containers we can address the challenges introduced by heterogeneous IoT networks. In this article, we have presented a Container-Based architecture to provide Mobility, Energy and Security as a Service (MESMaaS) at the core of network. Implementing such architecture at initial network level will ensure better communication by eliminating performance, communication and security anomalies. This will provide better communication within nodes, remove delays, re-transmissions, packet losses and will help the network to generate quality data. Thus, this will solve performance issues and will enable the efficient processing of data for fog and cloud based IoT systems.

References

1. Borgia, E.: The internet of things vision: key features, applications and open issues. Comput. Commun. **54**, 1–31 (2014)
2. Buratti, C., et al.: Testing protocols for the internet of things on the EuWIn platform. IEEE Internet Things J. **3**(1), 124–133 (2015)
3. Celesti, A., Mulfari, D., Fazio, M., Villari, M., Puliafito, A.: Exploring container virtualization in IoT clouds. In: 2016 IEEE International Conference on Smart Computing (SMARTCOMP), pp. 1–6. IEEE (2016)
4. da Cruz, M.A., Rodrigues, J.J., Sangaiah, A.K., Al-Muhtadi, J., Korotaev, V.: Performance evaluation of IoT middleware. J. Netw. Comput. Appl. **109**, 53–65 (2018)
5. Fotouhi, H., Moreira, D., Alves, M., Yomsi, P.M.: mRPL+: a mobility management framework in RPL/6LoWPAN. Comput. Commun. **104**, 34–54 (2017)

6. Krylovskiy, A.: Internet of Things gateways meet linux containers: performance evaluation and discussion. In: 2015 IEEE 2nd World Forum on Internet of Things (WF-IoT), pp. 222–227. IEEE (2015)
7. Lee, K., Kim, H., Kim, B., Yoo, C.: Analysis on network performance of container virtualization on IoT devices. In: 2017 International Conference on Information and Communication Technology Convergence (ICTC), pp. 35–37. IEEE (2017)
8. Lee, K., Kim, Y., Yoo, C.: The impact of container virtualization on network performance of IoT devices. Mob. Inf. Syst. **2018**, 1–6 (2018)
9. Merkel, D.: Docker: lightweight linux containers for consistent development and deployment. Linux J. **2014**(239), 2 (2014)
10. Mosenia, A., Jha, N.K.: A comprehensive study of security of Internet-of-Things. IEEE Trans. Emerg. Top. Comput. **5**(4), 586–602 (2016)
11. Mulfari, D., Fazio, M., Celesti, A., Villari, M., Puliafito, A.: Design of an IoT cloud system for container virtualization on smart objects. In: Celesti, A., Leitner, P. (eds.) ESOCC Workshops 2015. CCIS, vol. 567, pp. 33–47. Springer, Cham (2016). https://doi.org/10.1007/978-3-319-33313-7_3
12. Nishio, T., Shinkuma, R., Takahashi, T., Mandayam, N.B.: Service-oriented heterogeneous resource sharing for optimizing service latency in mobile cloud. In: Proceedings of the First International Workshop on Mobile Cloud Computing & Networking, pp. 19–26 (2013)
13. Sharma, V., Sharma, V., Mishra, N.: Internet of Things: concepts, applications, and challenges. In: Exploring the Convergence of Big Data and the Internet of Things, pp. 73–95. IGI Global (2018)
14. Taleb, T., Ksentini, A., Jantti, R.: "Anything as a service" for 5G mobile systems. IEEE Netw. **30**(6), 84–91 (2016)

Towards a Rule-Based Recommendation Approach for Business Process Modeling

Diana Sola[1,2(✉)]

[1] Intelligent Robotic Process Automation, SAP SE, Walldorf, Germany
[2] Data and Web Science Group, University of Mannheim, Mannheim, Germany
diana@informatik.uni-mannheim.de

Abstract. Business process modeling can be time-consuming and error-prone, especially for inexperienced users. For this reason, graphical editors for business process modeling should support users by providing suggestions on how to complete a currently developed business process model. We address this problem with a rule-based activity recommendation approach, which suggests suitable activities to extend the business process model that is currently edited at a user-defined position. Contrary to alternative approaches, rules provide an additional explanation for the recommendation, which can be useful in cases where a user might be torn between two alternatives. We plan to investigate how rule learning can be efficiently designed for the given problem setting and how a rule-based approach performs compared to alternative methods. In this paper we describe the basic idea, a first implementation and first results.

1 Introduction

A business process model is the graphical representation of an organization's business process and an important instrument for Business Process Management. When modeling a business process, it is essential to precisely label the individual elements such that the process is consistent and unambiguous. In the case of domain-specific processes, this might require using a specialized and sometimes technical vocabulary, which often turns out to be challenging. There are a lot of tools supporting the modeling of business processes in a graphical notation such as Business Process Model and Notation or Petri Nets. Usually, they are graphical editors providing the user with a repository of symbols, which represent the building blocks of the underlying modeling language [12]. However, business process modeling remains time-consuming and error-prone, especially for inexperienced users. The modeling task can be facilitated by providing features which assist users during modeling and make recommendations on how to complete a business process model that is being edited [8]. The basis for such a recommendation feature could be a repository of completed business process models.

Supervised by Heiner Stuckenschmidt (Data and Web Science Group, University of Mannheim, Mannheim, Germany), heiner@informatik.uni-mannheim.de.

© Springer Nature Switzerland AG 2021
H. Hacid et al. (Eds.): ICSOC 2020 Workshops, LNCS 12632, pp. 25–31, 2021.
https://doi.org/10.1007/978-3-030-76352-7_4

One possible recommendation approach in business process modeling is activity recommendation [17]. Given the business process model being worked on, the recommendation system makes suggestions regarding suitable activities to extend the model at a user-defined position. In other words, the system recommends proper activities to support the user modeling business processes in an iterative way. Figure 1 shows an example of a business process model that is currently developed. The user has just added the sequence flow with label 'Yes'. The task of the recommendation system is to find a suitable activity at this position. Since the business process model that has been developed so far depicts a version of the order-to-cash process, the recommender system suggests activities that have been used in similar order-to-cash processes in the repository: 'Submit purchase order', 'Analyze quotation' and 'Create and submit the quotation'.

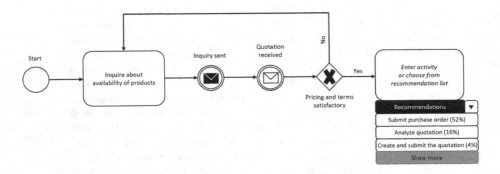

Fig. 1. A business process model under development

Structural patterns of activities (e.g. the order of activities) play an important role in the development of a recommendation method since some activities are more relevant at the current modeling phase than others. This poses a challenge in comparison to usual applications of recommender systems as they rarely need to consider structural patterns for the similarity of items or users.

Rules are a good option to model such patterns, therefore we want to address the activity recommendation problem by learning rules that capture the interrelationships between activities in processes of a repository. The rules can then be applied to a process under development. We intend to investigate how rule learning can be efficiently designed for the given problem setting. Moreover, we plan to analyse how a rule-based approach performs compared to alternative methods and what advantages can be derived in contrast to other methods. In this paper we propose a first, simple implementation and report about experiments where we compare it against one alternative approach.

2   Related Work

An overview of recommendation methods for business process modeling is presented in [11]. Kluza et al. distinguish between a subject-based classification, which concentrates on what is actually suggested, and a complementary position-based classification, which focuses on the position where the suggestion is to be placed. According to their categorization, our work falls into the category of full-name suggestion for an element. The evaluation method that we use in the experimental studies corresponds to a forward completion approach.

There are several works that abstract a business process model to a directed graph and use graph-mining techniques to extract structural patterns from the process repository. While Cao et al. [3] calculate the distance between patterns and the partial business process based on graph edit distance [2], Li et al. [14] propose an efficient string edit distance [13] based similarity metric which turns the graph-matching problem into a string matching problem. Different distance calculation strategies are compared by Deng et al. [7].

An approach that involves semantic information and patterns observed in other users' preferences is proposed by Koschmider et al. [12]. They present a business process modeling editor with two features. First, the editor allows the user to search for process model fragments via a query interface, which is based on semantic annotations. Second, the system recommends appropriate process model fragments to the model being edited, which is based on the combination of several aspects as the frequency a process part has been selected by other users or its process design quality. In two experiments, the authors prove the usefulness and efficiency of their editor. However, the evaluation was based on a comparably small repository of process fragments that were developed particularly for the study's modeling exercise.

In [17], Wang et al. present their embedding-based activity recommendation method RLRecommender which extracts relations between activities of the process models and embeds both activities and relations into a continuous low-dimensional space. The training model used is based on TransE [1]. The embedded vectors for activities and relations and their distances in the space are then used to recommend an appropriate activity.

Jannach et al. [10] propose different recommendation techniques to provide modeling support for users in the specific area of data analysis workflows and evaluate them using a pool of several thousand existing workflows. The user support consists of recommending additional operations to insert into the currently developed machine learning workflow and is hence similar to activity recommendation. In a laboratory study, the authors show that their recommendation tool helps users to significantly increase the efficiency of the modeling process.

3   A Rule-Based Recommendation Approach

Following [17], we frame the activity recommendation problem in terms of a knowledge graph completion task (sometimes also referred to as link prediction). Within this framework, the processes of the repository and the incomplete process are represented as a (large) graph consisting of triples (*head activity, relation, tail activity*) and the recommendation of an appropriate activity has to be understood as the completion task. Within the last decade, the knowledge graph completion task has received lots of attention and the majority of approaches uses embedding methods, where the knowledge graph is embedded into a low-dimensional space [18]. Thus, it is no surprise that this technique has also been used by Wang et al. in [17] to propose the embedding-based method RLRecommender as a solution for the activity recommendation problem.

While approaches that are based on embeddings dominate knowledge graph completion, more recently rule-based approaches, which have their origin in the field of inductive logic programming [5], have proven to be competitive [15]. As an additional benefit, these approaches offer an explanation for the given recommendation. Explainable recommendations have recently attracted more and more interest since they help improving the transparency, persuasiveness, effectiveness, trustworthiness, and satisfaction of recommendation systems [19].

We propose an approach which learns logical rules that describe how activities are used in the given process repository. These rules are used to give explainable recommendations for an appropriate activity at a given position. Our rule learner is based on the top-down search implemented in the association rule mining systems WARMR [6] and AMIE [9]. However, our implementation supports a specific language especially designed for predicting activities. The learned rules are Horn rules that predict the label u of an activity node X. In particular, they have the form $u(X) \leftarrow relation(X,Y), v(Y)$, where Y denotes another activity node in the process, v denotes the label of Y and $relation$ denotes the relation between the activities X and Y. For the relations between activities we make use of the definitions in [17]. The 'Direct After' relation depicts the connections of activities but makes it impossible to distinguish between an AND and an OR split. The 'Direct Causal' relations allow to capture the semantics of business process models more precisely. The concurrency of activities are described by 'Direct Concurrent' relations. The definition of these three relation families results in three rule learning strategies. Our first rule learning strategy (rules-after) is to only allow 'after' relations in the rule bodies. Analogously, we only allow 'causal' relations in the rule bodies for the second strategy (rules-causal). The third rule learning strategy (rules-concurrent) is to allow 'causal' and 'concurrent' relations in the rule body.

For the experiments, we made use of the two datasets that have also been used in the evaluation of RLRecommender in [17]. The first dataset (large dataset) consists of processes from the model collection of the Business Process Management Academic Initiative [16]. The second dataset (small dataset) consists of 221 processes collected from a district government in Hangzhou, China [4]. As evaluation metric we use the hit rate, which is the fraction of hits, where a hit is achieved if the generated recommendation list contains the activity that was actually chosen. We adopted the evaluation method from [17]. For the small dataset, we performed a fivefold cross-validation. For the large dataset, we chose the training and test split that we received from running the preprocessing code from RLRecommender on Github[1]. As in [17], we report the hit rate for recommendation list lengths 1–5 and for the large dataset additionally for length 10.

---

[1] https://github.com/THUBPM/RLRecommender .

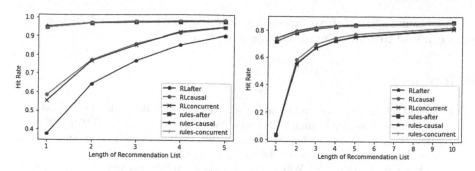

Fig. 2. Results on small (left) and large (right) dataset

The results of the experimental study in which we compared the embedding-based approach RLRecommender to our rule-based approach are depicted in Fig. 2. They show that our rule-based method outperforms the embedding-based approach on both datasets and for every recommendation list length.

4   Conclusion and Research Plans

This paper presents our ongoing work on a rule-based recommendation method for business process modeling which allows for explainable recommendations and outperforms an embedding-based approach.

In future work, we want to allow other forms of rules that involve more than one preceding activity of the process model. We also intend to conduct similar experimental evaluations with other existing methods and on other datasets. Jannach et al. [10] propose different approaches for predicting labels in the specific area of data analysis workflows. We plan to analyze whether these methods can also be used for the more general problem that we tackle, and if applicable, we will include their approach in a comprehensive experimental study.

The activity recommendation problem that we investigate is a multi-class classification problem. The learning of rules for multiple classes causes the problem of multiple rules firing. Until now we make use of a maximum strategy for the case that multiple rules make the same recommendation and assign the maximum confidence score of the rules to this recommendation. We plan to analyse other aggregation methods that are able to take entailment relations between rules into account.

In addition, we want to extend the problem to the possible case that there is no label in the process under development that has also been used in the process repository. In this case, we could try to match the labels to the labels in the repository and then apply the learned rules. However, this requires developing an approach that aggregates the confidence of generated mappings with the confidence scores of the rules to compute the ranking of the final recommendations.

Furthermore, we plan to investigate if a combined use of embeddings and rules can lead to further improvements in accuracy.

References

1. Bordes, A., Usunier, N., García-Durán, A., Weston, J., Yakhnenko, O.: Translating embeddings for modeling multi-relational data. In: NIPS, pp. 2787–2795 (2013)
2. Bunke, H.: On a relation between graph edit distance and maximum common subgraph. Pattern Recogn. Lett. **18**(8), 689–694 (1997)
3. Cao, B., Yin, J., Deng, S., Wang, D., Wu, Z.: Graph-based workflow recommendation: on improving business process modeling. In: Proceedings of the 21st ACM International Conference on Information and Knowledge Management, pp. 1527–1531. Association for Computing Machinery, New York (2012)
4. Dataset consisting of processes collected from a district government in Hangzhou, China. https://github.com/THUBPM/RLRecommender/tree/master/dataset/SRD
5. De Raedt, L.: Logical and Relational Learning. Springer, Berlin (2008). https://doi.org/10.1007/978-3-540-68856-3
6. Dehaspe, L., Toivonen, H.: Discovery of relational association rules. In: Džeroski, S., LavračN. (eds.) Relational data mining, pp. 189–212. Springer, Berlin (2001). https://doi.org/10.1007/978-3-662-04599-2_8
7. Deng, S., et al.: A recommendation system to facilitate business process modeling. IEEE Trans. Cybern. **47**(6), 1380–1394 (2017)
8. Fellmann, M., Zarvic, N., Metzger, D., Koschmider, A.: Requirements catalog for business process modeling recommender systems. In: Wirtschaftsinformatik, pp. 393–407 (2015)
9. Galárraga, L.A., Teflioudi, C., Hose, K., Suchanek, F.: AMIE: association rule mining under incomplete evidence in ontological knowledge bases. In: Proceedings of the 22nd International Conference on World Wide Web, pp. 413–422 (2013)
10. Jannach, D., Jugovac, M., Lerche, L.: Supporting the design of machine learning workflows with a recommendation system. ACM Trans. Interact. Intell. Syst. (TiiS) **6**(1), 1–35 (2016)
11. Kluza, K., Baran, M., Bobek, S., Nalepa, G.J.: Overview of recommendation techniques in business process modeling. In: KESE CEUR Workshop Proceedings, vol. 1070. CEUR-WS.org (2013)
12. Koschmider, A., Hornung, T., Oberweis, A.: Recommendation-based editor for business process modeling. Data Knowl. Eng. **70**(6), 483–503 (2011)
13. Levenshtein, V.I.: Binary codes capable of correcting deletions, insertions and reversals. Sov. Phys. Dokl. **10**(8), 707–710 (1966)
14. Li, Y., et al.: An efficient recommendation method for improving business process modeling. IEEE Trans. Ind. Inform. **10**(1), 502–513 (2014)
15. Meilicke, C., Chekol, M.W., Ruffinelli, D., Stuckenschmidt, H.: Anytime bottom-up rule learning for knowledge graph completion. In: Proceedings of the 28th International Joint Conference on Artificial Intelligence, pp. 3137–3143. AAAI Press (2019)
16. Model collection of the Business Process Management Academic Initiative. http://bpmai.org/

17. Wang, H., Wen, L., Lin, L., Wang, J.: RLRecommender: a representation-learning-based recommendation method for business process modeling. In: Pahl, C., Vukovic, M., Yin, J., Yu, Q. (eds.) ICSOC 2018. LNCS, vol. 11236, pp. 478–486. Springer, Cham (2018). https://doi.org/10.1007/978-3-030-03596-9_34
18. Wang, Q., Mao, Z., Wang, B., Guo, L.: Knowledge graph embedding: a survey of approaches and applications. IEEE Trans. Knowl. Data Eng. **29**(12), 2724–2743 (2017)
19. Zhang, Y., Chen, X.: Explainable recommendation: a survey and new perspectives. arXiv preprint arXiv:1804.11192 (2018)

Towards a Privacy Conserved and Linked Open Data Based Device Recommendation in IoT

Fouad Komeiha[1,2(✉)], Nasredine Cheniki[1], Yacine Sam[1], Ali Jaber[2],
Nizar Messai[1], and Thomas Devogele[1]

[1] University of Tours, Tours, France
{fouad.komeiha,nasredine.cheniki,yacine.sam,
nizar.messai,thomas.devogele}@univ-tours.fr
[2] Lebanese University, Beirut, Lebanon
{fouad.komeiha,ali.jaber}@ul.edu.lb

Abstract. Interconnecting Internet of Things (IoT) devices creates a network of services capable of working together to accomplish certain goals in different domains. The heterogeneous nature of IoT environments makes it critical to find devices that extend existing architectures and helps in reaching the desired goal; especially if we have to take into consideration data privacy. In this paper, we present a Linked Open Data (LOD) based approach to semantically annotate and recommend IoT devices while adding a layer of data security and privacy through implementing the SOLID (SOcial LInked Data) framework.

Keywords: IoT · Linked Open Data · Social Linked Data

1 Introduction

Internet of Things (IoT) as a paradigm, aims to connect pervasive devices to the internet and provides means for communication between them. The utilization of IoT architectures in different domains, led to an increase in the number of connected devices adapting different communication standards. This issue created an overhead for device discovery and recommendation.

The Web of Things (WoT) intends to increase device interoperability by using existing web standards such as HTTP[1], and REST[2] for data communication. Many protocol based device discovery frameworks have been proposed [13]. Although protocol based recommendations provide standard, reliable architectures, they lack interoperability [13]. Device representation in such protocols is still limited and more description expressivity for devices and their capabilities is needed. Semantic Web of Things (SWoT), an evolution of the WoT was introduced where domain ontologies are used to model devices and their capabilities in order to solve interoperability issues.

---

[1] https://www.w3.org/Protocols/.
[2] https://www.w3.org/2001/sw/wiki/REST.

© Springer Nature Switzerland AG 2021
H. Hacid et al. (Eds.): ICSOC 2020 Workshops, LNCS 12632, pp. 32–39, 2021.
https://doi.org/10.1007/978-3-030-76352-7_5

Semantic-based device recommendation systems incorporate semantics provided by ontologies within the recommendation process. These systems represent devices using designed ontologies for domain modeling and device annotation. This process helps in unifying devices' representation allowing systems to search for, and recommend devices based on their semantic similarities. Ontologies, however, are usually domain-specific and cover restricted knowledge domains. Linked Open Data, on the other hand, is a global data space containing assertions about diverse domains. This allows for a more expressive representation of devices. It also eliminates the need for ontology extensions or redesigns when recommendation in new domains is needed.

By annotating devices using LOD concepts, they become part of the LOD cloud, enabling us to use existing LOD-based similarity measures to perform similarity calculation and recommend devices to end-users. Besides, a recommendation method that provides trusted results by the user is necessary. Hence, device recommendations based on user-social relations could increase user confidence. In addition, semantic similarity calculation for device recommendations requires querying and access to semantic data representing devices. This means that any device recommending framework should respect the privacy and security of data (device) owners.

In this paper, we present an approach for user-user social device recommendation using LOD-based semantic annotation of devices. Our approach supports data privacy and access control by implementing the SOLID[3] architecture.

The paper is structured as follows: Sect. 2 describes our proposed device recommendation approach and its architecture. In Sect. 3 we explain the main challenges facing our proposal. Existing works for IoT device recommendation are presented in Sect. 4. Finally Sect. 5 provides a conclusion and future works.

2 Proposed Approach

In this section, we present our approach for device recommendation taking as a use case the smart home domain. We explain the semantic representation of devices, the recommendation process, and the implemented data privacy mechanism.

2.1 Semantic Representation

Our approach focuses on using Linked Open Data (LOD) cloud for both user profile and device annotations due to its expressivity and coverage of diverse domains. LOD is a global data space containing billions of semantic assertions. It's the result of the adaptation of Linked Data Principles by many publishers, both individuals and organizations.

Semantic similarity measures, on the other hand, are metrics that assess the similarity between concepts by measuring the degree of overlap between their

---

[3] https://solidproject.org/.

semantic representation. Many LOD-based similarity measures have been proposed in the literature [3, 7, 9, 11]. By connecting devices to the LOD cloud, annotated devices can be treated as LOD resources and existing LOD-based measures can be applied to compare the semantic similarity between them without the necessity to define new measures. By applying these similarity measures we can thus recommend new devices to a user based on his ecosystem.

2.2 Semantic and Social IoT Device Recommendation

Different recommendation algorithms have been proposed in the literature. User-based collaborative filtering and content-based filtering are two of the commonly used algorithms. User-based collaborative filtering recommends items based on the ratings of similar users to the target user. Content-based filtering method, on the other hand, recommends items that are similar to the ones rated by the user. In the domain of IoT, many approaches to recommend devices based on social relations were introduced [1, 2, 12].

In our approach, we opted towards using a hybrid system that combines different approaches. We will use a combination of social user-user relations and collaborative filtering to filter candidate devices. These devices will be candidates for a target user as they are owned by trusted friends. After clustering devices, a semantic content-based filtering method measures the semantic similarity between a target user's devices and the devices of his friends, thus avoiding the cold start problem. An overall semantic value representing the similarity between a target user's smart environment and his friend's environment will be obtained. Through receiving recommendations of devices from users trusted by the target user, the confidence in the results is higher. In our architecture, the control node (explained in Sect. 2.4) will allow a user to add other users to his list of friends by adding their WebIDs[4], thus maintaining a user-user relationship. Through this process, the control node can recommend devices to a user based on the ecosystem maintained by his friends.

2.3 Data Access and Privacy

Since device recommendation process needs to access data about user-profiles and device descriptions, a privacy policy that defines who has the right to access and use data is needed. The data owner should have full control over each piece of his semantic stored data.

In our approach, we adopt the SOLID framework to allow owners to control access to the data concerned in the recommendation process. SOLID is a personal data management framework that grants data owners full control over their personal data. The SOLID architecture is based on two main components: the Personal Online Data store (POD)/SOLID server where data is being stored, semantically annotated and accessed by third parties according to the data owner's preferences. To do this, a data POD provides each new user with

---

[4] https://www.w3.org/wiki/WebID.

a unique WebID, this id is a unique URL that will identify a user and provide access to all data authorized by him through HTTP requests. The second component, is the SOLID application that allows users to register/add, view, and edit data on PODs. By implementing the SOLID framework in our approach, we provide each device owner the ability to control access to his devices' data thus preserving his privacy.

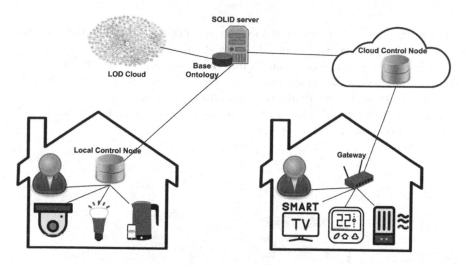

Fig. 1. Overview of the approach architecture in a smart home domain

2.4   Architecture

Fig. 1 represents an architecture of our approach in a smart home domain. It's made up of two components. The user will also have a role in the architecture by registering and providing initial information about his devices.

SOLID Server: The SOLID server will be available for all control nodes. It will be used to register and annotate users and devices as well as store all annotated data. To do this, the server will use a base ontology to describe and annotate devices with LOD resources. After registration, each user and device will be identified using a WebID. To allow the discovery of different devices, the server will act as a global directory for the control nodes. This directory will contain WebIDs of users and their owned devices. Using this global directory, a control node can get the WebID of a device through which semantic data describing this device can be accessed according to the device's owner permissions.

Control Node: The control node could be hosted locally or deployed remotely on a Cloud service. It has access to the SOLID server to perform the following operations:

1. Providing the user with an interface to register and create his profile thus obtaining a unique WebID for himself.
2. Allowing the user to register new devices and obtaining a unique WebID for each registered device via the SOLID server.
3. Tracking the availability of the devices in its managed network.
4. Performing similarity and recommendation of IoT devices. For this task, the control node will use LDS[5], our library of LOD-based similarity measures.

End User: The user registers devices in his network through the interface implemented in the control node. The registration process will allow him to create his profile, add/manage a list of his friends as well as provide information about IoT devices in his ecosystem. The user will also set the permissions for data availability and accessibility to other control nodes.

Figure 2 depicts detailed description about the components and functionalities of our proposed approach.

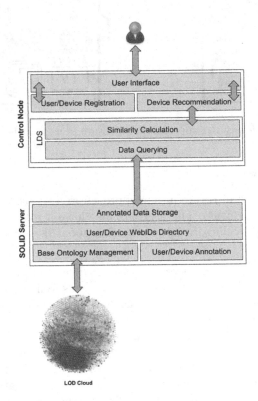

Fig. 2. Detailed representation of our SOLID-based device recommendation approach

---

[5] https://github.com/FouadKom/lds.

3   Challenges

Three major challenges face our approach, mainly in the recommendation process. These challenges stem from the SOLID architecture itself which aims to decentralize data storage. To do so, SOLID allows PODs to be data containers accessed from different applications. In our approach, the data POD used can be either hosted by the SOLID server or a different POD provider could be used to contain data about users and their devices. Our challenges come from the latter case:

- The first challenge may be faced as each POD provider may have different device description frameworks or ontologies. This means that different vocabularies might be used which will affect the ability of the system to perform similarity calculations and recommendations correctly. In this case, ontology alignment mechanisms should be utilized.
- The second challenge is the data distribution. The utilization of several POD providers to describe devices means data is distributed among several servers. This means that the recommendation process, in this case, will require accessing several providers to retrieve data of devices. Thus, the recommendation process could be time-consuming. This issue can be solved by introducing data caching and indexing of previously recommended devices at the control node level. This might help decrease the time needed for recommendations and limit the number of data access. This leads us to the third challenge.
- Caching user data by the recommendation system could violate user privacy policy when the user updates it. A user could permit our system to use for the first time a part of the data, then he decides to no longer share access. To solve this challenge, we must provide a privacy policy update mechanism triggered once the user updates his permissions.

4   Related Work

After the introduction of SWoT, many approaches relying on semantic device annotation were introduced.

In [4], authors propose a recommender system based on a service broker. They utilize the service-oriented architecture where devices are conceptualized as services for accessing their functionalities. The framework uses Agglomerative bottom-up Clustering to increase the velocity of recommendations. Similarity between services is calculated using the Normalized Google Distance (NGD).

In [6] authors propose a context-aware semantic-based discovery mechanism. The architecture is composed of three main levels. Directory level which uses a multi-proxy module for translation between devices using different protocols. Devices are then annotated semantically through ontological concepts. Constrained-network level where a concept directory containing entries of used concepts for annotations and the URL of annotated resources is defined. This facilitates the discovery of similar devices annotated using the same ontology concepts. Unconstrained network (Internet) level where resource directories communicate over a P2P overlay for a global device discovery.

In [8] authors introduce a distributed modular directory of service properties. Authors try to solve the problem of complexity and low processing of ontologies by providing distributed independent directories called search providers. Each search provider represents a semantic predicate in an ontology. The ontology can be extended by adding a new search provider. To discover services, the architecture enables complex querying of search providers through query federation. The service discovery mechanism is based on the Virtual State Layer (VSL) middleware which implements a security-by-design, and secure data access is ensured through role-based access model.

In [12], authors present a social architecture for IoT service recommendation. The architecture has 3 layers: a perception layer, a network layer, and an interoperability layer. The perception layer detects devices in the architecture. The network layer maps IoT data to communication protocols. The interoperability layer is based on oneM2M[6] and fiware[7] protocols for ensuring seamless communication between devices. The architecture also includes a recommender system that manages user-user, thing-thing, and user-thing relations based on data retrieved from the interoperability layer. Relations are then used for service recommendation.

To the best of our knowledge, our work is the first attempt to annotate smart devices using the LOD cloud for IoT device recommendation. Other works such as [5,10] benefit from LOD to annotate sensor data for reasoning and knowledge discovery. Overall in the device discovery domain, only [8] took into consideration the incapability of single domain ontologies to cover all necessary domain aspects. Besides, their approach was the only approach to provide a secure data access. However, contrary to their approach, by employing an LOD based annotation we are benefiting from the regularly updated data cloud without any additional update to the device annotation schema. In terms of privacy and security, our SOLID based approach is a more dynamic user-centric approach where users control access to their data in contrast to their role-based data access approach.

5   Conclusion and Future Work

In this paper, we presented our IoT device recommendation approach that annotates devices using the LOD resources. Our approach benefits from existing LOD-based semantic measures for similarity calculation between annotated devices. We utilized a user-user social relation network to recommend devices to a user based on his friend's ecosystems. Taking into consideration data privacy, we utilize the SOLID architecture providing device owners more control over their data and who can access it.

---

[6]   https://www.onem2m.org/.

[7]   https://www.fiware.org/developers/catalogue/.

In the future, we intend to start our experimentation and evaluation process in a smart-home use case. We also aim to further improve our approach and solve the previously mentioned and/or eventual new challenges.

References

1. Beltran, V., Ortiz, A.M., Hussein, D., Crespi, N.: A semantic service creation platform for social IoT (March 2014). https://doi.org/10.1109/WF-IoT.2014.6803173
2. Chen, Y., Zhou, M., Zheng, Z., Chen, D.: Time-aware smart object recommendation in social internet of things. IEEE Internet Things J. **7**(3), 2014–2027 (2020)
3. Cheniki, N., Belkhir, A., Sam, Y., Messai, N.: LODS: a linked open data based similarity measure. In: 2016 IEEE 25th International Conference on Enabling Technologies: Infrastructure for Collaborative Enterprises (WETICE), Paris, France, pp. 229–234 (June 2016)
4. Chirila, S., Lemnaru, C., Dinsoreanu, M.: Semantic-based IoT device discovery and recommendation mechanism. In: 2016 IEEE 12th International Conference on Intelligent Computer Communication and Processing (ICCP), pp. 111–116 (2016)
5. Gyrard, A.: An architecture to aggregate heterogeneous and semantic sensed data. In: Cimiano, P., Corcho, O., Presutti, V., Hollink, L., Rudolph, S. (eds.) ESWC 2013. LNCS, vol. 7882, pp. 697–701. Springer, Heidelberg (2013). https://doi.org/10.1007/978-3-642-38288-8_54
6. Mecibah, R., Djamaa, B., Yachir, A., Aissani, M.: A scalable semantic resource discovery architecture for the internet of things. In: Demigha, O., Djamaa, B., Amamra, A. (eds.) CSA 2018. LNNS, vol. 50, pp. 37–47. Springer, Cham (2019). https://doi.org/10.1007/978-3-319-98352-3_5
7. Meymandpour, R., Davis, J.G.: Enhancing recommender systems using linked open data-based semantic analysis of items. In: 3rd Australasian Web Conference (AWC 2015), Sydney, Australia (27–30 January 2015)
8. Pahl, M., Liebald, S.: A modular distributed IoT service discovery. In: 2019 IFIP/IEEE Symposium on Integrated Network and Service Management (IM), pp. 448–454 (2019)
9. Passant, A.: Measuring semantic distance on linking data and using it for resources recommendations. In: AAAI Spring Symposium: Linked Data Meets Artificial Intelligence, vol. 77, p. 123 (2010)
10. Pfisterer, D., et al.: Spitfire: toward a semantic web of things. IEEE Commun. Mag. **49**(11), 40–48 (2011)
11. Piao, G., Ara, S., Breslin, J.G.: Computing the semantic similarity of resources in DBpedia for recommendation purposes. In: Qi, G., Kozaki, K., Pan, J.Z., Yu, S. (eds.) JIST 2015. LNCS, vol. 9544, pp. 185–200. Springer, Cham (2016). https://doi.org/10.1007/978-3-319-31676-5_13
12. Saleem, Y., Crespi, N., Rehmani, M.H., Copeland, R., Hussein, D., Bertin, E.: Exploitation of social IoT for recommendation services. In: 2016 IEEE 3rd World Forum on Internet of Things (WF-IoT), pp. 359–364 (2016)
13. Zorgati, H., Djemaa, R.B., Amor, I.A.B.: Service discovery techniques in internet of things: a survey. In: 2019 IEEE International Conference on Systems, Man and Cybernetics (SMC), pp. 1720–1725 (2019)

Learning Performance Models Automatically

Runan Wang[(✉)]

Department of Computing, Imperial College London, London, UK
runan.wang19@imperial.ac.uk

Abstract. To ensure the quality of frequent releases in DevOps context, performance models enable system performance simulation and prediction. However, building performance models for microservice or serverless-based applications in DevOps is costly and error-prone. Thus, we propose to employ model discovery learning for performance models automatically. To generate basic models to represent the application, we first introduce performance-related TOSCA models as architectural models. Then we transform TOSCA models into layered queueing network models. A main challenge of performance model generation is model parametrization. We propose to learn parametric dependencies from monitoring data and systems analysis to capture the relationship between input data and resource demand. With frequent releases of new features, we consider employing detecting parametric dependencies incrementally to keep updating performance models in each iteration.

Keywords: Model discovery · Performance models · Model parameterization · Parametric dependencies

1 Introduction

DevOps has been widely adopted by in the industry, becoming an important part of software development methodologies. However, how to keep the rapid pace of deliveries and ensure the quality of the software at the same time is an open challenge in the context of DevOps.

Learning performance models is a model discovery process of building performance models with accurate specification of properties based on learning information through systems with testing and monitoring. Learning performance models for prediction is quite useful in DevOps practices because it can provide the possibility to answer a series of what-if questions about system performance. Besides, learning performance prediction models can also help to speculate the system structure without looking deep into each component development for both developers and cloud infrastructure providers. Developers can analyze and predict system performance with simulation results and calibrate the performance models in each DevOps iteration. In general, the providers are not

Supervised by: G. Casale and A. Filieri.

explicit about the internals of microservices (e.g., source codes). Thus, learning performance models through monitoring the deployed microservices can help to generate basic performance models with limited information about the development of each microservice. The infrastructure providers can infer the customer usage of microservices and achieve resource management with such performance prediction models.

In order to satisfy the requirement of automation in DevOps, high-degree automation is required in learning performance models. In this paper, we propose to learn performance models and their parameters from data-driven analysis on monitoring data. Compared with these existing works modelling with PCM or UML specifications [2,12] that requires deep understanding of internal components or manual definition, the generation of a basic TOSCA model can be done by analyzing network traffic, allowing automatic extraction and generation. With parameters of performance models specified in TOSCA models, it can be transformed into LQN models automatically. By learning parametric dependencies from monitoring data and code-level analysis, resource demand can be calibrated with additional dependencies and the topological structure can be updated to adapt to new changes.

This paper aims to provide an insight into learning performance models automatically for DevOps practices, which will combine model discovery, program analysis, and machine learning approaches to carry out the following research problems.

- P1: Generate and transform performance models automatically by learning monitoring data.
- P2: Accurate model parameters estimation.
- P3: Extract and learn parametric dependencies for performance models incrementally.
- P4: Iteratively update and enrich performance models in DevOps cycles.

2   Related Work

Performance Model Generation. Performance models are an abstraction of a real system, which can describe the system with a simplified representation and enable simulations and predictions. To learn performance models in the context of DevOps, it is important to involve both architectural and stochastic models. This is because generating architectural models can help adapt to new changes and involving with stochastic models can be solved with analytical solvers or simulations. Existing methods for generating architecture-level models like UML [12] and Palladio component model (PCM) [2] rely on manual analysis and domain knowledge, which cannot satisfy the requirement of high-degree automation in DevOps. In addition, the description languages of architectural models in previous works are independent of deployment, which brings complexity to frequent deployment and automatic calibration of performance models responding to new alternatives during this process.

Model Parameterization. Resource demand is a critical factor among model parameters that should be specified. The accuracy of resource demand estimation is decisive for the performance of performance models. Generally, regression-based algorithms have been employed to solve the resource demand estimation mostly based on utilization and response times [8,11]. With machine learning, there are existing studies that estimate resource demands (e.g., CPU utilization) for the ability to predict time series data [3] and can also help to select the optimal approaches for estimation on account of varieties of existing resource demand approaches [6].

Parametric Dependencies for Performance Models. Accurate resource demand estimation requires identifying appropriate input parameters, their dependencies, and their impact on the different elements of the performance model [4]. With current adoption of DevOps, parameterizing the performance models is more challenging for frequent releases and new features. To identify and specify the parametric dependencies of performance models, reverse engineering has been adopted in some works including static analysis and dynamic analysis of source codes [7,9]. To characterize the relationship between dependent and independent variables, [5] provides a machine learning-based feature selection approach to represent parametric dependencies from monitoring data. However, for microservice or serverless-based architecture, incrementally detecting new changes and features to update parametric dependencies is required in each DevOps cycle and the complexity of dependencies analysis should also be considered.

3   Methodology

For learning performance models, sock shop[1] has been selected as the system under test (SUT). Sock shop is a microservices-based application that has been widely used as benchmarks and study case for performance and quality of cloud service [10,13], which simulates an e-commerce website on which users can log in to the account, view and buy items. The application has been build with different language frameworks including Java Spring, GoLang and Node JS, along with MongoDB and MySQL as data storage, which is intended to provide a testbed for microservices and cloud orchestration. Sock shop provides several services that are packaged in Docker containers and all services communicate using REST over HTTP.

3.1   TOSCA Modelling

To generate architectural performance models, we propose to use TOSCA modelling with performance-related specification that can be further transformed the specification model into layered queueing networks for performance prediction.

---

[1] https://microservices-demo.github.io.

TOSCA [1] is an OASIS standard language that can standardize the specification of cloud applications and allow deployment and management based on TOSCA modelling. Components and topological structure of the application can be extracted by learning from monitoring data. With TOSCA, an application can be specified in a service template. The nodes representing application components are modelled within node templates by classifying different endpoints and the edges between different components are extracted by analyzing the network traffic between different services. After constructing the structure of the application, based on monitoring data, resource demand of each microservice can be estimated by regression approaches.

The application sock shop has been modelled with TOSCA based on RADON modeling profile[2]. RADON TOSCA provides an extended TOSCA modelling that helps to describe microservice and serverless orchestration. The representation of sock shop with RADON type hierarchy is shown in Fig. 1. As it can be observed in Fig. 1, the client node represents the close workload for the application. For these three target services, all of them are hosted on respective containers. The relationship 'routed_front' indicated that the requests made by the client could be sent to the frontend service and then frontend service would route requests to cart and catalogue service according to the requested endpoints with defined relationships 'catalogue_request' and 'cart_request'.

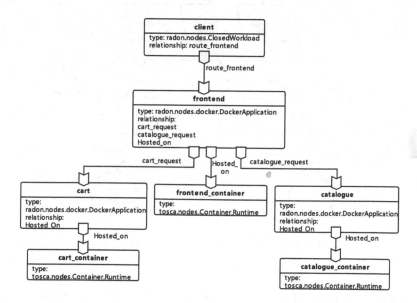

Fig. 1. Representation of modelling Sock Shop with RADON type hierarchy

[2] https://radon-h2020.eu/wp-content/uploads/2019/11/D4.3-RADON-Models-I.pdf.

3.2   Layered Queueing Network (LQN) Models

After specifying the application with TOSCA, RADON decomposition tool has been employed to generate LQN models, which can parse YAML file and transform the topology graph into LQN models automatically[3]. The generated LQN model is shown in Fig. 2. The task client simulated the concurrent users that made requests to different services. Each task in LQN model represented a certain microservice except to client. Entries in each task were defined as different classes of workload according to both endpoints the requests made to and different HTTP methods. The time consuming on the processor of each activity has been declared with the mean service demand. All of the mentioned parameters in LQN models have been specified with TOSCA in properties within node and relationship templates. By solving LQN models with simulation or analytical solvers, we can predict the performance of the application.

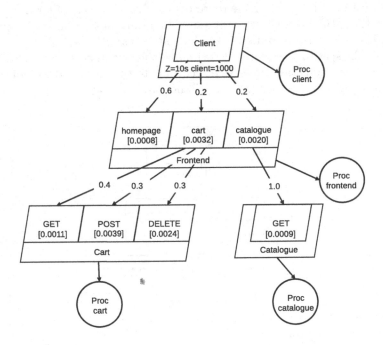

Fig. 2. The generated LQN model for Sock Shop

3.3   Learning Parametric Dependencies from Code Analysis

Static analysis of code allows to check, debug and obtain information of programs without actual execution of codes. In DevOps practices, after committing new version of code in the development phase, source code analysis could be

---

[3] https://radon-h2020.eu/wp-content/uploads/2020/01/D3.2-Decomposition-Tool-I.pdf.

employed to capture new changes that are related to performance, e.g., adding new functions and changing inside a method. TOSCA models can be updated according to these detected changes of source code, providing a more accurate specification in each iteration of committing source code. Continuous TOSCA model updating can keep the architectural performance model consistent with the latest development. In order to response to new changes in DevOps context, incremental detection and analysis of alternatives should be implemented, instead of analyzing and modelling the whole system every iteration.

Parametric dependencies can also be leveraged to optimize the resource demand estimation. By tracing the data flow and control flow that generated from source code and execution, the relations between input parameters and corresponding methods can be captured, including branches, loops, numbers of service invocations. Then the resource demand can be formulated with parametric dependencies as additional features to revise the estimated value that could be fed into TOSCA parameterization.

4   Conclusion and Future Work

In this paper, we propose an approach to learn performance models automatically that reach the objective of the research problems. To generate architectural models, we introduce TOSCA modelling by analyzing monitoring data. Then we transform performance-related TOSCA models into LQN models for simulation and prediction. To enrich and update performance models responding to new changes, we also present that learning parametric dependencies can provide a possible way to incrementally detect changes and also help to calibrate the estimation of resource demand.

In the future, we plan to integrate the performance models into DevOps pipeline and implement a framework enabling to update performance prediction models automatically.

References

1. Binz, T., Breitenbücher, U., Kopp, O., Leymann, F.: TOSCA: portable automated deployment and management of cloud applications. In: Bouguettaya, A., Sheng, Q., Daniel, F. (eds.) Advanced Web Services, pp. 527–549. Springer, New York (2014). https://doi.org/10.1007/978-1-4614-7535-4_22
2. Brosig, F., Kounev, S., Krogmann, K.: Automated extraction of palladio component models from running enterprise java applications. In: Proceedings of the Fourth International ICST Conference on Performance Evaluation Methodologies and Tools, pp. 1–10 (2009)
3. Duggan, M., Mason, K., Duggan, J., Howley, E., Barrett, E.: Predicting host CPU utilization in cloud computing using recurrent neural networks. In: 2017 12th International Conference for Internet Technology and Secured Transactions (ICITST), pp. 67–72 (2017)

4. Eismann, S., Walter, J., von Kistowski, J., Kounev, S.: Modeling of parametric dependencies for performance prediction of component-based software systems at run-time. In: 2018 IEEE International Conference on Software Architecture (ICSA), pp. 135–13509 (2018)
5. Grohmann, J., Eismann, S., Elflein, S., Kistowski, J.V., Kounev, S., Mazkatli, M.: Detecting parametric dependencies for performance models using feature selection techniques. In: 2019 IEEE 27th International Symposium on Modeling, Analysis, and Simulation of Computer and Telecommunication Systems (MASCOTS), pp. 309–322 (2019)
6. Grohmann, J., Herbst, N., Spinner, S., Kounev, S.: Using machine learning for recommending service demand estimation approaches-position paper. In: CLOSER, pp. 473–480 (2018)
7. Kappler, T., Koziolek, H., Krogmann, K., Reussner, R.: Towards automatic construction of reusable prediction models for component-based performance engineering. Software Engineering, **2008** (2008)
8. Kraft, S., Pacheco-Sanchez, S., Casale, G., Dawson, S.: Estimating service resource consumption from response time measurements. In: Proceedings of the Fourth International ICST Conference on Performance Evaluation Methodologies and Tools, pp. 1–10 (2009)
9. Krogmann, K., Kuperberg, M., Reussner, R.: Using genetic search for reverse engineering of parametric behavior models for performance prediction. IEEE Trans. Software Eng. **36**, 865–877 (2010)
10. Nguyen, C., Mehta, A., Klein, C., Elmroth, E.: Why cloud applications are not ready for the edge (yet). In: Proceedings of the 4th ACM/IEEE Symposium on Edge Computing, pp. 250–263 (2019)
11. Pérez, J.F., Pacheco-Sanchez, S., Casale, G.: An offline demand estimation method for multi-threaded applications. In: 2013 IEEE 21st International Symposium on Modelling, Analysis and Simulation of Computer and Telecommunication Systems, pp. 21–30. IEEE (2013)
12. Petriu, D.C., Shen, H.: Applying the UML performance profile: graph grammar-based derivation of LQN models from UML specifications. In: Field, T., Harrison, P.G., Bradley, J., Harder, U. (eds.) TOOLS 2002. LNCS, vol. 2324, pp. 159–177. Springer, Heidelberg (2002). https://doi.org/10.1007/3-540-46029-2_10
13. Rahman, J., Lama, P.: Predicting the end-to-end tail latency of containerized microservices in the cloud. In: 2019 IEEE International Conference on Cloud Engineering (IC2E), pp. 200–210. IEEE (2019)

Blockchain-Based Business Processes: A Solidity-to-CPN Formal Verification Approach

Ikram Garfatta[1,2]($\boxtimes$), Kaïs Klai[2], Mahamed Graïet[3,4], and Walid Gaaloul[5]

[1] University of Tunis El Manar, National Engineering School of Tunis, OASIS, Tunis, Tunisia
[2] University Sorbonne Paris North, LIPN UMR CNRS 7030, Villetaneuse, France
`ikram.garfatta@lipn.univ-paris13.fr`
[3] Higher Institute for Computer Science and Mathematics, University of Monastir, Monastir, Tunisia
[4] National School for Statistics and Information Analysis, Rennes, France
[5] Institut Mines-Télécom,Télécom SudParis, UMR 5157, SAMOVAR, Évry, France

Abstract. With its span of applications widening by the day, the technology of Blockchain has been gaining more interest in different domains. It has intrigued many investors, but also numerous malicious users who have put different Blockchain platforms under attack. It is therefore an inescapable necessity to guarantee the correctness of smart contracts as they are the core of Blockchain applications. Existing verification approaches, however, focus on targeting particular vulnerabilities, seldom supporting the verification of domain-specific properties.

In this paper, we propose a translation of Solidity smart contracts into CPNs (Coloured Petri nets) and investigate the capability of *CPN Tools* to verify CTL (Computation Tree Logic) properties.

Keywords: Blockchain · Formal verification · Smart contract · Solidity · Coloured Petri nets · CTL properties

1 Introduction

Within the span of the two last decades, many advances have been made in the world of Blockchain, allowing this technology to expand its reach to a myriad of application domains including Business Process Management (BPM) [13]. A Blockchain platform can indeed provide a reliable execution of business processes (BPs) even within a trustless network, especially thanks to the concept of smart

Supervised by Kaïs Klai, University Sorbonne Paris North, LIPN UMR CNRS 7030, Villetaneuse, France, and Mahamed Graïet Higher Institute for Computer Science and Mathematics, University of Monastir, Monastir, Tunisia and National School for Statistics and Information Analysis, Rennes, France.
Co-directed by Walid Gaaloul, Institut Mines-Télécom,Télécom SudParis, UMR 5157, SAMOVAR, Paris, France.

© Springer Nature Switzerland AG 2021
H. Hacid et al. (Eds.): ICSOC 2020 Workshops, LNCS 12632, pp. 47–53, 2021.
https://doi.org/10.1007/978-3-030-76352-7_7

contracts. In a BPM context, a smart contract can define business collaborations in general and inter-organizational BPs in particular. In fact, smart contracts are pieces of script code that act like autonomous software agents, used to enforce management rules on the execution of transactions on the Blockchain. They are stored on and executed by the Blockchain and therefore inherit its characteristics, particularly its immutability. This same feature can, however, turn into a weak spot for such contracts. In fact, as a smart contract cannot be altered once it has been deployed on the Blockchain, it cannot be corrected either, which makes verifying its correctness prior to its deployment an indispensable necessity. Furthermore, the correctness verification is an important aspect for the design of blockchain-based BPs. The assessment of such processes involves both requirements validation and consistency.

The main long-term objective of this thesis is therefore to develop an approach that allows to construct correct blockchain-based BPs. In this paper, we present our progress for the first milestone towards this goal, which we define as the verification of smart contracts in a general context. We are interested in Ethereum smart contracts as it is currently the second largest cryptocurrency platform after Bitcoin besides being the inaugurator of smart contracts, and particularly those written in Solidity [1] as it is the most popular language used by Ethereum. The contribution described herein is a first step towards a formal verification approach based on CPNs [7] for Solidity smart contracts.

Existing studies on the formal verification of smart contracts follow two main streams. The first group of studies are based on theorem proving [2–4]. In this case, the verification is not automated and requires the user's expertise in the manipulation of the used theorem prover as well as manual intervention in discharging proofs. The second group of studies are based on symbolic model checking coupled with complementary techniques such as symbolic execution and abstraction [5,8,11,12,14]. In order to use symbolic execution to generate the traces that would be used for the verification, the proposed approaches usually use under-approximation (e.g., in the form of loop bounds) which means that critical violations can be overlooked. This explains the presence of false negatives and/or positives in their reported results. We also note that most of the existing studies target specific vulnerabilities in smart contracts, and few are those that allow expressing customizable properties, in which case they are control flow-related properties. In fact, none of these studies target data-related properties. It is worth mentioning that most of the proposed approaches operate on the EVM bytecode rather than on the Solidity code because of the latter's lack of formal semantics. This, however, results in loss of contextual information, and consequently limits the range of properties that can be verified on the contract.

To overcome these shortcomings, we propose an algorithm for the translation of a Solidity smart contract into a hierarchical CPN model over which CTL properties can be verified. This work can easily be integrated as an extension layer into existing studies that rely on the translation of BP models into smart contracts as in [10] which generates Solidity code from models written in BPMN (Business Process Model and Notation), to verify their output and therefore check the correctness of the initial BP models.

2   Solidity-to-CPN Translation

To illustrate our approach and prove its feasibility, we adapt the Blind Auction example included in [1]. Figure 1 describes the workflow of the blind auction system. For a full description of the use case, we refer the reader to the *Solidity2CPN* document available at this repository[1].

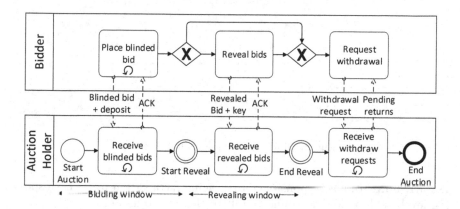

Fig. 1. Blind auction workflow

The general idea of our approach is to start from a CPN model representing the general workflow of the smart contract (level-0 model) and then to build on it by embedding it with submodels representing the smart contract functions (level-1 models). In a level-0 model, we distinguish the *user's behaviour* part which models the way users can interact with the system and the *smart contract's behaviour* part which represents the system. These two are linked via *communication places*. Figure 2 shows the level-0 model of the previously described blind auction use case.

We see a smart contract as a set of statements. A statement can be either a compound, a simple or a control one. A simple statement can be an assignment, a variable declaration, a sending or a returning statement. A control statement can be a requirement, a selection or a loop (a *for* or a *while* loop).

2.1   Translation Algorithm

Our proposed algorithms are structured as follows:

- EXTENDMODEL takes as input the level-0 CPN model and builds the extended hierarchical model by by calling INSERTSUBMODEL for each transition corresponding to a function in the Solidity smart contract.

---

[1] https://github.com/garfatta/solidity2cpn.

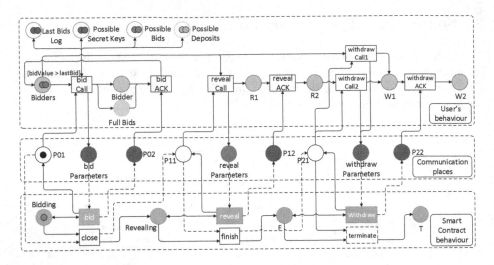

Fig. 2. Blind auction - level-0 model

- INSERTSUBMODEL is responsible for replacing a transition by its corresponding level-1 submodel and connecting it to the level-0 model.
- CREATESUBMODEL is the main algorithm. It generates the level-1 submodel for each transition by browsing the body of its corresponding function recursively and creates CPN patterns according to the type of the processed statement that interconnect to create the transition's submodel.

2.2   Application on the Blind Auction Use Case

The application of the algorithm on the level-0 model of the Blind Auction use case (see Fig. 2) yields a hierarchical CPN model whose level-1 submodels are created by the execution of CREATESUBMODEL. Figure 3 shows the submodel corresponding to transition *withdraw* in the level-0 model. The rest of the submodels can be found in the online Solidity2CPN document (See footnote 1).

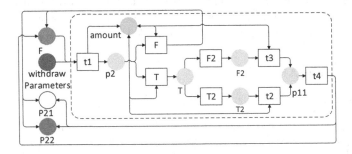

Fig. 3. SubModel of transition *withdraw*

3   Smart Contract Verification via CPN Tools

Having established the CPN model for a smart contract, verifying properties of the smart contract would come down to verifying properties on the CPN model. We have implemented the CPN model for our Blind Auction use case using *CPN Tools* which leverages explicit model checking techniques, and investigated its potential in the verification of behavioural and contract-specific properties.

In Table 1 we present state space analysis statistics for different initial marking values. We note that the unprovided values mean that the state space generation had not finished after several hours of execution. This is due to the infamous state space explosion problem associated with explicit state space exploration.

Table 1. State Space Analysis Results for different initial markings

Bidders		1	2	3	4	5	1	1	1
Possible bids		1	1	1	1	1	2	3	4
Possible secret keys		1	1	1	1	1	2	3	4
Possible deposits		1	1	1	1	1	2	3	4
State space	Without hierarchy	4 s	4 s	6 s	252 s	–	4 s	30 s	–
generation time	With hierarchy	5 s	5 s	10 s	1001 s	–	5 s	74 s	–
#Nodes	Without hierarchy	24	235	3118	47621	–	484	19984	–
	With hierarchy	44	583	9166	156117	–	1424	65513	–
#Arcs	Without hierarchy	26	378	7106	145062	–	555	22980	–
	With hierarchy	46	900	19784	446326	–	1545	70704	–
#Dead	Without hierarchy	3	10	35	124	–	49	1999	–
Markings	With hierarchy	3	10	35	124	–	49	1999	–

The state space report generated by *CPN Tools* allows the deduction of several general behavioural properties. For instance, in our use case application, the report confirms the boundedness of all the places of the modelled system. More specific properties can be verified by elaborating CTL properties. For instance, we can formulate a *termination* property to check the model's capability to always reach a terminal state (a dead marking) where certain conditions are met. We include the definition of such a property in the Solidity2CPN document[6].

4   Conclusion

The goal of our work is to propose a formal approach for the verification of smart contracts. In this context, we propose in this paper a translation algorithm that generates a hierarchical CPN model representing a given Solidity smart contract, including both its control-flow and data aspects. CTL properties are then verified on the CPN model to check corresponding properties on the smart contract, unrestrictedly to certain predefined vulnerabilities.

In view of the results presented in this paper, it may be concluded that *CPN Tools* does not hold much potential for the verification of properties on CPN models of smart contracts due to the state space explosion problem. We do prove, however, that the idea of using CPNs as a representation formalism is promising for it allows the consideration of the data aspect, and thus the formulation of contract-specific properties. To overcome the encountered limitations, we intend to investigate the potential of *Helena* [6] as an analyzer for High Level Nets. This tool offers on-the-fly verification of LTL properties, which unlike the verification of CTL properties offered by *CPN Tools*, does no always require the generation of the whole state space. To further improve the tool's performance, we also intend to work on *Helena*'s model checker by embedding it with an extension to an existing technique previously developed to deal with the state space explosion problem in regular Petri nets [9] and applying it on CPNs.

References

1. Solidity documentation. https://solidity.readthedocs.io/en/latest/
2. Formal verification for solidity contracts - ethereum community forum, October 2015. https://forum.ethereum.org/discussion/3779/formal-verification-for-solidity-contracts
3. Amani, S., Bégel, M., Bortin, M., Staples, M.: Towards verifying ethereum smart contract bytecode in isabelle/hol. In: Proceedings of the 7th ACM SIGPLAN International Conference on Certified Programs and Proofs, pp. 66–77 (2018)
4. Bhargavan, K., et al.: Formal verification of smart contracts: short paper. In: Proceedings of the 2016 ACM Workshop on Programming Languages and Analysis for Security, PLAS@CCS 2016, Vienna, Austria, 24 October 2016, pp. 91–96 (2016)
5. Chen, T., Li, X., Luo, X., Zhang, X.: Under-optimized smart contracts devour your money. In: IEEE 24th International Conference on Software Analysis, Evolution and Reengineering, SANER 2017, Austria, 20–24 February 2017, pp. 442–446 (2017)
6. Evangelista, S.: High level petri nets analysis with helena. In: Ciardo, G., Darondeau, P. (eds.) ICATPN 2005. LNCS, vol. 3536, pp. 455–464. Springer, Heidelberg (2005). https://doi.org/10.1007/11494744_26
7. Jensen, K., Kristensen, L.M.: Coloured Petri Nets: Modelling and Validation of Concurrent Systems, 1st edn. Springer, Heidelberg (2009). https://doi.org/10.1007/b95112 10.1007/b95112
8. Kalra, S., Goel, S., Dhawan, M., Sharma, S.: ZEUS: analyzing safety of smart contracts. In: 25th Annual Network and Distributed System Security Symposium, NDSS 2018, San Diego, California, USA, 18–21 February 2018 (2018)
9. Klai, K., Poitrenaud, D.: MC-SOG: an LTL model checker based on symbolic observation graphs. In: 29th International Conference Applications and Theory of Petri Nets, PETRI NETS 2008, China, 23–27 June, Proceedings. pp. 288–306 (2008)
10. López-Pintado, O., García-Bañuelos, L., Dumas, M., Weber, I., Ponomarev, A.: Caterpillar: a business process execution engine on the ethereum blockchain. Softw. Pract. Exp. **49**(7), 1162–1193 (2019)

11. Luu, L., Chu, D., Olickel, H., Saxena, P., Hobor, A.: Making smart contracts smarter. In: Proceedings of the 2016 ACM SIGSAC Conference on Computer and Communications Security, Vienna, Austria, 24–28 October 2016, p. 254–269 (2016)
12. Mavridou, A., Laszka, A., Stachtiari, E., Dubey, A.: Verisolid: correct-by-design smart contracts for ethereum. In: Financial Cryptography and Data Security - 23rd International Conference, St. Kitts and Nevis, 18–22 February 2019, p. 446–465 2019)
13. Mendling, J., Weber, I.: Blockchains for business process management - challenges and opportunities. EMISA Forum **38**(1), 22–23 (2018)
14. Torres, C.F., Schütte, J., State, R.: Osiris: hunting for integer bugs in ethereum smart contracts. In: Proceedings of the 34th Annual Computer Security Applications Conference, ACSAC 2018, PR, USA, 03–07 December 2018, p. 664–676 (2018)

Formal Quality of Service Analysis in the Service Selection Problem

Agustín Eloy Martinez-Suñé[✉]

Departamento de Computación, Universidad de Buenos Aires,
Buenos Aires, Argentina
aemartinez@dc.uba.ar

Abstract. The Service Selection problem has driven a lot of attention from the Service-Oriented community in the past few decades. Rapidly evolving cloud computing technologies foster the vision of a Service-Oriented Computing paradigm where multiple providers offer specific functionalities as services that compete against each other to be automatically selected by service consumers. We present a research program that focuses on *Quality of Service aware Service Selection*. We discuss our vision and research methodology in the context of the state of the art of the topic and review the main contributions of our approach.

1 Introduction

One of the main characteristics of the Service-Oriented Computing paradigm is the capability of building complex applications by composing web services as fundamental building blocks [3]. The emergence of this paradigm has been accompanied by an increasing number of globally available computational resources and communication infrastructure, and also by a deep transformation of the business models associated with the construction of software systems. Providers of cloud computing platforms such as *Amazon Web Services* and *Google Cloud* rely on this type of infrastructure to offer a wide variety of options to developers and web service providers, where a distinctive feature is the high level of granularity with which companies can optimize the use of these resources to better suit their business goals. Pricing schemes that depend on the amount of time a computational resource is used or the amount of memory available are some examples of the attributes used as knobs. Emerging technologies such as *serverless computing* or *Function as a Service (FaaS)* open up even more possibilities for the vision of a Service-Oriented Computing paradigm.

From an academic point of view, one of the issues that has driven a lot of attention from the Service-Oriented community in the past few decades is the

---

Research partly supported by the European Unions Horizon 2020 research and innovation program under the Marie Skodowska-Curie grant agreement No 778233, and by Universidad de Buenos Aires by grant UBACyT 20020170100544BA.
Supervised by Carlos G. López Pombo.

H. Hacid et al. (Eds.): ICSOC 2020 Workshops, LNCS 12632, pp. 54–60, 2021.
https://doi.org/10.1007/978-3-030-76352-7_8

Service Selection problem [3, pt. II]. Stated in general terms, this is the problem of selecting the most appropriate web service(s), from a pool of available ones, that best match(es) the functional and non-functional requirements and constraints specified by the client. Such a procedure is referred to as the determination of a Service Level Agreement (SLA) between the service consumer and the provider. In this research project we focus on *Quality of Service (QoS) aware service selection*: given multiple providers offering services that are satisfactory in terms of the functional requirements, how can we decide the best match in terms of their QoS, non-functional, characteristics.

The rest of this paper is structured as follows: Sect. 2 gives a clear statement of the problem and introduces the main questions that guide our research, Sect. 3 reviews the state of the art and puts our approach in context, Sect. 4 reviews the main contributions of our approach and, finally, Sect. 5 presents some concluding remarks.

2 Problem Definition

Our research problem can be stated more clearly in the following way:

Definition 1 (QoS-aware service selection problem). *Given a set* $S = \{s_1, s_2, \ldots, s_n\}$ *of QoS profiles each describing the Quality of Service aspects of different services being offered, and given a* QoS requirement *Rq describing what the client needs (or prefers) in terms of QoS:*

Choose a service $s \in S$ *that fulfils (or best suits) the client needs Rq.*

The reader might be familiar with different variations of this problem such as the *service ranking* problem where the goal is to obtain an ordering of S, instead of just one particular element of it; or the problem in the context of *web services composition* where the goal usually is to choose multiple services such that, if composed according to a given plan, the aggregated Quality of Service satisfies predefined needs or established criteria. Some variations of the problem will drive important research questions but the core challenges are captured by Definition 1. Some of the research questions that immediately arise from the problem statement are:

- What is the appropriate representation for the *QoS profile* and the *QoS requirements*?
- What is the appropriate analysis procedure to automatically find the right $s \in S$? Can this selection be performed in a computationally efficient way?
- How are the *QoS profiles* obtained? Are they manually designed or can they be automatically constructed?

Our vision for this research project is to explore, from both a theoretical and a practical point of view, the boundaries of what can be formally specified and automatically analyzed concerning Quality of Service and non-functional requirements. For this we leverage on the existing literature about the problem and approach the research from a Formal Methods [2,10] perspective.

3   Related Work

There has been extensive discussion on what constitutes non-functional require-ments [5,15]. Nevertheless, most of the research in service selection has focused on QoS attributes that allow some type of measurement. We follow this assump-tion since it is a consequence of having formal treatment and automatic analysis as goals and, according to the taxonomy of requirements presented in [5], we call this subset *quantitative attributes*. Some examples of these attributes are *response time, memory usage, price, reputation,* etc. For an extensive study of the state of the art in the service selection problem, we point the interested reader to [14] where Moghaddam et al. present a comparative review of existing approaches and to [7] where the authors perform a systematic literature review. A clear overview of the problem and its foundations can also be found in [8].

One way of classifying the literature is by how each work models QoS attributes and policies. This yields three broad categories: single-value attributes, range-valued attributes, and probabilistic attributes. In the single-value setting the QoS provided for a particular attribute is modelled as a constant value, an example of this can be seen in [18] where the initial quality model proposed by the authors is a vector with specific values of attributes such as price, dura-tion, reputation, etc. Instead, in the range-valued setting the model is usually an interval or a set of values (e.g. *resp. time* $= \{100, 200, 300, 400\}$ or *resp. time* $= [100, 400]$), a description of this kind of models can be found in [11, Section 2]. The latter perspective emerges by acknowledging the limitations of the former: there are scenarios in which the single value model becomes insufficient. Simi-larly, what has emerged in recent years is the proposal of models that capture the probabilistic nature of many quality of service attributes, a recent work in this direction can be found in [19].

In this context, one of the distinctive points of our research program is the perspective of treating QoS attributes from a general point of view while aiming to contribute with practical implementations of our proposals. Instead of study-ing the behavior of specific quality attributes such as *time, memory consumption,* or *availability* our approach makes no assumption on the attribute other than that they are interpreted as values. Such a homogeneous treatment of attributes enables the modeling of interdependences between them and a more holistic treatment of their space of values. Furthermore, this general point of view will also have an impact when we discuss our approach to the probabilistic setting.

4   Contributions

Under the light of Definition 1, having descriptions of s_i and Rq in a formal specification language turns the selection problem into the problem of checking whether a satisfaction judgment of the form $s_i \vdash Rq$ holds or not. Thus, a natural roadmap to follow starts with the search for an adequate logical language and the development of analysis methods for such language.

4.1   Towards a Formalization of Quality of Service Contracts

A relatively new approach to the analysis of hybrid systems' specification, due to Pappas et al. [17], integrates SMT-solving [4] with convex constraints [6], under the name of *SMC – Satisfiability Modulo Convex Optimization*. As a first step of our research program, we adopted *SMC* as a specification language and developed an efficient two-phase procedure for evaluating SLA based on SMC. A formula in this language is essentially a boolean combination of convex constraints and propositional variables, as it is shown in the following example.

Example 1 (QoS requirement in SMC). Consider a client interested in setting an upper bound for the execution time of the service with attributes that model the initialization time and the time it takes to process a kilobyte of information. A possible requirement can be stated like this:

$$initTime + 256 \cdot timePerKB \leq 50$$

The two-phase analysis procedure for deciding whether $s \vdash Rq$ is adapted to profit from the fact that specifications can be minimized in a preprocessing phase when the service is registered in the repository The expectation is that such preprocessing might produce an efficiency gain when, at runtime, $s \vdash Rq$ is checked to evaluate if an SLA is met. A detailed presentation, both theoretical and experimental, of the technique can be found in [12]. An interesting aspect of our proposal is that the minimization procedure is susceptible to be done incrementally by performing successive partial minimizations of a given specification before it is fully minimized. This exposes that most of the efficiency gain in the analysis is reached after investing a small portion of the total minimization time required.

4.2   Quality of Service Ranking

The aforementioned view guarantees the selection of a service satisfying the requirements of the executing application but it does not provide any insight when there is no service whose QoS profile fully complies with those requirements. In this scenario, an application would be pushed to abort its execution for there is no possible SLA on the QoS required.

A geometrical interpretation of the judgment $s \vdash Rq$ expresses that all satisfying values of real variables of s also satisfy Rq; on the other hand, a negative answer just means that there exist at least one value satisfying s and not satisfying Rq. This perspective opens up the possibility of evaluating partial compliance of QoS contracts by estimating what we call *inclusion ratio*, serving the purpose of ranking services by their degree of QoS compliance. A detailed explanation of the notion of inclusion ratio can be found in [13, Section 2.1]. In the following, we introduce the main concepts.

Given a formula α in a formal QoS specification language such as [13, Definition 1] we define the set of values satisfying the formula as $[\![\alpha]\!] = \{\ \bar{x} \mid \bar{x} \models \alpha\ \}$. Based on this definition, proving the formula $s \vdash Rq$ is to check whether

$[\![s]\!] \subseteq [\![Rq]\!]$ (i.e. whether all of the values of the quantitative attributes offered by the service are accepted by the requirements). Then, we propose to compute the volume of the intersection between $[\![s]\!]$ and $[\![Rq]\!]$ relative to the volume of $[\![s]\!]$, referred to as *inclusion ratio*. This indicator is essentially quantifying what is the percentage of the QoS values offered by the QoS profile that are actually accepted by the requirements contract. Under this interpretation we argue such indicator serves the purpose of quantifying the partial compliance of Rq by s and then, functionally compliant services can be QoS ranked by ordering them through inclusion ratio.

We developed a prototype tool based on a state-of-art convex polytope volume estimator to analyze partial compliance of Rq by s represented as sets of polytopes. The interested reader is pointed to [13] for further details of the technique, together with a nontrivial discussion about its implementation in the context of contracts consisting of sets of convex polytopes with non-empty intersection.

4.3   Probabilistic Treatment of Quality of Service

Recently there has been a lot of interest in the probabilistic treatment of QoS attributes in the context of web services [9, 16, 19]. This is a consequence of recognizing that: a) the behavior of many attributes is intrinsically probabilistic, and b) summarized metrics such as min, max or $average$ that give rise to single-value or range-valued models may not be sufficient to distinguish between different services. To the best of our knowledge, most of the research has focused on adopting QoS models based on discrete probabilistic distributions where each attribute is independent. To overcome these limitations we are currently developing an approach that considers probability density functions of continuous multivariate distributions as *QoS profiles* and probability bounds over regions of attribute values as *QoS requirements*. For the development of the procedure that verifies if a given QoS requirement holds we are studying state-of-art mathematical tools for integration and for operating with continuous distributions.

4.4   Validation and Experimental Evaluation

The methodology we propose for validating our work is both theoretical and experimental. On the one hand, rigorous mathematical proofs allow us to draw conclusions about the expressive power and limits of the formal languages we propose, and the correctness and computational complexity of the analysis methods we develop. On the other hand, extensive experimentation allows us to draw statistically meaningful conclusions to reason about the efficiency and scalability of the tools we develop. The difficulty of finding real-life scenarios where QoS specifications are formally described pushes us to construct case studies by automatically generating synthetic randomized specifications of different sizes and shapes. As it has been noted in [7], this is a common and accepted practice in the field. Nevertheless, to face the challenge of constructing more realistic case studies our approach for future work is the inclusion of data from publicly available datasets on QoS, such as QWS dataset [1] or the WS-DREAM dataset [20].

5 Summary

We presented a research program consisting of the development of formal languages and tools to analyze QoS attributes in the context of the Service Selection problem. In particular, our research focus on what is called quantitative attributes, the fragment of the QoS attributes that is considered to admit formal treatment. We discussed our vision and methodology in the context of the state of the art and briefly overviewed the main contributions of our approach. We believe the rapid development and radical transformation of cloud computing technologies give exciting possibilities for the research into Service Selection and the treatment of QoS in the context of Service-Oriented Computing.

References

1. Al-Masri, E., Mahmoud, Q.H.: QoS-based discovery and ranking of web services. In: 2007 16th International Conference on Computer Communications and Networks, pp. 529–534, August 2007. https://doi.org/10.1109/ICCCN.2007.4317873
2. Bjørner, D., Havelund, K.: 40 years of formal methods. In: Jones, C., Pihlajasaari, P., Sun, J. (eds.) FM 2014. LNCS, vol. 8442, pp. 42–61. Springer, Cham (2014). https://doi.org/10.1007/978-3-319-06410-9_4
3. Bouguettaya, A., Sheng, Q.Z., Daniel, F. (eds.): Web Services Foundations. Springer, New York (2014). https://doi.org/10.1007/978-1-4614-7518-7
4. De Moura, L., Bjørner, N.: Satisfiability modulo theories: introduction and applications. Commun. ACM **54**(9), 69–77 (2011). https://doi.org/10.1145/1995376.1995394
5. Glinz, M.: On non-functional requirements. In: 15th IEEE International Requirements Engineering Conference (RE 2007), pp. 21–26, October 2007. https://doi.org/10.1109/RE.2007.45
6. Grünbaum, B.: Convex Polytopes. Graduate Texts in Mathematics, Springer, New York (2003). https://doi.org/10.1007/978-1-4613-0019-9
7. Hayyolalam, V., Pourhaji Kazem, A.A.: A systematic literature review on QoS-aware service composition and selection in cloud environment. J. Netw. Comput. Appl. **110**, 52–74 (2018). https://doi.org/10.1016/j.jnca.2018.03.003
8. Ishikawa, F.: QoS-based service selection. In: Bouguettaya, A., Sheng, Q.Z., Daniel, F. (eds.) Web Services Foundations, pp. 375–397. Springer, New York (2014). https://doi.org/10.1007/978-1-4614-7518-7_15
9. Klein, A., Ishikawa, F., Bauer, B.: A probabilistic approach to service selection with conditional contracts and usage patterns. In: Baresi, L., Chi, C.-H., Suzuki, J. (eds.) ICSOC/ServiceWave 2009. LNCS, vol. 5900, pp. 253–268. Springer, Heidelberg (2009). https://doi.org/10.1007/978-3-642-10383-4_17
10. Kreiker, J., Tarlecki, A., Vardi, M.Y., Wilhelm, R.: Modeling, Analysis, and Verification - The Formal Methods Manifesto 2010 (Dagstuhl Perspectives Workshop 10482) (2011). https://doi.org/10.4230/DAGMAN.1.1.21. 20 pages

11. Martín-Díaz, O., Ruiz-Cortés, A., Durán, A., Benavides, D., Toro, M.: Automating the procurement of web services. In: Orlowska, M.E., Weerawarana, S., Papazoglou, M.P., Yang, J. (eds.) ICSOC 2003. LNCS, vol. 2910, pp. 91–103. Springer, Heidelberg (2003). https://doi.org/10.1007/978-3-540-24593-3_7

12. Martinez Suñé, A.E., Lopez Pombo, C.G.: Automatic quality-of-service evaluation in service-oriented computing. In: Riis Nielson, H., Tuosto, E. (eds.) COORDINATION 2019. LNCS, vol. 11533, pp. 221–236. Springer, Cham (2019). https://doi.org/10.1007/978-3-030-22397-7_13

13. Martinez Suñé, A.E., Lopez Pombo, C.G.: Quality of service ranking by quantifying partial compliance of requirements. In: Bliudze, S., Bocchi, L. (eds.) COORDINATION 2020. LNCS, vol. 12134, pp. 181–189. Springer, Cham (2020). https://doi.org/10.1007/978-3-030-50029-0_12

14. Moghaddam, M., Davis, J.G.: Service selection in web service composition: a comparative review of existing approaches. In: Bouguettaya, A., Sheng, Q.Z., Daniel, F. (eds.) Web Services Foundations, pp. 321–346. Springer, New York (2014). https://doi.org/10.1007/978-1-4614-7518-7_13

15. O'Sullivan, J., Edmond, D., ter Hofstede, A.: What's in a service? Distrib. Parallel Databases **12**(2), 117–133 (2002). https://doi.org/10.1023/A:1016547000822

16. Rosario, S., Benveniste, A., Haar, S., Jard, C.: Probabilistic QoS and soft contracts for transaction-based web services orchestrations. IEEE Trans. Serv. Comput. **1**(4), 187–200 (2008). https://doi.org/10.1109/TSC.2008.17

17. Shoukry, Y., Nuzzo, P., Sangiovanni-Vincentelli, A.L., Seshia, S.A., Pappas, G.J., Tabuada, P.: SMC: satisfiability modulo convex optimization. In: Proceedings of the 20th International Conference on Hybrid Systems: Computation and Control, HSCC 2017, New York, NY, USAm pp. 19–28. Association for Computing Machinery, April 2017. https://doi.org/10.1145/3049797.3049819

18. Zeng, L., Benatallah, B., Ngu, A., Dumas, M., Kalagnanam, J., Chang, H.: QoS-aware middleware for Web services composition. IEEE Trans. Software Eng. **30**(5), 311–327 (2004). https://doi.org/10.1109/TSE.2004.11

19. Zheng, H., Yang, J., Zhao, W.: Probabilistic QoS aggregations for service composition. ACM Trans. Web **10**(2), 12:1–12:36 (2016). https://doi.org/10.1145/2876513

20. Zheng, Z., Lyu, M.R.: WS-DREAM: a distributed reliability assessment mechanism for web services. In: 2008 IEEE International Conference on Dependable Systems and Networks With FTCS and DCC (DSN), pp. 392–397 (2008). https://doi.org/10.1109/DSN.2008.4630108

Software Demonstrations

A Crowdsourcing-Based Knowledge Graph Construction Platform

Xingkun Liu[✉], Zhiying Tu[✉], Zhongjie Wang[✉], Xiaofei Xu[✉], and Yin Chen[✉]

School of Computer Science and Technology, Harbin Institute of Technology, Harbin, China
18s003082@stu.hit.edu.cn, {tzy_hit,rainy,xiaofei,chenyin}@hit.edu.cn

Abstract. Nowadays, knowledge graphs are backbones of many information systems that require to have access to structured knowledge. While there are many openly available knowledge graphs, self-constructed knowledge graphs in specific domains are still in need, and the process of construction usually consumes a lot of manpower. In this paper, we present a novel platform that takes advantage of crowdsourcing to construct and manage knowledge graphs. The platform aims to provide knowledge graph automatic construction as a service and reduce the tenants' effort to construct knowledge graphs.

Keywords: Knowledge graph · Crowdsourcing · Ontology alignment

1 Introduction

Nowadays, knowledge graphs are backbones of many information systems that require to have access to structured knowledge. A large number of knowledge graphs such as YAGO [7], Freebase [1], DBPedia [4] have been constructed and applied to many real-world applications. A knowledge graph is composed of entities that present as nodes and relations which present as different types of edges between nodes. While there are many openly available knowledge graphs, self-constructed knowledge graphs in specific domains are still in need, and the process of construction usually consumes a lot of manpower.

In this paper, we present a novel platform that takes advantage of crowdsourcing to construct and manage knowledge graphs. The platform aims to provide knowledge graph automatic construction as a service and reduce the tenants' effort to construct knowledge graphs. Tenants are able to define knowledge graph schema according to the format of their resources for building knowledge graphs. While different tenants define knowledge graph schema differently, the platform will align and merge those schemas and generate a unified schema. Multiple tenants could contribute resources used for constructing knowledge graphs incrementally. Also, for convenience, the platform provides several general application interfaces for manipulating knowledge graphs such as querying interfaces

© Springer Nature Switzerland AG 2021
H. Hacid et al. (Eds.): ICSOC 2020 Workshops, LNCS 12632, pp. 63–66, 2021.
https://doi.org/10.1007/978-3-030-76352-7_9

and reasoning interfaces based on graph embedding models. Those application interfaces could be used to develop third-party applications.

The platform provides a process that has four main phases: *graph namespace applying, knowledge graph schema creating, resource uploading, graph embedding model training*. In the first phase, the platform handles requests from a tenant and creates a graph namespace, which is an environment for identifying one tenant's graph data and separating from others' graph data. In the second phase, tenants create knowledge graph schemas using GUI provided by the platform, after system manager examines and verifies those schemas, the platform aligns and merges those schemas. In the third phase, tenants upload resources using REST API provided by the platform, and those resources would be integrated into a knowledge graph in the graph namespace automatically. In the fourth phase, the platform will train and deploy several graph embedding models such as TransE [2], TransH [9], and TransD [3] according to tenants' needs.

The rest of the paper is organized as follows. In Sect. 2, we present an overview of the platform, while in Sect. 3 we describe our demonstration scenario. In Sect. 5, we conclude and outline future work.

2   System Overview

We provide the architecture of the platform, as shown in Fig. 1. In the architecture, the platform has five main components: *Web Frontend, Backend Service, Knowledge Graph Construction Service, Graph Embedding Model Management Service, Graph Database Management Service*.

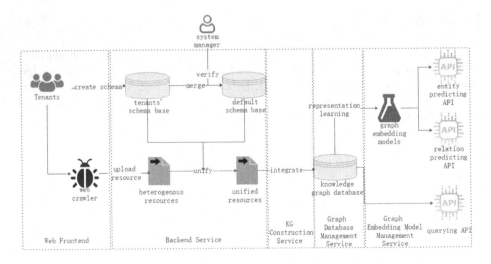

Fig. 1. The platform architecture

As shown in Fig. 1, tenants mainly interact with the web frontend. A tenant could apply for graph namespace, view graph namespace status, and delete graph namespace through interacting with web frontend. The tenant could create and edit graph schema, and after the schema gets verified, the tenant could get the REST API token for uploading resources. Also, the tenant could start a graph embedding model training job. The web frontend is developed using React and get data from backend service by REST API provided by backend service.

The backend service mainly deals with schema aligning, resource processing, graph embedding model training job scheduling, and service routing. The schema is based on OWL [5], and aligned in a way that combines linguistic-based strategy [8] and structure-based strategy [6]. After having a unified schema, heterogeneous resources could be transformed into a unified format. Then those resources will be put into a resource queue, which will be consumed by knowledge graph construction service. When a tenant requests for training a graph embedding model, the backend service will choose a graph embedding model management service registered on Zookeeper, which represents a service registry. The backend service is developed using Java and SpringBoot and interacts with other services by REST API or GRPC.

The knowledge graph construction service fetch resources from the resource queue and integrate those resources into the knowledge graph in the specified graph namespace. The knowledge graph construction service is written in Python, and the service is exported as REST API using the flask framework. We use RabbitMQ as a resource queue so that multiple knowledge graph construction services could subscribe to the queue to speed up the knowledge graph construction process, and it could be ensured that the resources won't get lost.

The graph embedding model management service mainly provides model training and model deploying service. When there comes a model training job from backend, the service exports graph data from graph database management service and starts training. After that, the service writes model parameters to the file system. And then, the deploying service would be notified and loads model parameters from the file system. The graph embedding model management service registers to Zookeeper once it gets start up so that the backend could discover the service. The service is developed using Python and Pytorch, and the service is exported using the flask framework.

The graph database management service uses Docker containers to run Neo4j graph databases. The service is developed using Python and Py2Neo, which is a client library for working with Neo4j and exported using flask framework.

3 Demonstration Scenario

We demonstrate the platform by constructing a real case of a knowledge graph. First, a tenant login to the platform and apply a graph namespace. Then several tenants create knowledge graph schemas according to their resource format. After the system manager examines and verifies the schemas, the schemas would be aligned and merged with a unified schema, and the tenants who create the schemas could get the tokens for uploading resources. While tenants upload their resources, the knowledge graph in the graph namespace gets constructed incrementally.

4   Where to Watch This Video

The demo file is titled "A Crowdsourcing-Based Knowledge Graph Construction Platform". It is a MP4 video format, no sound. It can be found in the following link: https://youtu.be/xBosip57XCs.

5   Conclusion and Future Work

This paper presented a novel platform that takes advantage of crowdsourcing to construct and manage knowledge graphs. The platform aims to provide knowledge graph automatic construction as a service and reduce the tenants' effort to construct knowledge graphs. Currently, the platform is used in different projects in order to construct knowledge graphs and make use of them. In the future, the platform will be enriched with more powerful features such as, knowledge graph refinement support with the help of third-party knowledge graphs, incrementally graph embedding model support to increase its applicability.

Acknowledgement. Research in this paper is partially supported by the National Key Research and Development Program of China (No 2018YFB1402500), the National Science Foundation of China (61802089, 61832004, 61772155, 61832014).

References

1. Bollacker, K., Evans, C., Paritosh, P., Sturge, T., Taylor, J.: Freebase: a collaboratively created graph database for structuring human knowledge. In: Proceedings of the 2008 ACM SIGMOD International Conference on Management of Data, pp. 1247–1250 (2008)
2. Bordes, A., Usunier, N., Garcia-Duran, A., Weston, J., Yakhnenko, O.: Translating embeddings for modeling multi-relational data. In: Advances in Neural Information Processing Systems, pp. 2787–2795 (2013)
3. Ji, G., He, S., Xu, L., Liu, K., Zhao, J.: Knowledge graph embedding via dynamic mapping matrix. In: Proceedings of the 53rd Annual Meeting of the Association for Computational Linguistics and the 7th International Joint Conference on Natural Language Processing (Volume 1: Long Papers), pp. 687–696 (2015)
4. Lehmann, J., et al.: Dbpedia-a large-scale, multilingual knowledge base extracted from wikipedia. Semant. Web 6(2), 167–195 (2015)
5. McGuinness, D.L., Van Harmelen, F., et al.: Owl web ontology language overview. W3C recommendation 10(10), 2004 (2004)
6. Melnik, S., Garcia-Molina, H., Rahm, E.: Similarity flooding: a versatile graph matching algorithm and its application to schema matching. In: Proceedings 18th International Conference on Data Engineering, pp. 117–128. IEEE (2002)
7. Suchanek, F.M., Kasneci, G., Weikum, G.: Yago: a core of semantic knowledge. In: Proceedings of the 16th International Conference on World Wide Web, pp. 697–706 (2007)
8. Tang, J., Li, J., Liang, B., Huang, X., Li, Y., Wang, K.: Using Bayesian decision for ontology mapping. J. Web Semant. 4(4), 243–262 (2006)
9. Wang, Z., Zhang, J., Feng, J., Chen, Z.: Knowledge graph embedding by translating on hyperplanes. In: Twenty-Eighth AAAI Conference on Artificial Intelligence (2014)

Data Interaction for IoT-Aware Wearable Process Management

Stefan Schönig[1]([✉]), Richard Jasinski[2], and Andreas Ermer[2]

[1] Institute for Management Information Systems, University of Regensburg,
Regensburg, Germany
stefan.schoenig@ur.de
[2] Maxsyma GmbH & Co. KG, Floß, Germany
{rjasinski,aermer}@maxsyma.de

Abstract. Process execution and monitoring based on Internet of Things (IoT) data can enable a more comprehensive view on processes. In our previous research, we developed an approach that implements an IoT-aware Business Process Management System (BPMS), comprising an integrated architecture for connecting IoT data to a BPMS. Furthermore, a wearable process user interface allows process participants to be notified in real-time at any location in case new tasks occur. In many situations operators must be able to directly influence data of IoT objects, e.g., to control industrial machinery or to manipulate certain device parameters from arbitrary places. However, a BPM controlled interaction and manipulation of IoT data has been neglected so far. In this demo paper, we extend our approach towards a framework for IoT data interaction by means of wearable process management. BPM technology provides a transparent and controlled basis for data manipulation within the IoT.

Keywords: Internet of Things · Process execution · Data interaction

1 Introduction and Relevance

Business process management (BPM) is considered as powerful technology to control, design, and improve processes. Processes are executed within systems that are part of the real world involving humans, computer systems as well as physical objects [1]. Internet of Things (IoT) as well as Cyber-Physical Systems (CPS), denoting the inter-networking of physical devices, have become very popular these days [2]. Process execution, monitoring and analytics based on IoT data can enable a more comprehensive view on processes. Embedding intelligence by way of real-time data gathering from devices and sensors and consuming them through BPM technology helps businesses to achieve cost savings and efficiency.

In our previous research [3–5], we developed an approach that implements an IoT-aware BPMS called *iot2flow*, comprising an architecture for connecting IoT data to a BPMS. Furthermore, we developed a wearable process user interface

© Springer Nature Switzerland AG 2021
H. Hacid et al. (Eds.): ICSOC 2020 Workshops, LNCS 12632, pp. 67–71, 2021.
https://doi.org/10.1007/978-3-030-76352-7_10

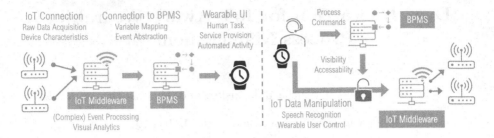

Fig. 1. Overview of wearable *IoT data provision* and *IoT data manipulation*

that allows operators to be notified in real-time at any location in case new tasks occur. In many situations operators must be able to directly influence data of IoT objects, e.g., to control industrial machinery or to manipulate certain device parameters from arbitrary places. However, a BPM controlled interaction and manipulation of IoT data has been neglected so far. In this demo paper, we extend our approach towards a fully implemented framework for IoT data interaction by means of wearable process management (cf. Fig. 1). BPM technology provides a transparent and controlled basis for data modification within the IoT. Additionally, we provide voice control for IoT data modification. The approach has been introduced in production processes of corrugation industry plants where paper is glued together to produce corrugated paper as raw material for cardboard boxes. Based on the presented approach operators productivity improved in terms of reduced stop times and increased production speed. The *iot2flow* framework is currently in use in several production plants. The toolset is continuously developed, enhanced and improved. A screencast presenting the complete tool is available at https://www.youtube.com/watch?v=gt9aJwTto2EE.

2 Involved Technology and Innovative Aspects

2.1 Bidirectional Connection of IoT Data Sources

We build upon the standard modelling notation BPMN 2.0 and use the Camunda BPMS (https://camunda.com). We communicate with the workflow engine by means of the Camunda Rest API. In order to connect IoT objects, we implemented an IoT middleware that supports IoT protocols like MQTT, TCP as well as PLC protocols such as OPC-UA and Simatic S7 (cf. Fig. 2). The IoT middleware specifies a mapping from IoT variables to process variables. Based on this, it keeps the BPMS updated with the latest IoT values. All running instances of a particular process receive the corresponding data value. The application cyclically acquires the IoT values and sends them to the BPMS. Given the current IoT data values, the engine calculates available activities.

2.2 Wearable Process Execution Interface

Participants are seamlessly notified when interaction is required, independent of where the user is located. This requires a real time notification on mobile devices

Fig. 2. IoT variable and voice command definition

of users. During process execution, available tasks for a specific participant are directly sent to mobile devices. As a mobile user interface we implemented an *Android* based smartwatch application. The IoT middleware cyclically requests the available user tasks from the Camunda API for each defined user and publishes them to the MQTT topic. The application allows users to start and complete tasks as well as to initiate new process instances. It is possible to start or complete tasks and processes. IoT and Industry 4.0 applications heavily depend on data modification during process execution. Production parameters and settings need to be changed and data needs to be fed into a system. Such data provision and manipulation can be controlled and scheduled with the help of the underyling process model. Therefore, we enhanced the existing architecture by means of wearable data interaction functionality that fills this gap towards a full fledged IoT-aware BPMS. The introduced data perspective is conceptually divided into *(i) Data Provision* and *(ii) Data Modification*. To classifiy the implemented concepts, we refer to the well-known Workflow Data Patterns [6].

2.3   Data Visibility and Provision

Sensor and machine data must be provided in real-time to trigger events or as a means to support task execution. Our tool supports the provision of IoT data to be referenced in process models and the running instances (Pattern P7 Workflow Data, Pattern P5 Case Data). IoT data can also be provisioned independently from any process (Pattern P8 Environmental Data). The interaction pattern is a push-oriented environment to workflow pattern (P25). Operators frequently need relevant information such documents or videos during the execution of tasks. The *iot2flow* tool implements the provision of media to wearables. The files to be provided are specified in the BPMN diagram. This function implements a task data visibility pattern (Pattern P1). The underlying implementation follows a pull-oriented environment to task interaction pattern (Pattern P16).

2.4   Data Modification

Process participants are able to actively influence environmental data, e.g. production parameters, in real-time from arbitrary locations. BPM technology

Fig. 3. Wearable process command and data manipulation interfaces

serves as a controlling instance ensuring that data access is restricted to specific situations, e.g., whole certain tasks are active, or to specific user groups. The *iot2flow* framework implements two different ways to manipulate IoT data: *(i) user input controls on wearables)* and *(ii) speech recognition directly on wearable devices.* The first option is to define a user input control either as string based or a numerical textfield that is bound to a task in the model. The textfield is shown on the device when the corresponding task is started. The inserted content is directly mapped either to an instance variable or to an external IoT variable.

The second option is based on speech recognition fueled by research in neural networks. We rely on an end-to-end (E2E) model approach that runs entirely on the device. *iot2flow* allows to specify terms combined with either boolean or numerical values that can be provided by means of speech recognition. The mapping from the term to the IoT variable is defined in the IoT middleware. Operators initiate the speech recognition mode during execution and declare both the term as well as the new value for the variable. This value is directly transfered to the IoT device. *iot2flow* provides the possibility to constrain the scope of commands to tasks, processes and/or user groups. This way, not the whole list of voice commands is applicable in every situation but restricted according to the visibility patterns P1 and P7 as well as to specific resources.

References

1. Schönig, S., Aires, A.P., Ermer, A., Jablonski, S.: Workflow support in wearable production information systems. In: Mendling, J., Mouratidis, H. (eds.) CAiSE 2018. LNBIP, vol. 317, pp. 235–243. Springer, Cham (2018). https://doi.org/10.1007/978-3-319-92901-9_20
2. Mosterman, P.J., Zander, J.: Industry 4.0 as a cyber-physical system study. Softw. Syst. Model. **15**(1), 17–29 (2016)
3. Schönig, S., Ermer, A., Market, M., Jablonski, S.: Sensor-enabled wearable process support in corrugation industry. BPM Industry Track, pp. 118–129 (2019)
4. Schönig, S., Ackermann, L., Jablonski, S., Ermer, A.: An integrated architecture for IoT-aware business process execution. In: Enterprise, Business-Process and Information Systems Modeling, pp. 19–34 (2018)

5. Schönig, S., Ackermann, L., Jablonski, S., Ermer, A.: IoT meets BPM: a bidirectional communication architecture for IoT-aware process execution. Softw. Syst. Model. 1–17 (2020)
6. Russell, N., ter Hofstede, A.H.M., Edmond, D., van der Aalst, W.M.P.: Workflow data patterns: identification, representation and tool support. In: Delcambre, L., Kop, C., Mayr, H.C., Mylopoulos, J., Pastor, O. (eds.) ER 2005. LNCS, vol. 3716, pp. 353–368. Springer, Heidelberg (2005). https://doi.org/10.1007/11568322_23

SiDD: The Situation-Aware Distributed Deployment System

Kálmán Képes(✉), Frank Leymann, Benjamin Weder, and Karoline Wild

Institute of Architecture of Application Systems, University of Stuttgart,
Stuttgart, Germany
{kepes,leymann,weder,wild}@iaas.uni-stuttgart.de

Abstract. Most of today's deployment automation technologies enable the deployment of distributed applications in distributed environments, whereby the deployment execution is centrally coordinated either by a central orchestrator or a master in a distributed master-workers architectures. However, it is becoming increasingly important to support use cases where several independent partners are involved. As a result, decentralized distributed deployment automation approaches are required, since organizations typically do not provide access to their internal infrastructure to the outside or leave control over application deployments to others. Moreover, the choice of partners can depend heavily on the current situation at deployment time, e.g. the costs or availability of resources. Thus, at deployment time it is decided which partner will provide a certain part of the application depending on the situation. To tackle these challenges, we demonstrate the situation-aware distributed deployment (SiDD) system as an extension of the OpenTOSCA ecosystem.

Keywords: Deployment · Choreography · Situation-aware system · TOSCA

1 Introduction and Motivation

Deployment technologies enable reusable and portable application deployments, making them key technologies for today's application management. A variety of technologies offer different capabilities and own domain-specific languages for modeling deployments. Many use declarative deployment models in which the desired state of an application can be specified by a structural description of the application with its components and their relationships among each other.

However, in recent years, deployment automation has focused primarily on centralized approaches, which allow the deployment of distributed applications, but the deployment execution is centrally coordinated either by a central orchestrator or a master in distributed master-workers architectures. Especially in industrial use cases, e. g., in a supply chain, or if specialized compute infrastructure is required, e. g., in quantum computing, several partners are involved each

© Springer Nature Switzerland AG 2021
H. Hacid et al. (Eds.): ICSOC 2020 Workshops, LNCS 12632, pp. 72–76, 2021.
https://doi.org/10.1007/978-3-030-76352-7_11

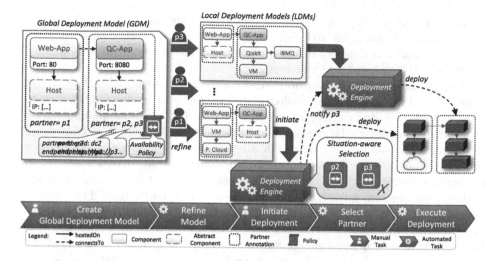

Fig. 1. Overview of the SiDD concept exemplary depicting the partner selection based on the available capacities of the partners (p2 and p3).

deploying a part of the overall application. Due to security concerns, organizations typically do not provide access to internal infrastructure to the outside or leave control over application deployments to others. Thus, centralized deployment approaches cannot be applied. Moreover, the involved partners can change depending on certain conditions, e.g., costs or availability. Thus, (i) the modeling of a distributed application involving multiple partners, (ii) the partner selection during deployment, and (iii) the decentralized execution of the overall deployment has to be enabled. To tackle these challenges, we present the *Situation-aware Distributed Deployment (SiDD) system* as an extension of the OpenTOSCA ecosystem [1], an open-source toolchain for modeling and executing application deployments using the Topology and Orchestration Specification for Cloud Applications (TOSCA).

2   Exemplary Application Scenario and SiDD Concept

In previous work, a concept for the *decentralized cross-organizational application deployment automation* [3] as well as the *situation-aware management* [2] has been introduced. In this work, we demonstrate how these concepts can be combined to enable the situation-aware partner selection for a distributed and decentralized deployment based on an exemplary scenario as shown in Fig. 1 on the left: Three partners *p1*, *p2*, and *p3* collaborate to run an application using a quantum algorithm and a visualization web application component. To reduce costs the partners share their infrastructure, p1 provides a classical private cloud, p2 a high-performance computing environment (HPCE) to run a quantum simulator, and p3 provides access to a quantum computer. At deployment time it has to be decided how the quantum algorithm shall be executed either using the quantum computer of p3 or running a quantum simulator in the HPCE of p2. This selection decision has to be made based on an availability policy, e. g., if there is no available time slot on the quantum computer, the quantum simulator in the HPCE is selected and vice versa.

The described scenario can be partially automated with our SiDD concept depicted in Fig. 1. First, a so-called *global deployment model (GDM)* has to be specified as a declarative deployment model that contains all application components and their relationships (see left in Fig. 1). This model only contains the application components on which all partners have agreed upon and not necessarily contain any infrastructure components such as virtual machines or application servers. In addition, a GDM contains abstract components that are replaced by concrete infrastructure components by each partner. Thus, in the second step, each partner refines the GDM into a so-called *local deployment model (LDM)* which specifies the needed infrastructure components of their own infrastructure, as other partners usually do not have or must not have knowledge about these. The refinement can be automated, e. g., using available refinement fragments [4]. After all partners have defined their LDMs, in the third step, any partner can initiate deployment by requesting the deployment engine to start a deployment. In the fourth step, it is decided which partners are actually involved in the deployment, e. g., in our scenario the infrastructure will be used from p1 and additionally either from p2, which can run a quantum simulator, or from p3, which has access to a quantum computer. This is based on the annotated policies, such as the availability policy in the GDM in Fig. 1. For example, if the quantum computer of p3 is not available at deployment time, a quantum simulator can be deployed in the HPCE of p2 if there are enough resources available. Finally, the selected partners are notified to start the deployment of their components. For the deployment each partner generates a workflow containing tasks to install components as well as tasks to exchange data between the partners, e. g., endpoint information to establish a connection to components of other partners.

3   The SiDD System

The SiDD System is an extension of the OpenTOSCA ecosystem that consists of *Winery*, a graphical TOSCA modeling tool, and *OpenTOSCA container*, a TOSCA deployment engine [1] (see video at https://youtu.be/A0JY9TW4ZFM). The architecture of the system with the relevant components is shown in Fig. 2. *Winery* can be used to graphically model a declarative deployment model as a TOSCA *topology template* by using defined types and attaching the executable artifacts, e. g., a WAR for running a web application. Winery is used to model the GDM, which is then passed to the involved partners (a). The *CSAR Importer/Exporter* enables the export of a standardized *Cloud Service Archive (CSAR)* that can be consumed by a TOSCA deployment engine. The *Substitution Mapping* Component can be used to refine abstract components, called node templates, by concrete ones using refinement fragments. This is used

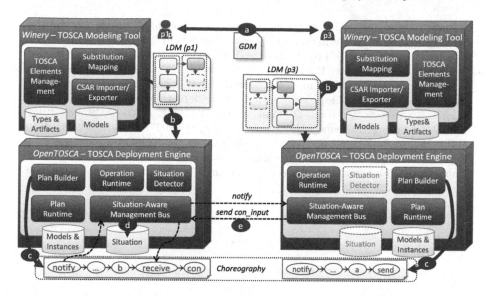

Fig. 2. Architecture of the SiDD system exemplary shown with two partners.

by each partner to refine the abstract node templates in the GDM and to obtain the LGM that can be processed by the deployment engine *OpenTOSCA container* (b). For deployment execution, a BPEL workflow is generated based on the declarative deployment model by the *Plan Builder* (c). The workflows of the partners form a choreography by sending and receiving messages to share deployment data. The *Plan Runtime* runs the plan when the deployment is instantiated. All operations that have to be executed and which are not provided as a service run in the *Operation Runtime*. For the situation-aware selection, the required information has to be provided by an external application which is then used by the *Situation Detector* to determine which situations are active. When an application is instantiated, the *Situation-Aware Management Bus* is responsible for the partner selection based on the current situation (d) and to exchange messages during deployment (e).

Acknowledgments. This work was partially funded by the DFG project DiStOPT (252975529), the BMWi project *PlanQK* (01MK20005N), and the DFG's Excellence Initiative project *SimTech* (EXC 2075 - 390740016).

References

1. Breitenbücher, U., et al.: The OpenTOSCA ecosystem - concepts & tools. In: European Space project on Smart Systems, Big Data, Future Internet - Towards Serving the Grand Societal Challenges - Volume 1: EPS Rome 2016, pp. 112–130, December 2016

2. Képes, K., et al.: Situation-aware management of cyber-physical systems. In: Proceedings of the 9th International Conference on Cloud Computing and Services Science (CLOSER 2019). pp. 551–560. SciTePress, May 2019
3. Wild, K., Breitenbücher, U., Képes, K., Leymann, F., Weder, B.: Decentralized cross-organizational application deployment automation: an approach for generating deployment choreographies based on declarative deployment models. In: Dustdar, S., Yu, E., Salinesi, C., Rieu, D., Pant, V. (eds.) CAiSE 2020. LNCS, vol. 12127, pp. 20–35. Springer, Cham (2020). https://doi.org/10.1007/978-3-030-49435-3_2
4. Wild, K., et al.: TOSCA4QC: two modeling styles for TOSCA to automate the deployment and orchestration of quantum applications. In: 2020 IEEE 24th International Enterprise Distributed Object Computing Conference (EDOC). IEEE Computer Society (2020)

AuraEN: Autonomous Resource Allocation for Cloud-Hosted Data Processing Pipelines

Sunil Singh Samant[1]($\boxtimes$), Mohan Baruwal Chhetri[1,2], Quoc Bao Vo[1],
Ryszard Kowalczyk[1,3], and Surya Nepal[2]

[1] Swinburne University of Technology, Melbourne, Australia
{ssamant,bvo,rkowalczyk}@swin.edu.au
[2] CSIRO Data61, Sydney, Australia
{mohan.baruwalchhetri,surya.nepal}@data61.csiro.au
[3] Systems Research Institute, Polish Academy of Sciences, Warsaw, Poland

Abstract. Ensuring cost-effective end-to-end QoS in an IoT data processing pipeline (DPP) is a non-trivial task. A key factor that affects the overall performance is the amount of computing resources allocated to each service in the pipeline. In this demo paper, we present **AuraEN**, an **Au**tonomous resource allocation **EN**gine that can proactively scale the resources of each individual service in the pipeline in response to predicted workload variations so as to ensure end-to-end QoS while optimizing the associated costs. We briefly describe the AuraEN system architecture and its implementation and demonstrate how it can be used to manage the resources of a DPP hosted on the Amazon EC2 cloud.

Keywords: Data processing pipeline · Resource optimization · End-to-end QoS · Cloud resource orchestration · Resource scaling

1 Introduction

In the IoT paradigm, objects or 'things' with sensing, effecting and communication capabilities generate massive volumes of data, mostly as streaming data. This data is typically *ingested*, *processed*, and *stored*, before being *consumed* by end-user applications. A data processing pipeline (DPP) is essentially a composite service that comprises specialised atomic software services for ingestion, processing, and storage. Each of these atomic services can be fulfilled by a number of different big data processing software platforms, e.g., Apache Kafka (https://kafka.apache.org) and Apache Pulsar (https://pulsar.apache.org/) for ingestion; Apache Spark (https://spark.apache.org/), Apache Storm (http://storm.apache.org/) and Apache Flink (https://flink.apache.org/) for processing; and Cassandra (https://cassandra.apache.org/), MongoDB (https://www.mongodb.com/) and HBase (https://hbase.apache.org/) for storage. Due to these platforms' distributed design and native support for horizontal scalability, cloud computing

© Springer Nature Switzerland AG 2021
H. Hacid et al. (Eds.): ICSOC 2020 Workshops, LNCS 12632, pp. 77–80, 2021.
https://doi.org/10.1007/978-3-030-76352-7_12

is the default choice for running IoT DPPs. The cloud provides a scalable, elastic, easily accessible and inexpensive way of meeting the processing needs of the constituent services in a DPP, even under varying workloads.

IoT applications that provide real-time actionable insights based on the streaming data have stringent QoS requirements that must be fulfilled at all times, e.g., traffic accident detection. At the same time, cloud cost optimisation is a key challenge faced by most consumers of cloud infrastructure. Therefore, a key research challenge related to the adaptive management of cloud resources for IoT DPPs is to *ensure cost-effective, end-to-end QoS fulfilment, even under varying workload conditions.* The end-to-end QoS of a DPP depends upon the performance of its constituent software services; their performance, in turn, depends upon several factors including their individual configurations, the inter-dependencies between adjacent services, the amount of computing resources allocated to each service, and, the data ingestion rate. This makes autonomous adaptive resource management for IoT DPPs a non-trivial task.

We have previously proposed a systematic approach for building a *sustainable QoS profile* for constituent services that can be used to inform resource allocation decisions for a DPP in response to varying workloads [2]. We have also proposed an approach for end-to-end QoS and cost-aware resource allocation that uses the sustainable QoS profile for decision-making [1]. In this demo paper, we present a proof-of-concept implementation of **AuraEN**, an **Au**tonomous resource allocation **EN**gine for the adaptive management of computing resources for cloud-deployed IoT DPPs. We present the conceptual system architecture of *AuraEN* and briefly describe its implementation details. We demonstrate how it can be used to (a) deploy a custom DPP on the Amazon EC2 cloud, and (b) autonomously manage the allocated computing resources for each service in the pipeline in response to the varying workload.

2   System Architecture

As shown in Fig. 1, AuraEN has two key components - *Resource Optimizer* (ROpt) and *Resource Orchestrator* (ROrch). The ROpt takes the following as input: (a) the one-step-ahead workload prediction, (b) the sustainable QoS profiles for the candidate cloud instance types, (c) their pricing information, (d) the current resource allocation for each DPP service, and (e) the end-to-end QoS constraints that need to be satisfied. A DPP specific workload transformation function is used to compute the input workload for each DPP service based on the forecast workload. The sustainable QoS profile is obtained through performance benchmarking as discussed in [2]; pricing information for the candidate cloud instance types is obtained by querying the cloud provider API; the one-step ahead workload can be estimated from the historical workload using standard forecasting algorithms. Based on these inputs, the ROpt (a) computes the cost-optimal resource allocation for each service in the DPP for the next period, and (b) determines how these resources should be provisioned. It uses the following three resource scaling strategies:

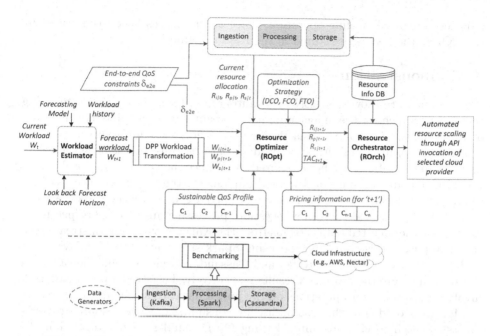

Fig. 1. The architecture of DPP resource management system (AuraEN)

- *Delta-capacity optimization (DCO)* finds the cost-optimal allocation of resources for each DPP service based on the *delta* between the existing capacity and the predicted workload. *DCO for scale-out* finds the optimal allocation for the delta increase in the workload while retaining existing compute instances for each DPP service. Similarly, *DCO for scale-in* finds the optimal set of running compute instances that can be removed for each DPP service in response to the delta decrease in the workload.
- *Full-capacity optimization (FCO)* finds the cost-optimal resource allocation per DPP service based on the predicted workload while ensuring the end-to-end QoS requirements. It does not consider the existing resource allocation in its decision-making. And can potentially replace all of the existing compute instances if a more cost-optimal allocation can be found.
- *Fault-tolerant optimization (FTO)* aims at making the DPP fault-tolerant. This can be done in a number of different ways including by (a) over provisioning resources for each service in the pipeline, (b) using heterogeneous resources for each service, and (c) using heterogeneous contracting options for the computing resources.

Depending upon the scaling strategy chosen, the resource allocation for the next time period will be different. The ROrch takes the output of ROpt and manages the creation/termination of resources on the target cloud platform. It uses the target cloud platform API to send requests to launch new resources or terminate existing ones; it also re-configures/restarts the DPP services as required after each scaling operation. Once the resources have been scaled by the ROrch, the results are stored in the Resource Info DB.

3   Demonstration

We demonstrate how AuraEN can autonomously manage the computing resources for each service in a custom data stream processing pipeline[1] deployed on the Amazon EC2 cloud in Asia-Pacific ap-southeast-2 zone. We use a synthetic workload generator application[2] to produce data streams representing the vehicular streaming data. It can be configured to produce the data streams at different rates to simulate varying workload and implements the Kafka producer API for sending data streams to the Kafka service. The DPP uses Apache Kafka, Apache Spark and Apache Cassandra for ingestion, processing and storage respectively. To process the ingested data streams from the Kafka service, a data stream processor application is[3] implemented using Spark streaming APIs for Java along with the connector APIs for Kafka and Cassandra service to pull the raw data streams and store the processed results respectively. The stream processing task involves map and filter operations on the pulled data streams and the processed results are stored into the Cassandra for consumption. The resource requirements for each service are fulfilled using On-Demand EC2 instances. For the purpose of demonstration we use *t2.micro, t2.small* and *t2.medium* instances running the Ubuntu operating system.

In the demonstration, we show how the DPP is deployed and managed on the EC2 cloud using AuraEN. Initially, resources are allocated for the pipeline services based on a pre-defined default input workload and end-to-end latency requirements. Once the DPP has been successfully deployed and all the services are running, the workload generated by the workload generator application is varied to trigger resource scale-in and scale-out using the different scaling strategies. A video demonstrating the different scenarios can be found at: [https://youtu.be/r8S1PcbsCsU].

Acknowledgement. This research is supported by a PhD Scholarship from CSIRO Data61, an Australian federal government agency responsible for scientific research.

References

1. Samant, S.S., Baruwal Chhetri, M., Vo, Q.B., Kowalczyk, R., Népal, S.: Towards end-to-end QoS and cost-aware resource scaling in cloud-based IoT data processing pipelines. In: 2018 IEEE International Conference on Services Computing (SCC), pp. 287–290. IEEE (2018)
2. Samant, S.S., Baruwal Chhetri, M., Vo, Q.B., Nepal, S., Kowalczyk, R.: Benchmarking for end-to-end QoS Sustainability in Cloud-hosted Data Processing Pipelines. In: 2019 IEEE 5th International Conference on Collaboration and Internet Computing (CIC), pp. 39–48. IEEE (2019)

---

[1] https://github.com/samantsunil/AuraEN.
[2] https://github.com/samantsunil/data-generator.
[3] https://github.com/samantsunil/data-processor-app.

Artificial Intelligence for IT Operations
(AIOPS 2020)

International Workshop on Artificial Intelligence for IT Operations (AIOps 2020)

Large-scale systems of all types, such as data centres, cloud computing environments, edge clouds, IoT, and embedded environments, are characterized by extreme complexity. The large number of processes and their complex interactions make the successful management of such systems an incrementally harder task. To deal with this, the operators are increasingly relying on employing artificial intelligence and data analytic tools against the observational data from the IT system. As a result, the research field of Artificial Intelligence for IT operations (AIOps) is increasingly important. It holds the promise to develop and utilize AI-based learning methods against the data from the systems to aid the operators in successful operation. The high number of submissions to AIOps 2020 reflects the existing high interest of the research area.

The accepted papers address several important aspects of AIOps related to the issues of reliability, security, and scalability for the operation of IT systems, such as cloud systems, achieved via data analysis. The numerous experimental evaluations and use-cases presented deliver an insight into the behaviour of the methods in controlled test-beds and real-world scenarios, showing their usability. But more importantly, they point out further open research questions for investigation. Each submission was reviewed by at least three senior reviewers. From the 28 submitted papers, we selected 14 high-quality papers. Contextually, the papers are divided into three groups and are briefly summarized in the following.

The first group, consisting of eight papers, is concerned with the task of detecting anomalous behaviour of metrics and events represented as textual or numerical data (including streams). From a modeling perspective, the techniques range from more traditional data analysis approaches, such as rules, through to deep learning methods, with a greater proclivity for the latter.

The second group, consisting of four papers, is focused on the problem of fault localization from metric, event, and alert data. From a modeling perspective, the proposed methods utilize various approaches from causal discovery to reconstruct the causal graph of event relations, or to discover frequently co-occurring events.

The third group consists of three papers on novel topics in the area of AIOps, such as issues arising from sharing the data and efficient resource utilization. The third paper of this group is of great importance for the community since it depicts the landscape of what constitutes AIOps, what are the historical trends, which are the related tasks, and the types of data sources as well as the most important further directions in the field.

Two keynote speeches were presented. The first by Dan Pei from Tsingua University on "Towards Autonomous IT Operations through Artificial Intelligence" and the second by Michael R. Lyu on "Software Reliability Engineering for Resilient Cloud Operations". Both keynotes discussed open challenges, possible solutions, and future directions that can guide the field of AIOps.

Organization

Workshop Organizers

Odej Kao   Technische Universität Berlin, Germany
Jorge Cardoso   Huawei Munich Research Center, Germany

Workshop Co-chairs

Jasmin Bogatinovski   Technische Universität Berlin, Germany
Sasho Nedelkoski   Technische Universität Berlin, Germany
Alexander Acker   Technische Universität Berlin, Germany
Thorsten Wittkopp   Technische Universität Berlinv Germany
Soeren Becker   Technische Universität Berlin, Germany
Li Wu   Technische Universität Berlin, Germany
Florian Schmidt   Technische Universität Berlin, Germany

Program Committee

Ivona Brandic   Vienna University of Technology, Austria
Ana Juan   Universitat Oberta de Catalunya, Spain
Dan Pei   Tsinghua University, China
Johan Tordsson   Umeå University, Sweden
Feng Liu   Huawei European Research Center, Belgium
Rama Akkiraju   IBM, USA
Filipe Araujo   University of Coimbra, Portugal
Samuel Kounev   University of Wuerzburg, Germany
Domenico Cotroneo   University of Naples Federico II, Italy
Roberto Natella   University of Naples Federico II, Italy
Michael R. Lyu   The Chinese University of Hong Kong, Hong Kong
Jonathan Maces   Max Planck Institute for Software Systems, Germany
Stefan Schulte   Vienna University of Technology, Austria
Stefan Tai   Technische Universität Berlin, Germany
Dragi Kocev   Jozef Stefan Institute, Slovenia
Vladimir Podolskiy   Technical University of Munich, Germany
Shenglin Zhang   Nankai University, China
Gjorgji Madjarov   University of Skopje, North Macedonia
Matej Petkovic   Jozef Stefan Institute, Slovenia

Ljupco Todorovski   University of Ljubljana, Slovenia
Cesare Pautasso   University of Lugano, Switzerland
Martin Breskvar   Jozef Stefan Institute, Slovenia

We would like to take the opportunity of thanking the authors who submitted a contribution, as well as our sponsor for the conference, Huawei, and the external Program Committee members, whose extensive collaboration made this event possible.

Performance Diagnosis in Cloud Microservices Using Deep Learning

Li Wu[1,2](✉), Jasmin Bogatinovski[2], Sasho Nedelkoski[2], Johan Tordsson[1,3], and Odej Kao[2]

[1] Elastisys AB, Umeå, Sweden
{li.wu,johan.tordsson}@elastisys.com
[2] Distributed and Operating Systems Group, TU Berlin, Berlin, Germany
{jasmin.bogatinovski,nedelkoski,odej.kao}@tu-berlin.de
[3] Department of Computing Science, Umeå University, Umeå, Sweden

Abstract. Microservice architectures are increasingly adopted to design large-scale applications. However, the highly distributed nature and complex dependencies of microservices complicate automatic performance diagnosis and make it challenging to guarantee service level agreements (SLAs). In particular, identifying the culprits of a microservice performance issue is extremely difficult as the set of potential root causes is large and issues can manifest themselves in complex ways. This paper presents an application-agnostic system to locate the culprits for microservice performance degradation with fine granularity, including not only the anomalous service from which the performance issue originates but also the culprit metrics that correlate to the service abnormality. Our method first finds potential culprit services by constructing a service dependency graph and next applies an autoencoder to identify abnormal service metrics based on a ranked list of reconstruction errors. Our experimental evaluation based on injection of performance anomalies to a microservice benchmark deployed in the cloud shows that our system achieves a good diagnosis result, with 92% precision in locating culprit service and 85.5% precision in locating culprit metrics.

Keywords: Performance diagnosis · Root cause analysis · Microservices · Cloud computing · Autoencoder

1 Introduction

The microservice architecture design paradigm is becoming a popular choice to design modern large-scale applications [3]. Its main benefits include accelerated development and deployment, simplified fault debugging and recovery, and producing a rich software development technique stacks. With microservices, monolithic application can be decomposed into (up to hundreds of) single-concerned, loosely-coupled services that can be developed and deployed independently [12].

As microservices deployed on cloud platforms are highly-distributed across multiple hosts and dependent on inter-communicating services, they are prone

© Springer Nature Switzerland AG 2021
H. Hacid et al. (Eds.): ICSOC 2020 Workshops, LNCS 12632, pp. 85–96, 2021.
https://doi.org/10.1007/978-3-030-76352-7_13

to performance anomalies due to the external or internal issues. Outside factors include resource contention and hardware failure or other problems e.g., software bugs. To guarantee the promised service level agreements (SLAs), it is crucial to timely pinpoint the root cause of performance problems. Further, to make appropriate decisions, the diagnosis can provide some insights to the operators such as where the bottleneck is located, and suggest mitigation actions. However, it is considerably challenging to conduct performance diagnosis in microservices due to the large scale and complexity of microservices and the wide range of potential causes.

Microservices running in the cloud have monitoring capabilities that capture various application-specific and system-level metrics, and can thus understand the current system state and be used to detect service level objective (SLO) violations. These monitored metrics are externalization of the internal state of the system. Metrics can be used to infer the failure in the system and we thus refer to them as *symptoms* in anomaly scenarios. However, because of the large number of metrics exposed by microservices (e.g., Uber reports 500 million metrics exposed [14]) and that faults tend to propagate among microservices, many metrics can be detected as anomalous, in addition to the true root cause. These additional anomalous metrics make it difficult to diagnose performance issues manually (research problems are stated in Sect. 2).

To automate performance diagnosis in microservices effectively and efficiently, different approaches have been developed (briefly discussed in Sect. 6). However, they are limited by either coarse granularity or considerable overhead. Regarding granularity, some work focus on locating the service that initiates the performance degradation instead of identifying the real cause with fine granularity [8,9,15] (e.g., resource bottleneck or a configuration mistake). We argue that the coarse-grained fault location is insufficient as it cannot give us more details to the root causes, which makes it difficult to recover the system timely. As for considerable overhead, to narrow down the fault location, several systems can pinpoint the root causes with fine granularity. But they need to instrument application source code or runtime systems, which brings considerable overhead to a production system and/or slows down development [4].

In this paper, we adopt a two-stage approach for anomaly detection and root cause analysis (system overview is described in Sect. 3). In the first stage, we model the service that causes the failure following a graph-based approach [16]. This allows us to pinpoint the potential faulty service that initiates the performance degradation, by identifying the root cause (anomalous metric) that contributes to the performance degradation of the faulty service. The second stage, inference of the potential failure, is based on the assumption that the most important symptoms for the faulty behaviour have a significant deviation from their values during normal operation. Measuring the individual contribution to each of the symptoms at any time point, that leads to the discrepancy between observed and normal behaviour, allows for localization of the most likely symptoms that reflect the fault. Given this assumption, we aim to model the symptoms values under normal system behaviour. To do this we adopt an autoencoder method

(Sect. 4). Assuming a Gaussian distribution of the reconstruction error from the autoencoder, we can suggest interesting variations in the data points. We then decompose the reconstruction error assuming each of the symptoms as equally important. Further domain and system knowledge can be adopted to re-weight the error contribution. To deduce the set of possible symptoms as a preference rule for the creation of the set of possible failure we consider the symptom with a maximal contribution to the reconstruction error. We evaluate our method in a microservice benchmark named Sock-shop[1], running in a Kubernetes cluster in Google Cloud Engine (GCE)[2], by injecting two types of performance issues (CPU hog and memory leak) into different microservices. The results show that our system can identify the culprit services and metrics well, with 92% and 85.5% in precision separately (Sect. 5).

2   Problem Description

Given a collection of loosely coupled microservices S, we collect the relevant performance metrics over time for each service $s \in S$. We use $m^{(s,t)}$ to denote the metrics for service s at time t. Furthermore, $m_i^{(s,t)}$ denotes the individual metric (e.g., response time, container cpu utilization, etc.) for service s, collected at time t.

Based on above definition, the performance diagnosis problem is formulated as follows: given metrics m of a cloud microservice, assuming anomalies are detected from metric m_i of a set of services s_a at time t, where i is the index of response time, how can we identify the culprit metrics that cause the anomaly? Furthermore, we break down the research problem as following two sub-problems:

1. How to pinpoint the culprit service s_{rc} that initiates the performance degradation in microservices?;
2. Given the culprit service s_{rc}, how to pinpoint the culprit metric m_{rc} that contributes to its abnormality?

3   System Overview

To address the culprit services and metrics diagnosis problems, we propose a performance diagnosis system shown in Fig. 1. In overall, there are four components in our system, namely data collection, anomaly detection, culprit service localization (CSL) and culprit metric localization (CML). Firstly, we collect metrics from multiple data resources in the microservices, including run-time operating system, the application and the network. In particular, we continuously monitor the response times between all pairs of communicating services. Once the anomaly detection module identifies long response times from services, it triggers the system to localize the culprit service that the anomaly originates from.

---

[1] Sock-shop - https://microservices-demo.github.io/.
[2] Google Cloud Engine - https://cloud.google.com/compute/.

After the culprit service localization, it returns a list of potential culprit services, sorted by probability of being the source of the anomaly. Next, for each potential culprit service, our method identifies the anomalous metrics which contribute to the service abnormality. Finally, it outputs a list of (service, metrics list) pairs, for the possible culprit service, and metrics, respectively. With the help of this list, cloud operators can narrow down the causes and reduce the time and complexity to get the real cause.

3.1 Data Collection

We collect data from multiple data sources, including the application, the operating system and the network, in order to provide culprits for performance issues caused by diverse root causes, such as software bugs, hardware issues, resource contention, etc. Our system is designed to be application-agnostic, requiring no instrumentation to the application to get the data. Instead, we collect the metrics that reported by the application and the run-time system themselves.

3.2 Anomaly Detection

In the system, we detect the performance anomaly on the response times between two interactive services (collected by service mesh) using a unsupervised learning method: BIRCH clustering [6]. When a response time deviates from their normal status, it is detected as an anomaly and trigger the subsequent performance diagnosis procedures. Note that, due to the complex dependency among services and the properties of fault propagation, multiple anomalies could be also detected from services that have no issue.

3.3 Culprit Service Localization (CSL)

After anomalies are detected, the culprit service localization is triggered to identify the faulty service that initiates the anomalies. To get the faulty services, we use the method proposed by Wu, L., et al. [16]. First, it constructs an attributed graph to capture the anomaly propagation among services through not only the service call paths but also the co-located machines. Next, it extracts an anomalous subgraph based on detected anomalous services to narrow down the root cause analysis scope from the large number of microservices. Finally, it ranks the faulty services based on the personalized PageRank, where it correlates the anomalous symptoms in response times with relevant resource utilization in container and system levels to calculate the transition probability matrix and Personalized PageRank vector. There are two parameters for this method that need tuning: the anomaly detection threshold and the detection confidence. For the detail of the method, please refer to [16].

With the identified faulty services, we further identify the culprit metrics that make the service abnormal, which is detailed in Sect. 4.

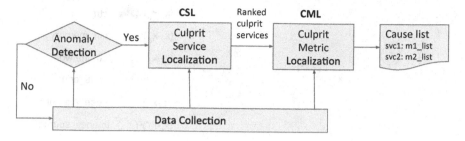

Fig. 1. Overview of the proposed performance diagnosis system.

4   Culprit Metric Localization (CML)

The underlying assumption of our culprit metric localization of the root cause lies in the observation that the underlying symptoms for the faulty behaviour differ from their expected values during normal operation. For example, when there is an anomaly of type "memory leak" it is expected that the memory in the service increases drastically, as compared to the normal operation. In the most general case, it is not known in advance which metric is contributing the most and it is the most relevant for the underlying type of fault in an arbitrary service. Besides, there may exist various inter-relationships between the observed metrics that manifest differently in normal or abnormal scenarios. Successful modelling of this information may improve the anomaly detection procedure and also better pinpoint the potential cause for the anomaly. For example in "CPU hog" we experience not only CPU increase but also a slight memory increase. Thus, some inter-metric relationships may not manifest themselves in same way during anomalies as normal operation.

To tackle these challenges we adopt the autoencoder architecture. An autoencoder is an approach that fits naturally under stressed conditions. The first advantage of the method is that one can add an arbitrary number of input metrics. Thus it can include many potential symptoms as potential faults to be considered at once. The second advantage is that it can correlate arbitrary relationships within the observed metric data with various complexity based on the depth and applied nonlinearities.

An autoencoder [5] is a neural network architecture that learns a mapping from the input to itself. It is composed of an encoder-decoder structure of at least 3 layers: input, hidden and output layer. The encoder provides a mapping from the input to some intermediate (usually lower-dimensional) representation, while the decoder provides an inverse mapping from the intermediate representation back to the input, Thus the cost function being optimized is given as in:

$$\mathcal{L}(X, X) = ||\phi(X) - UU^T\phi(X)||_2^2 \qquad (1)$$

where U can be seen as weights of the encoder-decoder structure learned using the backpropagation learning algorithm. While there exist various ways how the

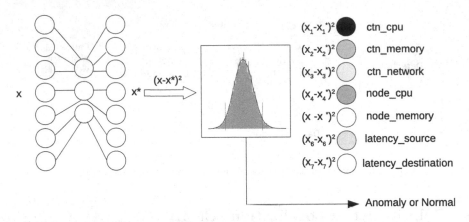

Fig. 2. The inner diagram of the culprit metric localization for anomaly detection and root cause inference. The Gaussian block produces decision that a point is an anomaly if it is below a certain probability threshold. The circle with the greatest intensity of black contributes the most to the error and is pointed as an root cause symptom.

mapping from one instance to another can be done, especially interesting is the mapping when the hidden layer is of reduced size. This allows to compress the information from the input and enforce it to learn various dependencies. During the training procedure, the parameters of the autoencoder are trained using just normal data from the metrics. This allows us to learn the normal behaviour of the system. In our approach, we further penalize the autoencoder to enforce sparse weights and discourage propagation of information that is not relevant via the L_1 regularization technique. This acts in discouraging the modeling of non-relevant dependancies between the metrics.

Figure 2 depicts the overall culprit metric localization block. The approach consists of three units: the autoencoder, anomaly detection and root-cause localization part. The root-cause localization part produces an ordered list of most likely cause given the current values of the input metrics. There are two phases of operation: the offline and online phase. During the offline phase, the parameters of the autoencoder and the gaussian distribution part are tuned. During the online phase, the input data is presented to the method one point at the time. The input is propagated through the autoencoder and the anomaly detection part. The output of the latter is propagated to the root-cause localization part that outputs the most likely root-cause.

After training the autoencoder, the second step is to learn the parameters of a Gaussian distribution of the reconstruction error. The underlying assumption is that the data points that are very similar (e.g., lie within 3σ (standard deviations) from the mean) are likely to come from a Gaussian distribution with the estimated parameters. As such they do not violate the expected values for metrics. The parameters of the distribution are calculated on a held-out validation set from normal data points. As each of the services in the system is run in a separate container and we have the metric for each of them, the autoencoder can be utilized as an additional anomaly detection method on a service level. As the culprit service localization module exploits the global dependency graph of the overall architecture, it suffers from the eminent noise propagated among the services. While unable to exploit the structure of the architecture, the locality property of the autoencoder can be used to fine-tune the results from the culprit service localization module. Thus, with a combination of the strengths of the two methods, we can produce better results for anomaly detection.

The decision for the most likely symptom is done such that we calculate the individual errors between the input and the corresponding reconstructed output. As the autoencoder is constrained to learn normal state, we hypothesize change of the underlying symptom when an anomaly arises to occur. Hence, for a given anomaly as a most likely cause, we report the symptom that contributes to the final error the most.

5   Experimental Evaluation

In this section, we present the experimental setup and evaluate the performance of our system in identifying the culprit metrics and services.

5.1   Testbed and Evaluation Metrics

To evaluate our system, we set up a testbed on Google Cloud Engine (see footnote 2), where we run the Sock-shop (see footnote 1) microservice benchmark consisting of seven microservices in a Kubernetes cluster, and the monitoring infrastructures, including the Istio service mesh[3], node-exporter[4], Cadvisor[5], Prometheus[6]. Each worker node in the cluster has 4 virtual CPUs, 15 GB of memory with Container-Optimized OS. We also developed a workload generator to send requests to different services.

To inject the performance issues in microservices, we customize the Docker images of the services by installing the fault injection tools. We inject two types of faults: CPU hog and memory leak, by exhausting the resource CPU and memory in the container, with stress-ng[7], into four different microservices. For each anomaly, we repeated the experiments 6 times in the duration of at least 3 min. To train the autoencoder, we collect data of 2 h in normal status.

To quantify the performance of our system, we use the following two metrics:

- *Precision at top* k denotes the probability that the root causes are included in the top k of the results. For a set of anomalies A, $PR@k$ is defined as:

$$PR@k = \frac{1}{|A|} \sum_{a \in A} \frac{\sum_{i<k}(R[i] \in v_{rc})}{(min(k, |v_{rc}|))} \qquad (2)$$

where $R[i]$ is the rank of each cause and v_{rc} is the set of root causes.

---

[3] Istio - https://istio.io/.

[4] Node-exporter - https://github.com/prometheus/node_exporter.

[5] Cadvisor - https://github.com/google/cadvisor.

[6] Prometheus - https://prometheus.io/.

[7] stress-ng - https://kernel.ubuntu.com/~cking/stress-ng/.

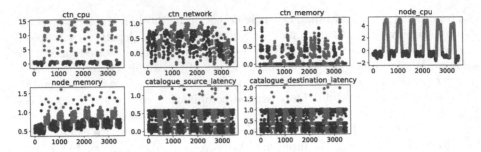

Fig. 3. Collected metrics when CPU hog is injected to microservice catalogue. (Color figure online)

– *Mean Average Precision* (MAP) quantifies the overall performance of a method, where N is the number of microservices:

$$MAP = \frac{1}{|A|} \sum_{a \in A} \sum_{1 \le k \le N} PR@k. \tag{3}$$

5.2 Effectiveness Evaluation

For each anomaly case, we collect the performance metrics from the application (suffixed with latency) and run-time system, including containers (prefixed with ctn) and worker nodes (prefixed with node). Figure 3 gives an example of the collected metrics when the "CPU hog" anomaly fault is injected to the catalogue microservice, repeated six times within one hour. The data collected during the fault injection is marked in red. The CPU hog fault is expected to be reflected by the ctn_cpu metric. We can see that (i) there are obvious spikes in metrics ctn_cpu and node_cpu. The spike of node_cpu is caused by the spike of ctn_cpu as container resource usage is correlated to node resource usage; (ii) metrics ctn_memory and node_memory also have some deviations; (iii) the fault CPU hog causes spikes in service latency. Therefore, we can conclude that the fault injected to the service manifests itself with a significant deviation from normal status. Meanwhile, it also affects some other metrics.

For each fault injected service, we train the autoencoder with normal data and test with the anomalous data. Figure 4 shows the reconstruction errors from autoencoder for each metric. We can see that the metric ctn_cpu has a large error comparing with other metrics, which indicates it has a higher probability to be the cause of the anomaly of service catalogue. The second highest reconstruction error is in the node_cpu metric, which is due to its strong correlation with the container resource usage. Hence, we conclude that ctn_cpu is the culprit metric.

Table 1 demonstrates the results of our method on different microservices and faults, in terms of PR@1, PR@3 and MAP. We observe that our method achieve a good performance with 100% in PR@1 in different services and faults, except for the service orders and carts with the fault memory leak. This is because (i)

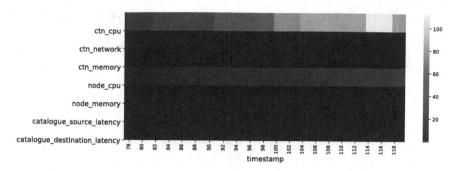

Fig. 4. Reconstruction errors for each metric when CPU hog is injected to microservice catalogue.

Table 1. Performance of identifying culprit metrics.

Service	Orders	Catalogue	Carts	User	Average
CPU hog					
PR@1	1.0	1.0	1.0	1.0	1.0
PR@3	1.0	1.0	1.0	1.0	1.0
MAP	1.0	1.0	1.0	1.0	1.0
Memory leak					
PR@1	0.83	1.0	0	1.0	0.71
PR@3	0.83	1.0	1.0	1.0	0.96
MAP	0.88	1.0	0.83	1.0	0.93

orders and carts are computation-intensive services; (ii) we exhaust their resource memory heavily in our fault injection; (iii) fault memory leak issues manifest as both high memory usage and high CPU usage. As our method target root cause that manifests itself with a significant deviation of causal metric, the accuracy decreases when the root cause manifests in multiple metrics. On average, our system achieves 85.5% in precision and 96.5% in MAP.

Furthermore, we apply the autoencoder to all of the pinpointed faulty services by the culprit service localization (CSL) module and analyze its performance of identifying the culprit services. For example, in an anomaly case where we inject a CPU hog into service catalogue, the CSL module returns a ranked list and the real cause service catalogue is ranked as the third. The other two services with higher rank are service orders and front-end. We leverage autoencoder to these three services, and the results show (i) autoencoder of service order returns Normal, which means it is a false positive and can be removed from the ranked list; (ii) autoencoder of service front-end returns Anomaly, and the highest ranked metric is the latency, which indicates that the abnormality of front-end is caused by an external factor, which is the downstream service catalogue. In this case, we conclude that it is not a culprit service and remove it from the ranked list; (iii)

Table 2. Comparisons of identifying culprit services.

Metrics	CSL	CSL + CML	Improvement(%)
PR@1	0.57	0.92	61.4
PR@3	0.83	0.98	18.1
MAP	0.85	0.97	14.1

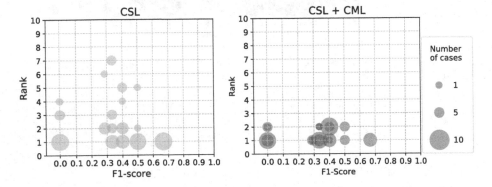

Fig. 5. Calibration of culprit service localization with autoencoder.

autoencoder of service catalogue returns Anomaly and the top-ranked metric is ctn_cpu. Therefore, with autoencoder, we can reduce the number of potential faulty services from 3 to 1.

Figure 5 shows the rank of culprit services identified by CSL and the calibration results with CSL and culprit metric localization module (CSL + CML) against the F1-score (the harmonic mean of precision and recall) of anomaly detection for all anomaly cases. We observe that applying autoencoder on the service relevant metrics can significantly improve the accuracy of culprit service localization by ranking the faulty service within the top two. Table 2 shows the overall performance of the above two methods for all anomaly cases. It shows that complementing culprit service localization with autoencoder can achieve a precision of 92%, which outperforms 61.4% than the results of CSL only.

6   Related Work

To diagnose the root causes of an issue, various approaches have been proposed in the literature. Methods and techniques for root cause analysis have been extensively studied in complex system [13] and computer networks [7].

Recent approaches for cloud services typically focus on identifying coarse-grained root causes, such as the faulty services that initiate service performance degradation [8,9,15]. In general, they are graph-based methods that construct a dependency graph of services with knowledge discovery from metrics or provided service call graph, to show the spatial propagation of faults among services; then they infer the potential root cause node which results in the abnormality of other nodes in the graph. For example, Microscope [8] locates the faulty service by building a service causality graph with the service dependency and service interference in the same machine. Then it returns a ranked list of potential culprit services by traversing the causality graph. These approaches can help operators narrow down the services for investigation. However, the causes set for an abnormal service are of a wide range, hence it is still time-consuming to get the real cause of faulty service, especially when the faulty service is low-ranked in the results of the diagnosis.

Some approaches identify root causes with fine granularity, including not only the culprit services but also the culprit metrics. Seer [4] is a proactive online performance debugging system that can identify the faulty services and the problematic resource that causes service performance degradation. However, it requires instrumentation to the source code; Meanwhile, its performance may decrease when re-training is frequently required to follow up the updates in microservices. Loud [10] and MicroCause [11] identify the culprit metrics by constructing the causality graph of the key performance metrics. However, they require anomaly detection to be performed on all gathered metrics, which might introduce many false positives and decrease the accuracy of causes localization. Álvaro Brandón, et al. [1] propose to identify the root cause by matching the anomalous graphs labeled by an expert. However, the anomalous patterns are supervised by expert knowledge, which means it can only detect previously known anomaly types. Besides, the computation complexity of graph matching is exponential to the size of the previous anomalous patterns. Causeinfer [2] pinpoints both the faulty services and culprit metrics by constructing a two-layer hierarchical causality graph. However, this system uses a lag correlation method to decide the causal relationship between services, which requires the lag is obviously included in the data. Compared to these methods, our proposed system leverages the spatial propagation of the service degradation to identify the culprit service and the deep learning method, which can adapt to arbitrary relationships among metrics, to pinpoint the culprit metrics.

7 Conclusion and Future Work

In this paper, we propose a system to help cloud operators to narrow down the potential causes for a performance issue in microservices. The localized causes are in a fine-granularity, including not only the faulty services but also the culprit metrics that cause the service anomaly. Our system first pinpoints a ranked list of potential faulty services by analyzing the service dependencies. Given a faulty service, it applies autoencoder to its relevant performance metrics and leverages the reconstruction errors to rank the metrics. The evaluation shows that our system can identify the culprit services and metrics with high precision.

The culprit metric localization method is limited to identify the root cause that reflects itself with a significant deviation from normal values. In the future, we would like to develop methods to cover more diverse root causes by analyzing the spatial and temporal fault propagation.

Acknowledgment. This work is part of the FogGuru project which has received funding from the European Union's Horizon 2020 research and innovation programme under the Marie Skłodowska-Curie grant agreement No 765452. The information and views set out in this publication are those of the author(s) and do not necessarily reflect the official opinion of the European Union. Neither the European Union institutions and bodies nor any person acting on their behalf may be held responsible for the use which may be made of the information contained therein.

References

1. Brandón, Á., et al.: Graph-based root cause analysis for service-oriented and microservice architectures. J. Syst. Softw. **159**, 110432 (2020)
2. Chen, P., Qi, Y., Hou, D.: Causeinfer: automated end-to-end performance diagnosis with hierarchical causality graph in cloud environment. IEEE Trans. Serv. Comput. **12**(02), 214–230 (2019)
3. Di Francesco, P., Lago, P., Malavolta, I.: Migrating towards microservice architectures: an industrial survey. In: ICSA, pp. 29–2909 (2018)
4. Gan, Y., et al.: Seer: leveraging big data to navigate the complexity of performance debugging in cloud microservices. In: Proceedings of the Twenty-Fourth International Conference on Architectural Support for Programming Languages and Operating Systems, ASPLOS 2019, pp. 19–33 (2019)
5. Goodfellow, I., Bengio, Y., Courville, A.: Deep Learning. MIT Press, Cambridge (2016). http://www.deeplearningbook.org
6. Gulenko, A., et al.: Detecting anomalous behavior of black-box services modeled with distance-based online clustering. In: 2018 IEEE 11th International Conference on Cloud Computing (CLOUD), pp. 912–915 (2018)
7. łgorzata Steinder, M., Sethi, A.S.: A survey of fault localization techniques in computer networks. Sci. Comput. Program. **53**(2), 165–194 (2004)
8. Lin, J., et al.: Microscope: pinpoint performance issues with causal graphs in microservice environments. In: Service-Oriented Computing, pp. 3–20 (2018)
9. Ma, M., et al.: Automap: diagnose your microservice-based web applications automatically. In: Proceedings of the Web Conference 2020, WWW 2020, pp. 246–258 (2020)
10. Mariani, L., et al.: Localizing faults in cloud systems. In: ICST, pp. 262–273 (2018)
11. Meng, Y., et al.: Localizing failure root causes in a microservice through causality inference. In: 2020 IEEE/ACM 28th International Symposium on Quality of Service (IWQoS), pp. 1–10. IEEE (2020)
12. Newman, S.: Building Microservices. O'Reilly Media Inc., Newton (2015)
13. Solé, M., Muntés-Mulero, V., Rana, A.I., Estrada, G.: Survey on models and techniques for root-cause analysis (2017)
14. Thalheim, J., et al.: Sieve: actionable insights from monitored metrics in distributed systems. In: Proceedings of the 18th ACM/IFIP/USENIX Middleware Conference, pp. 14–27 (2017)
15. Wang, P., et al.: Cloudranger: root cause identification for cloud native systems. In: CCGRID, pp. 492–502 (2018)
16. Wu, L., et al.: MicroRCA: root cause localization of performance issues in microservices. In: NOMS 2020 IEEE/IFIP Network Operations and Management Symposium (2020)

Anomaly Detection at Scale: The Case for Deep Distributional Time Series Models

Fadhel Ayed[1]([✉]), Lorenzo Stella[2], Tim Januschowski[2], and Jan Gasthaus[2]

[1] University of Oxford, Oxford, UK
[2] Amazon Research, Berlin, Germany
{stellalo,tjnsch,gasthaus}@amazon.de

Abstract. This paper introduces a new methodology for detecting anomalies in time series data, with a primary application to monitoring the health of (micro-) services and cloud resources. The main novelty in our approach is that instead of modeling time series consisting of real values or vectors of real values, we model time series of probability distributions. This extension allows the technique to be applied to the common scenario where the data is generated by requests coming in to a service, which is then aggregated at a fixed temporal frequency. We show the superior accuracy of our method on synthetic and public real-world data.

Keywords: Anomaly detection · Recurrent neural networks · Time series analysis

1 Introduction

In large-scale distributed systems or cloud environments, the detection of anomalous events allows operators to detect and understand operational issues and facilitates swift troubleshooting. Undetected anomalies can result in potentially significant losses and can impact customers of these systems and services negatively. In this work we focus on anomaly detection in the context of our target application of monitoring compute systems and cloud resources, where main object of interest are metrics emitted by these systems; we refer to this setting as *cloud monitoring*. We refer the reader to detailed overviews [2,11] on other application areas for anomaly detection.

In the setting of cloud monitoring, an anomaly detection system needs to be able to efficiently detect anomalous events in a streaming fashion. The fundamental difficulties that any anomaly detection system has to face are threefold. First, due to the number and diversity of the monitored metrics (often millions) and the streaming nature of the data, it is uncommon to have sufficient amounts of labeled data. Even if labels are available, due to the subjectivity of the task, labels may not represent an "objective" ground truth. This raises the need for *unsupervised* techniques. Second, the monitoring systems have to track the evolution of a large number of time series simultaneously, which often leads to a

© Springer Nature Switzerland AG 2021
H. Hacid et al. (Eds.): ICSOC 2020 Workshops, LNCS 12632, pp. 97–109, 2021.
https://doi.org/10.1007/978-3-030-76352-7_14

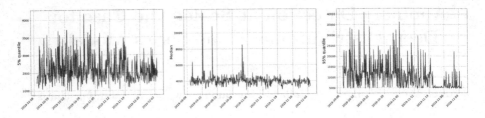

Fig. 1. Latency metric monitoring with temporal aggregation using different summary statistics. The three panels show the same underlying event data, aggregated into five-minute intervals using three summary statistics: (left) 5% quantile; (center) median; (right) 95% quantile. The anomalous region occuring at the end of November is clearly visible in the 95% quantile, but harder (or impossible) to detect in the other two.

considerable flow of data to process in near real-time, so the models have to *scale efficiently* to the amount of data available. Here, scalability comes not only in the traditional flavor of computational scalability, but also in terms of the need to involve experts to tune the systems. Finally, the methods have to be *flexible* in order to handle time series and anomalies of different nature (e.g. CPU usage, latency, error rate).

The main contribution of the present work is a novel anomaly detection method based on distributional time series models that addresses all three challenges. To the best of our knowledge it is the first anomaly detection methodology that builds on a predictive model for a distributional time series representation. It employs an autoregressive LSTM-based recurrent neural network to provide flexibility while still being statistically sound. Our model scales well at inference time and has a compact model state making it deployable in low-latency, streaming application scenarios. Finally, our methodology can detect collective anomalies[1], which most non-distributional techniques are unable to detect. We evaluate our method on a number of data sets including synthetic, publicly available, and AWS-internal data sets, and show that our method compares favorably to the state of the art. While we develop our methodology for the cloud monitoring setting, we further show that our method is competitive in classical anomaly detection settings.

We proceed by first discussing a motivating example for our method and provide background in Sect. 2, introduce the model formally in Sect. 3, evaluate it empirically in Sect. 4, discuss related work in Sect. 5, before concluding in Sect. 6.

---

[1] A collective anomaly consists of a subset of points that deviates from the rest of the dataset even though individually each point may appear normal.

2   Motivation

In the following we motivate our *distributional* time series modeling approach from two angles: the data generation process of request-driven metrics, and high-frequency time series.

In a typical (micro-) service monitoring setup, a metric datum is emitted for each request handled by the service. The raw monitoring data is thus a stream of events, where each event is a tuple consisting of a timestamp and a set of measurements. As a measurement is triggered for each incoming request, the time stamps are not equally spaced, and for large services one may collect hundreds of thousands of events per minute. To facilitate further processing, the typical anomaly detection pipeline starts with a temporal aggregation step, where the event data is aggregated into fixed-sized time intervals (e.g. one minute), recovering the classical, equally-spaced time series setting. This aggregation of events requires choosing a meaningful statistic which summarizes all measurements within a given time interval, while allowing detection of abnormal behaviors. Commonly used summary statistics are the mean, the median, or extreme percentiles. However, the summary statistics chosen ultimately determine the range of anomalies one can detect, and one risks missing anomalies if the statistics are chosen inappropriately (see Fig. 1 for illustration from internal services). The method we propose here embraces this event-based data generation process by considering the entire *distribution* of measurements within each time interval. This means

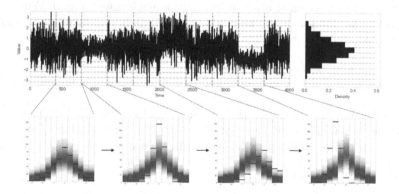

Fig. 2. Illustration of our approach. The undelying signal (top panel) is grouped into fixed-size time intervals (vertical red dashed lines) of size n_t (here $n_t = 400$). We estimate the probability distribution $\tilde{F}_t$ of the values within the interval using a histogram (blue horizontal bars in the bottom panels), with bin edges (dashed grey lines) chosen according to a global strategy (e.g. based the marginal distribution, top right panel). For each time interval, the model predicts a probability distribution over probability distributions (yellow-red heatmaps in the bottom panel) using a RNN. For "normal" periods (e.g. bottom left panel), the observed data (blue lines) aligns with the model's prediction, i.e. the blue lines fall into the shaded area. For "anomalous" periods (three rightmost bottom panels), the observed histogram falls outside the high-probability region predicted by the model. (Color figure online)

considering time series of equally spaced "points" in time, but where each "point" is a probability distribution, called a *distributional data point*. This is in contrast to most classical anomaly detection approaches that do not explicitly model the temporal data aggregation step.

Even though the proposed method was originally designed for the particular nature of the data described above, we demonstrate highly competitive performance even in the "classical" setting, where the starting point are time series of real values sampled at a regular frequency. We discuss this via the example of high-frequency time series, arising for example from measuring the CPU utilization or temperature of a compute node every second. Our approach solves several difficulties specific to such metrics.

The main challenge one faces when monitoring high-frequency data is that the temporal dynamics governing the data evolve at a slower pace than the frequency of observation. In typical application settings, meaningful variations are expected to occur from one hour to the next, but not every second. The underlying dynamics can thus often be adequately described by using one hour or half an hour time granularity, with seasonal patterns that are daily, or weekly. However, both classical and deep-learning-based time series models are commonly unable to model long range dependencies (measured in number of observations), so that if high-frequency data is modeled directly, these models commonly fail to capture medium and long term patterns. Our approach allows modeling the temporal evolution at a more appropriate frequency by aggregating the observations, while retaining the ability to detect anomalies at the original frequency by modeling the distribution of observations within each time interval. Within each aggregated time interval t, we treat the high-frequency data point as samples from this distribution.

3   Model

In the following, we introduce the necessary notation and tools used in the rest of the paper. We start by recalling that a generic strategy for anomaly detection using probabilistic models is to mark an observation as anomalous if its probability under the model is low. More details can be found for example in [7]. Our method builds on this approach and is summarized in Fig. 2.

3.1   Distributional Time Series

Let $F_{1:T} = F_1, \ldots, F_T$ be a time series of univariate probability distributions, represented by their cumulative distribution functions (CDFs). We assume that the support for all F_t is the interval $\mathbb{Y} = [y_{\min}, y_{\max}]$. Even though these distributions are the objects of interest, we usually do not to have access to them directly. Because of this, we also consider the scenario where we observe F_t only indirectly through samples, i.e. at each time t a set $\mathcal{Y}_t = \{Y_{t1}, \ldots, Y_{tn_t}\}$ of n_t iid samples from F_t is observed. We can differentiate three real-world use cases:

1. **Monitoring services with frequent requests:** This corresponds to the setting described in Sect. 2, where for each time interval (e.g. each minute), the number of measurements n_t is large, e.g. on the order of 10^5 or more. The underlying distributions F_t can then be estimated with a high enough precision for us to consider that they are directly observed. We will also refer to this as the *asymptotic settting*, since it corresponds $n_t \to +\infty$.
2. **High-frequency time series:** This corresponds to the setting where the temporal resolution of the original time series is higher than the scale at which meaningful temporal variation occurs, e.g. $n_t = 60$ when aggregating from seconds to minutes.
3. **Low-frequency time series:** We also consider the $n_t = 1$ setting, where our model reduces to a classical probabilistic time series model over real-values observations. Even though this is not the setting for which our approach was originally designed, we will show that it still yields competitive results.

Our model handles all three settings. We will refer to the last two scenarios as the *finite n_t scenarios*, in contrast to the first one. In the asymptotic setting, the distributions F_t are observed directly. In the finite n_t settings, we only observe samples from them. Therefore, we need to be able to assess the likelihood of F_t for the asymptotic regime, and the likelihood of $\mathcal{Y}_t$, where F_t is marginalized out, for the finite n_t regimes.

3.2 Probabilistic Model on Binned Densities

A common approach to modeling distributional data is to represent the functions of interest (e.g. the CDFs or PDFs) by a point in a carefully chosen finite-dimensional space. In this work, we will consider the space of piece-wise linear functions to approximate the CDFs, or equivalently, the space of binned (piecewise-constant) distributions to approximate the PDFs. Specifically, we chose to approximate each CDF F_t by a piece-wise linear function $\widetilde{F}_t$, composed of d linear pieces. A given function in this class is specified by two sets of parameters: the start and end points of linear pieces (the *knot positions*), and the slopes in each segment. While it is possible to adapt the knot positions dynamically (as done in [8]), we keep the knot positions fixed and only model the temporal evolution of the slopes within each segment. We emphasize the fact that one can approximate any F_t arbitrarily well as the grid becomes finer (d becomes larger). Therefore, we will assume that the F_t themselves are piece-wise linear.

We divide $\mathbb{Y}$ into d bins using the grid $y_{\min} = a_0 < ... < a_d = y_{\max}$. Suppose that the CDF F_t is piece-wise linear, interpolating the points $(a_k, F_t(a_k))_{k=0,...,d}$. Then the probability of falling into one of the bins $[a_{k-1}, a_k)$ is $p_{tk} = F_t(a_k) - F_t(a_{k-1})$. Given F_t, a sufficient statistic of the set of observations $\mathcal{Y}_t$ is the count vector $m_t = (m_{t1}, ..., m_{td})$, where m_{tk} denotes the number of elements of $\mathcal{Y}_t$ that fell into the bin $[a_{k-1}, a_k)$. It follows a Multinomial distribution with n_t trials and outcome probabilities p_t.

Specifying a distribution on the d dimensional probability vector $p_t = (p_{t1}, ..., p_{td})$ entails a distribution over the piece-wise linear CDFs F_t. We model this distribution over probability vectors using a Dirichlet distribution, i.e. $p_t \sim \text{Dir}(\alpha_t)$, where $\alpha_t \in \mathbb{R}_+^d$ denotes the concentration parameter whose temporal evolution is modeled using an RNN. With this choice of prior, p_t can be marginalized out and we have a closed form probability mass function for the observations m_t. More precisely, m_t follows a Dirichlet-Multinomial distribution with n_t number of trials and concentration vector α_t.

To summarize, given α_t, the likelihood of the observation is:

$$\mathcal{L}_t = \mathcal{L}(p_t; \alpha_t) = \text{Dir}(p_t; \alpha_t) \qquad \text{(Asymptotic setting)}$$

$$\mathcal{L}_t = \mathcal{L}(m_t; n_t, \alpha_t) = \text{Dir-Mult}(m_t; n_t, \alpha_t), \qquad \text{(Finite } n_t \text{ setting)}$$

where as explained previously in the asymptotic regime we suppose that we directly observe p_t which is equal to the normalized counts $\frac{1}{n_t} m_t$.

3.3   RNN Temporal Dynamics Model

In both settings, the temporal evolution of the data is described through the time-varying parameter α_t, and it is this dynamic behavior that we aim to learn. In order to do so, we will use an autoregressive LSTM-based recurrent neural network, whose architecture follows the one described in [23].

Recurrent neural networks (RNNs) form a class of artificial neural networks designed to handle sequential data. One of the key benefits of RNNs is their ability to handle sequences of varying lengths. RNNs sequentially update a hidden state h: at every time step t, the next hidden state h_t is computed by using the previous h_{t-1} and the next input (the next observation y_t and other covariates). A crucial detail is that the weights of the network are shared across time steps, which makes the RNN *recurrent*, and capable of handling sequences of varying length. This compact representation makes them amenable to streaming settings. Here, we mainly rely on long short-term memory networks (LSTM), the arguably most popular subclass of RNNs.

Let $z_{1:T}$ be the sequence of observations, either $p_{1:T}$ or $m_{1:T}$ depending on the setting. Denote ϕ the parameters of the RNN model. Given a horizon τ, the aim is to predict the probability distribution of future trajectories $z_{T+1:T+\tau}$, with the potential use of observed covariates $x_{1:T+\tau}$.

The parameter α_t is function of the output h_t of an autoregressive recurrent neural network with

$$h_t = r_\phi(h_{t-1}, z_{t-1}, x_t) \qquad (1)$$

$$\alpha_t = \theta_\phi(h_t) \qquad (2)$$

where r_ϕ is a multi-layer recurrent neural network with LSTM cells. The model is autoregressive and recurrent in the sense that it uses respectively the observation at the last time step z_{t-1} and the previous hidden state h_{t-1} as input. Then a layer θ_ϕ projects the output h_t to $\mathbb{R}_+^d$, the domain of α_t. The parameters ϕ of the model are chosen to minimize the negative log likelihood $\text{L} = -\sum_{t=1}^{T} \log(\mathcal{L}_t)$. Finally, we note that when dealing with anomaly detection we only require a time horizon $\tau = 1$.

3.4   Anomaly Detection with Level Sets

Once we forecast α_{T+1}, we can assess whether the observation z_{T+1} is a potential anomaly. Indeed, given α_{T+1}, we know the distribution of the random variable Z_{T+1}, of which z_{T+1} should be a sample if no anomaly happened. Therefore, we can compute the threshold η_{T+1} such that the probability of $\mathcal{L}(Z_{T+1})$ being smaller than η_{T+1} is smaller than a given level ε (for example $\varepsilon = 5\%$). Hence, the observation z_{T+1} will be considered anomalous if its likelihood is smaller than η_{T+1}. The remaining difficulty is to compute η_{T+1}. When the number of possible outcomes for Z_{T+1} is relatively small, this can be done exactly by computing the likelihoods of all outcomes. Otherwise, we will use a Monte Carlo method, following an idea that goes back to [13]: if we consider the univariate random variable defined as $\mathcal{L}_{T+1}(Z_{T+1})$, we remark that η_{T+1} can be interpreted as the ε quantile of that distribution.

For the high frequency setting, we use a two-stage approach, described on the following illustrative example. Suppose that we observe a minute frequency time series, and we are interested in hourly aggregation. From the forecasting module, we predict α_{T+1} and hence the distribution of the observations for the hour to come. In the first stage, before the hour is over, we assess every minute whether the current observation is anomalous. Once the hour is over, we assess whether the past 60 observations jointly constitute a collective anomaly. If we want to detect collective anomalies that are shorter, we can add an intermediate stage. Finally, as explained in the experiment section, we will need to give an anomaly score to each time point to evaluate the models. The score used is the logarithm of the p-value, which is the smallest ε for which a given point is considered as an anomaly. For the two-stage strategy, we simply add the two scores.

4   Experiments

Our implementation[2] is based on GluonTS [1] which in turn is based on MXNet [5]. We learn a global model (across all metrics) which takes roughly 3mins per 100 metrics. For such models, we do not have to re-train often, so we may disregard the training time for the production scenario. Inference scales embarrassingly parallel. Scoring of a single data point take 1 ms for 1 minutely aggregated data (note that we do not perform the costly Monte-Carlo estimates at every time point). We can limit memory consumption of the models to a fixed size of 80kb per metric. For all the experiments we learn the parameters of the model on the learning time range $\{0, ..., T\}$, and we perform anomaly detection on the detection time range $\{T + 1, ..., T + D\}$. We consider two different grids to define the bins. The first one is the simple regular grid, $a_k = k/d$. The second grid is obtained using $d + 1$ regularly spaced quantile levels of the marginal distribution. Depending on the problem, the regular or the quantiles grid can be better.

---

[2] The code is available at https://github.com/awslabs/gluon-ts/tree/distribution_anomaly_detection/distribution_anomaly_detection.

4.1 Evaluation Metric

For comparing the different models we will use the area under the receiver operating characteristic curve (ROC-AUC). It is a metric commonly used for classification problems to compare algorithms which performances depend on selecting a threshold. This measure quantifies how much a model is able to distinguish between the two classes. It takes values between 0 and 1, the higher the better. This score is independent of the threshold chosen since it only considers the ranking of the observations by the model in terms of how much abnormal it looks. Therefore it allows to quantify the maximum potential of a method.

4.2 Synthetic Data

Let $\mu_t = \sin(\frac{2\pi t}{P})$ and $\sigma_t = 1$, where $P = 24$ is a period length and $\epsilon_t \sim \mathcal{N}(0, 0.1)$ are iid noise. We will consider the two following dynamics:

1. **DS1:** $F_t = \mathcal{N}(\mu_t + \epsilon_t, \sigma_t)$
2. **DS2:** $F_t = \mathcal{N}(\mu_t, \sigma_t + \epsilon_t)$

We consider $T = 1500$ learning time points and a detection time horizon of $D = 2000$. In the detection time range, we add an anomaly with probability 3% at each location independently. For each experiment, we use one of two types of anomalies: a sudden distributional shift (by adding 1 to μ_t), or a distributional collapse (removing $1/2$ to the standard deviation σ_t). We therefore get four different settings, we will denote them respectively DS1 μ, DS1 σ, DS2 μ and DS2 σ.

We set the threshold for anomaly detection to be 95%. For the Monte Carlo approximation, we take $M = 1000$ samples from the predictive distribution of the log likelihoods. An observation can then be considered anomalous in two cases. In the first case, the generated noise term ϵ_t falls outside of a 95% confidence interval of the $\mathcal{N}(0, 0.1)$: these are false positives, and if the model perfectly captures the generating process, this should happen 5% of the times on average. The second case corresponds to the anomalies that are artificially added, considered as malfunctions, or true positives.

Asymptotic Setting. This setting mimics popular services on AWS; we have access to a grid of a thousand quantiles of F_t at each time step. We take a regular grid of $d = 30$ bins. We report the results in Table 1.

Finite n_t Setting. In this setting, we observe $n_t = 60$ samples from evey distribution F_t (hourly aggregation). We take a quantile grid of $d = 10$ bins. In most practical settings, we are able to take d much larger since we can make use of multiple time series simultaneously, even though they represent different metrics (CPU usage, Latency, Number of connected users, etc.).

We compare the performance of our approach to the standard one of monitoring an aggregated statistic. We use two state-of-the-art open source algorithms, namely Luminol and TwitterAD, as competitors. These algorithms are run on the appropriate aggregated statistics (empirical mean and standard deviation

Table 1. Anomaly detection for synthetic datasets in the asymptotic setting. Results are expressed in percent. When the name of the dataset is followed by μ (resp. σ), it corresponds to distributional shifts malfunctions (resp. distributional collapse). Otherwise, no malfunctions are introduced. We expect the FPR to be 5% in all cases.

	False positive rate	Recall
DS1	5.73 ± 0.61	–
DS1 μ	5.60 ± 0.99	99.7 ± 0.67
DS2 σ	5.43 ± 0.11	100
DS2	4.96 ± 1.0	–
DS2 μ	5.15 ± 1.5	100
DS2 σ	4.98 ± 0.72	99.8 ± 0.5

Table 2. Comparative evaluation of anomaly detection methods on the synthetic high frequency data. When the name of the dataset is followed by μ (resp. σ), it corresponds to distributional shifts malfunctions (resp. distributional collapse).

	Distribution	TwitterAD	Luminol
DS1 μ	0.9928	**0.9998**	0.9400
DS1 σ	**0.9864**	0.5010	0.9691
DS2 μ	0.9973	**0.9999**	0.9596
DS2 σ	**0.9797**	0.4990	0.9456

per hour). We note that in a practical setting, we don't know which statistics is most appropriate to monitor. The results are reported in Table 2.

4.3   Yahoo Webscope Dataset

Yahoo Webscope is an open dataset often used as a benchmark for anomaly detection since it is labeled. It is composed of 367 time series, varying in length from 700 to 1700 observations. Some of these time series come from real traffic to Yahoo services and some are synthetic. The dataset is divided into 4 sub-benchmarks, from $A1$ to $A4$. The time frequency of all the time series is one hour. Since the frequency is relatively low, and since there are no collective anomalies in this dataset, we take $n_t = 1$, which corresponds to the classical anomaly detection setting. We report the results of [19] to compare the performance of our approach with the state of the art anomaly detection algorithms. We report the results per sub-benchmark, since they contain different patterns. We use 40% of each time series for training. We learn a single model for all the series of a same sub-benchmark, which means that we train the model on all the time series simultaneously. The results are given in Table 3. Here, since $n_t = 1$, the total number of possible outcomes is equal to $d = 100$. Therefore, we do not need Monte Carlo estimates.

Table 3. Comparative evaluation of state-of-the-art anomaly detection methods on the Yahoo Webscope dataset. Average AUC per benchmark.

Benchmark	iForest	OCSVM	LOF	PCA	TwitterAD	DeepAnT	FuseAD	Distribution
A1	0.8888	0.8159	0.9037	0.8363	0.8239	0.8976	**0.9471**	0.9435
A2	0.6620	0.6172	0.9011	0.9234	0.5000	0.9614	0.9993	**0.9999**
A3	0.6279	0.5972	0.6405	0.6278	0.6176	0.9283	**0.9987**	0.9988
A4	0.6327	0.6036	0.6403	0.6100	0.6534	0.8597	0.9657	**0.9701**

4.4 AWS Data

Finally, we consider three benchmark datasets of high frequency time series, collected from AWS. These datasets are often used internally at Amazon to compare models. The benchmark **B1** has a 1 min time frequency, it is composed of 55 time series. The benchmarks **B2** and **B3** have a 5 min time frequency. They are composed of 100 time series each. All datasets are composed of different metrics, among them CPU usage, latency, number of users, etc. Each time series of all three benchmarks have approximately 17 000 time points. We use 60% of the time range for training and the remainder for detection. We set $d = 100$ and aggregate all time series to a 30 min frequency, so $n_t = 30$ for **B1** and $n_t = 6$ for **B2** and **B3**. However the quality of the labeling is heterogeneous, **B1** being the most reliable one. We find that the anomalies identified by our method are false positives under the labels but should probably be counted as true positives. We perform a two stage anomaly detection, the first stage gives scores for the single observations, the second for the collection of observations within half-hours. We again compare to Luminol and TwitterAD. The results are reported in Table 4 which show the dominance of our method on this data set.

Table 4. Comparative evaluation of anomaly detection methods on AWS data. Average ROC-AUC per benchmark.

	Distribution	TwitterAD	Luminol
B1	**0.8183**	0.7134	0.6467
B2	**0.7534**	0.5895	0.5804
B3	**0.6860**	0.5889	0.5860

5 Related Work

Anomaly detection is a rich field with many applications and solutions available (see for example [2,11,12,21]). We focus our discussion of related work on unsupervised anomaly detection models. A first approach for anomaly detection are the so called outlier detection methods, which quantify how much every point is different from its neighborhood. In [21] for example, a saliency score is computed for every observation using Fourier Transforms. In [15], an outlier score is computed using Recurrent Autoencoder Ensembles.

Most related to our approach are anomaly detection methods using (probabilistic) forecasting models. Overview of forecasting models can be found in recent tutorials [7]. These approaches have the advantage that we can obtain an interpretable and normalized score. Other approaches, e.g., [10,27] do not allow for this and hence we do not consider them further. The most common forecasting models are the classical ARIMA (see for example [18] for applications) and increasingly deep learning based approaches as we apply it here [16,19]. Even though in the more general area of sequence learning, attention-based models [26] have become the state of the art, our choice of an RNN similar to [8,23] is motivated by the streaming setting that we consider here. The compact model state of an RNN is well suited to a streaming setting, whereas attention-based models have a prohibitively large state. Building on extreme value theory, the authors of [24] propose a noteworthy approach that can be seen as non-deep forecasting based method, but where only the tail of the distribution is modeled.

Functional Time Series (FTS) models are related to our problem. In that setting, the learner observes a time sequence of functional data, and tries to forecast the next functions. This framework and the resulting methods have many applications ranging from the study of demographic curves (for example [14]) to electricity price forecasting (see [9]). While these models allow for more general functions than distributions as data points, the restriction to distributions has led to further models [4,20] and Bayesian variants have also been proposed [3,22].

Instead of time series models, it is also possible to disregard time dependencies and case the time series prediction as a regression problem. This would allow for employing distribution regression models such as [25]. Given the strong auto-correlation in metrics data as we observe in our application, time dependence should not be disregarded and we therefore do not consider this approach further.

We can also cite the field of anomaly detection from log data ([6,17] for example), which also aim at detecting abnormal events in large-scale distributed systems. The setting is however quite different from the one considered here, as these methods are designed for unstructured data, which is not the case here as we observe real valued time series.

6   Conclusion

We presented the first anomaly detection method based on deep distributional time series models. The development of this model was motivated by real-world anomaly detection data and use-cases that we commonly find in monitoring cloud services. In the experiments, we show that on synthetic, public and AWS-internal data, our method compares favorably to other anomaly detection offerings. Our method was designed for streaming scenarios as they occur in monitoring compute metrics and it is fully elastic. While labels for anomalies are sparse, imbalanced and noisy, they nevertheless exists. Future work should consider how to improve the algorithms described here by incorporating labels during learning and acquiring them during production runs to lead to a continuously improving anomaly detection system.

References

1. Alexandrov, A., et al.: Gluonts: probabilistic time series models in python. arXiv preprint arXiv:1906.05264 (2019)
2. Bendre, S.: Outliers in statistical data (1994)
3. Caron, F., Davy, M., Doucet, A., Duflos, E., Vanheeghe, P.: Bayesian inference for linear dynamic models with dirichlet process mixtures. IEEE Trans. Signal Process. **56**(1), 71–84 (2007)
4. Chang, Y., Kaufmann, R.K., Kim, C.S., Miller, J.I., Park, J.Y., Park, S.: Evaluating trends in time series of distributions: a spatial fingerprint of human effects on climate. J. Econom. **214**(1), 274–294 (2020)
5. Chen, T., et al.: Mxnet: a flexible and efficient machine learning library for heterogeneous distributed systems. arXiv preprint arXiv:1512.01274 (2015)
6. Du, M., Li, F., Zheng, G., Srikumar, V.: Deeplog: anomaly detection and diagnosis from system logs through deep learning. In: Proceedings of the 2017 ACM SIGSAC Conference on Computer and Communications Security, pp. 1285–1298 (2017)
7. Faloutsos, C., Gasthaus, J., Januschowski, T., Wang, Y.: Forecasting big time series: old and new. Proc. VLDB Endow. **11**(12), 2102–2105 (2018)
8. Gasthaus, J., et al.: Probabilistic forecasting with spline quantile function RNNs. In: The 22nd International Conference on Artificial Intelligence and Statistics, pp. 1901–1910 (2019)
9. González, J.P., San Roque, A.M., Perez, E.A.: Forecasting functional time series with a new hilbertian armax model: application to electricity price forecasting. IEEE Trans. Power Syst. **33**(1), 545–556 (2017)
10. Guha, S., Mishra, N., Roy, G., Schrijvers, O.: Robust random cut forest based anomaly detection on streams. In: International Conference on Machine Learning, pp. 2712–2721 (2016)
11. Hawkins, D.M.: Identification of Outliers, vol. 11. Springer, Heidelberg (1980)
12. Hochenbaum, J., Vallis, O.S., Kejariwal, A.: Automatic anomaly detection in the cloud via statistical learning. arXiv preprint arXiv:1704.07706 (2017)
13. Hyndman, R.J.: Computing and graphing highest density regions. Am. Stat. **50**(2), 120–126 (1996)
14. Hyndman, R.J., Ullah, M.S.: Robust forecasting of mortality and fertility rates: a functional data approach. Comput. Stat. Data Anal. **51**(10), 4942–4956 (2007)
15. Kieu, T., Yang, B., Guo, C., Jensen, C.S.: Outlier detection for time series with recurrent autoencoder ensembles (2019)
16. Malhotra, P., Vig, L., Shroff, G., Agarwal, P.: Long short term memory networks for anomaly detection in time series. In: Proceedings, vol. 89, pp. 89–94. Presses universitaires de Louvain (2015)
17. Meng, W., et al.: Loganomaly: unsupervised detection of sequential and quantitative anomalies in unstructured logs. In: IJCAI, pp. 4739–4745 (2019)
18. Moayedi, H.Z., Masnadi-Shirazi, M.: Arima model for network traffic prediction and anomaly detection. In: 2008 International Symposium on Information Technology, vol. 4, pp. 1–6. IEEE (2008)

19. Munir, M., Siddiqui, S.A., Chattha, M.A., Dengel, A., Ahmed, S.: Fusead: unsupervised anomaly detection in streaming sensors data by fusing statistical and deep learning models. Sensors **19**(11), 2451 (2019)
20. Park, J.Y., Qian, J.: Functional regression of continuous state distributions. J. Econom. **167**(2), 397–412 (2012)
21. Ren, H., et al.: Time-series anomaly detection service at microsoft. In: Proceedings of the 25th ACM SIGKDD International Conference on Knowledge Discovery & Data Mining, pp. 3009–3017 (2019)
22. Rodriguez, A., Ter Horst, E., et al.: Bayesian dynamic density estimation. Bayesian Anal. **3**(2), 339–365 (2008)
23. Salinas, D., Flunkert, V., Gasthaus, J., Januschowski, T.: Deepar: probabilistic forecasting with autoregressive recurrent networks. Int. J. Forecast. **36**(3), 1181–1191 (2019)
24. Siffer, A., Fouque, P.A., Termier, A., Largouet, C.: Anomaly detection in streams with extreme value theory. In: Proceedings of the 23rd ACM SIGKDD International Conference on Knowledge Discovery and Data Mining, pp. 1067–1075 (2017)
25. Szabó, Z., Sriperumbudur, B.K., Póczos, B., Gretton, A.: Learning theory for distribution regression. J. Mach. Learn. Res. **17**(1), 5272–5311 (2016)
26. Vaswani, A., et al.: Attention is all you need. In: Advances in Neural Information Processing Systems, pp. 5998–6008 (2017)
27. Yeh, C.C.M., et al.: Matrix profile i: all pairs similarity joins for time series: a unifying view that includes motifs, discords and shapelets. In: 2016 IEEE 16th International Conference on Data Mining (ICDM), pp. 1317–1322. IEEE (2016)

A Systematic Mapping Study in AIOps

Paolo Notaro[1,2(✉)] ⓘ, Jorge Cardoso[2,3], and Michael Gerndt[1]

[1] Chair of Computer Architecture and Parallel Systems,
Technical University of Munich, Munich, Germany
`paolo.notaro@tum.de, gerndt@in.tum.de`
[2] Ultra-scale AIOps Lab, Huawei Munich Research Center, Munich, Germany
`{paolo.notaro,jorge.cardoso}@huawei.com`
[3] Department of Informatics Engineering/CISUC,
University of Coimbra, Coimbra, Portugal

Abstract. IT systems of today are becoming larger and more complex, rendering their human supervision more difficult. Artificial Intelligence for IT Operations (AIOps) has been proposed to tackle modern IT administration challenges thanks to AI and Big Data. However, past AIOps contributions are scattered, unorganized and missing a common terminology convention, which renders their discovery and comparison impractical. In this work, we conduct an in-depth mapping study to collect and organize the numerous scattered contributions to AIOps in a unique reference index. We create an AIOps taxonomy to build a foundation for future contributions and allow an efficient comparison of AIOps papers treating similar problems. We investigate temporal trends and classify AIOps contributions based on the choice of algorithms, data sources and the target components. Our results show a recent and growing interest towards AIOps, specifically to those contributions treating failure-related tasks (62%), such as anomaly detection and root cause analysis.

Keywords: AIOps · Operations and Maintenance · Artificial Intelligence

1 Introduction

Modern society is increasingly dependent on large-scale IT infrastructures. At the same time, the latest IT challenges impose higher levels of reliability and efficiency on computer systems. Because of the large increase in size and complexity of these systems, IT operators are increasingly challenged while performing tedious administration tasks manually. This has sparked in recent years much interest towards the study of self-managing and autonomic computing systems to improve efficiency and responsiveness of IT services. While many static algorithmic solutions have been proposed, these automated solutions often show limitations in terms of adaptiveness and scalability. The presence of large data volumes in different modalities motivates the investigation of intelligent learning systems, able to adapt their behavior to new observations and situations.

H. Hacid et al. (Eds.): ICSOC 2020 Workshops, LNCS 12632, pp. 110–123, 2021.
https://doi.org/10.1007/978-3-030-76352-7_15

Artificial Intelligence for IT Operations (AIOps) investigates the use of Artificial Intelligence (AI) for the management and improvement of IT services. AIOps relies Machine Learning, Big Data, and analytic technologies to monitor computer infrastructures and provide proactive insights and recommendations to reduce failures, improve mean-time-to-recovery (MTTR) and allocate computing resources efficiently [3]. AIOps offers a wide, diverse set of tools for several applications, from efficient resource management and scheduling to complex failure management tasks such as failure prediction, anomaly detection and remediation [13,23]. However, being a recent and cross-disciplinary field, AIOps is still a largely unstructured research area. The existing contributions are scattered across different conferences and apply different terminology conventions. Moreover, the high number of application areas renders the search and collection of relevant papers difficult. Some previous systematic works only treat single tasks or subareas inside AIOps [20,31]. This motivates the need for a complete and updated study of AIOps contributions.

In this paper, we present in-depth analysis of AIOps to cover for these limitations. We have identified and extracted over 1000 AIOps contributions through a systematic mapping study, enabling us to delineate common trends, problems and tools. First, we provide an in-depth description of the methodology followed in our mapping study (Sect. 2), reporting and motivating our planning choices regarding problem definition, search, selection and mapping. Then, we present and discuss the results drawn from our study, including the identification of most common topics, data sources, and target components (Sect. 3). Finally, Sect. 4 summarizes the outcomes and conclusions treated in this work.

2   Methodology

2.1   Systematic Mapping Studies

A systematic mapping study (SMS) is a research methodology widely adopted in many research areas, including software engineering [34]. The ultimate goal of a SMS is to provide an overview of a specific research area, to obtain a set of related papers and to delineate trends present inside such area. Relevant papers are collected via predefined search and selection techniques and research trends are identified using categorization techniques across different aspects of the identified papers, e.g. topic or contribution type. We choose to perform a SMS because we are interested in gathering contributions and obtaining statistical insights about AIOps, such as the distribution of works in different subareas and the presence of temporal trends for particular topics. SMSs have also been shown to increase the effectiveness of follow-up systematic literature reviews [34]. To this end, we have also used our systematic mapping study to collect references for a survey on failure management in AIOps separately published.

2.2   Planning

According to the step outline followed in [34], a systematic mapping study is composed of:

– **Formulation,** i.e. express the goals intended for the study through research
 questions. Equally important is to clearly define the scope of investigation;
– **Search,** i.e. define strategies to obtain a sufficiently high number of papers
 within the scope of investigation. This comprises the selection of one or more
 search strategies (database search, manual search, reference search, etc.);
– **Selection** (or screening), i.e. define and apply a set of inclusion/exclusion
 criteria for identifying relevant papers inside the search result set;
– **Data Extraction and Mapping,** i.e. gather the information required to
 map the selected papers into predefined categorization scheme(s). Finally,
 results are presented in graphical form, such as histograms or bubble plots.

The next sections illustrate and motivate our choices regarding these four steps
for our systematic mapping study in AIOps.

2.3   Formulation

The main goal of this mapping study is to identify the extent of past research
in AIOps. In particular, we would like to identify a representative set of AIOps
contributions which can be grouped based on the similarity of goals, employed
data sources and target system components. We also wish to understand the
relative distribution of publications within these categories and the temporal
implications involved. Formally, we articulate the following research questions:

**RQ1. What categories can be observed while classifying AIOps con-
tributions in scientific literature?**
RQ2. What is the distribution of papers in such categories?
RQ3. Which temporal trends can be observed for the field of AIOps?

In terms of scope, we express the boundaries of AIOps as the union of goals
and problems in IT Operations when dealt with AI techniques. To circumvent
ambiguity about the term AI, we adopt an inclusive convention where we consider AI both date-driven approaches, such as Machine Learning and data mining, as well as goal-based approaches, such as reasoning, search and optimization
approaches. However, we mostly concentrate our efforts on the first category due
to its stronger presence and connection to AIOps methodologies (e.g. data collection).

2.4   Search and Selection

Selection Criteria. We start illustrating the selection principles beforehand,
so that the discussion will appear clearer when we describe our result collection strategy, composed of search and selection altogether. In terms of inclusion
criteria, we define only one relevance criterion, based on the main topic of the
document. Following from our discussion on scoping such inclusion criterion
comprises two necessary conditions:

Table 1. The two keyword sets obtained via PICO used for database search.

AI Keywords	IT Operations Keywords
("AI" OR "artificial intelligence")	
"classification"	("DevOps" OR "site reliability engineering"
"clustering"	OR "SRE") ("IT operations")
"logistic regression"	("anomaly detection" OR "outlier detection")
"regression"	("cloud computing")
("DL" OR "deep learning")	("cloud")
("ML" OR "Machine Learning")	("fault detection" OR "failure detection")
("inference" OR "logic" OR "reasoning)	("fault localization" OR "failure localization")
("supervised" OR "unsupervised" OR	("fault prediction" OR "failure prediction")
"semi-supervised" OR "reinforcement") AND ("learning")	("fault prevention" OR "failure prevention")
("support vector machine" OR "SVM")	("log" OR "logs" OR "log analysis")
("tree" OR "tree-based" OR "trees" OR "forest")	("metrics" OR "KPI" OR "key performance indicator")
(("bayesian" OR "neural") AND "network")	("remediation" OR "recovery")
((("hidden" AND "markov") OR ("gaussian"	("root-cause analysis" OR "root cause analysis")
AND "mixture")) AND "model")	("service desk automation")
(("datacenter" OR "data center") AND "management")	("tracing" OR "trace" OR "traces")

- The document references one or more AI methods. These mentions can either be part of the implementation or as part of its discussion/analysis (e.g. in a survey). Any mention to AI algorithms employed by others (i.e. mentioned in the related work section or as baseline comparison) that is not strictly the focus of the document, is not considered valid;
- The document applies its concepts to some kind of IT system management. We therefore exclude papers with no specific target domain or with a target domain outside of IT Operations.

In terms of exclusion criteria, we define the following as exclusion rules:

- The language of the document is not English;
- The document is not accessible online;
- The document does not belong to the following categories: scientific article (conference paper, journal article), book, white paper;
- The main topic of the document is one of the following: cybersecurity, industrial process control, cyber-physical systems, and optical sensor networks.

For the special case of survey and review papers, we consider them relevant as long while carrying out our mapping study, but we then exclude them from our final result set, as these articles are useful to find other connected works through references, but they do not constitute novel contributions to the field.

Database Search. For the search process, database search represents the first and most important step, as it aims to provide the highest number of results and perform an initial screening of irrelevant papers. We perform database search in three steps: keywording, query construction and result polling. For keywording we use the PICO technique presented in [34] to derive a set of keywords for AI and a set of keywords for IT Operations. The keywords are listed in Table 1. Then, following our scoping considerations, we construct queries so that they return results where both AI and IT Operations are present. In particular, we apply logic conjunction of keywords across all combinations of the two keyword sets (e.g. "logistic regression" and "cloud computing"). This helps enforcing precision in our search results. For keywords with synonyms and abbreviations, we allow all equivalent expressions via OR disjunction. We also perform general search queries, related to the topic as a whole (e.g. "AIOps"). Finally, we group some queries with common terms to reduce the number of queries.

We select three online search databases that are appropriate for the scope of investigation: IEEE Xplore, ACM Digital Library and arXiv. For each query we restrict our analysis to the top 2000 results returned. We aggregate results from all searches in one large set of papers, removing duplicates and annotating for each item corresponding search metadata (e.g. number of hits, index position in corresponding searches, etc.). The result from this step consists of 83817 unique articles. For each item we collect the title, authors, year, publication venue, contribution type and citation count (from Google Scholar).

Preliminary Filtering and Ranking-Based Selection. In the filtering step we start improving the quality of our selection of papers. First, papers are automatically excluded based on publication venue, for those venues that are clearly irrelevant for topic reasons (e.g. meteorology). We also exclude based on the year of publication (year <1990) as it precedes the advent of large-scale IT services. By doing so, we can exclude approximately 8000 elements.

Usually at this point, a full-text analysis would be performed on all the available papers to screen relevant contributions using the above cited selection rules. Although we partly filtered results, it is still not feasible to perform an exhaustive selection analysis, even as simple as filtering by title. It is also impractical to attempt an automated selection by content, as it is not clear how to perform an efficient, high-recall, high-precision text classification without supervision. Therefore, before proceeding with the rest of the search and selection steps, we apply a ranking procedure on these intermediate results, so that we can prioritize investigation of more relevant papers. We apply the exclusion and inclusion rules of Sect. 2.4 to the papers examined in ranking order.

This approximate procedure however raises the question of when it is convenient to stop our selection and discard the remaining items. To solve this, we develop a new approach from our observations of ranked items. We base the method on the following assumption: a considerable ratio of relevant papers can be identified by ranking and selecting top results using different relevance criteria (conference, position index in the query result set, number of hits in all queries, etc.), but in this sorting scenario we also observe a long-tail distribution of relevant documents, i.e. some relevant papers appear in the last positions even after sorting with our relevance heuristics (see Fig. 1). This is coherent with the known impossibility of performing exhaustive systematic literature reviews and mapping studies, as completing the long tail provides less results at the expense of a larger research effort. We assume the ratio of relevant papers in the long tail to be constant and comparable in magnitude to the number of relevant papers when sampled randomly from the result set. Based on this assumption, we proceed as follows:

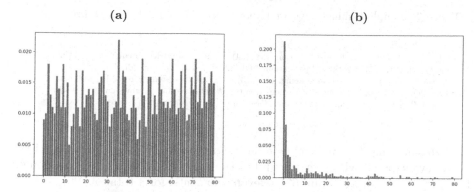

Fig. 1. Estimated relevance probability for collected papers (y-axis), as a function of the index in the result set (x-axis, in thousands), with paper arranged: (a) in random order (b) using a relevance heuristic based on search hits. We can observe how, thanks to the heuristic (b), the majority of relevant papers can be identified by examining only a small fraction of the set (the top results on the left side).

- We start screening all papers in the result set, ranked according to different relevance heuristics (e.g. number of hits in queries), and we observe the ratio of relevant papers identified over time;
- We examine the same papers in random order, and measure the same ratio;
- When the two ratios are comparable, we assert we reached the tail of the distribution of relevant papers and stop examining and selecting new papers.

As sorting criteria, we use the number of hits in the search performed in the previous step, as well as other more complex heuristics, taking into account the index position in result sets and the number of citations. When examining a paper, we look into the full content to identify concepts related to our selection criteria previously illustrated. As done previously with search results, we gather relevant papers in one unique group. Using this stopping criterion, we conclude this selection step when we have identified 430 relevant papers.

2.5   Additional Search Techniques

The "early stopping" criterion previously described, while allowing a feasible and comprehensive selection strategy across thousands of contributions, has a natural tendency towards discarding relevant papers. We also expect to miss other relevant papers, not present in the initial set of 83817, because they were not identified by our database search. To cover for these limitations, we apply other search techniques in addition to database search. Differing from before, we here apply our selection criteria exhaustively for each document retrieved.

Reference Search. For each of the 430 relevant papers identified in the previous step, we search inside their cited references. In particular, we adopt backward

Table 2. Sample k-shingles with relevance probability (and total occurrences).

k=1	k=2	k=3
tpc-w, 1.00 (13)	defect prediction, 1.00 (34)	software defect prediction, 1.00 (22)
log-based, 0.92 (11)	(work)load prediction, 1.00 (32)	disk failure prediction, 1.00 (8)
sla, 0.84 (48)	software aging, 1.00 (13)	failure prediction model, 1.00 (7)
stragglers, 0.83 (10)	resource allocation, 1.00 (6)	cloud resource provisioning, 1.00 (5)
vm, 0.83 (59)	hardware failures, 0.89 (8)	automatic anomaly detection, 0.88 (7)

snowball sampling [18]: we include in our relevant set all papers previously cited by a relevant paper whenever they fulfill the selection requirements mentioned above. By doing so, we obtain 631 relevant elements, for a total of 1061.

Conference Search. Reference search allows to identify prominent contributions frequently mentioned by other authors. A drawback is the introduction of bias towards specific research groups and authors. We also observed how reference search rewards specific tasks and research fields as they are typically more cited. We therefore apply other search techniques to compensate for these facts. We perform a manual search by inspecting papers published in relevant conferences. These relevant conferences are identified via correlation with other relevant papers and have also been confirmed by experts in the field. We look at the latest 3 editions of each conference, in an effort to compensate the sampling of dated papers performed by reference search. We obtain 5 more papers with this method.

Iterative Search Improvement. To conclude our search, we attempt at improving our initial guess on IT Operations keywords via analysis of the available text content (text and abstract). Using our relevant paper set as positive samples, we perform a statistical analysis to identify k-shingles (sets of k consecutive tokens) that appear often in relevant documents (Table 2). In particular, we measure the document relevance probability given the set of shingles observed in the available text content. We choose $k = 1, \ldots, 5$. We use these shingles as keywords to construct new queries along with previously used AI keywords. We here limit the collection to 20 results per query. Thanks to this step, we identify 20 new relevant papers. As a by-product, we get in contact with frequently cited concepts and keywords in AIOps, later useful for taxonomy and classification.

2.6   Data Extraction and Categorization

After obtaining the result set of relevant papers (counting 1086 contributions), we analyze the available information to draw quantitative results and answer our research questions. We describe here the data extraction process and the analysis techniques employed to gather insights and trends for the AIOps field.

First, we classify the relevant papers according to target components and data sources. Target components indicate a high-level piece of software or hardware in an IT system that the document tries to enhance (e.g. hard drive for hard

Table 3. Selection of result papers grouped by data sources, targets and (sub)categories.

Ref.	Source Code	Testing Resources	System Metrics	KPIs/SLO data	Network Traffic	Topology	Incident Reports	Event Logs	Execution Traces	Source Code	Application	Hardware	Network	Datacenter	Cat.
[27]	●									●					1.1
[32]	●	●								●					1.2
[16]			●								●			●	1.3
[41]			●	●							●			●	1.3
[29]			●								●			●	1.4
[47]			●									●			2.1
[14]			●								●			●	2.1
[12]			●					●			●				2.1
[46]								●			●	●			2.1
[8]	●										●			●	2.2
[11]	●	●									●				2.2
[17]	●	●									●			●	2.2
[35]	●					●					●			●	2.2
[24]								●			●			●	2.2
[37]								●			●	●	●	●	2.2
[45]								●			●				2.2
[43]	●							●						●	3.1
[42]			●								●				3.1
[40]			●	●							●			●	3.1
[21]						●	●						●		3.1
[22]						●	●				●		●		3.1
[15]							●				●				3.1
[10]							●	●			●			●	3.1
[6]							●	●			●			●	3.1
[28]							●				●				3.1
[30]						●							●	●	3.2
[49]	●							●						●	3.3
[1]	●	●										●	●		4.1
[33]		●					●						●	●	4.1
[5]			●										●	●	4.1
[44]	●							●			●				4.2
[4]	●		●	●							●				4.2
[19]					●	●							●		4.2
[9]									●		●			●	4.2
[36]		●					●				●				4.3
[7]	●	●									●			●	4.3
[26]							●				●			●	4.3
[2]								●			●			●	4.3
[39]								●			●			●	5.1
[48]								●			●			●	5.2
[25]							●	●			●			●	5.2
[38]	●	●									●			●	5.3

(Sub)Category Legend					
1.1	Software Defect Prediction	2.2	System Failure Prediction	4.2	Root Cause Diagnosis
1.2	Fault Injection	3.1	Anomaly Detection	4.3	RCA - Others
1.3	Software Rejuvenation	3.2	Internet Traffic Classification	5.1	Incident Triage
1.4	Checkpointing	3.3	Log Enhancement	5.2	Solution Recommendation
2.1	Hardware Failure Prediction	4.1	Fault Localization	5.3	Recovery

disk failure prediction). We group components in five high-level categories: code, application, hardware, network and datacenter. Data sources provide an indication of the input information of the algorithm (such as logs, metrics, or execution traces). Data sources are categorized in source code, testing resources, system metrics, key performance indicators (KPIs), network traffic, topology, incident reports, logs and traces. the "AI Method" axis denotes the actual algorithm employed, with similar methods aggregated in bigger classes to avoid excessive fragmentation (e.g. 'clustering' may contain both k-means and agglomerative hierarchical clustering approaches). Table 3 presents a selection of papers from the result set with the corresponding target, source and category annotation.

Then, we use the result set to infer a taxonomy based on tasks and target goals. The taxonomy is depicted in Fig. 2. We divide in AIOps contributions in failure management (FM), the study on how to deal with undesired behavior

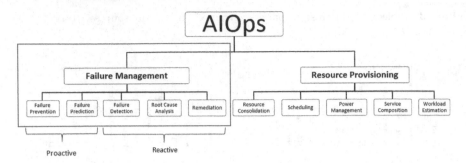

Fig. 2. Taxonomy of AIOps as observed in the identified contributions

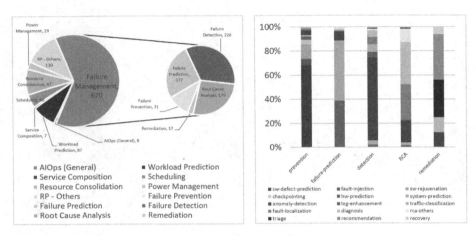

Fig. 3. Left: distribution of AIOps papers in macro-areas and categories. Right: percent distribution of failure management papers by category in corresponding sub-categories.

in the delivery of IT services; and resource provisioning, the study of allocation of energetic, computational, storage and time resources for the optimal
delivery of IT services. Within each of these macro-areas, we further distinguish
approaches in categories based on the similarity of goals. In failure management,
these categories are failure prevention, online failure prediction, failure detection,
root cause analysis (RCA) and remediation. In resource provisioning, we divide
contributions in resource consolidation, scheduling, power management, service
composition, and workload estimation. We further choose to expand our analysis
of FM (red box of Fig. 2) by applying for this macro-area an additional subcate-
gorization based on specific problems. Examples of subcategories are checkpointing for failure prevention, or fault localization for root cause analysis (see also
Table 3).

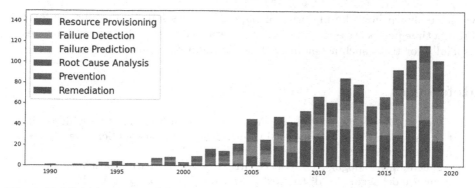

Fig. 4. Published papers in AIOps by year and categories from the described taxonomy.

3   Results

We now discuss the results of our mapping study. We first analyze the distribution of papers in our taxonomy. The left side of Fig. 3 visualizes the distribution of identified papers by macro-area and category. Excluding papers treating AIOps in general (8), we observe that more the majority of items (670, 62.1%) are associated with failure management (FM), with most contributions concentrated in online failure prediction (26.4%), failure detection (33.7%), and root cause analysis (26.7%); the remaining resource provisioning papers support in large part resource consolidation, scheduling and workload prediction. On the right side, we can observe that the most common problems in FM are software defect prediction, system failure prediction, anomaly detection, fault localization and root cause diagnosis. To analyze temporal trends present inside the AIOps field, we measured the number of publications in each category by year of publication. The corresponding bar plot is depicted in Fig. 4. Overall, we observe a large, on-growing number of publications in AIOps. We can observe how failure detection has gained particular traction in recent years (71 publications for the 2018–2019 period) with a contribution size larger than the entire resource provisioning macro-area (69 publications in the same time frame). Failure detection is followed by root cause analysis (39) and online failure prediction (34), while failure prevention and remediation are the areas with the smallest number of attested contributions (11 and 5, respectively).

4   Conclusion

In this paper, we presented our contribution towards better structuring the AIOps field. We planned and conducted a systematic mapping study by means of pre-established formulation, search, selection, and categorization techniques, thanks to which we collected more than 1000 contributions and grouped into several categories thanks to our proposed taxonomy, and differing substantially in terms of goals, data sources and target components. In our result section, we have shown how the majority of papers address failures in different forms. From a time perspective, we observed a generalized on-growing research interest, espcially for tasks such as anomaly detection and root cause analysis.

References

1. Abreu, R., Zoeteweij, P., Gemund, A.J.V.: Spectrum-based multiple fault localization. In: IEEE/ACM International Conference on Automated Software Engineering, November 2009. https://doi.org/10.1109/ase.2009.25
2. Aguilera, M.K., Mogul, J.C., Wiener, J.L., Reynolds, P., Muthitacharoen, A.: Performance debugging for distributed systems of black boxes. ACM SIGOPS Oper. Syst. Rev. **37**(5), 74–89 (2003). https://doi.org/10.1145/1165389.945454
3. Lerner, A.: AIOps Platforms, August 2017. https://blogs.gartner.com/andrew-lerner/2017/08/09/aiops-platforms/
4. Attariyan, M., Chow, M., Flinn, J.: X-ray: automating root-cause diagnosis of performance anomalies in production software. In: Proceedings of the 10th USENIX Conference on Operating Systems Design and Implementation, OSDI 2012, Hollywood, CA, USA, pp. 307–320, October 2012. https://doi.org/10.5555/2387880.2387910
5. Bahl, P., Chandra, R., Greenberg, A., Kandula, S., Maltz, D.A., Zhang, M.: Towards highly reliable enterprise network services via inference of multi-level dependencies. In: Proceedings of the 2007 Conference on Applications, Technologies, Architectures, and Protocols for Computer Communications - SIGCOMM (2007). https://doi.org/10.1145/1282380.1282383
6. Barham, P., Isaacs, R., Mortier, R., Narayanan, D.: Magpie: online modelling and performance-aware systems. In: Proceedings of the 9th Conference on Hot Topics in Operating Systems, HOTOS 2003, Lihue, Hawaii, vol. 9, p. 15, May 2003. https://doi.org/10.5555/1251054.1251069
7. Bodik, P., Goldszmidt, M., Fox, A., Woodard, D.B., Andersen, H.: Fingerprinting the datacenter: automated classification of performance crises. In: Proceedings of the 5th European Conference on Computer Systems - EuroSys 2010 (2010). https://doi.org/10.1145/1755913.1755926
8. Chalermarrewong, T., Achalakul, T., See, S.C.W.: Failure prediction of data centers using time series and fault tree analysis. In: IEEE 18th International Conference on Parallel and Distributed Systems, December 2012. https://doi.org/10.1109/icpads.2012.129
9. Chen, M., Kiciman, E., Fratkin, E., Fox, A., Brewer, E.: Pinpoint: problem determination in large, dynamic Internet services. In: Proceedings of IEEE International Conference on Dependable Systems and Networks (2002). https://doi.org/10.1109/dsn.2002.1029005
10. Chow, M., Meisner, D., Flinn, J., Peek, D., Wenisch, T.F.: The mystery machine: end-to-end performance analysis of large-scale internet services. In: OSDI 2014: Proceedings of the 11th USENIX Conference on Operating Systems Design and Implementation, pp. 217–231 (2014). https://doi.org/10.5555/2685048.2685066
11. Cohen, I., Goldszmidt, M., Kelly, T., Symons, J., Chase, J.S.: Correlating instrumentation data to system states: a building block for automated diagnosis and control. In: Proceedings of the 6th USENIX Conference on Symposium on Operating Systems Design & Implementation, OSDI 2004 (2004). https://doi.org/10.5555/1251254.1251270

12. Costa, C.H., Park, Y., Rosenburg, B.S., Cher, C.Y., Ryu, K.D.: A system software approach to proactive memory-error avoidance. In: SC 2014: International Conference for High Performance Computing, Networking, Storage and Analysis, November 2014. https://doi.org/10.1109/sc.2014.63

13. Dang, Y., Lin, Q., Huang, P.: AIOps: real-world challenges and research innovations. In: IEEE/ACM 41st International Conference on Software Engineering: Companion, May 2019. https://doi.org/10.1109/icse-companion.2019.00023

14. Davis, N.A., Rezgui, A., Soliman, H., Manzanares, S., Coates, M.: FailureSim: a system for predicting hardware failures in cloud data centers using neural networks. In: IEEE 10th International Conference on Cloud Computing (CLOUD), Jun 2017. https://doi.org/10.1109/cloud.2017.75

15. Du, M., Li, F., Zheng, G., Srikumar, V.: DeepLog: anomaly detection and diagnosis from system logs through deep learning. In: Proceedings of ACM SIGSAC Conference on Computer and Communications Security (2017). https://doi.org/10.1145/3133956.3134015

16. Garg, S., van Moorsel, A., Vaidyanathan, K., Trivedi, K.: A methodology for detection and estimation of software aging. In: Proceedings Ninth International Symposium on Software Reliability Engineering (Cat. No.98TB100257). IEEE Computer Society (1998). https://doi.org/10.1109/issre.1998.730892

17. Islam, T., Manivannan, D.: Predicting application failure in cloud: a machine learning approach. In: 2017 IEEE International Conference on Cognitive Computing (ICCC), Jun 2017. https://doi.org/10.1109/ieee.iccc.2017.11

18. Jalali, S., Wohlin, C.: Systematic literature studies: database searches vs. backward snowballing. In: Proceedings of the 2012 ACM-IEEE International Symposium on Empirical Software Engineering and Measurement, pp. 29–38, September 2012. https://doi.org/10.1145/2372251.2372257

19. Kandula, S., Katabi, D., Vasseur, J.P.: Shrink: a tool for failure diagnosis in IP networks. In: Proceedings of the 2005 ACM SIGCOMM Workshop on Mining Network Data - MineNet 2005 (2005). https://doi.org/10.1145/1080173.1080178

20. Kobbacy, K.A.H., Vadera, S., Rasmy, M.H.: AI and OR in management of operations: history and trends. J. Oper. Res. Soc. **58**(1), 10–28 (2007). https://doi.org/10.1057/palgrave.jors.2602132

21. Lakhina, A., Crovella, M., Diot, C.: Diagnosing network-wide traffic anomalies. In: Proceedings of the 2004 Conference on Applications, Technologies, Architectures, and Protocols for Computer Communications - SIGCOMM 2004. ACM Press (2004). https://doi.org/10.1145/1015467.1015492

22. Lakhina, A., Crovella, M., Diot, C.: Mining anomalies using traffic feature distributions. In: Proceedings of the 2005 Conference on Applications, Technologies, Architectures, and Protocols for Computer Communications - SIGCOMM 2005. ACM Press (2005). https://doi.org/10.1145/1080091.1080118

23. Li, Y., et al.: Predicting node failures in an ultra-large-scale cloud computing platform: an AIOps solution. ACM Trans. Software Eng. Methodol. **29**(2), 1–24 (2020). https://doi.org/10.1145/3385187

24. Liang, Y., Zhang, Y., Xiong, H., Sahoo, R.: Failure prediction in IBM BlueGene/L event logs. In: Seventh IEEE International Conference on Data Mining (ICDM) (2007). https://doi.org/10.1109/icdm.2007.46

25. Lin, F., Beadon, M., Dixit, H.D., Vunnam, G., Desai, A., Sankar, S.: Hardware remediation at scale. In: 2018 48th Annual IEEE/IFIP International Conference on Dependable Systems and Networks Workshops (DSN-W), June 2018. https://doi.org/10.1109/dsn-w.2018.00015

26. Lin, Q., Zhang, H., Lou, J.G., Zhang, Y., Chen, X.: Log clustering based problem identification for online service systems. In: Proceedings of the 38th ACM International Conference on Software Engineering Companion (ICSE) (2016). https://doi.org/10.1145/2889160.2889232

27. Menzies, T., Greenwald, J., Frank, A.: Data mining static code attributes to learn defect predictors. IEEE Trans. Software Eng. **33**(1), 2–13 (2007). https://doi.org/10.1109/TSE.2007.256941

28. Chen, M.Y., Accardi, A., Kiciman, E., Lloyd, J., Patterson, D., Fox, A., Brewer, E.: Path-based failure and evolution management. In: Proceedings of the 1st Conference on Symposium on Networked Systems Design and Implementation, NSDI 2004, San Francisco, California, vol. 1, p. 23, March 2004. https://doi.org/10.5555/1251175.1251198

29. Moody, A., Bronevetsky, G., Mohror, K., Supinski, B.R.D.: Design, modeling, and evaluation of a scalable multi-level checkpointing system. In: 2010 ACM/IEEE International Conference for High Performance Computing, Networking, Storage and Analysis, November 2010. https://doi.org/10.1109/sc.2010.18

30. Moore, A.W., Zuev, D.: Internet traffic classification using Bayesian analysis techniques. In: Proceedings of the 2005 ACM International Conference on Measurement and Modeling of Computer Systems - SIGMETRICS 2005 (2005). https://doi.org/10.1145/1064212.1064220

31. Mukwevho, M.A., Celik, T.: Toward a smart cloud: a review of fault-tolerance methods in cloud systems. IEEE Trans. Serv. Comput. 1 (2018). https://doi.org/10.1109/tsc.2018.2816644

32. Natella, R., Cotroneo, D., Duraes, J.A., Madeira, H.S.: On fault representativeness of software fault injection. IEEE Trans. Software Eng. **39**(1), 80–96 (2013). https://doi.org/10.1109/tse.2011.124

33. Nguyen, H., Shen, Z., Tan, Y., Gu, X.: FChain: toward black-box online fault localization for cloud systems. In: IEEE 33rd International Conference on Distributed Computing Systems, July 2013. https://doi.org/10.1109/icdcs.2013.26

34. Petersen, K., Vakkalanka, S., Kuzniarz, L.: Guidelines for conducting systematic mapping studies in software engineering: an update. Inf. Softw. Technol. **64**, 1–18 (2015). https://doi.org/10.1016/j.infsof.2015.03.007

35. Pitakrat, T., Okanović, D., van Hoorn, A., Grunske, L.: Hora: architecture-aware online failure prediction. J. Syst. Softw. **137**, 669–685 (2018). https://doi.org/10.1016/j.jss.2017.02.041

36. Podgurski, A., et al.: Automated support for classifying software failure reports. In: Proceedings of IEEE 25th International Conference on Software Engineering (2003). https://doi.org/10.1109/icse.2003.1201224

37. Salfner, F., Malek, M.: Using hidden semi-Markov models for effective online failure prediction. In: 26th IEEE International Symposium on Reliable Distributed Systems (SRDS), October 2007. https://doi.org/10.1109/srds.2007.35

38. Samir, A., Pahl, C.: A controller architecture for anomaly detection, root cause analysis and self-adaptation for cluster architectures. In: International Conference on Adaptive and Self-Adaptive Systems and Applications (2019). 10993/42062

39. Shao, Q., Chen, Y., Tao, S., Yan, X., Anerousis, N.: Efficient ticket routing by resolution sequence mining. In: Proceeding of the 14th ACM SIGKDD International Conference on Knowledge Discovery and Data Mining - KDD (2008). https://doi.org/10.1145/1401890.1401964

40. Sharma, A.B., Chen, H., Ding, M., Yoshihira, K., Jiang, G.: Fault detection and localization in distributed systems using invariant relationships. In: 43rd Annual IEEE/IFIP International Conference on Dependable Systems and Networks (DSN), June 2013. https://doi.org/10.1109/dsn.2013.6575304

41. Vaidyanathan, K., Trivedi, K.: A comprehensive model for software rejuvenation. IEEE Trans. Dependable Secure Comput. **2**(2), 124–137 (2005). https://doi.org/10.1109/tdsc.2005.15

42. Xu, H., et al.: Unsupervised anomaly detection via variational auto-encoder for seasonal KPIs in web applications. In: Proceedings of the 2018 World Wide Web Conference on World Wide Web (2018). https://doi.org/10.1145/3178876.3185996

43. Xu, W., Huang, L., Fox, A., Patterson, D., Jordan, M.I.: Detecting large-scale system problems by mining console logs. In: Proceedings of the ACM SIGOPS 22nd Symposium on Operating Systems Principles - SOSP 2009 (2009). https://doi.org/10.1145/1629575.1629587

44. Yuan, D., Mai, H., Xiong, W., Tan, L., Zhou, Y., Pasupathy, S.: SherLog: error diagnosis by connecting clues from run-time logs. In: ACM SIGARCH Computer Architecture News, vol. 38, no. 1, pp. 143–154 (2010). https://doi.org/10.1145/1735970.1736038

45. Zhang, K., Xu, J., Min, M.R., Jiang, G., Pelechrinis, K., Zhang, H.: Automated IT system failure prediction: a deep learning approach. In: IEEE International Conference on Big Data (2016). https://doi.org/10.1109/bigdata.2016.7840733

46. Zhang, S., et al.: Syslog processing for switch failure diagnosis and prediction in datacenter networks. In: IEEE/ACM 25th International Symposium on Quality of Service (IWQoS), June 2017. https://doi.org/10.1109/iwqos.2017.7969130

47. Zheng, S., Ristovski, K., Farahat, A., Gupta, C.: Long short-term memory network for remaining useful life estimation. In: IEEE International Conference on Prognostics and Health Management (ICPHM) (2017). https://doi.org/10.1109/icphm.2017.7998311

48. Zhou, W., Tang, L., Li, T., Shwartz, L., Grabarnik, G.Y.: Resolution recommendation for event tickets in service management. In: IFIP/IEEE International Symposium on Integrated Network Management (IM) (2015). https://doi.org/10.1109/inm.2015.7140303

49. Zhu, J., He, P., Fu, Q., Zhang, H., Lyu, M.R., Zhang, D.: Learning to log: helping developers make informed logging decisions. In: IEEE/ACM 37th IEEE International Conference on Software Engineering, May 2015. https://doi.org/10.1109/icse.2015.60

An Influence-Based Approach for Root Cause Alarm Discovery in Telecom Networks

Keli Zhang[✉], Marcus Kalander, Min Zhou, Xi Zhang, and Junjian Ye

Noah's Ark Lab, Huawei Technologies, Shenzhen, China
zhangkeli1@huawei.com

Abstract. Alarm root cause analysis is a significant component in the day-to-day telecommunication network maintenance, and it is critical for efficient and accurate fault localization and failure recovery. In practice, accurate and self-adjustable alarm root cause analysis is a great challenge due to network complexity and vast amounts of alarms. A popular approach for failure root cause identification is to construct a graph with approximate edges, commonly based on either event co-occurrences or conditional independence tests. However, considerable expert knowledge is typically required for edge pruning. We propose a novel data-driven framework for root cause alarm localization, combining both causal inference and network embedding techniques. In this framework, we design a hybrid causal graph learning method (HPCI), which combines Hawkes Process with Conditional Independence tests, as well as propose a novel Causal Propagation-Based Embedding algorithm (CPBE) to infer edge weights. We subsequently discover root cause alarms in a real-time data stream by applying an influence maximization algorithm on the weighted graph. We evaluate our method on artificial data and real-world telecom data, showing a significant improvement over the best baselines.

Keywords: Network management · Root cause analysis · Alarm correlation analysis · Influence maximization

1 Introduction

Recent years have seen rapid development in cellular networks, both in increasing network scale and complexity coupled with increasing network performance demands. This growth has made the quality of network management an even greater challenge and puts limits on the analysis methods that can be applied. In cellular networks, anomalies are commonly identified through alarms. A large-scale network can generate millions of alarms during a single day. Due to the interrelated network structure, a single fault can trigger a flood of alarms from

---

K. Zhang and M. Kalander—These authors contributed equally to this work.

H. Hacid et al. (Eds.): ICSOC 2020 Workshops, LNCS 12632, pp. 124–136, 2021.
https://doi.org/10.1007/978-3-030-76352-7_16

multiple devices. Traditionally, to recover after a failure, an operator will analyze all relevant alarms and network information. This can be a slow and time-consuming process. However, not all alarms are relevant. There exists a subset of alarms that are the most significant for fault localization. We denote these as root cause alarms, and our main goal is to intelligently identify these alarms.

There exist abundant prior research in the areas of Root Cause Analysis (RCA) and fault localization. However, most proposed methods are highly specialized and take advantage of specific properties of the deployed network, either by using integrated domain knowledge or through particular design decisions [2–4]. A more general approach is to infer everything from the data itself.

In our proposed alarm RCA system, we create an influence graph to model alarm relations. Causal inference is used to infer an initial causal graph, and then we apply a novel Causal Propagation-Based Embedding (CPBE) algorithm to supplement the graph with meaningful edge weights. To identify the root cause alarms, we build upon ideas in how influence propagates in social networks and view the problem as an influence maximization problem [10], i.e., we want to discover the alarms with the largest influence. When a failure transpires, our system can automatically perform RCA based on the sub-graph containing the involved alarms and output the top-K most probable root cause alarms.

In summary, our main contributions are as follows:

- We design a novel unsupervised approach for root cause alarm localization that integrates casual inference and influence maximization analysis, making the framework robust to causal analysis uncertainty without requiring labels.
- We propose HPCI, a Hawkes Process-based Conditional Independence test procedure for causal inference.
- We further propose CPBE, a Causal Propagation-Based Embedding algorithm based on network embedding techniques and vector similarity to infer edge weights in causality graphs.
- Extensive experiments on a synthetic and a real-world citywide dataset show the advantages and usefulness of our proposed methods.

2   Related Work

Root Cause Alarms. There are various ways to discover alarm correlations and root cause alarms. Rules and experience of previous incidents are frequently used. In more data-driven approaches, pattern mining techniques that compress alarm data can assist in locating and diagnosing faults [24]. Abele et al. [1] propose to find root cause alarms by combining knowledge modeling and Bayesian networks. To use an alarm clustering algorithm that considers the network topology and then mine association rules to find root cause alarms was proposed in [21].

Graph-Based Root Cause Analysis. Some previous works depend on system dependency graphs, e.g., Sherlock [2]. A disadvantage is the requirement of exact conditional probabilities, which is impractical to obtain in large networks. Other systems are based on causality graphs. G-RCA [4] is a diagnosis system, but its causality graph is configured by hand, which is unfeasible in large scale, dynamic environments. The PC algorithm [19] is used by both CauseInfer [3] and [11] to estimate DAGs, which are then used to infer root causes. However, such graphs can be very unreliable. Co-occurrence and Bayesian decision theory are used in [13] to estimate causal relations, but it is mainly based on log event heuristics and is hard to generalize. Nie et al. [16] use FP-Growth and lag correlation to build a causality graph with edge weights added with expert feedback.

3   Preliminaries

In this section, we shortly review the two key concepts that our proposed method depends upon, Hawkes process [7] and the PC algorithm [20].

Hawkes Process. This is a popular method to model continuous-time event sequences where past events can excite the probability of future events. The keystone of Hawkes process is the conditional intensity function, which indicates the occurrence rate of future events conditioned on past events, denoted by $\lambda_d(t)$, where $u \in \mathcal{C} = \{1, 2, ..., U\}$ is an event type. Formally, given an infinitely small time window $[t, t + \Delta t)$, the probability of a type-u event occurring in this window is $\lambda_d(t)\Delta t$. For U-dimensional Hawkes process with event type set $\mathcal{C}$, each dimension u has a specific form of conditional intensity function defined as

$$\lambda_u(t) = \mu_u + \sum_{v \in \mathcal{C}} \sum_{t_i < t} k_{uv}(t - t_i),$$ (1)

where $\mu_u \geq 0$ is the background intensity for type-u events and $k_{uv}(t) \geq 0$ is a kernel function indicating the influence from past events. An exponential kernel is most frequently used, i.e., $k_{uv}(t) = \alpha_{uv}e^{-\beta_{uv}(t)}$, where α_{uv} captures the degree of influence of type-v events to type-u events and β_{uv} controls the decay rate. The parameters are commonly learned by optimizing a log-likelihood function. Let $\mu = (\mu_u) \in R^U$ be the background intensities, and $A = (\alpha_{uv}) \in R^{U \times U}$ the influence matrix reflecting the certain causality between event types. For a set of event sequences $\mathcal{S} = \{S_1, S_2, ..., S_m\}$, where each event sequence $S_i = \{(a_{ij}, t_{ij})\}_{j=1}^{n_i}$ is observed during a time period of $[0, T_i]$, and each pair $((a_{ij}, t_{ij}))$ represents an event of type a_{ij} that occurred at time t_{ij}. The log-likelihood of a Hawkes process model with parameters $\Theta = \{\mu, A\}$ can then be expressed as

$$\mathcal{L}(\mu, A) = \sum_{i=1}^{m} (\sum_{j=1}^{n_i} \log \lambda_{a_{ij}}(t_{ij}) - \sum_{u=1}^{U} \int_0^{T_i} \lambda_u(t)dt).$$ (2)

The influence matrix A is generally sparse or low-rank in practice, hence, adding penalties into $\mathcal{L}(\mu, A)$ is common. For instance, Zhou [25] used a mix of Lasso and nuclear norms to constrain A to be both low-rank and sparse by using

$$\min_{\mu \geq 0, A \geq 0} -\mathcal{L}(\mu, A) + \rho_1 ||A||_1 + \rho_2 ||A||_*,$$ (3)

where $|| \cdot ||_1$ is the L_1-norm, and $|| \cdot ||_* = \sum_{i=1}^{rankA} \sigma_i$ is the nuclear norm. The parameters ρ_1 and ρ_2 controls their weights. A number of algorithms can be applied to solve the above learning problem, more details can be found in [22].

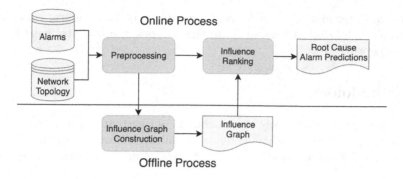

Fig. 1. Architecture of the proposed system.

PC Algorithm. This algorithm is frequently used for learning directed acyclic graphs (DAGs) due to its strong causal structure discovery ability [23]. Conditional Independence (CI) tests play a central role in the inference. A significance level p is used as a threshold to determine if an edge should be removed or retained. Formally, given a variable set Z, if X is independent of Y, denoted as $X \perp\!\!\!\perp Y | Z$, the edge between X and Y will be removed, otherwise it will be kept in the causal graph. A rigorous description can be found in [9].

The G-square test and Fisher-Z test are two common realizations for conditional independence testing in causal inference [18]. The G-square test is used for testing independence of categorical variables using conditional cross-entropy while the Fisher-Z test evaluates conditional independence based on Pearson's correlation. CI tests assume that the input is independent and identically distributed. In our alarm RCA scenario, the size of the time window depends on the network characteristics and needs to be selected to ensure that causal alarms exist in one window and the data between different windows are independent.

4   System Overview

Our proposed framework consists of two main procedures: influence graph creation and alarm ranking. A system overview can be found in Fig. 1. The alarm preprocessing module is shared and handles alarm filtering and aggregation with consideration to the network topology.

The influence graph is constructed using historical alarm transactions and is periodically recreated. It is comprised of alarm types as nodes and their inferred relations as edges. To create the graph, we first exploit causal inference methodology to infer an initial alarm causality graph structure by applying HPCI, a hybrid method that merges Hawkes process and conditional independence tests. We further apply a network embedding technique, CPBE, to infer the edge weights. The alarm stream is monitored in real-time. When a failure transpires, the system attempts to discover the underlying root cause alarms. The related alarms are aggregated with the created influence graph and are ranked by their influence to determine the top-K most probable root cause alarm candidates. The alarm candidates are then given to the network operators to assist in handling the network issue.

5   Methodology

This section introduces the key components in our system; alarm preprocessing, the influence graph construction, and how the influence ranking is done. We start by presenting the data and its required preprocessing and aggregation steps.

5.1   Data and Alarm Preprocessing

Network Topology. This is the topological structure of the connections between network devices. Connected network devices will interact with each other. If a failure occurs on one device, then any connected devices can be affected, triggering alarms on multiple network nodes.

Alarms. Network alarms are used to identify faults in the network. Each alarm record contains information about occurrence time, network device where the alarm originated, alarm type, etc. In practice, any alarms with missing key information are useless and removed. Furthermore, alarm types that are either systematic or highly periodical are also removed. These types of alarms are irrelevant for root cause analysis since they will be triggered regardless if a fault occurred or not.

Alarm Preprocessing. We partition the raw alarm data into alarm sequences and alarm transactions in three steps as follows.

1. Devices in connected sub-graphs of the network can interact, i.e., alarms from these devices can potentially be related to the same fault. Consequently, we first aggregate alarms from the same sub-graph together.
2. Alarms related to the same failure will generally occur together within a short time interval. We thus further partition the alarms based on their occurrence times. Alarms that occurred within the same time window w_i are grouped and sorted by time. The window size can be adjusted depending on network characteristics. We define each group as an alarm sequence, denoted as $S_i = \{(a_{ij}, t_{ij})\}_{j=1}^{n_i}$, where w_i is the window, $a_{ij} \in A$ is the alarm type, $t_{ij} \in w_i$ is occurrence time, and n_i the number of alarms.
3. Each alarm sequence is transformed into an alarm transaction denoted by $T_i = \{(a, t, n) | a \in A_i\}$, where a, t, n indicates the alarm type, the earliest occurrence time and the number of occurrences, respectively. Different from S_i, A_i contains a single element for each alarm type in window w_i.

5.2   Alarm Influence Graph Construction

In this section, we elaborate on the construction of the alarm influence graph. The graph has the alarm types as nodes and their relation as the edges. First, an initial causal structure DAG is inferred by a hybrid causal structure learning method (HPCI). Subsequently, edge weights are inferred using a novel network embedding method (CPBE).

HPCI. A multi-dimensional Hawkes process can capture certain causalities behind event types, i.e., the transpose of the influence matrix A can be seen as the adjacency matrix of the causal graph for event types. However, redundant or indirect edges tend to be discovered since the conditional intensity function can not perfectly model real-world data and due to the difficulty in capturing the instantaneous causality.

To reduce this weakness, we propose a hybrid algorithm HPCI that is based on Hawkes process and the PC algorithm. HPCI is used to discover the causal structure for the alarm types in our alarm RCA scenario. The main procedure can be expressed in three steps. (1) Use multi-dimensional Hawkes process without penalty to capture the influence intensities among the alarm types. We use the alarm sequences $\mathcal{S} = \{S_i\}$ as input and obtain an initial weighted graph. The weights on an edge (u, v) is the influence intensity $\alpha_{uv} > 0$, reflecting the expectation of how long it takes for a type-u event to occur after an type-v event. All edges with positive weights are retained. (2) Any redundant and indirect causal edges are removed using CI tests. We use the alarm transactions $\mathcal{T} = \{T_i\}$ as input and for each alarm a_i the sequence of alarm occurrences $N_i = \{n | T_k \in \mathcal{T}, (a, t, n) \in T_k, a = a_i\}$ is extracted. Note that n can be 0 if an alarm type is not present in a window w_i. For each pair of alarm types (a_i, a_j), the CI test of their respective occurrence sequences is used to test for independence and remove edges. The output is a graph with unwanted edges removed. (3) Finally, we iteratively remove the edge with the smallest intensity until the graph is a DAG. Our final causal graph is denoted as G^C.

We select CI tests to enforce sparsity in the causal graph in the second step. Compared to adding penalty terms such as $L1$-norm, the learning procedure is more interpretable, and our experiments show more robust results.

Edge Weights Inference. The causal graph G^C learned by HPCI is a weighted graph, however, the weights do not account for global effects on the causal intensities. Hence, to encode more knowledge into the graph, we propose a novel network embedding-based weight inference method, Causal Propagation-Based Embedding (CPBE). CPBE consists mainly of two steps; (1) For each node u, we obtain a vector representation $Z_u \in \mathcal{R}^L$ using a novel network embedding technique. (2) Use vector similarity to compute edge weights between nodes.

The full CPBE algorithm is shown in Algorithm 1. CPBE uses a new procedure to generate a context for the skip-gram model [15] (lines 1–9). This procedure is also illustrated in Fig. 2. In essence, for each historical alarm transaction $T_i \in \mathcal{T}$, we use the learned causality graph G^C and extract a causal propagation graph G_i^{PC}, where only the nodes corresponding to alarm types in T_i are

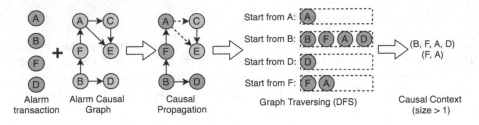

Fig. 2. Context generation procedure for CPBE.

Algorithm 1. Causal Propagation-Based Embedding (CPBE)

Input: Alarm Transactions $T = \{T_i\}$; Causal Graph $G^C = (V, E)$;

1: $C = \{\}, E_w = \{\}$;
2: **for** $T_i \in T$ **do:**
3:     $G_i^{PC} = \text{ConstructPropagationGraph}(T_i, G^C)$
4:     **for** $alarm\_node \in T_i$ **do:**
5:         $C_{alarm\_node} = \text{GraphTraversing}(G_i^{PC}, alarm\_node)$
6:         $C = C \cup C_{alarm\_node}$
7:     **end for**
8: **end for**
9: $Z \leftarrow$ skip-gram(C) to map nodes to embedding vectors
10: **for** $(u, v) \in E$ **do:**
11:     $w = Cosine(Z_u, Z_v)$
12:     $E_w = E_w \cup (u, v, w)$
13: **end for**

Output: Alarm Influence Graph $G^I = (V, E_w)$;

---

retained. Starting from each node in G_i^{PC}, we traverse the graph to generate a node-specific causal context. During the traversal for a node u, only nodes that have a causal relation with u are considered. There are various possible traversing strategies, e.g., depth-first search (DFS) and RandomWalk [6]. The skip-gram model is applied to the generated contexts to obtain an embedding vector $Z_u \in \mathcal{R}^L$ for each node u. Finally, the edge weight between two nodes is set to be the cosine similarity of their associated vectors. We denote the final weighted graph as the alarm influence graph G^I.

5.3 Root Cause Alarm Influence Ranking

This section describes how the alarm influence graph G^I is applied to an alarm transaction to identify the root cause alarms. For each alarm transaction $T_i \in \mathcal{T}$, an alarm propagation graph G_i^{PI} is created with the relevant nodes $v \in T_i$ and applicable edges $\{(u, v, w) | u, v \in T_i\}$. Any nodes corresponding to alarms not present in T_i are removed. The process is equivalent to how G_i^{PC} is created from the causal graph G^C. The alarms in each propagation sub-graph are then ranked independently. The process is illustrated in Fig. 3.

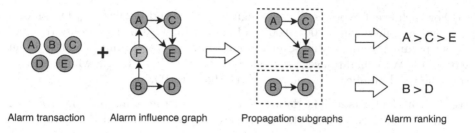

Fig. 3. Processing flow from alarm transaction to ranked alarms.

We consider the problem of finding the root cause alarm as an influence maximization problem [10]. We want to discover a small set of K *seed nodes* that maximizes the influence spread under an influence diffusion model. A suitable model is the independent cascade model, which is widely used in social network analysis. Following this model, each node v is activated by each of its neighbors independently based on an influence probability $p_{u,v}$ on each edge (u, v). These probabilities directly correspond to the learned edge weights. Given a seed set S_0 to start with at $t = 0$, at step $t > 0$, $u \in S_{t-1}$ tries to activate its outgoing inactivated neighbors $v \in \mathcal{N}^{out}(u)$ with probability $p_{u,v}$. Activated nodes are added to S_t and the process terminates when $|S_t| = 0$, i.e., when no nodes further nodes are activated. The influence of the seed set S_0 is then the expected number of activated nodes when applying the above stochastic activation procedure.

There are numerous algorithms available to solve the influence maximization problem [12]. In our scenario, each graph G_i^{PI} is relatively small and the actual algorithm is thus less important. We directly select the Influence Ranking Influence Estimation algorithm (IRIE) [8] for this task. IRIE estimates the influence $r(u)$ for each node u by deriving a system of n linear equations with n variables. The influence of a node u comprises of its own influence, 1, and the sum of the influences it propagates to its neighbors.

6   Evaluation

In this section, we present the experimental setup and evaluation results. We perform two main experiments, one to verify the correctness of our causal graph and a second experiment to evaluate the root cause identification accuracy. The first experiment is performed on both synthetic and real-world data, while the second is completed on the real-world dataset. The datasets and code are available at https://github.com/shaido987/alarm-rca.

Synthetic Data Generation. The synthetic event sequences are generated in four steps. (1) We randomly generate a DAG G with an average out-degree d with N event types. We set d to 1.5 to emulate the sparsity property of our real-world dataset. (2) For each edge (u, v), a weight α_{uv} is assigned by uniform random sampling from a range $r \in [(0.01, 0.05), (0.05, 0.1), (0.1, 0.5), (0.5, 1.0)]$.

(3) For each event type $u \in U$, we assign a background intensity μ_u by uniform random sampling from $(0.001, 0.005)$. (4) Following Ogata [17], we use α_{uv} and μ_u as parameters of a Multi-dimensional Hawkes process and simulate event sequences. We generate event sequences of length $T = 14$ days while ensuring that the total number of events is greater than 10,000.

Real-World Dataset. The dataset was collected from a major cellular carrier in a moderate-sized city in China between Aug 4th, 2018 and Oct 24th, 2018. After preprocessing, it consists of 672,639 alarm records from 3,818 devices with 78 different alarm types. Due to the difficulty of labeling causal relations, we only have the ground-truth causal relations for a subset of 15 alarm types, 44 directed edges in the graph. Furthermore, we have also obtained the ground-truth root cause alarms in a random sample of 6,000 alarm transactions. These are used to evaluate the root cause localization accuracy.

6.1 Causal Graph Structure Correctness

We evaluate our proposed HPCI method and the accuracy of the discovered causal graphs. We use four frequently used causal inference methods for sequential data as baselines.

- PC-GS: PC algorithm with G-square CI test.
- PC-FZ: PC algorithm with Fisher-Z CI test.
- PCTS: Improved PC algorithm for causal discovery in time series [14].
- HPADM4: Multi-dimensional Hawkes process with exponential parameterization of the kernels and a mix of $L1$ and nuclear-norm [25].

The significance level p in the conditional independence tests included in the methods are all set to 0.05. The size of time window w for aggregating event sequences is set to $300\,s$, the maximum lag $\tau_{max} = 2$ in PCTS, and the penalization level in HPADM4 is set to the default 1,000. Furthermore, the decay parameter β in Hawkes process is set to 0.1, and we select Fisher-Z as the CI test in our HPCI algorithm. For evaluation, we define three metrics as follows.

$$Precision = \frac{|P \cap S|}{|P|}, \quad Recall = \frac{|P \cap S|}{|S|}, \quad F1\text{-}score = 2 \cdot \frac{Precision \cdot Recall}{Precision + Recall},$$

where P is the set of all directed edges in the learned causal graph G^C and S is the set of ground-truth edges.

Results. The F1-scores using synthetic data with $N \in [10, 15, 20]$ are shown in Table 1. As shown, HPCI outperforms the baselines for nearly all settings of N and α. However, HPADM4 obtains the best result for $N = 10$ and low α, this is due to the distribution of event occurrence intervals being sparse which makes the causal dependency straightforward to capture using a Hawkes process. However, for higher N or α the events will be denser. Thus, Hawkes process has trouble distinguishing instantaneous causal relations, especially when events co-occur. The use of CI tests in HPCI helps to distinguish these instantaneous

Table 1. F1-scores on the synthetic dataset for different number of event types and α.

α	N	PS-GS	PS-FZ	PCTS	HPADM4	HPCI
(0.01,0.05)	10	0.286	0.174	0.283	**0.566**	0.200
	20	0.133	0.321	0.156	0.306	**0.604**
	30	0.216	0.267	0.104	0.229	**0.357**
(0.05,0.1)	10	0.500	0.529	0.283	0.500	**0.867**
	20	0.367	0.585	0.155	0.323	**0.806**
	30	0.265	0.484	0.227	0.227	**0.756**
(0.1,0.5)	10	0.467	0.811	0.278	0.517	**0.933**
	20	0.621	0.889	0.151	0.306	**0.984**
	30	0.495	0.845	0.103	0.227	**0.967**
(0.5,1.0)	10	0.800	0.722	0.272	0.517	**0.929**
	20	0.708	0.906	0.151	0.302	**0.983**
	30	0.433	0.845	0.103	0.227	**0.967**

Table 2. Results of the causal graph structure evaluation on the real-world dataset.

Method	Precision	Recall	F1-score
PC-GS	0.250	0.159	0.194
PC-FZ	0.452	0.432	0.442
PCTS	0.220	**0.864**	0.350
HPADM4	0.491	0.614	0.545
HPCI	**0.634**	0.591	**0.612**

causal relations by taking another perspective in which causality is discovered based on distribution changes in the aggregated data without considering the time-lagged information among events. HPCI thus achieves better results. The use of time aggregation is disadvantageous for PCTS due to its focus on time series, which can partly explain its comparatively worse results.

The results on the real-world data are shown in Table 2. HPCI performs significantly better than all baselines in precision and F1-score, while PTCS obtains the highest recall. PTCS also has significantly lower precision, indicating more false positives. PCTS is designed for time series, however, those may be periodic, which can give higher lagged-correlation values leading to more redundant edges. HPCI instead finds a good balance between precision and recall. The competitive result indicates that the causality behind the real alarm data conforms to the assumptions of HPCI to a certain extent.

6.2   Root Cause Alarm Identification

We evaluate the effectiveness of CPBE and the root cause alarm accuracy on the real-world dataset. We use the causal graph structure created by HPCI as the

Table 3. Root cause alarm identification accuracy using different edge weight inference strategies together with IRIE for alarm ranking at different K.

Method	$K=1$	$K=2$	$K=3$	$K=4$	$K=5$
IT	0.576	0.590	0.672	0.810	0.900
Pearson	0.407	0.435	0.456	0.486	0.486
CP	0.474	0.640	0.730	0.790	0.840
ST	0.439	0.642	0.750	0.785	0.814
CPBE	**0.618**	**0.752**	**0.851**	**0.929**	**0.961**

base and augment it with the 44 known causal ground-truths. The causal graph is thus as accurate as possible. CPBE is compared with four baseline methods, all used for determining edge weights.

- IT, directly use the weighted causal graph discovered by HPCI with the learned influence intensities as edge weights.
- Pearson, uses the aligned Pearson correlation of each alarm pair [16].
- CP, the weights of an edge (u, v) is set to $\frac{A_{uv}}{A_u}$ where A_{uv} is the number of times u and v co-occur in a window, and A_u is the total number of u alarms.
- ST, a static model with maximization likelihood estimator [5]. It is similar to CP, but A_{uv} represents the number of times u occurs before v.

For each method, IRIE is used to find the top-K most likely root cause alarms in each of the 6,000 labeled alarm transactions. For IRIE, we use the default parameters. We attempt to use RandomWalk, BFS, and DFS for traversal in CPBE, as well as different Skip-gram configurations with $w \in [1, 5]$ and vector length $L \in [10, 30]$. However, there is no significant difference in the outcome, indicating that CPBE is insensitive to these parameter choices on our data. The results for different K when using RandomWalk are shown in Table 3. As shown, CPBE outperforms the baselines for all K. For $K = 1$, CPBE achieves an accuracy of 61.8% which, considering that no expert knowledge is integrated into the system, is an excellent outcome. Moreover, the running time of CPBE is around 10 s and IRIE takes 325 s for all 6,000 alarm transactions. This is clearly fast enough for system deployment.

7   Conclusion

We present a framework to identify root cause alarms of network faults in large telecom networks without relying on any expert knowledge. We output a clear ranking of the most crucial alarms to assist in locating network faults. To this end, we propose a causal inference method (HPCI) and a novel network embedding-based algorithm (CPBE) for inferring network weights. Combining the two methods, we construct an alarm influence graph from historical alarm data. The learned graph is then applied to identify root cause alarms through a flexible ranking method based on influence maximization. We verify the correctness of the learned graph using known causal relation and show a significant improvement over the best baseline on both synthetic and real-world data. Moreover, we demonstrate that our proposed framework beat the baselines in identifying root cause alarms.

References

1. Abele, L., Anic, M., et al.: Combining knowledge modeling and machine learning for alarm root cause analysis. IFAC Proc. Volumes **46**(9), 1843–1848 (2013)
2. Bahl, P., Chandra, R., et al.: Towards highly reliable enterprise network services via inference of multi-level dependencies. In: ACM SIGCOMM Computer Communication Review, vol. 37, pp. 13–24. ACM (2007)
3. Chen, P., Qi, Y., et al.: Causeinfer: automatic and distributed performance diagnosis with hierarchical causality graph in large distributed systems. In: INFOCOM, 2014 Proceedings IEEE, pp. 1887–1895. IEEE (2014)
4. Ge, Z., Yates, J., et al.: GRCA: a generic root cause analysis platform for service quality management in large ISP networks. In: ACM ACM Conference on Emerging Networking Experiments and Technologies (2010)
5. Goyal, A., Bonchi, F., et al.: Learning influence probabilities in social networks. In: Proceedings of the third ACM International Conference on Web Search and Data Mining, pp. 241–250. ACM (2010)
6. Grover, A., Leskovec, J.: node2vec: scalable feature learning for networks. In: Proceedings of the 22nd ACM SIGKDD International Conference on Knowledge Discovery and Data Mining, pp. 855–864. ACM (2016)
7. Hawkes, A.G.: Spectra of some self-exciting and mutually exciting point processes. Biometrika **58**(1), 83–90 (1971)
8. Jung, K., Heo, W., et al.: IRIE: scalable and robust influence maximization in social networks. In: 2012 IEEE 12th International Conference on Data Mining (ICDM), pp. 918–923. IEEE (2012)
9. Kalisch, M., Bühlmann, P.: Estimating high-dimensional directed acyclic graphs with the PC-algorithm. J. Mach. Learn. Res. **8**, 613–636 (2007)
10. Kempe, D., Kleinberg, J., et al.: Maximizing the spread of influence through a social network. In: Proceedings of the Ninth ACM SIGKDD International Conference on Knowledge Discovery and Data Mining, pp. 137–146. ACM (2003)
11. Kobayashi, S., Otomo, K., et al.: Mining causality of network events in log data. IEEE Trans. Netw. Serv. Manag. **15**(1), 53–67 (2018)
12. Li, Y., Fan, J., et al.: Influence maximization on social graphs: a survey. IEEE Trans. Knowl. Data Eng. **30**(10), 1852–1872 (2018)
13. Lou, J.G., Fu, Q., et al.: Mining dependency in distributed systems through unstructured logs analysis. SIGOPS Oper. Syst. Rev. **44**(1), 91–96 (2010)
14. Meng, Y., et al.: Localizing failure root causes in a microservice through causality inference. In: 2020 IEEE/ACM 28th International Symposium on Quality of Service (IWQoS), pp. 1–10. IEEE (2020)
15. Mikolov, T., Sutskever, I., et al.: Distributed representations of words and phrases and their compositionality. In: Advances in Neural Information Processing Systems, pp. 3111–3119 (2013)
16. Nie, X., Zhao, Y., et al.: Mining causality graph for automatic web-based service diagnosis. In: 2016 IEEE 35th International Performance Computing and Communications Conference (IPCCC), pp. 1–8 (2016)

17. Ogata, Y.: On lewis' simulation method for point processes. IEEE Trans. Inf. Theory **27**(1), 23–31 (1981)
18. Peters, J., Mooij, J.M., et al.: Causal discovery with continuous additive noise models. J. Mach. Learn. Res. **15**(1), 2009–2053 (2014)
19. Spirtes, P., Glymour, C.: An algorithm for fast recovery of sparse causal graphs. Soc. Sci. Comput. Rev. **9**(1), 62–72 (1991)
20. Spirtes, P., Glymour, C.N., et al.: Causation, Prediction, and Search. MIT Press, Cambridge (2000)
21. Su, C., Hailong, Z., et al.: Association mining analysis of alarm root-causes in power system with topological constraints. In: Proceedings of the 2017 International Conference on Information Technology, pp. 461–468. ACM (2017)
22. Veen, A., Schoenberg, F.P.: Estimation of space-time branching process models in seismology using an EM-type algorithm. J. Am. Stat. Assoc. **103**(482), 614–624 (2008)
23. Wang, P., Xu, J., et al.: Cloudranger: root cause identification for cloud native systems. In: 2018 18th IEEE/ACM International Symposium on Cluster, Cloud and Grid Computing (CCGRID), pp. 492–502. IEEE (2018)
24. Zhang, X., Bai, Y., et al.: Network alarm flood pattern mining algorithm based on multi-dimensional association. In: Proceedings of the 21st ACM International Conference on Modeling, Analysis and Simulation of Wireless and Mobile Systems, pp. 71–78. ACM (2018)
25. Zhou, K., Zha, H., et al.: Learning social infectivity in sparse low-rank networks using multi-dimensional hawkes processes. In: Artificial Intelligence and Statistics, pp. 641–649 (2013)

Localization of Operational Faults in Cloud Applications by Mining Causal Dependencies in Logs Using Golden Signals

Pooja Aggarwal[✉], Ajay Gupta[✉], Prateeti Mohapatra[✉], Seema Nagar[✉], Atri Mandal[✉], Qing Wang[✉], and Amit Paradkar[✉]

IBM Research AI, New Delhi, India
{aggarwal.pooja,ajaygupta,pramoh01,senagar3,atri.mandal}@in.ibm.com
qing.wang1@ibm.com, paradkar@us.ibm.com

Abstract. Cloud based microservice architecture has become a powerful mechanism in helping organizations to scale operations by accelerating the pace of change at minimal cost. With cloud based applications being accessed from diverse geographies, there is a need for round-the-clock monitoring of faults to prevent or to limit the impact of outages. Pinpointing source(s) of faults in cloud applications is a challenging problem due to complex interdependencies between applications, middleware, and hardware infrastructure all of which may be subject to frequent and dynamic updates. In this paper, we propose a light-weight fault localization technique, which can reduce human effort and dependency on domain knowledge for localizing observable operational faults. We model multivariate error-rate time series using minimal runtime logs to infer causal relationship among the golden signal errors (error rates) and micro-service errors to discover ranked list of possible faulty components. Our experimental results show that our system can localize operational faults with high accuracy (F1 = 88.4%) underscoring the effectiveness of using golden signal error rates in fault localization.

Keywords: Hybrid cloud · Fault localization · Golden signals · Causal modeling · PageRank

1 Introduction

Run time failures in running software systems are unavoidable. Recovering systems from such failures is an important aspect of incident management process in IT Operations. In order to identify the mitigating actions that will lead to system recovery, the failure needs to be triaged to a faulty system component (along with the reason for component failure). This triaging process, known as fault localization, can consume significant operator resources in a complex system environment. In cloud based distributed micro-services applications, localizing

© Springer Nature Switzerland AG 2021
H. Hacid et al. (Eds.): ICSOC 2020 Workshops, LNCS 12632, pp. 137–149, 2021.
https://doi.org/10.1007/978-3-030-76352-7_17

fault is a challenging task due to the following reasons: a) heterogeneity of infrastructure components, b) difficulty to trace the execution paths, c) overhead of instrumentation, and d) lack of a synchronized clock across a distributed system to interpret the dependency among the components. Therefore, localizing faults in cloud-based distributed applications using artificial intelligence is an active area of research.

There are three types of approaches present in the literature for fault localization in distributed systems: trace-based, metric-based and log-based. Trace and metric based approaches require instrumentation of code and metrics respectively, thus adding extra overhead and may not always be feasible to do. On the other hand, logs are easily available with any distributed system making log-based approaches more practical to use as compared to trace and metric based approaches.

Using causal inference techniques for trace and metric based approaches has been thoroughly explored [11,15,18]. Jia et al. have used causal inference techniques for log-based approaches [9], but they make the assumption that the abnormal or the failed service is known and that the time order of the log templates within a service is maintained. However, this is not always true, making it difficult for the approaches to generalize well. The above cited state-of-the art approaches also don't take into account the causality relationships across templates of multiple services in the system.

We address the limitations of existing log based approaches by taking a holistic view of causality mining among the micro-services level as well as the template level across micro-services, capturing the fine-grained causality of the various components involved. Also, our fault-localization approach does not assume an abnormal or failed service. We use Golden Signal errors also known as gateway errors [4] that the users face or observe when a system fails. These errors are very critical and have to be handled in real time. The faults are characterized by the PIE model [23] which states that the fault should be *executed*, the error state/fault then *infects* the system and the errors are *propagated* to the output state. We term these faults as observable operational faults. Our approach does not depend on the time order of log templates like [9] within a micro-service to localize the fault at the template level, since we build causality at the template level and then use PageRank-based centrality metrics to find the most important log template responsible for the fault. Unlike previous approaches [12,13], we do not need data spanning days or weeks to compute causality. Our work in this paper uses very less runtime data (in the order of minutes) to infer causal relationships.

1.1   Main Contributions

In this paper, we propose a transactional error log based fault localization technique which has the following key characteristics:

1. **Multi-variate Time Series Modeling of Logs**: We transform the logs to multivariate timeseries data by counting the number of error logs in each time

bin for the impacted services. This transformation from log space to metric space helps us in running causal dependency models directly on log data.
2. **Causal dependency model** which is mined from golden signals viz. error logs in application and gateway at the log template level. Unlike state-of-the-art approaches which need weeks of training data our system can compute causal relationship among anomalous micro-services from only a few minutes of logs.
3. **Personalized PageRank** algorithm that uses the extracted anomalous subgraph from the dependency and causal graphs and the golden signal errors to provide a ranked order of faulty nodes.

We experiment with two Granger causal based techniques: Regression and Conditional Independence [7, 22]. The experiments are run on a simulated micro-service system, the TrainTicket application that contains 41 micro-services [2]. Our empirical results demonstrate the accuracy of our proposed approach in identifying and localizing operational faults.

The rest of the paper is organized as follows. We provide a brief summary of related works in Sect. 2. Section 3 describes the proposed fault localization system in detail. Section 4 gives an overview of the dataset generation technique. In Sect. 5, we present our experimental results and highlight the key takeaways from our experiments. Finally, we conclude in Sect. 6 with some future directions of our work.

2 Related Work

In recent years, different solutions have been proposed for fault localization in distributed systems, networks, clouds and micro-services. Those proposed solutions typically fall into three categories, i.e. trace-based, metric-based and log-based approaches [26].

Trace-based approaches need to gather information through complete tracing of the execution path, and then detect the potential faults through the outlier analysis along the execution path [5, 16, 18]. These methods are able to identify the root causes of the underlying problems with high accuracy, but they require significant domain knowledge. These approaches also require system code logic which might not be available.

Metric-based methods collect the metrics from both application and infrastructure components and construct the causality graph among components to infer the root causes of faults [11, 14, 15, 24, 26]. These approaches often treat the applications as a black box and only collect the system metrics related to the applications such as CPU utilization, memory usage, etc. The challenging task in these methods is in how to build the accurate causality graph using the huge amount of collected metrics data.

Log-based approaches [8, 9, 21, 25, 27] parse the system logs first, and then identify the faulty components based on parsed logs. Though log-based methods are capable of identifying informational causes, the difficulties in log parsing and abnormal information locating from large scale of logs pose great challenges in practice. The log-based approaches can further classified in three categories, a) machine learning based approach [27], b) domain dependent [21,25] and b) causal inference based approach [9]. Jia et al. [9] present a fault localization system using causal inference techniques to build a dependency graph among the services first and then within a service at log template level. Using causal inference on logs from a distributed system is still in its infancy. The existing approaches suffer from lack of benchmark data, assumptions such as failed or abnormal service is given as well as ignoring the causality across the log templates of multiple micro-services. We address these limitations in the proposed approach.

3   Central Idea

Our proposed fault localization technique using golden signals addresses the problem of localizing the faults that are characterized by the PIE model [23]. There are four types of golden signals viz. **Latency, Error, Traffic**, and **Saturation** [4]. These signals can help monitor the health of systems by identifying faults and triaging of issues. In this paper, we focus on one type of golden signal viz. *Error* to localize faults.

Figure 1 illustrates the predictive and reactive modes of our approach. In the *predictive* mode, the logs are analyzed in real time. Our fault localization technique is triggered when the number of golden signal errors reach above a threshold value in a given time window, indicating that there is a fault in the system which is repeatable. To localize the fault, causal relationships among the micro-services, emitting error signals, and the service emitting golden signal errors are inferred after modeling the logs as multivariate time series data (*time series modeling*). We identify the services which are causing golden signal errors and extract a subgraph of those services from the dependency graph.

Generally, the faulty node has high causality score with golden signal errors. However, it is highly possible that nodes that have no relation with the golden signal errors can also have high causal scores. To avoid such false positives, we explore graph centrality indices (e.g., *PageRank*) to find the micro-service which best characterizes the golden signal errors. The node with the highest centrality scores is likely the faulty service. It is possible that the nature of fault is such that the actual faulty node does not emit any error signals. In such scenarios, we analyse the error messages emitted by a node in order to detect faulty service. We refer to this technique as *Last Mile Fault Localization*. We propose 3 variants of our approach: (1) All the nodes causing golden signal error are considered faulty, (2) We run Personalized PageRank on the static/dependency graph, and (3) We run Personalized PageRank on causal graph.

During the *reactive* mode, the fault localization model is triggered whenever the system raises an alert/alarm. In this scenario, we extract fault timing by parsing the alert messages and retrieve the corresponding logs from log database. Subsequent process is same as described in the *predictive* mode.

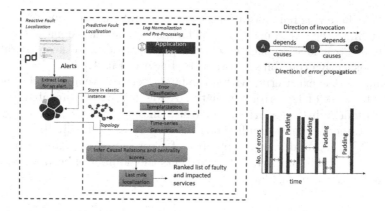

Fig. 1. (Left:) Architecture of proposed fault localization approach. **(Upper Right:)** Graph showing direction of cause and depends-on relationship. **(Lower Right)** Time series data - Each bar corresponds to an error signal, each color represents a microservice, and the height indicates the frequency of the error

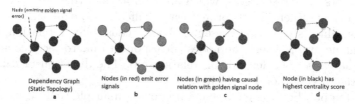

Fig. 2. A sample fault localization flow using causal inference and PageRank.

3.1 Fault Localization

Figure 2 shows the flow of our fault localization approach. We start with the dependency graph where nodes in the graph represent various services and the edges indicate the requests flow direction (Fig. 2a). Dependency graphs can be inferred from architecture diagrams, or can be discovered from logs [9]. We focus on a subgraph extracted from the dependency graph consisting of only those nodes which emit error signals (Fig. 2b). Next, we use two Granger causality techniques: regression based and independence testing based to infer the causal relationship among micro-services. Causal dependencies indicate the strength of the correlation between the errors in various micro-services. We run PageRank based centrality index to find the faulty node (Fig. 2d) among the nodes which have causal relationship with the node emitting golden signal errors (Fig. 2c). In the following subsections, we will describe each sub-component of our approach in detail.

Multivariate Time Series Modeling. We model the logs as multivariate time series data. We count the number of error logs in each time bin to obtain a time series corresponding to each impacted micro-service for a given time window. The resulting data representation can be viewed as a multidimensional array

$M \in R^{n \times t}$, where n is number of micro-services emitting error logs and t is the number of time steps. It is possible that for some time bins, none of the micro-services emit error logs. In such cases, we do zero padding to keep the interval between consecutive time bins constant as shown in Fig. 1.

To pinpoint the exact error within the faulty service, we also model the time series at the level of log error templates. A log *template* is an abstraction of a print statement in the source code, which manifests itself in raw logs with different parameter values in different runs [17]. It is represented as $T \in R^{e \times t}$, where $e = \sum_{i=1}^{n} x_i$, and x_i is the number of error templates emitted by micro-service n_i. Each column of array, T, represents the number of occurrences of the error template for a particular time bin.

Causal Inference. We infer causal relationships among the error signals emitted by individual micro-services and the golden signal errors, after modeling the log data as multiple time series. We assume that the anomalous behavior of a faulty component is likely to result in error signals being emitted by neighboring components (micro-services), which are components that interact with the faulty component either directly or indirectly. Different from association and correlation, causality is used to represent a direct "cause-effect" relation. Figure 1 shows a sample graph where the nodes correspond to micro-services and edges represent the cause and depends on relationship. The direction of causality is reverse of the direction of dependency.

Mining temporal dependency structure among multiple time series has been extensively studied. As mentioned in Sect. 2, the Granger causality framework [7] is used to infer the causal dependencies among time series data. In this paper, we use two types of Granger causality techniques: (1) Regression Based and, (2) Conditional Independence Testing. One of the classic approaches for a Granger causal test is to linearly regress B_t on $A_{t-1:t-p}, B_{t-1:t-p}$ for some lag p and compare the residue with that of regression of B_t on $B_{t-1:t-p}$ alone [6], where A and B are two time-series. In this paper, we refer to this approach as *BLinear*. In order to track the causal dependencies among time series instantly, [28] developed a novel Bayesian Lasso-Granger method, *BLasso*, which conducts the causal inference from the Bayesian perspective [19] in a sequential online mode. We use *BLinear* and *BLasso* regression based methods to infer the causality graph of micro-services. For conditional independence based causal inference, we use the $PC - Algorithm$ [10,20]. The algorithm starts from a complete, undirected graph and deletes recursively the edges based on conditional independence decisions. We leverage a cross entropy based metric, namely G^2, to test whether two services are dependent on one another or not. The micro-services which cause golden signal errors are identified as potential source of fault.

Personalized PageRank Algorithm. As described above, the nodes which cause golden signal errors are useful for finding the actual root cause of the fault. However, it is highly possible that nodes that have no relation with the golden signal errors can have high causal scores. To avoid such false positives, the nodes (micro-services) causing golden signal errors are considered as candidate nodes.

We experiment with both dependency and causal graphs and use the extracted anomalous sub-graph (nodes having causal scores with golden signal and their connections) to rank the nodes using the *Personalized PageRank* method proposed in [11]. We assign higher weights to the nodes which cause golden signal error.

The inputs to the PageRank algorithm are the graphs (causal and dependency), the golden signal errors and the causal score of each node with the golden signal errors. Let CS_i define the causal score of node i with respect to the golden signal errors. We derive the anomalous sub-graph from both dependency and causal graphs by preserving the nodes that cause golden signal errors (candidate nodes) and their direct connections. Considering that the request flow is from node i to node j, the weight of each edge e_{ij} is assigned to the value of CS_j, the weight of each added self-edge e_{ii} is assigned to the value of CS_i, and the weight of each added backward edge e_{ji} is assigned to the value of ρCS_i, where $\rho \in [0, 1]$. We set ρ to a high value if the causal graph represents the true dependency graph. As error propagation happens in the opposite direction of request flow, we reverse the direction of the edges when applying the PageRank algorithm.

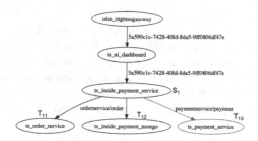

Fig. 3. Last mile fault localization on TrainTicket system.

Last Mile Fault Localization. The output from the previous step namely, *Personalized PageRank*, provides a ranked list of potential nodes (or microservices) which are faulty. However, this information is not sufficient to identify the precise location of the fault. To get the precise location, we use the *Last Mile Fault Localization* technique which examines the service emitting the error as well as it's neighbourhood to identify the correct fault location.

Let us assume that the nodes localized by PageRank are given by:

$$L_p = [S_i], \quad where \; i \in \{1..n\} \tag{1}$$

For last mile fault localization, we inspect the following set of nodes:

$$L_m = [S_i, T_{i_1}, ..., T_{i_j}, ...T_{i_k}], \quad where \; i \in \{1..n\} \; and \; j \in \{1...k\} \tag{2}$$

Herein, $\{T_{i_1}, ..., T_{ik}\}$ are nodes which are one hop away from S_i returned by Page Rank. To detect the correct fault location, we use the application topology obtained using execution behaviour model [17] along with trace information obtained using discriminating parameters [3]. The last mile localization technique is illustrated in Fig. 3 which shows the service call flow mined from execution logs. For this instance, PageRank returns the faulty service as *ts-inside-payment-service* which has three directly connected edges. As such we inspect the nodes *[ts-inside-payment-service, ts-order-service, ts-inside-payment-mongo, ts-payment-service]*. With the use of discriminating parameter values in the call flow viz. *5a590c1c-7428-408d-8da5-9ff0806df47e*) and *paymentservice/payment* we are able to correctly localize the fault to *ts-payment-service* as shown in the figure.

4   Dataset Details

We use the TrainTicket application [2], an open-source micro-service application, to inject faults and generate log data to evaluate the effectiveness of our proposed approach. The application contains 41 micro-services. Service *ts-ui-dashboard* acts as the gateway service which records the status of each incoming and outgoing service. The error signals emitted by this service and capturing the failure of a request are considered as *golden signals*. We use Istio [1] to inject HTTP abort fault in multiple services. In abort fault, incoming request is intercepted by Istio and returns 500 error status. Users can also specify the percentage of requests that should be failed. We injected faults in 17 services that cover the main flow of the train ticket application. We generated data by running a scenario where **100%** incoming requests are failed. To measure the observability of faults, the user flow was executed **20** times after injecting the fault. For each service, the average running time for 20 iterations was **19 min** and the average number of log lines were **164,740**. We found that the average number of error messages were **270** while the minimum and maximum number of error messages were **40** and **1,436** respectively. We observed that the maximum number of error are generated by the *ts-station-service* fault. This is because it is one of the farthest node from the gateway node *ts-ui-dashboard* and also since all nodes in the path emit errors.

5   Experiments

In this section, we present an empirical study conducted to show the effectiveness of our proposed approach for fault localization.

The experiments are designed to show that causal graph in conjunction with golden signals delivers superior results over just using dependency graph (static topology) among the micro-services emitting error templates. Localizing operation faults has two aspects, 1) localizing the faulty micro-service and 2) further, within the faulty micro-service, which template (error message) is responsible. In order to evaluate both the aspects, we first evaluate fault localization at micro-service level then further localizing at finer granularity of template level within the faulty micro-service.

For evaluating the first aspect, we divide the experiments into two groups, $G1$ and $G2$. In group $G1$, topology is static, while in group $G2$, topology is dynamic and is computed using causality based techniques. To generate the causal graph, we use 3 approaches: PC, $Blasso$, $Blinear$. We show that the dynamic topology, called causal graph, computed using causality inference techniques are outperforming static topology in all the variations. Finally, we perform last mile fault localization to localize the faulty error template.

Group 1: Group G1 covers 3 approaches, $G1_A$, $G1_B$ and $G1_C$. In $G1_A$, we use error signals emitted by each micro-service and using them as the starting node, we traverse the dependency graph to the leaf nodes. The leaf nodes are picked as the faulty nodes. In $G1_B$, instead of traversing till the leaf nodes, we run PageRank over the subgraph consisting of all micro-services emitting error signals to get a ranked list of faulty nodes. In $G1_C$, we consider only those micro-services nodes which cause the golden signal errors and run *Personalized PageRank* algorithm on static topology to get ranked list of faulty nodes.

Group 2: In Group G2, we have two variations $G2_A$ and $G2_B$ for fault localization using dynamic causal graphs. In $G2_A$ the nodes which cause golden signal error are consider as potential faulty nodes. In $G2_B$, micro-service nodes causal to golden signal node are identified and are assigned weights based on the causality scores. Then, *Personalized PageRank* is used on causal graph factoring in weights on the causal nodes to do first level of fault localization.

5.1   Results

We use a time bin of size $10\,\text{ms}$ as the inter-arrival time between error logs in this dataset. We have considered threshold for the number of golden signals errors as 15. To calculate precision and recall we use a graph-based approach. If the localized node does not exactly match the ground truth node we calculate match based on the distance (in number of hops) of the returned node n according to the following equation:

$$S = 1 - h_n/(H+1) \tag{3}$$

Here S is the final match score for the returned node n, h_n is the distance(in hops) of this node from the ground truth node and H is a pre-configured threshold for the maximum number of hops allowed. For our experiments we use H = 3. In all the PageRank based methods, we measure precision and recall for $Top3$.

Figure 4 shows the results of experiments conducted to evaluate accuracy of our approach at micro-service level. We observe, low precision and recall for $G1_A$, indicating that the errors emitted by the leaf nodes are not the potential source of fault. In $G1_B$, we see 7% point increase in F1 score as PageRank helps in identifying most impact-ful node. However, the performance is low as all the nodes emitting error signals are considered. In $G2_A$, where golden signal errors are factored in, we observe an increase of 42% points in F1 score (with *Blinear* causal technique), signifying the usefulness of golden signals to localize the fault. The error signals often start from the faulty micro-services and then propagate to other non-faulty micro-services via inter-micro-service interactions. Due to this, sometimes the non faulty micro-services might show causal relationship which golden signal errors, resulting in low precision for $G2_A$. Out of the 17 faults, six faults are injected in the services which directly interacts with *ts-ui-dashboard* service. When fault is ingested in any of these 6 services, only the error signals emitted by *ts-ui-dashboard* service has evidence of failure. In $G2_A$ approach, we consider only those micro-services, for fault localization, which have causal relationship with *ts-ui-dashboard*, therefore the failure of these 6 services is not captured by $G2_A$ approach, hence the recall is low. To further improve the performance, we use *Personalized PageRank*. The results clearly indicates that our approach $G2_C$, using *Blinear* causal technique, outperforms all other approaches with F1 score of 0.88 and with casual techniques *Blasso* and *PC*, we have F1-scores of 0.83 and 0.87 respectively. *PC* based technique has highest recall of 0.96. We observe that with the dependency graph, $G2_B$ we get low F1-score viz. 0.49 (*Blasso*), 0.51 (*PC*) and 0.58 (*Blinear*). The possible explanation could be that the dependency graph is not a true indication of runtime behavior. A micro-service might interact with multiple services but in a user scenario all the flows need not execute. *Personalized PageRank* algorithm on the static dependency graph does not assign high centrality scores to the faulty nodes as non-faulty nodes have many incoming and out going edges resulting in high centrality score. Whereas, the causal graph indicates the runtime interactions among various micro-services. Therefore, the causal relationship among various micro-services captures the error propagation across various micro-services in the application well.

Method	Precision	Recall	F1 Score
$G1_A$	0.25	0.31	0.28
$G1_B$	0.32	0.39	0.35
$G2_{A-PC}$	0.57	0.59	0.58
$G2_{A-Blasso}$	0.58	0.83	0.68
$G2_{A-Blinear}$	0.62	0.88	0.73
$G1_{C-PC}$	0.49	0.55	0.51
$G1_{C-Blasso}$	0.46	0.52	0.49
$G1_{C-Blinear}$	0.54	0.63	0.58
$G2_{B-PC}$	0.73	**0.96**	0.83
$G2_{B-Blasso}$	**0.84**	0.90	0.87
$G2_{B-Blinear}$	**0.84**	0.93	**0.88**

Fig. 4. Performance results with different fault localization methods

Method	Precision	F1 Score
$G2_{A-PC}$	0.57	0.58
$G2_{A-Blasso}$	0.62	0.71
$G2_{A-Blinear}$	0.66	0.76
$G1_{C-PC}$	0.55	0.54
$G1_{C-Blasso}$	0.48	0.50
$G1_{C-Blinear}$	0.56	0.59
$G2_{B-PC}$	0.75	0.84
$G2_{B-Blasso}$	0.85	0.87
$G2_{B-Blinear}$	0.85	0.89

Fig. 5. Performance results with *Last Mile Fault Localization*

We apply our *LastMilefaultlocalization* (*LMFL*) technique to analyze the error templates emitted by the top 3 potential faulty micro-services. We observe that usually there are 3–4 unique error templates emitted by each micro-service. Instead of analyzing all the error templates of the top 3 micro-services, we narrow down to the error templates which have high causal relationship with golden signal errors and high centrality score. This reduces the number of potential root cause error templates by 70%. On these error templates, we do further analysis to find out whether the service emitting error template is at fault or one of its children nodes. Figure 5 shows the improvement in precision and F1 for $G1_C$, $G2_A$ and $G2_B$ approaches. The recall remains same as $LMFL$ is applied on the output of previous step.

6   Conclusion

In this paper, we present a golden signal based fault localization approach, which is based on inferring the causal relationship among services emitting error signals and the one emitting golden signal error. PageRank based graph centrality approach is used to efficiently localize the faults. The proposed approach improves state-of-the art techniques by (i) using golden signal error rate to localize the operational faults (ii) time series modeling of the error rate from log data (iii) using only the positive samples to do last mile fault localization and (iv) using regression and conditional independence based causal techniques. Our experimental results show the effectiveness of this approach. This technique can easily be extended to use other golden signals such as latency, saturation, and traffic. In future, we plan to perform more experiments with real world dataset and explore other types of golden signals.

References

1. Istio service mesh. https://istio.io/. Accessed 16 Aug 2020
2. Train ticket: a benchmark microservice system. https://github.com/FudanSELab/train-ticket/. Accessed 16 Aug 2020
3. Aggarwal, P., Atreja, S., Dasgupta, G., Mandal, A.: System anomaly detection using parameter flows, December 2019
4. Beyer, B., Jones, C., Petoff, J., Murphy, N.R.: Site Reliability Engineering: How Google Runs Production Systems. O'Reilly Media Inc., Newton (2016)
5. Chow, M., Meisner, D., Flinn, J., Peek, D., Wenisch, T.F.: The mystery machine: end-to-end performance analysis of large-scale internet services. In: 11th {USENIX} Symposium on Operating Systems Design and Implementation ({OSDI} 2014), pp. 217–231 (2014)
6. Geweke, J.F.: Measures of conditional linear dependence and feedback between time series. J. Am. Stat. Assoc. **79**(388), 907–915 (1984)
7. Granger, C.W.J.: Investigating causal relations by econometric models and cross-spectral methods. Econometrica **37**(3), 424–438 (1969)
8. Gupta, M., Mandal, A., Dasgupta, G., Serebrenik, A.: Runtime monitoring in continuous deployment by differencing execution behavior model. In: Pahl, C., Vukovic, M., Yin, J., Yu, Q. (eds.) ICSOC 2018. LNCS, vol. 11236, pp. 812–827. Springer, Cham (2018). https://doi.org/10.1007/978-3-030-03596-9_58
9. Jia, T., Chen, P., Yang, L., Li, Y., Meng, F., Xu, J.: An approach for anomaly diagnosis based on hybrid graph model with logs for distributed services. In: 2017 IEEE International Conference on Web Services (ICWS), pp. 25–32. IEEE (2017)

10. Kalisch, M., Bühlmann, P.: Estimating high-dimensional directed acyclic graphs with the PC-algorithm. J. Mach. Learn. Res. **8**, 613–636 (2007)
11. Kim, M., Sumbaly, R., Shah, S.: Root cause detection in a service-oriented architecture. ACM SIGMETRICS Perform. Eval. Rev. **41**(1), 93–104 (2013)
12. Kobayashi, S., Otomo, K., Fukuda, K.: Causal analysis of network logs with layered protocols and topology knowledge. In: 2019 15th International Conference on Network and Service Management (CNSM), pp. 1–9 (2019)
13. Kobayashi, S., Otomo, K., Fukuda, K., Esaki, H.: Mining causality of network events in log data. IEEE Trans. Netw. Serv. Manag. **15**(1), 53–67 (2018)
14. Lin, J., Chen, P., Zheng, Z.: Microscope: pinpoint performance issues with causal graphs in micro-service environments. In: Pahl, C., Vukovic, M., Yin, J., Yu, Q. (eds.) ICSOC 2018. LNCS, vol. 11236, pp. 3–20. Springer, Cham (2018). https://doi.org/10.1007/978-3-030-03596-9_1
15. Mariani, L., Monni, C., Pezzé, M., Riganelli, O., Xin, R.: Localizing faults in cloud systems. In: 2018 IEEE 11th International Conference on Software Testing, Verification and Validation (ICST), pp. 262–273 (2018)
16. Mi, H., Wang, H., Zhou, Y., Lyu, M.R.T., Cai, H.: Toward fine-grained, unsupervised, scalable performance diagnosis for production cloud computing systems. IEEE Trans. Parallel Distrib. Syst. **24**(6), 1245–1255 (2013)
17. Nandi, A., Mandal, A., Atreja, S., Dasgupta, G.B., Bhattacharya, S.: Anomaly detection using program control flow graph mining from execution logs. In: Proceedings of the 22nd ACM SIGKDD International Conference on Knowledge Discovery and Data Mining, pp. 215–224 (2016)
18. Nguyen, H., Shen, Z., Tan, Y., Gu, X.: Fchain: toward black-box online fault localization for cloud systems. In: 2013 IEEE 33rd International Conference on Distributed Computing Systems, pp. 21–30 (2013)
19. Park, T., Casella, G.: The Bayesian lasso. J. Am. Stat. Assoc. **103**(482), 681–686 (2008)
20. Spirtes, P., Glymour, C.: An algorithm for fast recovery of sparse causal graphs. Soc. Sci. Comput. Rev. **9**(1), 62–72 (1991)
21. Tan, J., Pan, X., Marinelli, E., Kavulya, S., Gandhi, R., Narasimhan, P.: Kahuna: problem diagnosis for mapreduce-based cloud computing environments. In: 2010 IEEE Network Operations and Management Symposium-NOMS 2010, pp. 112–119. IEEE (2010)
22. Valdés-Sosa, P., et al.: Estimating brain functional connectivity with sparse multivariate autoregression. Philos. Trans. R. Soc. Lond. Ser. B Biol. Sci. **360**, 969–81 (2005)
23. Voas, J.M.: Pie: a dynamic failure-based technique. IEEE Trans. Software Eng. **18**(8), 717 (1992)
24. Wang, P., et al.: Cloudranger: root cause identification for cloud native systems. In: 2018 18th IEEE/ACM International Symposium on Cluster, Cloud and Grid Computing (CCGRID), pp. 492–502. IEEE (2018)
25. Weber, I., Li, C., Bass, L., Xu, X., Zhu, L.: Discovering and visualizing operations processes with pod-discovery and pod-viz. In: 2015 45th Annual IEEE/IFIP International Conference on Dependable Systems and Networks, pp. 537–544. IEEE (2015)
26. Wu, L., Tordsson, J., Elmroth, E., Kao, O.: Microrca: root cause localization of performance issues in microservices. In: NOMS 2020–2020 IEEE/IFIP Network Operations and Management Symposium, pp. 1–9. IEEE (2020)

27. Xu, J., Chen, P., Yang, L., Meng, F., Wang, P.: Logdc: problem diagnosis for declartively-deployed cloud applications with log. In: 2017 IEEE 14th International Conference on e-Business Engineering (ICEBE), pp. 282–287. IEEE (2017)
28. Zeng, C., Wang, Q., Wang, W., Li, T., Shwartz, L.: Online inference for time-varying temporal dependency discovery from time series. In: 2016 IEEE International Conference on Big Data (Big Data), pp. 1281–1290. IEEE (2016)

Using Language Models to Pre-train Features for Optimizing Information Technology Operations Management Tasks

Xiaotong Liu[✉], Yingbei Tong[✉], Anbang Xu[✉], and Rama Akkiraju[✉]

IBM Research Almaden, San Jose, USA
{xiaotong.liu,yingbei.tong}@ibm.com, {anbangxu,akkiraju}@us.ibm.com

Abstract. Information Technology (IT) Operations management is a vexing problem for most companies that rely on IT systems for mission-critical business applications. While IT operators are increasingly leveraging analytical tools powered by artificial intelligence (AI), the volume, the variety and the complexity of data generated in the IT Operations domain poses significant challenges in managing the applications. In this work, we present an approach to leveraging language models to pretrain features for optimizing IT Operations management tasks such as anomaly prediction from logs. Specifically, using log-based anomaly prediction as the task, we show that the machine learning models built using language models (embeddings) trained with IT Operations domain data as features outperform those AI models built using language models with general-purpose data as features. Furthermore, we present our empirical results outlining the influence of factors such as the type of language models, the type of input data, and the diversity of input data, on the prediction accuracy of our log anomaly prediction model when language models trained from IT Operations domain data are used as features. We also present the run-time inference performance of log anomaly prediction models built using language models as features in an IT Operations production environment.

Keywords: AI for IT operations · Language modeling · Anomaly detection

1 Introduction

Information Technology (IT) Operations management is a vexing problem for most companies that rely on IT systems for mission critical business applications. Despite best intentions, designs, and development practices followed, software and hardware systems are susceptible to outages, resulting in millions of dollars in labor, revenue loss, and customer satisfaction issues. IT downtime costs an estimated $26.5 billion in lost revenue each year based on a survey of

© Springer Nature Switzerland AG 2021
H. Hacid et al. (Eds.): ICSOC 2020 Workshops, LNCS 12632, pp. 150–161, 2021.
https://doi.org/10.1007/978-3-030-76352-7_18

200 companies across North America and Europe [4]. The best of the analytical tools fall short of detecting incidents early, predicting when incidents may occur, offering timely and relevant guidance on how to resolve incidents quickly and efficiently and helping avoid them from recurring. This can be attributed to the complexity of the problem at hand. IT applications, the infrastructure that they run on and the networking systems that support that infrastructure, all produce large amounts of structured and unstructured data in the form of logs and metrics. The volume and the variety of data generated in real-time poses significant challenges for analytical tools in processing them for detecting genuine anomalies, correlating disparate signals from multiple sources, and raising only those alerts that need IT Operations management teams' attention. To add to this, data volumes continue to grow rapidly as companies move to modular microservices-based architectures, further compounding the problem. Furthermore, the heterogeneous nature of environments, where companies' IT applications can run on a mix of traditional bare metal, virtual machines, and public or private clouds operated by different parties, adds to the diversity of formats, platforms and scale that IT Operations management solutions must deal with. These complex and dynamic environments demand a new approach to IT Operations management that is fast, real-time, adaptive, customizable, and scalable.

The rise of Artificial Intelligence (AI) powered by the advancements in hardware architectures, cloud computing, natural language processing (NLP), and machine learning (ML), has opened up new opportunities for optimizing various industries and business processes. Operations management of IT systems is one such an area that is prime for optimization. AI can help IT Operations management personnel in detecting issues early, predicting them before they occur, locating the specific application or infrastructure component that is the source of the issue, and recommending relevant and timely recommendation actions based on mining prior issue records. All these analytics help reduce the mean time to resolve (MTTR) an incident, which in turn, saves millions of dollars by preventing direct costs (lost revenue, penalties, opportunity costs, etc.) and indirect costs (customer dissatisfaction, lost customers, lost references, etc.). Many IT Operations management vendors are starting to embed AI capabilities into their products. An advanced IT Operations management system needs to take all kinds of data as inputs, detect anomalies early, predict when incidents may occur, offer timely and relevant guidance on how to resolve incidents quickly and efficiently, automatically apply resolutions when applicable, and proactively avoid them from recurring by enforcing the required feedback loops into the various software development lifecycles. This can increase the productivity of IT Operations personnel or Site Reliability Engineers (SREs) and thereby improve the mean times to detect, identify and resolve incidents.

In this work, we present an approach to leveraging language models to pre-train features for optimizing IT Operations management tasks such as anomaly prediction from logs. Specifically, using log-based anomaly prediction as the task, we show that the machine learning models built using language models (embeddings) trained with the IT Operations domain data as features outperform those AI models built using language models with general-purpose data as features. Furthermore, we present our empirical results outlining the influence of factors such as the type of language models, the type of input data, and the diversity of input data, on the prediction accuracy of our log anomaly prediction model when language models trained from IT Operations domain data are used as features. We also present the run-time inference performance of log anomaly prediction models built using language models as features in an IT Operations production environment.

The structure of the rest of this paper is as follows: Sect. 2 discusses the related works and techniques. Section 3 presents our method to pre-train features using language models for IT Operations, followed by our case study on log anomaly prediction in Sect. 4. We conclude the paper in Sect. 5.

2   Related Works

The notion of application of AI to optimize IT is often referred to as AI Operations (or AIOps in short) in the industry. Coined by Gartner [6], the field of AIOps is a specific space, stretching across several markets including Application Performance Management (APM), IT Operations Management (ITOM), IT Automation and Configuration Management (ITACM), and IT Service Management (ITSM) with a specific focus on AI-infusion. Research problems in this field include anomaly detection and prediction [10], incident management [16], fault localization [21], root cause analysis [20], and so on. Our work uses log-based anomaly prediction as the task to study the effects of pre-trained features from language modeling. Prior work in this space mainly relies on parsing stable logs, where the set of distinct log events is known and will not change over time [22]. However, in practice log data often contains previously unseen log events or log sequences. Furthermore, it can be challenging for conventional log parsers to adapt to different microservices since logs in each microservice may have their own context information. In this work, we leverage language models to pre-train features from IT Operations log data, which makes it easier to parse dynamically evolving logs in a cloud environment for anomaly detection and prediction.

Language modeling and embedding representation learning has been an active area of research in NLP, starting from word embeddings that map words in a vocabulary to vectors of real numbers so that words of similar semantic meaning are close to each other in the embedding space. Two language models are typically used to learn the word embedding representations: Continuous Bag of Words Model (CBOW) and Skip-gram. In the CBOW model, the distributed representations of context are combined to predict the word in the middle, while in the Skip-gram model, the distributed representation of the input word is used to predict the context. Word2vec [12] was the first popular embedding method for NLP tasks. The embeddings were derived from a Skip-gram model represented as a neural network with a single hidden layer. GloVe [13] learned the embeddings through dimensionality reduction on the co-occurrence counts matrix. FastText [2] introduced the concept of subword-level embeddings, based on the Skip-gram model. Each word is represented as a bag of character n-grams, and their embeddings are the sum of vector representations associated with each character n-gram. Recent research in language modeling and deep learning has advanced contextualized embedding learning to address the issue of polysemous and the context-dependent nature of words. ELMo [14] extracts context-sensitive features from a bidirectional LSTM language model and provides additional features for a task-specific architecture. ULMFiT [7] advocates discriminative fine-tuning and slanted triangular learning rates to stabilize the fine-tuning process with respect to end tasks. OpenAI GPT [15] builds on multi-layer transformer [17] decoders instead of LSTM to achieve effective transfer while requiring minimal changes to the model architecture. BERT [5] uses bidirectional transformer encoders to pre-train a large corpus, and fine-tunes the pre-trained model that requires almost no specific architecture for each end task. In this work, we empirically investigate the impact of embeddings and language models pre-trained using IT Operations domain data for optimizing the domain-specific tasks.

3 Approach

In this section, we describe our approach to pre-training features using language models for optimizing different IT Operations management tasks.

3.1 Motivation

IT Operations environment generates many kinds of data. These include metrics, alerts, events, logs, tickets, application and infrastructure topology, deployment configurations, and chat conversations. Of these, metrics tend to be structured in nature while logs, alerts, and events are semi-structured, and tickets are unstructured data types. Also, among all the data types, logs and metrics sometimes can be leading indicators of problems, while alerts, tickets and chat conversations tend to be lagging indicators. The volume, the variety and the complexity of these data offers both challenges and opportunities in developing AI-infused analytical tools to optimize IT Operations management tasks. Leveraging language models to pre-train features from IT Operations domain data, we present a transfer learning approach that serves as strong a foundation to build AI models for various different IT Operations management tasks, such as log anomaly detection, fault localization, named entity extraction, similar incident analysis, and event grouping (as illustrated in Fig. 1).

3.2 Embedding and Language Modeling

Embeddings, also known as distributed vector representations, have been widely used in NLP and natural language understanding. In a pretrained embedding

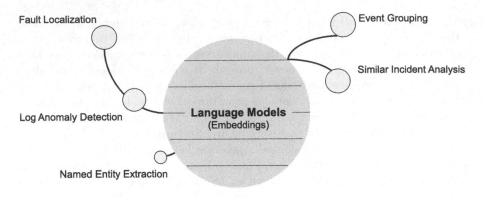

Fig. 1. An illustration of language models for different IT Operations management tasks.

space, words or phrases from the vocabulary are mapped to vectors of real numbers, and each is associated with a feature vector of a fixed dimension. Embeddings are pre-trained on large text corpus using a language modeling task, which assigns a probability distribution over sequences of words that matches the distribution of a language. After that, embeddings can be extracted from the pre-trained language model.

Typically, we can categorize embeddings into two types based on the language modeling approach in use: context-free embeddings and contextualized embeddings. Static word embeddings (e.g., Word2Vec, Glove, fastText) are context-free as the language models generate the same embedding for the same word even in different context. Deep pre-trained language models (e.g., ELMo, ULMFiT, BERT) can generate contextualized embeddings where the representation of each word also depends on the other words in a sentence (i.e., the context of the word). For the scope of this work, we select two representative embeddings from each type: fastText for context-free embedding and BERT for contextualized embedding.

3.3   Pre-training Language Models for IT Operations Management

Most existing embeddings available in the literature were created from text corpus in natural language such as Wikipedia pages and news articles. However, the text data generated in the IT Operations domain are different from natural language texts, as the vocabulary of the IT Operations domain is quite unique. For example, logs can contain a mix of the date, the time, the pod id, the level of logging, the component where the system runs, and the content of log message.

To pre-train language models in the IT Operations domain, we first process the input text data into a normalized format using predefined rules, extracting the most informative texts such as log messages, ticket descriptions and so on. We also remove duplicates of texts, which may be auto-generated multiple times by the system for the same event. Next, we randomly sample data from each data source, and use the data samples to learn the vocabulary of the whole IT Operations domain. For fastText the vocabulary of words are learned when pre-training the language model. For BERT since the vocabulary has to be predetermined prior to the neural model training, we use sentencepiece [9] to learn the vocabulary of subwords. After that, we pre-train the language model using the sampled data, and tune the parameters based on model evaluation. An overview of the pre-training pipeline is shown in Fig. 2.

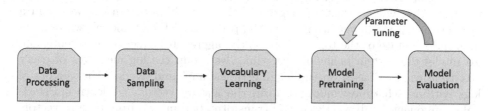

Fig. 2. The pipeline of pre-training language models using IT Operations domain data.

4   Case Study: Log Anomaly Detection

To demonstrate the feasibility of our approach, we describe a case study of using language models to pre-train features for building log anomaly detection and prediction models in IT Operations management.

4.1   Problem Statement

Anomaly detection from logs is one fundamental IT Operations management task, which aims to detect anomalous system behaviors and find signals that can provide clues to the reasons and the anatomy of a system's failure. Log messages are inherently unstructured, since system events can be recorded by developers using any text for the purpose of convenience and flexibility. Traditionally, log parsing is usually applied as a first step towards down-stream log analysis tasks to convert unstructured textual log messages into a structured format that enables efficient searching, filtering, grouping, counting, and sophisticated mining of logs. In particular, log templates are extracted from logs to represent an abstraction of log messages by masking system parameters recorded in logs. However, existing log parsing approaches are unable to adapt to evolving logs, making it challenging for continuous model improvement and customization. Furthermore, it is difficult to capture the semantic information in log messages with log templates.

4.2 System Overview

To tackle these challenges, we develop a system to perform anomaly detection and prediction based on pre-trained features from language models. An overview of our system is shown in Fig. 3. Our system consists of two subsystems: Off-line Training and Runtime Inference. The Off-line Training subsystem focuses on log parsing, embedding extraction and anomaly detector training. The input to this subsystem are randomly sampled log data and a pre-trained language model (fastText or BERT). We generate a vector representation for each log message: from fastText embeddings, we aggregate the embeddings from the words in a sentence using tf-idf weighting; from the pre-trained BERT model, we take its final hidden state of the token [CLS] as the aggregate sequence representation. To model the system behavior over time, we group the log messages of every 10 s time window based on the logging timestamp, and average the vectors of logs within each time window to form the feature vector. We learn a Principal Component Analysis (PCA) [18] transformation matrix from feature vectors in training data collected when the system was running in normal condition. The output of this subsystem are trained log anomaly detection models, which are saved to the storage repository Data Lake. The Runtime Inference subsystem checks if an anomaly occurs for a given time window during runtime. The input to this subsystem are the trained models from the Off-line Training subsystem in the Data Lake, as well as new logs in a streaming fashion through Kafka [8], a unified, high-throughput, low-latency platform for handling real-time data streams. Our system will predict anomaly if the feature vector of a new time window is sufficiently different from the normal space constructed by PCA. The output of this subsystem are anomalies detected from the logs.

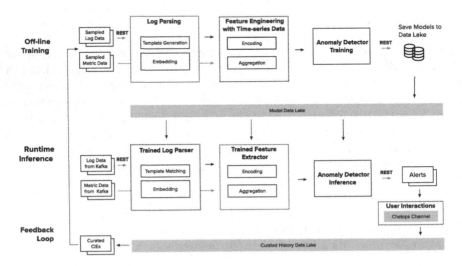

Fig. 3. An overview of our anomaly detection system built with embeddings from pre-trained language models.

Feedback on the anomaly detection results will be used to continuously improve the Off-line Training subsystem.

4.3   Evaluation

In our experiments, we explore the effects of three factors that may affect language model pre-training for IT Operations management: the type of language models (context-free or contextualized embeddings), the type of pretrained data (domain-specific data or general-purpose data), and the diversity of pre-trained data.

Datasets and Benchmarks. For pre-training the language models, we take 3 million sampled logs from 64 different applications: 48 cloud microservices in IBM Watson Assistant (WA) [1] (1 million logs), and 15 applications in Loghub collection [22] (2 million logs), ranging from distributed systems, supercomputers, operating systems to server applications. To test the accuracy of our trained anomaly detector models, we collect logs with ground truth from two IBM Watson Assistant microservices (denoted as *WA-1 and WA-2*) when the system was running in normal or abnormal condition, along with the HDFS ground truth data from Loghub. We compare the predictions with ground truth and compute the per-class accuracy as the percentage of correct predictions in the normal or abnormal test data set, respectively.

Accuracy Testing. We trained the following variants of anomaly detection models using different pre-trained features in our experiments:

- **Baseline**. Our baseline uses count vectors of log templates as feature vectors to build the model [19].
- **fastText-origin**. The model was built using features from the original pre-trained fastText embedding using general-purpose data [11].
- **fastText-wa**. The model was built using features from our fastText embedding pre-trained using IBM Watson Assistant data.
- **fastText-wa-loghub**. The model was built using features from our fastText embedding pre-trained using IBM Watson Assistant and Loghub data.
- **BERT-origin**. The model was built using features from the original pre-trained BERT using general-purpose data [5].
- **BERT-wa**. The model was built using features from our pre-trained BERT using IBM Watson Assistant data.
- **BERT-wa-loghub**. The model was built using features from our pre-trained BERT using IBM Watson Assistant and Loghub data.

In Table 1, we report the accuracy results of anomaly prediction on the benchmark datasets across various models. Firstly, we can see that the models trained with context-free embeddings consistently outperform our baseline as well as the models trained with contextualized embeddings. While our pre-trained fastText embedding using the WA data is comparable to the original general-purpose fast-Text embedding, we get a significant boost in accuracy of 8% with our pre-trained fastText embedding using both the WA data and the Loghub data, compared with the general-purpose fastText embeddings. This demonstrates the effectiveness of using diversified domain-specific data for pre-training embedding features to enhance the log anomaly detection task.

On the other hand, we can see that our pre-trained BERT embedding with the WA data is also comparable to the original general-purpose BERT embedding, though both are slightly worse than the baseline. Surprisingly, when we introduce more diverse data from both the WA data and the Loghub data to pre-train BERT, the accuracy drops considerably. One possible reason is that log data are much simpler than the general purpose text data such as Wikipedia articles or news articles, and the context in log data is not rich enough for BERT to learn during model pre-training. Besides, since log anomaly detection is essentially an unsupervised learning task, it is difficult to fine-tune pre-trained BERT models as in supervised learning tasks.

Overall, our experiments show that context-free embeddings are more robust and effective than contextualized embeddings to pre-train features for the log anomaly detection task, and it is even better when pre-training the embeddings with the IT Operations domain data of diverse applications.

Table 1. Accuracy results on log anomaly detection with pre-trained features from embeddings.

Model	HDFS		WA-1		WA-2	Average
	Normal	Abnormal	Normal	Abnormal	Normal	
Baseline	99.99%	44.9%	93.3%	66.7%	95%	80%
fastText-origin	98.8%	66%	93.3%	100%	98.3%	91%
fastText-wa	98.8%	59.7%	93.3%	100%	98.3%	90%
fastText-wa-loghub	99.6%	99.9%	96.7%	100%	98.3%	99%
BERT-origin	2%	99%	93.3%	100%	98.3%	79%
BERT-wa	97.1%	52.7%	93.3%	100%	95%	78%
BERT-wa-loghub	96.8%	47.5%	93.3%	100%	40%	67%

Performance Testing. We test the performance of our log anomaly prediction models built using language models as features in an IT Operations production environment. The production environment is a cluster with 2 CPUs and 4Gb memory. The test data consists of 10,000 randomly sampled logs from the WA data. We consider pre-trained BERT models of different numbers of layers, as well as the one-layer fastText model.

In Table 2, we report the total time (in seconds) and the average speed (the number of log lines per second). We observed that as the BERT model gets more complex with more layers, the average speed of embedding inference from the pre-trained model drops significantly. The fastText model is over 40 times faster than the one-layer BERT model since it is basically perform a lookup once the pre-trained embeddings are loaded into memory.

Table 2. Performance testing results on log anomaly detection with pre-trained features from embeddings.

	fastText	BERT		
	1 layer	1 layer	3 layers	6 layers
Total time	1.4 s	60 s	450 s	1700 s
Average speed	7000 lines/s	166 lines/s	22 lines/s	6 lines/s

5   Conclusion and Future Work

As IT complexity grows and the use of AI technologies expands, enterprises are looking to bring in the power of AI to transform how they develop, deploy and operate their IT. Pre-training a language model can accelerate the development of text-based AI models for optimizing IT Operations management tasks at a large scale. We investigate the effects of this language model pre-training approach for IT Operations management through a series of experiments on different language model types, data domains and data diversities. We present the empirical results on the prediction accuracy of our log anomaly prediction model and its run-time inference performance using language models as features in an IT Operations production environment. We show that the machine learning models built using context-free embeddings trained with diverse IT Operations domain data as features outperform those AI models built using language models with general-purpose data. Our pre-trained language models for IT Operations will be released soon. We hope that the insights gained from these experiments will help researchers and practitioners develop solutions and tools that enable better scalability, integration and management in the IT Operations domain. In the future, we will continue to explore the effects of pre-trained features using language models on different IT Operations management tasks such as fault localization and similar incident analysis. Besides, we plan to extend the studies to use more advanced language models such as GPT-3 [3].

References

1. I.W. Assistant (2020). https://www.ibm.com/cloud/watson-assistant/
2. Bojanowski, P., Grave, E., Joulin, A., Mikolov, T.: Enriching word vectors with subword information. Trans. Assoc. Comput. Linguist. **5**, 135–146 (2017)

3. Brown, T.B., et al.: Language models are few-shot learners. arXiv preprint arXiv:2005.14165 (2020)
4. CA-Technologies: The avoidable cost of downtime (2010)
5. Devlin, J., Chang, M.W., Lee, K., Toutanova, K.: Bert: pre-training of deep bidirectional transformers for language understanding. In: Proceedings of the 2019 Conference of the North American Chapter of the Association for Computational Linguistics: Human Language Technologies, Volume 1 (Long and Short Papers), pp. 4171–4186 (2019)
6. Gartner (2017). https://www.gartner.com/en/newsroom/press-releases/2017-04-11-gartner-says-algorithmic-it-operations-drives-digital-business
7. Howard, J., Ruder, S.: Universal language model fine-tuning for text classification. arXiv preprint arXiv:1801.06146 (2018)
8. Kafka (2020): https://kafka.apache.org/
9. Kudo, T., Richardson, J.: Sentencepiece: a simple and language independent subword tokenizer and detokenizer for neural text processing. arXiv preprint arXiv:1808.06226 (2018)
10. Meng, W., et al.: Loganomaly: Unsupervised detection of sequential and quantitative anomalies in unstructured logs. In: IJCAI, pp. 4739–4745 (2019)
11. Mikolov, T., Grave, E., Bojanowski, P., Puhrsch, C., Joulin, A.: Advances in pre-training distributed word representations. arXiv preprint arXiv:1712.09405 (2017)
12. Mikolov, T., Sutskever, I., Chen, K., Corrado, G.S., Dean, J.: Distributed representations of words and phrases and their compositionality. In: Advances in Neural Information Processing Systems, pp. 3111–3119 (2013)
13. Pennington, J., Socher, R., Manning, C.D.: Glove: global vectors for word representation. In: Proceedings of the 2014 Conference on Empirical Methods in Natural Language Processing (EMNLP), pp. 1532–1543 (2014)
14. Peters, M.E., Neumann, M., Iyyer, M., Gardner, M., Clark, C., Lee, K., Zettlemoyer, L.: Deep contextualized word representations. arXiv preprint arXiv:1802.05365 (2018)
15. Radford, A., Narasimhan, K., Salimans, T., Sutskever, I.: Improving language understanding by generative pre-training (2018). https://s3-us-west-2.amazonaws.com/openai-assets/researchcovers/languageunsupervised/languageunderstanding paper.pdf
16. Sarnovsky, M., Surma, J.: Predictive models for support of incident management process in it service management. Acta Electrotechnica et Informatica **18**(1), 57–62 (2018)
17. Vaswani, A., et al.: Attention is all you need. In: Advances in Neural Information Processing Systems, pp. 5998–6008 (2017)
18. Wold, S., Esbensen, K., Geladi, P.: Principal component analysis. Chemom. Intell. Lab. Syst. **2**(1–3), 37–52 (1987)
19. Xu, W., Huang, L., Fox, A., Patterson, D., Jordan, M.I.: Detecting large-scale system problems by mining console logs. In: Proceedings of the ACM SIGOPS 22nd symposium on Operating Systems Principles, pp. 117–132 (2009)
20. Zhang, Y., Rodrigues, K., Luo, Y., Stumm, M., Yuan, D.: The inflection point hypothesis: a principled debugging approach for locating the root cause of a failure. In: Proceedings of the 27th ACM Symposium on Operating Systems Principles, pp. 131–146 (2019)

21. Zhou, X., et al.: Latent error prediction and fault localization for microservice applications by learning from system trace logs. In: Proceedings of the 2019 27th ACM Joint Meeting on European Software Engineering Conference and Symposium on the Foundations of Software Engineering, pp. 683–694 (2019)
22. Zhu, J., et al.: Tools and benchmarks for automated log parsing. In: 2019 IEEE/ACM 41st International Conference on Software Engineering: Software Engineering in Practice (ICSE-SEIP), pp. 121–130. IEEE (2019)

Towards Runtime Verification via Event Stream Processing in Cloud Computing Infrastructures

Domenico Cotroneo, Luigi De Simone, Pietro Liguori$^{(\boxtimes)}$, Roberto Natella, and Angela Scibelli

DIETI, University of Naples Federico II, Naples, Italy
{cotroneo,luigi.desimone,pietro.liguori,roberto.natella}@unina.it,
ang.scibelli@studenti.unina.it

Abstract. Software bugs in cloud management systems often cause erratic behavior, hindering detection, and recovery of failures. As a consequence, the failures are not timely detected and notified, and can silently propagate through the system. To face these issues, we propose a lightweight approach to runtime verification, for monitoring and failure detection of cloud computing systems. We performed a preliminary evaluation of the proposed approach in the OpenStack cloud management platform, an "off-the-shelf" distributed system, showing that the approach can be applied with high failure detection coverage.

Keywords: Runtime verification · Runtime monitoring · Cloud computing systems · OpenStack · Fault injection

1 Introduction

Nowadays, the cloud infrastructures are considered a valuable opportunity for running services with high-reliability requirements, such as in the telecom and health-care domains [13,35]. Unfortunately, residual software bugs in cloud management systems can potentially lead to high-severity failures, such as prolonged outages and data losses. These failures are especially problematic when they are *silent*, i.e., not accompanied by any explicit failure notification, such as API error codes, or error entries in the logs. This behavior hinders the timely detection and recovery, lets the failures to silently propagate through the system, and makes the traceback of the root cause more difficult, and recovery actions more costly (e.g., reverting a database state) [11,12].

To face these issues, more powerful means are needed to identify these failures at runtime. A key technique in this field is represented by *runtime verification strategies*, which perform redundant, end-to-end checks (e.g., after service API calls) to assert whether the virtual resources are in a valid state. For example, these checks can be specified using temporal logic and synthesized in a runtime monitor [7,14,30,36], e.g., a logical predicate for a traditional OS can assert that

© Springer Nature Switzerland AG 2021
H. Hacid et al. (Eds.): ICSOC 2020 Workshops, LNCS 12632, pp. 162–175, 2021.
https://doi.org/10.1007/978-3-030-76352-7_19

a thread suspended on a semaphore leads to the activation of another thread [2]. Runtime verification is now a widely employed method, both in academia and industry, to achieve reliability and security properties in software systems [4]. This method complements classical exhaustive verification techniques (e.g., model checking, theorem proving, etc.) and testing.

In this work, we propose a lightweight approach to runtime verification tailored for the monitoring and analysis of cloud computing systems. We used a non-intrusive form of tracing of events in the system under test, and we build a set of lightweight monitoring rules from correct executions of the system in order to specify the desired system behavior. We synthesize the rules in a runtime monitor that verifies whether the system's behavior follows the desired one. Any runtime violation of the monitoring rules gives a timely notification to avoid undesired consequences, e.g., non-logged failures, non-fail-stop behavior, failure propagation across sub-systems, etc. Our approach does not require any knowledge about the internals of the system under test and it is especially suitable in the multi-tenant environments or when testers may not have a full and detailed understanding of the system. We investigated the feasibility of our approach in the OpenStack cloud management platform, showing that the approach can be easily applied in the context of an "off-the-shelf" distributed system. In order to preliminary evaluate the approach, we executed a campaign of fault-injection experiments in OpenStack. Our experiments show that the approach can be applied in a cloud computing platform with high failure detection coverage.

In the following of this paper, Sect. 2 discusses related work; Sect. 3 presents the approach; Sect. 4 presents the case study; Sect. 5 experimentally evaluates the approach; Sect. 6 concludes the paper.

2 Related Work

Promptly detecting failures at runtime is fundamental to stop failure propagation and mitigate its effects on the system. In this work, we exploit runtime verification to state the correctness of a system execution according to specific properties. In literature, some studies refer to runtime verification as runtime monitoring or dynamic analysis. Runtime monitoring consists of the observation of behaviors of the target system during its operation instead of verifying the system according to a specific model.

Over the last decades, several efforts have been spent on methodologies and tools for debugging and monitoring distributed systems. *Aguilera et al.* [1] proposed an approach to collect black-box network traces of communications between nodes. The objective was to infer causal paths of the requests by tracing call pairs and by analyzing correlations. Magpie [3] and Pinpoint [8] reconstruct causal paths by using a tracing mechanism to record events at the OS-level and the application server level. The tracing system tags the incoming requests with a unique *path identifier* and links resource usage throughout the system with that identifier. *Gu et al.* [21] proposes a methodology to extract knowledge on distributed system behavior of request processing without source code or prior knowledge. The authors construct the distributed system's component architecture in request processing and discover the heartbeat mechanisms of target distributed systems. Pip [31] is a system for automatically checking the behavior of a distributed system against programmer-written expectations about the system. Pip provides a domain-specific expectations language for writing declarative descriptions of the expected behavior of large distributed systems and relies on user-written annotations of the source code of the system to gather events and to propagate path identifiers across chains of requests. OSProfiler [25] provides a lightweight but powerful library used by fundamental components in OpenStack cloud computing platform [24]. OSProfiler provides annotation system that can be able to generate traces for requests flow (RPC and HTTP messages) between OpenStack subsystems. These traces can be extracted and used to build a tree of calls which can be valuable for debugging purposes. To use OSProfiler, it is required deep knowledge about OpenStack internals, making it hard to use in practice.

Research studies on runtime verification focused on formalisms for describing properties to be verified. Typically, a runtime verification system provides a Domain Specification Language (DSL) for the description of properties to be verified. The DSL can be a stand-alone language or embedded in an existing language. Specification languages for runtime verification can be regular, which includes temporal logic, regular expressions, and state machines, but also non-regular, which includes rule systems, stream languages.

In the runtime verification literature, there is an established set of approaches for the specification of temporal properties, which include Linear Temporal Logic (LTL) [28], Property Specification Patterns (PSP) [16], and Event Processing Language (EPL) [18]. Linear Temporal Logic is the most common family of specification languages. This approach supports logical and temporal operators. LTL is extensively used as specification language in many model checkers [6,9,22]. The Property Specification Patterns consist of a set of recurring temporal patterns. Several approaches use PSP and/or extend original patterns used in [5]. Event Processing Language is used to translate event patterns in queries that trigger event listeners whether the pattern is observed in the event stream of a Complex Event Processing (CEP) environment [33]. The most interesting characteristic of CEP systems is that can be used in *Stream-based Runtime Verification* or *Stream Runtime Verification* (SRV) tools. SRV is a declarative formalism to express monitors using streams; the specifications are used to delineate the dependencies between streams of observations of the target systems and the output of the monitoring process.

In [36], *Zhou et al.* propose a runtime verification based trace-oriented monitoring framework for cloud computing systems. The requirements of the monitoring can be specified by formal specification language, i.e. LTL, Finite State Machine (FSM). The tracing adopted in this approach is fine-grained, in which traces are a collection of events and relationships: every event records the details of one execution step in handling the user request (function name, duration), every relationship records the causal relation between two events. Using both the events and the relationships, it is possible to represent a trace into a so-called *trace tree*. In a trace tree, a node represents an event and an edge represents a relationship between events. This approach is generalizable at the cost of accessing the target source code to get the knowledge needed for instrumenting the code and gaining information about events relationships. However, this is not always the case, leading this approach difficult to exploit in practice. In [29], *Power and Kotonya* propose Complex Patterns of Failure (CPoF), an approach that provides reactive and proactive Fault-Tolerance (FT) via Complex Event Processing and Machine Learning for IoT (Internet of Things). Reactive-FT support is used to train Machine Learning models that proactively handle imminent future occurrences of known errors. Even if CPoF is intended for IoT systems, it inspired us in the use of Complex Event Processing to build the monitor.

The proposed approach presents several points of novelty compared to state-of-the-art studies and tools in runtime verification literature. In particular, the proposed methodology relies on *black-box tracing*, instead of regular tracing, avoiding knowing about system internals and the collection of information about the relationships between events (i.e., uncorrelated events). Further, we provide a new set of monitoring rules that well fit distributed systems and cloud computing infrastructure requirements, in which we need to face peculiar challenges like multi-tenancy, complex communication between subsystems, lack of knowledge of system internals. Based on the analysis of the events collected during system operation, we can specify the normal behavior of the target system and perform *online anomaly detection*.

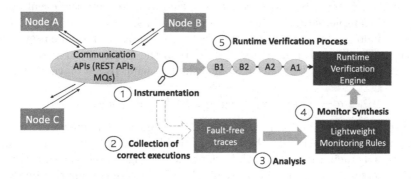

Fig. 1. Overview of the proposed approach.

3   Proposed Approach

Figure 1 shows an overview of the proposed approach. Firstly, we instrument the system under test to collect the events exchanged in the system during the experiments (step ①). Our instrumentation is a form of *black-box tracing* since we consider the distributed system as a set of black-box components interacting via public service interfaces. To instrument the system, we do not require any knowledge about the internals of the system under test, but only basic information about the communication APIs being used. This approach is especially suitable when testers may not have a full and detailed understanding of the entire cloud platform. Differently from traditional distributed system tracing [25], this lightweight form of tracing does not leverage any propagation of the event *IDs* to discriminate the events generated by different users or sessions.

In the step ②, we collect the *correct executions* of the system. To define its normal (i.e., correct) behavior, we exercise the system in "fault-free" conditions, i.e., without injecting any faults. Moreover, to take into account the variability of the system, we repeat several times the execution of the system, collecting different *"fault-free traces"*, one per each execution. We consider every fault-free trace a *past correct execution* of the system.

Step ③ analyzes the collected fault-free traces to define a set of *failure monitoring rules*. These rules encode the expected, correct behavior of the system, and detect a failure if a violation occurs. This step consists of two main operations. Firstly, the approach extracts only the attributes useful for expressing the monitoring rules (e.g., the name of the method, the name of the target system, the timestamp of the event, etc.). Then, we define the failure monitoring rules by extracting *"patterns"* in the event traces. We define a *"pattern"* as a recurring sequence of (not necessarily consecutive) events, repeated in every fault-free trace, and associated with an operation triggered by a workload. In this work, we identify patterns by manually inspecting the collected traces. In future work, we aim to develop algorithms to identify patterns using statistical analysis techniques, such as invariant analysis [17,20,34].

In general, we can express a monitoring rule by observing the events in the traces. For example, suppose there is an event of a specific type, say A, that is eventually followed by an event of a different type, say B, in the same user session (i.e., same ID). The term *event type* refers to all the events related to a specific API call. This rule can be translated into the following pseudo-formalism.

$$a \to b \; and \; id(a) = id(b), \quad with \; a \in A, \; b \in B \tag{1}$$

The rules can be applied in the multi-user scenario and concurrent systems as long as the information on the IDs is available. However, introducing an ID in distributed tracing requires both in-depth knowledge about the internals and intrusive instrumentation of the system. Therefore, to make our runtime verification approach easier to apply, we propose a set of coarse-grained monitoring rules (also known as *lightweight monitoring rules*) that do not require the use of any propagation ID. To apply the rules in a multi-user scenario, we define two different sets of events, A and B, as in the following.

$$A = \{all \; distinct \; events \; of \; type \; "A" \; happened \; in \; [t, \; t + \Delta]\}$$
$$B = \{all \; distinct \; events \; of \; type \; "B" \; happened \; in \; [t, \; t + \Delta]\} \tag{2}$$

with $|A| = |B| = n$. Our monitoring rule for the multi-user case then asserts that there should exist a binary relation R over A and B such that:

$$R = \{(a, b) \in A \times B \mid a \to b,$$

$$\nexists\, a_i, a_j \in A,\ b_k \in B \mid (a_i, b_k), (a_j, b_k), \qquad (3)$$

$$\nexists\, b_i, b_j \in B,\ a_k \in A \mid (a_k, b_i), (a_k, b_j)\ \}$$

with $i, j, k \in [1, n]$. That is, every event in A has an event in B that follows it, and every event a is paired with exactly one event b, and viceversa. These rules are based on the observation that, if a group of users performs concurrent operations on shared cloud infrastructure, then a specific number of events of type A is eventually followed by the same number of events of type B. The idea is inspired by the concept of flow conservation in network flow problems. Without using a propagation ID, it is not possible to define the couple of events a_i and b_i referred to the same session or the same user i, but it is possible to verify that the total number of events of type A is equal to the total number of events of type B in a pre-defined time window. We assume that the format of these rules can detect many of the failures that appear in cloud computing systems: if at least one of the rules is violated, then a failure occurred.

Finally, we synthesize a monitor from failure monitoring rules, expressed according to a specification language (step ④). The monitor takes as inputs the events related to the system under execution, and it checks, at runtime, whether the system's behavior follows the desired behavior specified in the monitoring rules (step ⑤). Any (runtime) violation of the defined rules alerts the system operator of the detection of a failure.

4   Case Study

In this paper, we investigated the feasibility of the proposed approach in the context of a large-scale, industry-applied case study. In particular, we applied the approach in the OpenStack project, which is the basis for many commercial cloud management products [26] and is widespread among public cloud providers and private users [27]. Moreover, OpenStack is a representative real-world large software system, which includes several sub-systems and orchestrates them to deliver rich cloud computing services. The most fundamental services of OpenStack [15,32] are (i) the **Nova** subsystem, which provides services for provisioning instances (VMs) and handling their life cycle; (ii) the **Cinder** subsystem, which provides services for managing block storage for instances; and (iii) the **Neutron** subsystem, which provides services for provisioning virtual networks for instances, including resources such as *floating IPs, ports* and *subnets*. Each subsystem includes several components (e.g., the Nova sub-system includes *nova-api, nova-compute*, etc.), which interact through message queues internally to OpenStack. The Nova, Cinder, and Neutron sub-systems provide external REST API interfaces to cloud users. To collect the messages (i.e., the events) exchanged in the system, we instrumented the *OSLO Messaging library*, which uses a message queue library and it is used for communication among OpenStack subsystems, and the *RESTful API libraries* of each OpenStack subsystem, which are used are used for communication between OpenStack and its clients. In total, we instrumented only 5 selected functions of these components (e.g., the `cast` method of OSLO to broadcast messages), by adding very simple annotations only at the beginning of these methods, for a total of 20 lines of code. We neither added any further instrumentation to the subsystems under test nor used any knowledge about OpenStack internals.

We collected one hundred correct executions by running the same workload in fault-free conditions. This workload configures a new virtual infrastructure from scratch, by stimulating all of the target subsystems (i.e., Nova, Neutron, and Cinder) in a balanced way. The workload creates VM instances, along with key pairs and a security group; attaches the instances to an existing volume; creates a virtual network consisting in a subnet and a virtual router; assigns a floating IP to connect the instances to the virtual network; reboots the instances, and then deletes them. We implemented this workload by reusing integration test cases from the *OpenStack Tempest* project [23], since these tests are already designed to trigger several subsystems and components of OpenStack and their virtual resources.

After the fault-free traces collection, we extract the information associated with every event within the trace. In particular, we record the time at which the communication API has been called and its duration, the component that invoked the API (*message sender*), and the remote service that has been requested through the API call (*called service*). Internally, the approach associates an *event name* to every collected event within a trace, so that two events of the same type are identified by the same name. In particular, we assign a unique name to every distinct pair *<message sender, called service>* (e.g., *<Cinder, attach volume>*).

4.1   Monitoring Rules

To determine the monitoring rules, we manually identified common patterns, in terms of events, in the fault-free traces. In particular, we determined a set of patterns for all the operations related to the workload execution (e.g., operations related to the instances, volumes, and networks). For example, the analysis of the events related to the attach of a volume to an instance pointed out common three different patterns in the fault-free traces. We derived failure monitoring rules for each pattern. Listing 1.1 shows three different monitoring rules, expressed in a pseudo-formalism, related to the `Volume Attachment` operation.

Listing 1.1. `Volume Attachment` monitoring rules

```
Rule#1:  event(name = "compute_reserve_block_device_name") is eventually
         followed by event( name = "compute_attach_volume")

Rule#2:  event( name = "compute_attach_volume") is eventually followed by
         event( name = "cinder-volume.localhost.␣
         localdomain@lvm_initialize_connection")

Rule#3:  Pattern of Rule#2 is eventually followed by event(name="cinder-volume
         .localhost.localdomain@lvm_attach_volume")
```

We derived the first rule by observing that, during the attachment of a volume, the `<compute, reserve_block_device_name>` event is always followed by the `<compute, attach_volume>` event in every fault-free trace. Indeed, to perform such an operation, the `reserve_block_device_name` method, a synchronous RPC call, is called before the `attach_volume` nova-compute API to get and reserve the next available device name. `Rule#2` follows the same structure of the `Rule#1`. `Rule#3` shows the possibility to write more complex rules, involving more than just two events.

In the same way, we derived rules for all further operations related to the volumes and the instances. Instead, the identification of the rules for network operations is different and more complex. Indeed, network operations are performed by the Neutron sub-system in an asynchronous way, such as by exchanging periodic and concurrent status polls among agents deployed in the datacenter and the Neutron server. This behavior leads to more non-deterministic variations of the events in the traces. Given the high source of non-determinism affecting the network operations, it is not possible to create rules based on event ordering. Therefore, to find these patterns, we observed the repetitions of the Neutron-associated events both in the fault-free and the faulty traces (i.e., with a fault injected in the Neutron subsystem). We found that, when the injected fault experiences a failure in the Neutron component, some network-related events occurred a much higher number than their occurrences in the fault-free traces.

For example, in several experiments targeting the Neutron component and that experienced a failure during the network operations, we found cases in which the event `<q-plugin, release_dhcp_port>` occurred more than 500 times. However, analyzing all the fault-free executions, this specific event occurred at most 3 times. Indeed, in such faulty experiments, the system repeatedly performed the same operation since it was unable to complete the request. Based on these observations, we defined the monitoring rules related to the network operations by checking, at runtime, if a specific event type occurred in a number higher than a threshold. Thus, for each event type, we defined a threshold as the higher number of times that the event type occurred during the fault-free executions. Figure 2 shows the logic adopted for the rules related to the network operations.

After identifying the rules, it is necessary to translate the rules in a particular specification language. We select EPL (*Event Processing Language*) as specification language. EPL is a formal language provided by the Esper software [18], that allows expressing different types of rules, such as temporal or statistical rules. It is a SQL-standard language with extensions, offering both typical SQL clauses (e.g., `select`, `from`, `where`) and additional clauses for event processing (e.g., `pattern`, `output`). In EPL, streams are the source of data and events are the basic unit of data. The typical SQL clause `insert into` is used to forward events to other streams for further downstream processing. We use the `insert into` clause for translating network operation rules using three interconnected statements, as shown in Listing 1.2.

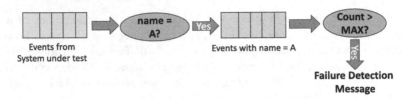

Fig. 2. Example of Neutron SSH failure monitoring rules.

Listing 1.2. EPL rules for the network operations

```
@name('S1') insert into EventNetworkStream select *
from Event where name='q-plugin_release_dhcp_port';
@name('S2') insert into countInfoStream select count(*) as count1 from
    EventNetworkStream;
@name('NetworkRule#1') select * from countInfoStream where count1 > maxEvent1
output when OutputTriggerVar1 = true then set OutputTriggerVar1 = false;
```

The first statement (S1) extracts <q-plugin, release_dhcp_port> events and forwards them to the stream NetworkEventStream. The second statement (S2) counts the number of events in NetworkEventStream and passes this information to the stream CountInfoStream. Finally, the third statement (NetworkRule#1) produces an output if the value in CountInfoStream is bigger than the maximum value. To avoid that the third statement outputs anytime it receives a new <q-plugin, release_dhcp_port> event after the first output, we use the OutputTriggerVar boolean variable, initialized to true and set to false after the first time the rule is verified.

The monitor synthesis is automatically performed once EPL rules are compiled. The Esper Runtime acts like a container for EPL statements which continuously executes the queries (expressed by the statements) against the data arriving as inputs. For more detailed information on Esper, we refer the reader to the official documentation [19].

4.2   Multi User Case

We applied the EPL statements, derived from the monitoring rules, also in the multi-user scenario. Since we do not collect an event ID, we use a *counter* to take into account multi-user operations. Indeed, we use the counter as an event ID to relate couples of events. We associate a different counter to each event type: when an event of a specific type occurs, we increment its counter. In particular, the translation of rules described in Listing 1.1 uses the clause of pattern, useful for finding time relationships between events. Pattern expressions usually consist of filter expressions combined with pattern operators. We use the pattern operators every, followed-by ($\rightarrow$), and timer:interval. The operator every defines that every time a pattern subexpression connected to this operator turns true, the Esper Runtime starts a new active subexpression. Without this operator, the subexpression stops after the first time it becomes true. The operator $\rightarrow$ operates on events order, establishing that the right-hand expression is evaluated only after that the left-hand expression turns true. The operator `timer:interval` establishes the duration of the time-window during which to observe the arriving events (it starts after that the left-hand expression turns true). The value of the counter is sent, along with the event name, to the Esper Runtime. Listing 1.3 shows the EPL translation of the rule **Rule#1** in the multi-user case.

Listing 1.3. EPL rule in the multi-user scenario

```
@name('Rule#1') select * from pattern [every a = Event(name="
    compute_reserve_block_device_name") -> (timer:interval(secondsToWait
    seconds) and not b=Event(name="compute_attach_volume", countEvent = a.
    countEvent))];
```

Every time the Esper Runtime observes an event <compute,reserve_ block_device_name> with its counter value, it waits for the receive of the event <compute, attach_volume> with the same counter value within a time window of `secondsToWait` seconds. If this condition is not verified, the approach generates a failure detection message.

5   Preliminary Experiments

To preliminary evaluate our approach, we performed a campaign of fault injection experiments in the OpenStack platform. In our experiments, we targeted OpenStack version 3.12.1 (release *Pike*), deployed on Intel Xeon servers (E5-2630L v3 @ 1.80 GHz) with 16 GB RAM, 150 GB of disk storage, and Linux CentOS v7.0, connected through a Gigabit Ethernet LAN. In particular, our tool [10] injected the following fault types:

- **Throw exception**: An exception is raised on a method call, according to pre-defined, per-API list of exceptions;
- **Wrong return value**: A method returns an incorrect value. In particular, the returned value is corrupted according to its data type (e.g., we replace an object reference with a null reference, or replace an integer value with a negative one);
- **Wrong parameter value**: A method is called with an incorrect input parameter. Input parameters are corrupted according to the data type, as for the previous fault type;
- **Delay**: A method is blocked for a long time before returning a result to the caller. This fault can trigger timeout mechanisms inside OpenStack or can cause a stall.

Before every experiment, we clean-up any potential residual effect from the previous experiment, in order to ensure that the potential failure is only due to the current injected fault. To this end, we re-deploy the cloud management system, remove all temporary files and processes, and restore the OpenStack database to its initial state.

In-between calls to service APIs, our workload generator performs *assertion checks* on the status of the virtual resources, in order to reveal failures of the cloud management system. In particular, these checks assess the connectivity of the instances through SSH and query the OpenStack API to ensure that the status of the instances, volumes, and the network is consistent with the expectation of the tests. In the context of our methodology, assertion checks serve as *ground truth* about the occurrence of failures during the experiments (i.e., a reference for evaluating the accuracy of the proposed approach).

We evaluated our approach in terms of the *failure detection coverage* (FDC), defined as the number of experiments identified as failed over the total number of experiments that experienced a failure. We focused only on the experiments that experienced a failure, for a total of 481 faulty traces, one per each fault-injection experiment. We define an experiment as failed if at least one API call returns an error (**API error**) or if there is at least one assertion check failure (**assertion check failure**). Also, to evaluate the most interesting cases, we focused on the experiments in which the target system was not able to timely notify the failure (i.e., failure notified with a long delay or not notified at all), as described in our previous work [12].

The coverage provided by our runtime verification approach is compared with the coverage provided by OpenStack API Errors by design. API Errors notifies the users that the system is not able to perform a request, thus they work as a failure detection mechanism. Table 1 shows the FDC of both approaches considering different failure cases (related to different operations). The results show that our approach is able to identify a failure in the 79.38% of the failures, showing significantly better performance of the OpenStack failure coverage mechanism. In particular, the table highlights how our rules are able to identify failures that were never notified by the system (Instance Creation and SSH Connection). The RV approach shows lower performance only in the Volume Creation case failure: this suggests the need to add further monitoring rules or to improve the existing ones for this specific case.

Table 1. Comparison with API errors coverage

Failure case	OpenStack FDC %	RV FDC %
Volume Creation	29.67	28.57
Volume Attachment	25.33	92.00
Volume Deletion	100	100
Instance Creation	0.00	90.96
SSH Connection	0.00	38.46
Total	**23.96**	**79.38**

We evaluated our approach also in a simulated multi-user scenario. To simulate concurrent requests, 10 traces (5 fault-free and 5 faulty) are "mixed-together" by alternating the events of all the traces but without changing the

Table 2. Average FDC in the multi-user scenario

Failure case	Avg FDC %
Volume Creation	32.00 ∓ 12.42
Volume Attachment	45.33 ∓ 13.82
Volume Deletion	36.00 ∓ 12.20
Total	37.78 ∓ 13.88

relative order of the events within every single trace. The faulty traces are related to the same failure type (e.g., Volume Creation). For each failure type, we performed the analysis 30 times by randomly choosing both the fault-free and the faulty traces. Table 2 shows the average FDC and the standard deviation of our monitoring rules for all the failure volume cases. The preliminary results can be considered promising. However, the high standard deviation indicates that the average FDC is very sensitive to the randomity of the analyzed traces.

6   Conclusion and Future Work

In this paper, we propose an approach to runtime verification via stream processing in cloud computing infrastructures. We applied the proposed approach in the context of the OpenStack cloud computing platform, showing the feasibility of the approach in a large and complex "off-the-shelf" distributed system. We performed a preliminary evaluation of the approach in the context of the fault-injection experiments. The approach shows promising results, both in the single-user and simulated multi-user cases.

Future work includes the development of algorithms able to automatically identify patterns using statistical analysis techniques, such as invariant analysis. We also aim to conduct fault-injection campaigns by using a multi-tenant workload in order to perform an evaluation in a real multi-user scenario and to analyze the overhead introduced by the approach.

Acknowledgements. This work has been supported by the COSMIC project, U-GOV 000010–PRD-2017-S-RUSSO_001_001.

References

1. Aguilera, M.K., Mogul, J.C., Wiener, J.L., Reynolds, P., Muthitacharoen, A.: Performance debugging for distributed systems of black boxes. ACM SIGOPS Oper. Syst. Rev. **37**(5), 74–89 (2003)
2. Arlat, J., Fabre, J.C., Rodríguez, M.: Dependability of cots microkernel-based systems. IEEE Trans. Comput. **51**(2), 138–163 (2002)
3. Barham, P., Isaacs, R., Mortier, R., Narayanan, D.: Magpie: online modelling and performance-aware systems. In: Proceedings of the HotOS, pp. 85–90 (2003)

4. Bartocci, E., Falcone, Y.: Lectures on Runtime Verification: Introductory and Advanced Topics, vol. 10457. Springer, Cham (2018). https://doi.org/10.1007/978-3-319-75632-5
5. Bianculli, D., Ghezzi, C., Pautasso, C., Senti, P.: Specification patterns from research to industry: a case study in service-based applications. In: Proceedings of the ICSE, pp. 968–976. IEEE (2012)
6. Blom, S., van de Pol, J., Weber, M.: LTSMIN: distributed and symbolic reachability. In: Touili, T., Cook, B., Jackson, P. (eds.) CAV 2010. LNCS, vol. 6174, pp. 354–359. Springer, Heidelberg (2010). https://doi.org/10.1007/978-3-642-14295-6_31
7. Chen, F., Roşu, G.: Mop: an efficient and generic runtime verification framework. In: Proceedings of the 22nd Annual ACM SIGPLAN Conference on Object-Oriented Programming Systems and Applications, pp. 569–588 (2007)
8. Chen, Y.Y.M., Accardi, A.J., Kiciman, E., Patterson, D.A., Fox, A., Brewer, E.A.: Path-based failure and evolution management. In: Proceedings of the NSDI, pp. 309–322 (2004)
9. Cimatti, A., et al.: NuSMV 2: an OpenSource tool for symbolic model checking. In: Brinksma, E., Larsen, K.G. (eds.) CAV 2002. LNCS, vol. 2404, pp. 359–364. Springer, Heidelberg (2002). https://doi.org/10.1007/3-540-45657-0_29
10. Cotroneo, D., De Simone, L., Liguori, P., Natella, R.: Profipy: programmable software fault injection as-a-service. In: 2020 50th Annual IEEE/IFIP International Conference on Dependable Systems and Networks (DSN), pp. 364–372 (2020)
11. Cotroneo, D., De Simone, L., Liguori, P., Natella, R., Bidokhti, N.: Enhancing failure propagation analysis in cloud computing systems. In: 2019 IEEE 30th International Symposium on Software Reliability Engineering (ISSRE), pp. 139–150. IEEE (2019)
12. Cotroneo, D., De Simone, L., Liguori, P., Natella, R., Bidokhti, N.: How bad can a bug get? an empirical analysis of software failures in the openstack cloud computing platform. In: Proceedings of the ESEC/FSE, pp. 200–211 (2019)
13. Dang, L.M., Piran, M., Han, D., Min, K., Moon, H., et al.: A survey on internet of things and cloud computing for healthcare. Electronics 8(7), 768 (2019)
14. Delgado, N., Gates, A.Q., Roach, S.: A taxonomy and catalog of runtime software-fault monitoring tools. IEEE Trans. Software Eng. 30(12), 859–872 (2004)
15. Denton, J.: Learning OpenStack Networking. Packt Publishing Ltd. (2015)
16. Dwyer, M.B., Avrunin, G.S., Corbett, J.C.: Patterns in property specifications for finite-state verification. In: Proceedings of the ICSE, pp. 411–420 (1999)
17. Ernst, M.D., et al.: The daikon system for dynamic detection of likely invariants. Sci. Comput. Program. 69(1–3), 35–45 (2007)
18. EsperTech: ESPER HomePage (2020). http://www.espertech.com/esper
19. EsperTech: Esper Reference (2020). http://esper.espertech.com/release-8.5.0/reference-esper/html_single/index.html
20. Grant, S., Cech, H., Beschastnikh, I.: Inferring and asserting distributed system invariants. In: Proceedings of the ICSE, pp. 1149–1159 (2018)
21. Gu, J., Wang, L., Yang, Y., Li, Y.: Kerep: experience in extracting knowledge on distributed system behavior through request execution path. In: Proceedings of the ISSREW, pp. 30–35. IEEE (2018)
22. Holzmann, G.J.: The model checker spin. IEEE Trans. Software Eng. 23(5), 279–295 (1997)
23. OpenStack: Tempest Testing Project (2018). https://docs.openstack.org/tempest
24. OpenStack: OpenStack HomePage (2020). https://www.openstack.org/
25. OpenStack: OSProfiler HomePage (2020). https://github.com/openstack/osprofiler

26. OpenStack project: The OpenStack marketplace (2018). https://www.openstack.org/marketplace/distros/
27. OpenStack project: User stories showing how the world #RunsOnOpenStack (2018). https://www.openstack.org/user-stories/
28. Pnueli, A.: The temporal logic of programs. In: Proceedings of the SFCS, pp. 46–57. IEEE (1977)
29. Power, A., Kotonya, G.: Providing fault tolerance via complex event processing and machine learning for IoT systems. In: Proceedings of the IoT, pp. 1–7 (2019)
30. Rabiser, R., Guinea, S., Vierhauser, M., Baresi, L., Grünbacher, P.: A comparison framework for runtime monitoring approaches. J. Syst. Software **125**, 309–321 (2017)
31. Reynolds, P., Killian, C.E., Wiener, J.L., Mogul, J.C., Shah, M.A., Vahdat, A.: PIP: detecting the unexpected in distributed systems. Proc. NSDI. **6**, 9 (2006)
32. Solberg, M.: OpenStack for Architects. Packt Publishing (2017)
33. Wu, E., Diao, Y., Rizvi, S.: High-performance complex event processing over streams. In: Proceedings of the SIGMOD/PODS, pp. 407–418 (2006)
34. Yabandeh, M., Anand, A., Canini, M., Kostic, D.: Finding almost-invariants in distributed systems. In: Proceedings of the SRDS, pp. 177–182. IEEE (2011)
35. Yin, Z., Yu, F.R., Bu, S., Han, Z.: Joint cloud and wireless networks operations in mobile cloud computing environments with telecom operator cloud. IEEE Trans. Wirel. Commun **14**(7), 4020–4033 (2015)
36. Zhou, J., Chen, Z., Wang, J., Zheng, Z., Dong, W.: A runtime verification based trace-oriented monitoring framework for cloud systems. In: Proceedings of the ISS-REW, pp. 152–155. IEEE (2014)

Decentralized Federated Learning Preserves Model and Data Privacy

Thorsten Wittkopp$^{(\boxtimes)}$ and Alexander Acker$^{(\boxtimes)}$

Technische Universität Berlin, Berlin, Germany
{t.wittkopp,alexander.acker}@tu-berlin.de

Abstract. The increasing complexity of IT systems requires solutions, that support operations in case of failure. Therefore, Artificial Intelligence for System Operations (AIOps) is a field of research that is becoming increasingly focused, both in academia and industry. One of the major issues of this area is the lack of access to adequately labeled data, which is majorly due to legal protection regulations or industrial confidentiality. Methods to mitigate this stir from the area of federated learning, whereby no direct access to training data is required. Original approaches utilize a central instance to perform the model synchronization by periodical aggregation of all model parameters. However, there are many scenarios where trained models cannot be published since its either confidential knowledge or training data could be reconstructed from them. Furthermore the central instance needs to be trusted and is a single point of failure. As a solution, we propose a fully decentralized approach, which allows to share knowledge between trained models. Neither original training data nor model parameters need to be transmitted. The concept relies on teacher and student roles that are assigned to the models, whereby students are trained on the output of their teachers via synthetically generated input data. We conduct a case study on log anomaly detection. The results show that an untrained student model, trained on the teachers output reaches comparable F1-scores as the teacher. In addition, we demonstrate that our method allows the synchronization of several models trained on different distinct training data subsets.

Keywords: AIOps · Federated learning · Knowledge representation · Anomaly detection · Transfer learning

1 Introduction

IT systems are expanding rapidly to satisfy the increasing demand for a variety of applications and services in areas such as content streaming, cloud computing or distributed storage. This entails an increasing number of interconnected devices, large networks and growing data centres to provide the required infrastructure [1]. Additionally, awareness for data privacy and confidentiality is rising

---

T. Wittkopp and A. Acker—Equal contribution.

especially in commercial industry. Big- and middle-sized companies are relying on private cloud, network and storage providers to deploy and maintain according solutions. Except for severe problems that require local access, the operation and maintenance is done remotely. Remote access together with the growing system complexity puts extreme pressure on human operators especially when problems occur. To maintain control and comply with defined service level agreements, operators are in need of assistance. Therefore, monitoring solutions are combined with methods from machine learning (ML) and artificial intelligence to support the operation and maintenance of those systems - usually referred to as AIOps. Examples of concrete solutions are early detection of system anomalies [2,3], root cause analysis [4], recommendation and automated remediation [5].

The majority of ML and AI methods relies on training data. In case of anomaly detection a common approach is to collect monitoring data such as logs, traces or metrics during normal system operation and utilize them to train models. Representing the normal system state, these models are utilized to detect deviations from the learned representation which are labeled as anomalies. Therefore, AIOps systems require preliminary training phases to adjust to the target environment until they can be utilized for detection. This is known as cold start problem. Although adjusted to customer requirements, deployed systems at different sites are very similar (e.g. private cloud solutions based on OpenStack, storage systems based on HDFS or network orchestration via ONAP). An obvious mitigation of the cold start problem would be to use training data from existing sites to train models and fine tune them after deployment within a target customer site. Furthermore, training data used from a variety of sites increases the holisticity of models allowing them to perform generally better. However, sharing data or model parameters, even if indirectly related with company business cases, is usually not possible due to confidentiality or legal restrictions [6].

Federated learning as a special form of distributed learning is gaining increased attention since it allows access to a variety of locally available training data and aims to preserve data privacy [6,7]. Utilizing this concept, we propose a method that allows different deployments of the same system to synchronize their anomaly detection models without exchanging training data or model parameters. It does not require a central instance for model aggregation and thus, improves scalability. We introduce a concept of student and teacher roles for models whereby student models are learning from teachers. As input, vectors that are randomly generated within a constrained value range are used as input to both, student and teacher models. Student models are trained on the output of teachers. We conduct a case study based on log anomaly detection for the Hadoop File System (HDFS). In a first experiment it is shown that our solution can mitigate the cold start problem. A second experiment reveals that the proposed method can be utilized to holistically train distributed models. Models that were trained by our method achieve comparable results to a model that was trained on the complete training dataset.

The rest of this paper is structured as follows. First, Sect. 2 gives an overview of federated learning and applied federated learning in the field of AIOps. Second, Sect. 3 describes the our proposed method together with relevant preliminaries. Third, Sect. 4 presents the conducted case study and experiment results. Finally, Sect. 5 concludes our work.

2 Related Work

Federated learning is a form of distributed machine learning method. Thereby, model training is done locally within the environment of the data owner without sending training data to a central server. Locality is defined within the boundaries of the data owner's private IT environment. Initially this concept was proposed by McMahan et al. [8]. Instead of training data, either model weights or gradients from clients are sent to a central instance which aggregated them to a holistic model during model training. Updated models are sent back to the clients. Yang et al. [9] provide a categorization, which are vertical federated learning, horizontal federated learning and federated transfer learning. Despite preventing direct data exchange, publications revealed possibilities to restore training data from constantly transmitted weights or gradients, which violates the data privacy requirement [10]. For an adversary it is possible to recover the training dataset by using a model inversion attack [11] or determine whether a sample is part of the training dataset by using a membership inference attack [12]. Since reconstruction of original training data is possible when observing model changes [11,12] different privacy preserving methods are introduced. These are focused on obfuscation of input data [13] or model prediction output [14]. Geyer et al. [15] applied differential privacy preserving on client side to realize privacy protection. Shokri and Shmatikov [6] select small subsets of model parameters to be shared in order to prevent data reconstruction. However, model parameters or gradients still need to be shared with a central instance. Furthermore, the requirement of a central instance is a major bottleneck for scalability [16].

Application of federated learning in the field of AIOps is mainly focused around anomaly [17–19] and intrusion detection [20,21]. Liu et al. [18,19] propose a deep time series anomaly detection model which is trained locally on IoT devices via federated learning. Although not directly applied on an AIOps related problem, the proposed method could be applied to perform anomaly detection on time series like CPU utilization or network traffic statistics of the device itself. Nguyen et al. [17] propose their system DÏoT for detecting compromised devices in IoT networks. It utilizes an automated technique for device-type specific anomaly detection. The unsupervised training is done via federated learning individually on each device in the IoT environment. Preuveneers et al. [20] develop an intrusion detection system based on autoencoder models. The model parameter exchange is coupled with a permissioned blockchain to provide integrity and prevent adversaries to alter the distributed training process.

3   Decentralized Federated Learning

In this chapter we present our method for decentralized federated learning that aims at preservation of model and data privacy. Thereby, models that were trained on individual and partly distinct training sets are synchronized. Beside preserving data privacy, the novelty of our method is the dispensability of model parameter sharing. We illustrate the entire process of local training and global knowledge distribution. To realize latter, the communication process between a set of entities is described.

3.1   Problem Definition and Preliminaries

To apply our proposed federated learning method we assume the following setup. Let $\Phi = \{\phi_1, \phi_2, \ldots\}$ be a set of models and $E = \{e_1, e_2, \ldots\}$ a set of environments. We define a model deployed in a certain environment as a tuple (ϕ_i, e_j). All models that perform the same task T in their environment are combined into a set of workers $W_T = \{(\phi_i, e_j)\}$. Each model ϕ_i has access to locally available training data but cannot directly access training data from other environments. Furthermore, neither gradients nor model parameters can be shared outside of their environment. Having a function $P(T, \phi_i, X_{e_j})$ that measures how well a model ϕ_i is performing the task T on data X_{e_j} from environment e_j, the goal is to synchronize the model training in a way that all models can perform well on data from all environments:

$$P(T, \phi_1, X_{e_1}) \approx P(T, \phi_1, X_{e_2}) \approx \ldots \approx P(T\phi_2, X_{e_1}) \approx P(T, \phi_2, X_{e_2}) \approx \ldots \quad (1)$$

Each model ϕ_i is defined as a transformation function $\phi : X^{d_1} \to Y^{d_2}$ of a given input $x \in X^{d_1}$ into an output $y \in Y^{d_2}$, where d_1 and d_2 are the corresponding dimensions. Since no original training data can be shared between environments, we define an input data range $\tilde{X}$. It allows to draw data samples $\tilde{x} \sim \tilde{X}^{d_1}$ that are restricted to the range of the original training data but otherwise are not related to samples of the original training data set X^{d_1}. Models can adopt the role of teachers $\phi_i^{(t)}$ and students $\phi_i^{(s)}$. Student models are directly trained on the output of teachers. We refer to a training set that is generated by a teacher as knowledge representation, formally defined as a set of tuples:

$$r = \{(\tilde{x}, \gamma(\phi_i^{(t)}(\tilde{x})) : \tilde{x} \sim \tilde{X}^{d_1}\}. \quad (2)$$

Thereby γ is a transformation function that is applied on the teacher model output. Student models $\phi_i^{(s)}$ are trained on the knowledge representations of teachers. The objective is to minimize the loss between the output of teacher and student models

$$\underset{\theta_j^{(s)}}{\arg\min} \, \mathcal{L}(\gamma(\phi_i^{(t)}(\tilde{x})), \phi_j^{(s)}(\tilde{x})). \quad (3)$$

where $\theta_j^{(s)}$ are parameters of the student model $\phi_j^{(s)}(\tilde{x})$.

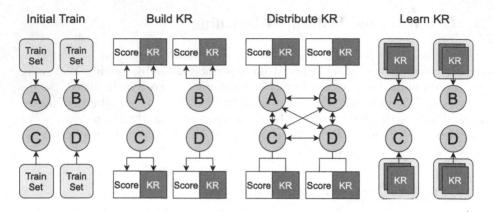

Fig. 1. The process of multi-cluster learning

3.2   The Concept of Teachers and Students

For the training process itself we introduce the teacher student concept. Every model can adopt the role of a student or teacher. A teacher model trains student models by providing a knowledge representation. First, we assume a set of models, each performing the same task in their own local environment. These models are trained on the same objective but only with the locally available training data. To prevent the sharing of training data or model parameter between environments, models are adapting roles of teachers to train student models within other environments. Overall, this process is realized by four steps: (1) Initial Train, (2) adapt teacher role and build knowledge representation, (3) distribute knowledge representation and (4) adapt student role and train on teacher knowledge representation. Figure 1 visualizes these steps the overall layout of each phase. The example shows four environments A-D with locally available training sets. Initially, all models are trained on the locally available training data. After that, models adapt the role of teachers and respectively generate the knowledge representations. Thereby, auxiliary input data are generated from the value range of locally available training data. This range must be synchronized across all environments. Otherwise, there is no connection to original local training data that was used to train models within their environments. Additionally, a score is calculated that reflects how well a model is performing on the task. Next, the knowledge representations are distributed. After receiving a knowledge representation, a model checks the attached score and compares it with its local score. Representations with lower scores are dropped. When receiving higher or equal scores, the models adapts the role of a student and is trained on the received knowledge representation. Through this process each model will be retrained and updated. Note that the loss function used during the initial training can differ from the loss function used for knowledge representation learning.

3.3 Loss Function

During training of student models the objective is to directly learn the transformed outputs of a teacher for a given input. We utilize the *tanh* as the transformation function to restrict teacher model outputs to the range $[-1, 1]$. This restriction should reduce the output value range and thus, stabilize the training process and accelerate convergence. Therefore, the student needs a loss function that can minimize the loss for every element from it's own output against the output vector of the teacher. This requires a regressive loss function. We utilize the smooth L1 loss which calculates the absolute element-wise distance. If this distance falls below 1 additionally a square operation is applied on it. It is less sensitive to outliers than the mean square error loss and prevents exploding gradients.

4 Evaluation

The evaluation of our method is done on the case study of log anomaly detection by utilizing a LSTM based neural network, called DeepLog [2]. DeepLog is trained on the task of predicting the next log based on a sequence of preceding log entries. However our decentralized federated learning method can be applied to any other machine learning model that is trainable via gradient descent. We utilize the labeled HDFS data set, which is a log data set collected from the Hadoop distributed file system deployed on a cluster of 203 nodes within the Amazon EC2 platform [22]. This data set consists of 11,175,629 log lines that form a total of 570,204 labeled log sequences. However, raw log entries are highly variant which entails a large number of prediction options. Thus, a preprocessing is applied to reduce the number of possible prediction targets. Log messages are transformed into templates consisting of a constant and variable part. The constant templates are used as discrete input and prediction target. The task of template parsing is executed by Drain [23]. We refer to the set of templates as $\mathbb{T}$.

4.1 Auxiliary Sample Generation

As described, student models are trained on auxiliary samples together with teacher model outputs. These samples are drawn from a restricted range but otherwise independent from the training data that was used to train the teacher. In the conducted use case study of log anomaly detection auxiliary samples are randomly generated as follows. Having $\mathbb{T}$ as the set of unique templates, an auxiliary sample is defined as $\tilde{x} = (t_i \sim \mathbb{T} : i = 1, 2, \dots, w + 1)$, where w is the input sequence length. Template t_{w+1} will be used as the prediction target. Note that templates are randomly sampled from the unique template set $\mathbb{T}$. Thus, auxiliary input samples for DeepLog are random template sequences.

4.2   Experiment 1: Training of an Untrained Model

In our first experiment we investigate the ability of the proposed method to mitigate the cold-start problem. Therefore, a DeepLog model is trained on the HDFS
data. Out of the 570,204 labeled log sequences of the HDFS data set, we use the
first 4,855 normal sequences as training and the remaining 565,349 as test set.
The test set contains 16,838 anomalies and 553,366 normal sequences. The two
hyper-parameters of DeepLog were set as follow: $w = 10$ and $k = 9$, where w is
the window size that determines the input sequence length and is required to generate auxiliary samples. Further we use cross-entropy loss to learn a distribution
when predicting possible next log lines. The teacher model uses a batch size of 16
and we trained it over 20 epochs on all 4,855 normal log sequences. The therewith
trained DeepLog model is utilized as the teacher while a completely untrained
model with the same architecture and parametrization adopts the role of a stu-
dent. The teacher performs following transformation: $\phi : X^{w \times |\mathbb{T}|} \rightarrow Y^{|\mathbb{T}| \times \mathbb{R}^{[-1,1]}}$
to generate the knowledge representation. It takes a sequence with w one-hot
encoded templates and outputs a *tanh* transformed probability distribution over
all w templates. Different amounts of knowledge representation tuples are tested:
$\{10, 50, 100, 500, 1000, 5000\}$. The student model uses a batch size of 16 and we
trained it over 10 epochs for every knowledge representation size. Auxiliary input
samples are generated as described in Sect. 4.1. Due to this randomness of sample generation, we repeat the experiment five times. Figure 2 shows the results as
a boxplot, which illustrated the F1-scores for the teacher and students. Bottom
and top whiskers reflect the minimum and maximum non-outlier values. The
line in the middle of the box represents the median. The box boundaries are
the first quartile and third quartile of the value distribution. The most left bar
shows the F1-score for the teacher model, which is 0.965. This bar is a flat line,
because it represents a single value. Remaining bars are visualizing the F1-score
for each knowledge representation size, from 10 to 5,000. First, it can be observed
that a knowledge representation size of at least 100 is required to have a decent
F1-score on the student model. As expected, the score of the student increases
with the number of used knowledge representation tuples. Due to the random
sampling F1-scored of student models trained with 100 and 500 knowledge representation tuples underlie high uncertainty. The 0.95 confidence interval ranges
from 0.153 to 0.558 for size 100 and from 0.313 to 0.805 for size 500. The student
model's F1-score becomes increasingly stable with higher knowledge representation sizes and reaches 0.961 within the 0.95 confidence interval of $[0.943, 0.958]$
for size 5000. Compared to the teacher model's F1-score of 0.965 we conclude
that our method can be utilized to mitigate the cold-start problem by training
an untrained model via knowledge representations of trained models.

4.3   Experiment 2: Federated Learning

In this experiment we investigate how multiple DeepLog models behave as teachers and students. It allows to train distributed models on locally available training data and subsequently synchronize the knowledge of models. Therefore, we

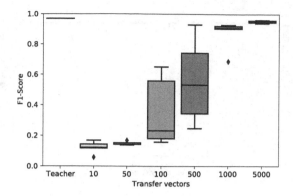

Fig. 2. Comparison between F1-scores of teacher model and student models that were trained on different knowledge representation sizes.

simulate 8 distributed HDFS systems by creating a set of unique sequences. Out of the 570,204 labeled log sequences of the HDFS data set, we again use the first 4,855 normal sequences to generate unique sequences of size w. This results in 4,092 unique training samples. These 4,092 training samples were evenly and randomly split into 8 sets, hence every set contains 511 training samples. Therefore, 8 DeepLog models are respectively trained. The two hyper-parameters of each DeepLog were set as follow: $w = 10$ and $k = 9$. To evaluate the model performance on predicting potentially unseen samples, remaining 565,349 sequences are used as a joint test set.

We initially trained the 8 DeepLog models over 10 epochs with a batch size of 16 and cross-entropy-loss as a loss function. The therewith trained DeepLog model are utilized as teacher models for each other. Hence, all 8 nodes are also students and where trained with the knowledge representation from the teachers with a batch size of 1 over 5 epochs. Again, we test different amounts of knowledge representation per model: $\{10, 50, 100, 500, 1000\}$. Note that a student model is trained on the representations of all teacher models that have a higher or equal score than itself.

Auxiliary input samples are generated as described in Sect. 4.1. Due to this random generation of auxiliary samples, we repeat the experiment 7 times. Figure 3 shows the results as a boxplot, which illustrates the F1-scores of all 8 models (marked as A-H) for different amounts of knowledge representation tuples over 7 experiment executions. The properties of the boxplot are the same as described in Sect. 4.2. The most left bars in the category *Before* for each model after initially trained on the locally available unique sequence set. No federated learning is applied here. It can be seen that their F1-scores ranging from 0.520 to 0.938. The 0.95 confidence interval for all 8 nodes in this section ranges from 0.875 to 0.901.

The Fig. 4 shows a zoomed-in version of the same boxplot on the F1-score range of between 0.90 and 0.94. As expected, an increasing amount of knowledge representations leads to overall higher F1-scores. Furthermore, it is visible that

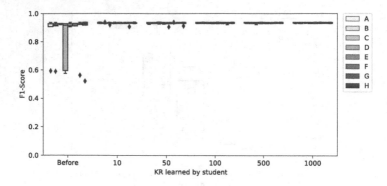

Fig. 3. Comparison of F1-scores of each node before and after the teacher student process with different knowledge representation sizes.

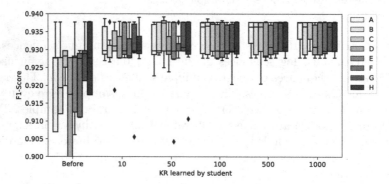

Fig. 4. Comparison of F1-scores between 0.90 and 0.94 of each node before and after the teacher student process with different knowledge representation sizes.

already 10 knowledge representations are improving the F1-scores significantly. After 10 knowledge representation the lowest F1-score is 0.906 and the 0.95 confidence interval for this section ranges from 0.923 to 0.933. In the category of 50 knowledge representations, the number of existing outliers is the same, but the 0.95 confidence interval ranges from 0.930 to 0.933, which is an improvement compared to 10 knowledge representations. Also the highest F1-score of 0.939 could be observed in this category. The lowest deviations of the F1-score for all nodes occur at 1000 knowledge representations with a 0.95 confidence interval from 0.933 to 0.935. In this category the highest F1-score is 0.938 and the lowest 0.928. Compared to the 0.95 confidence interval of 0.875 to 0.901 before the teacher student process, we observe an improvement in every category.

This experiment indicates, that it is able to train different models with the same configuration in a distributed system. The method preserves data privacy by using auxiliary samples to transfer knowledge between models. A central instance for synchronization of the training process is not required. Neither model parameters nor gradients need to be transferred.

5   Conclusion

In this work we proposed a federated learning solution for synchronizing distributed models trained on locally available data. Out method does not required a sharing of original training data or model parameter. Training is done via assignment of teacher and student roles to existing models. Students are trained on the output of teachers via auxiliary samples and respective outputs, referred to as knowledge representations. We evaluated our approach in a case study of log anomaly detection. DeepLog models were trained on distinct and unique log sequences from the HDFS data set. After that the teacher and student roles were applied to the models in order to test the ability of synchronizing them. In our first experiment we could show that this approach can mitigate the cold start problem. For this experiment we setup a trained teacher DeepLog model and an untrained DeepLog model as a student. We investigated how well they student adapts the teacher with different amounts of knowledge representations. After applying the proposed method, the student model achieved a comparable F1-score of 0.96 while the teacher achieved 0.97. In the second experiment, we demonstrated that our method allows the synchronization of several models trained on different distinct training data sets through the proposed decentralized federated learning process. Therefore, we split the training set into 8 equal and unique log sequence subsets and distributed these among 8 DeepLog models. With this training data all nodes could perform an initial training step before they entered the role of teachers and students. Even with distributing small amount of 10 knowledge representations all nodes could improve to a F1-score 0.95 confidence interval between 0.923 and 0.933 compared to their initial trained models, which reached a 0.95 confidence interval of 0.875 to 0.901. After 1,000 the models archive a 0.95 confidence interval of between 0.933 and 0.935.

For future work we plan to investigate more datasets in order to verify the general applicability of our approach. Furthermore, generating random sequences from a relatively small set of discrete elements represents a comparably limited search space for auxiliary sample generation. We expect the knowledge transfer to be harder within larger discrete sets or even within continuous space. Another goal is to research methods and heuristics to stabilize and accelerate the process of knowledge transfer with increasingly complex auxiliary samples.

References

1. Acker, A., Wittkopp, T., Nedelkoski, S., Bogatinovski, J., Kao, O.: Superiority of simplicity: A lightweight model for network device workload prediction. arXiv preprint arXiv:2007.03568 (2020)
2. Du, M., Li, F., Zheng, G., Srikumar, V.: DeepLog: anomaly detection and diagnosis from system logs through deep learning. In: Proceedings of the 2017 ACM SIGSAC Conference on Computer and Communications Security, pp. 1285–1298 (2017)
3. Nedelkoski, S., Bogatinovski, J., Acker, A., Cardoso, J., Kao, O.: Self-supervised log parsing. arXiv preprint arXiv:2003.07905 (2020)

4. Wu, L., Tordsson, J., Elmroth, E., Kao, O.: Microrca: root cause localization of performance issues in microservices. In: IEEE/IFIP Network Operations and Management Symposium (NOMS) (2020)
5. Lin, F., Beadon, M., Dixit, H.D., Vunnam, G., Desai, A., Sankar, S.: Hardware remediation at scale. In: 2018 48th Annual IEEE/IFIP International Conference on Dependable Systems and Networks Workshops (DSN-W), pp. 14–17. IEEE (2018)
6. Shokri, R., Shmatikov, V.: Privacy-preserving deep learning. In: Proceedings of the 22nd ACM SIGSAC Conference on Computer and Communications Security, pp. 1310–1321 (2015)
7. Hard, A., et al.: Federated learning for mobile keyboard prediction. arXiv preprint arXiv:1811.03604 (2018)
8. McMahan, B., Moore, E., Ramage, D., Hampson, S., Arcas, B.A.Y.: Communication-efficient learning of deep networks from decentralized data. In: Artificial Intelligence and Statistics, pp. 1273–1282 (2017)
9. Yang, Q., Liu, Y., Chen, T., Tong, Y.: Federated machine learning: concept and applications. ACM Trans. Intell. Syst. Technol. (TIST) 10(2), 1–19 (2019)
10. Papernot, N., Abadi, N.M., Erlingsson, U., Goodfellow, I., Talwar, K.: Semi-supervised knowledge transfer for deep learning from private training data. arXiv preprint arXiv:1610.05755 (2016)
11. Fredrikson, M., Jha, S., Ristenpart, T.: Model inversion attacks that exploit confidence information and basic countermeasures. In: Proceedings of the 22nd ACM SIGSAC Conference on Computer and Communications Security, pp. 1322–1333 (2015)
12. Shokri, R., Stronati, M., Song, C., Shmatikov, V.: Membership inference attacks against machine learning models. In: 2017 IEEE Symposium on Security and Privacy (SP), pp. 3–18. IEEE (2017)
13. Hu, R., Gong, Y., Guo, Y.: Sparsified privacy-masking for communication-efficient and privacy-preserving federated learning. arXiv preprint arXiv:2008.01558 (2020)
14. Bonawitz, K., et al.: Practical secure aggregation for federated learning on user-held data. arXiv preprint arXiv:1611.04482 (2016)
15. Geyer, R.C., Klein, T., Nabi, M.: Differentially private federated learning: a client level perspective. arXiv preprint arXiv:1712.07557 (2017)
16. Kairouz, P., et al.: Advances and open problems in federated learning. arXiv preprint arXiv:1912.04977 (2019)
17. Nguyen, T.D., Marchal, S., Miettinen, M., Fereidooni, H., Asokan, N., Sadeghi, A.-R.: Dïot: a federated self-learning anomaly detection system for IoT. In: 2019 IEEE 39th International Conference on Distributed Computing Systems (ICDCS), pp. 756–767. IEEE (2019)
18. Liu, Y., Kumar, N., Xiong, Z., Lim, W.Y.B., Kang, J., Niyato, D.: Communication-efficient federated learning for anomaly detection in industrial internet of things
19. Liu, Y., et al.: Deep anomaly detection for time-series data in industrial IoT: a communication-efficient on-device federated learning approach. IEEE Internet of Things J. (2020)
20. Preuveneers, D., Rimmer, V., Tsingenopoulos, I., Spooren, J., Joosen, W., Ilie-Zudor, E.: Chained anomaly detection models for federated learning: an intrusion detection case study. Appl. Sci. 8(12), 2663 (2018)
21. Nguyen, T.D., Rieger, P., Miettinen, M., Sadeghi, A.-R.: Poisoning attacks on federated learning-based IoT intrusion detection system (2020)

22. Xu, W., Huang, L., Fox, A., Patterson, D., Jordan, M.I.: Detecting large-scale system problems by mining console logs. In: Proceedings of the ACM SIGOPS 22nd Symposium on Operating Systems Principles, pp. 117–132 (2009)
23. He, P., Zhu, J., Zheng, Z., Lyu, M.R.: Drain: an online log parsing approach with fixed depth tree. In: 2017 IEEE International Conference on Web Services (ICWS), pp. 33–40. IEEE (2017)

Online Memory Leak Detection
in the Cloud-Based Infrastructures

Anshul Jindal[1]([✉])(iD), Paul Staab[2], Jorge Cardoso[2](iD), Michael Gerndt[1](iD),
and Vladimir Podolskiy[1](iD)

[1] Chair of Computer Architecture and Parallel Systems,
Technical University of Munich, Garching, Germany
{anshul.jindal,v.podolskiy}@tum.de, gerndt@in.tum.de
[2] Huawei Munich Research Center, Huawei Technologies Munich, Munich, Germany
{paul.staab,jorge.cardoso}@huawei.com

Abstract. A memory leak in an application deployed on the cloud can affect the availability and reliability of the application. Therefore, to identify and ultimately resolve it quickly is highly important. However, in the production environment running on the cloud, memory leak detection is a challenge without the knowledge of the application or its internal object allocation details.

This paper addresses this *challenge of online detection of memory leaks in cloud-based infrastructure without having any internal application knowledge* by introducing a novel machine learning based algorithm Precog. This algorithm solely uses one metric i.e. the system's memory utilization on which the application is deployed for the detection of a memory leak. The developed algorithm's accuracy was tested on 60 virtual machines manually labeled memory utilization data provided by our industry partner Huawei Munich Research Center and it was found that the proposed algorithm achieves the accuracy score of 85% with less than half a second prediction time per virtual machine.

Keywords: Memory leak · Online memory leak detection · Memory leak patterns · Cloud · Linear regression

1 Introduction

Cloud computing is widely used in the industries for its capability to provide cheap and on-demand access to compute and storage resources. Physical servers resources located at different data centers are split among the virtual machines (VMs) hosted on it and distributed to the users [5]. Users can then deploy their applications on these VMs with only the required resources. This allows the efficient usage of the physical hardware and reducing the overall cost. However, with all the advantages of cloud computing there exists the drawback of detecting a fault or an error in an application or in a VM efficiently due to the layered virtualisation stack [1,4]. A small fault somewhere in the system can impact the performance of the application.

© Springer Nature Switzerland AG 2021
H. Hacid et al. (Eds.): ICSOC 2020 Workshops, LNCS 12632, pp. 188–200, 2021.
https://doi.org/10.1007/978-3-030-76352-7_21

An application when deployed on a VM usually requires different system resources such as memory, CPU and network for the completion of a task. If an application is mostly using the memory for the processing of the tasks then this application is called a memory-intensive application [8]. It is the responsibility of the application to release the system resources when they are no longer needed. When such an application fails to release the memory resources, a **memory leak** occurs in the application [14]. Memory leak issues in the application can cause continuous blocking of the VM's resources which may in turn result in slower response times or application failure. In software industry, memory leaks are treated with utmost seriousness and priority as the impact of a memory leak could be catastrophic to the whole system. In the development environment, these issues are rather easily detectable with the help of static source code analysis tools or by analyzing the heap dumps. But in the production environment running on the cloud, memory leak detection is a challenge and it only gets detected when there is an abnormality in the run time, abnormal usage of the system resources, crash of the application or restart of the VM. Then the resolution of such an issue is done at the cost of compromising the availability and reliability of the application. Therefore it is necessary to monitor every application for memory leak and have an automatic detection mechanism for memory leak before it actually occurs. However, it is a challenge to detect memory leak of an application running on a VM in the cloud without the knowledge of the programming language of the application, nor the knowledge of source code nor the low level details such as allocation times of objects, object staleness, or the object references [10]. Due to the low down time requirements for the applications running on the cloud, detection of issues and their resolutions is to be done as quickly as possible. Therefore, this challenge is addressed in this paper *by solely using the VM's memory utilization as the main metric and devising a novel algorithm called **Precog** to detect memory leak.*

The main contribution of this paper are as follows:

- **Algorithm**: We propose an online novel machine learning based algorithm **Precog** for accurate and efficient detection of memory leaks by solely using the VM's memory utilization as the main metric.
- **Effectiveness**: Our proposed algorithm achieves the accuracy score of 85% on the evaluated dataset provided by our industry partner and accuracy score of above 90% on the synthetic data generated by us.
- **Scalability**: Precog's predict functionality is linearly scalable with the number of values and takes less than a second for predicting in a timeseries with 100,000 values.

Reproducibility: our code and synthetic generated data are publicly available at: https://github.com/ansjin/memory_leak_detection.

2   Related Work

Memory leak detection has been studied over the years and several solutions have been proposed. Sor et al. reviewed different memory leak detection approaches based on their implementation complexity, measured metrics, and intrusiveness and a classification taxonomy was proposed [11]. The classification taxonomy broadly divided the detection algorithms into *(1) Online detection, (2) Offline detection and (3) Hybrid detection.* The *online detection* category uses either staleness measure of the allocated objects or their growth analysis. *Offline detection* category includes the algorithms that make use of captured states i.e. heap dumps or use a visualization mechanism to manually detect memory leaks or use static source code analysis. *Hybrid detection* category methods combine the features offered by online and offline methods to detect memory leaks. Our work falls in the category of online detection therefore, we now restrict our discussion to the approaches related to the online detection category only.

Based on the staleness measure of allocated objects, Rudaf et al. proposed "LeakSpot" for detecting memory leaks in web applications [9]. It locates JavaScript allocation and reference sites that produce and retain increasing numbers of objects over time and uses staleness as a heuristic to identify memory leaks. Vladimir Šor et al. proposes a statistical metric called *genCount* for memory leak detection in Java applications [12]. It uses the number of different generations of the objects grouped by their allocation sites, to abstract the object staleness - an important attribute indicating a memory leak. Vilk et al. proposed a browser leak debugger for automatically debugging memory leaks in web applications called as "BLeak" [13]. It collects heap snapshots and analyzes these snapshots over time to identify and rank leaks. BLeak targets application source code when detecting memory leaks.

Based on the growth analysis objects, Jump et al. proposes "Cork" which finds the growth of heap data structure via a directed graph *Type Points-From Graph* - TPFG, a data structure which describes an object and its outgoing reference [6]. To find memory leaks, TPFG's growth is analyzed over time in terms of growing types such as a list. FindLeaks proposed by Chen et al. tracks object creation and destruction and if more objects are created than destroyed per class then the memory leak is found [2]. Nick Mitchell and Gary Sevitsky proposed "LeakBot", which looks for heap size growth patterns in the heap graphs of Java applications to find memory leaks [7]. "LEAKPOINT" proposed by Clause et al. uses dynamic tainting to track heap memory pointers and further analyze it to detect memory leaks [3].

Most of the online detection algorithms that are proposed focus either on the programming language of the running application or on garbage collection strategies or the internals of the application based on the object's allocation, references, and deallocation. To the best of our knowledge, there is no previous work that solely focuses on the detection of memory leaks using just the system's memory utilization data on which application is deployed. The work in this paper, therefore, focuses on the detection of a memory leak pattern irrespective of the programming language of the application or the knowledge of application's source code or the low-level details such as allocation times of objects, object staleness, or the object references.

Table 1. Symbols and definitions.

Symbol	Interpretation
t	A timestamp
x_t	The percentage utilization of a resource (for example memory or disk usage) of a virtual machine at time t
N	Number of data points
$x = \{x_1, x_2, ..., x_N\}$	A VM's memory utilization observations from the Cloud
T	Time series window length
$x_{t-T:t}$	A sequence of observations $\{x_{t-T}, x_{t-T+1}, ..., x_t\}$ from time $t - T$ to t
U	Percentage memory utilization threshold equal to 100
C	Critical time

3 Methodology for Memory Leak Detection

In this section, we present the problem statement of memory leak detection and describes our proposed algorithm's workflow for solving it.

3.1 Problem Statement

Table 1 shows the symbols used in this paper.

We are given $x = \{x_1, x_2, ..., x_N\}$, an $N \times 1$ dataset representing the memory utilization observations of the VM and an observation $x_t \in R$ is the percentage memory utilization of a virtual machine at time t. The objective of this work is to determine whether or not there is a memory leak on a VM such that an observation x_t at time t reaches the threshold U memory utilization following a trend in the defined critical time C. Formally:

Problem 1 (Memory Leak Detection)

- **Given**: a univariate dataset of N time ticks, $x = \{x_1, x_2, ..., x_N\}$, representing the memory utilization observations of the VM.
- **Output**: an anomalous window for a VM consisting of a sequence of observations $x_{t-T:t}$ such that these observations after following a certain trend will reach the threshold U memory utilization at time $t + M$ where $M \leq C$.

Definition 1 *(Critical Time). It is the maximum time considered relevant for reporting a memory leak in which if the trend line of memory utilization of VM is projected, it will reach the threshold U.*

3.2 Illustrative Example

Figure 1 shows the example memory utilization of a memory leaking VM with the marked anomalous window between t_k and t_n. It shows that the memory utilization of the VM will reach the defined threshold ($U = 100\%$) within the defined critical time C by following a linearly increasing trend (shown by the trend line) from the observations in the anomalous window. Therefore, this VM is regarded as a memory leaking VM.

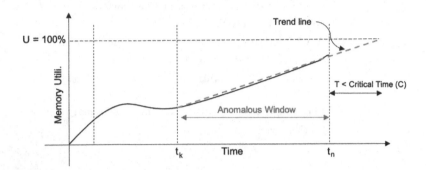

Fig. 1. Example memory utilization of a memory leaking VM with the marked anomalous window.

Our developed approach can be applied for multiple VMs as well. We also have conducted an experiment to understand the memory usage patterns of memory leak applications. We found that, if an application has a memory leak, usually the memory usage of the VM on which it is running increases steadily. It continues to do so until all the available memory of the system is exhausted. This usually causes the application attempting to allocate the memory to terminate itself. Thus, usually a memory leak behaviour exhibits a linearly increasing or "sawtooth" memory utilization pattern.

3.3   Memory Leak Detection Algorithm: Precog

The Precog algorithm consists of two phases: offline training and online detection. Figure 2 shows the overall workflow of the Precog algorithm.

Offline Training: The procedure starts by collecting the memory utilization data of a VM and passing it to *Data Pre-processing* module, where the dataset is first transformed by resampling the number of observations to one every defined resampling time resolution and then the time series data is median smoothed over the specified smoothing window. In *Trend Lines Fitting* module, firstly, on the whole dataset, the change points $P = \{P_1, P_2, ..., P_k\}$, where $k \leq n - 1$, are detected. By default, two change points one at the beginning and other at the end of time series data are added. If the change points are not detected, then the algorithm will have to go though each data point and it will be compute intensive, therefore these points allows the algorithm to directly jump from one change point to another and selecting all the points in between the two change points. *Trend Lines Fitting* module selects a sequence of observations $x_{t-L:t}$

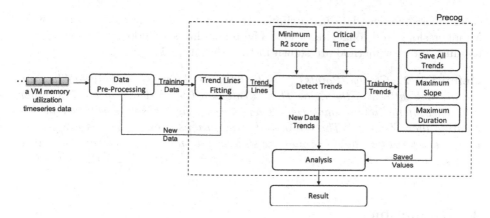

Fig. 2. Overall workflow of Precog algorithm.

between the two change points: one fixed P_1 and other variable P_r where $r \leq k$ and a line is fitted on them using the linear regression. The R-squared score, size of the window called as *duration*, time to reach threshold called *exit time* and slope of line are calculated. This procedure is repeated with keeping the fixed change point the same and varying the other for all other change points. Out of all the fitted lines, the best-fitted line based on the largest duration and highest slope is selected for the fixed change point. If this best-fitted lines' time to reach threshold falls below the critical time then its slope and duration are saved as historic trends.

This above procedure is again repeated by changing the fixed change point to all the other change points. At the end of this whole procedure, we get for each change point, a best-fitted trend if it exists. Amongst the captured trends, maximum duration and the maximum slope of the trends are also calculated and saved. This training procedure can be conducted routinely, e.g., once per day or week. The method's pseudocode is shown in the Algorithm's 1 *Train function*.

Online Detection: In the Online Detection phase, for a new set of observations $\{x_k, x_k+1, x_k+2, ..., x_k+t-1 x_k+t\}$ from time k to t where $t-k \geq P_{min}$ belonging to a VM after pre-processing is fed into the *Trend Lines Fitting* module. In *Trend Lines Fitting* module, the change points are detected. A sequence of observations $x_{t-L:t}$ between the last two change points starting from the end of the time series are selected and a line is fitted on them using the linear regression. The R-squared score, slope, duration and exit time to reach threshold of the fitted line is calculated. If its slope and duration are greater than the saved maximum counter parts then that window is marked anomalous. Otherwise, the values are compared against all the found training trends and if fitted-line's slope and duration are found to be greater than any of the saved trend then, again that window will be marked as anomalous. This procedure is further repeated by analyzing the observations between the last change point P_k and the previous next change point until all the change points are used. This is done for the cases where the new data has a similar trend as the historic data but now with a higher slope and longer duration. The algorithm's pseudo code showing the training and test method are shown in the Algorithm 1.

Definition 2 *(Change Points). A set of time ticks which deviate highly from the normal pattern of the data. This is calculated by first taking the first-order difference of the input timeseries. Then, taking their absolute values and calculating their Z-scores. The indexes of observations whose Z-scores are greater than the defined threshold (3 times the standard deviation) represents the change points. The method's pseudocode is shown in the Algorithm's 1 CPD function.*

4   Evaluation

We design experiments to answer the questions:

- **Q1. Memory Leak Detection Accuracy**: how accurate is Precog in the detection of memory leaks?
- **Q2. Scalability**: How does the algorithm scale with the increase in the data points?
- **Q3. Parameter Sensitivity**: How sensitive is the algorithm when the parameters values are changed?

We have used F1-Score (denoted as F1) to evaluate the performance of the algorithms. Evaluation tests have been executed on a machine with 4 physical cores (3.6 GHz Intel Core i7-4790 CPU) with hyperthreading enabled and 16 GB of RAM. These conditions are similar to a typical cloud VM. It is to be noted that the algorithm detects the cases where there is an ongoing memory leak and assumes that previously there was no memory leak. For our experiments, hyperparameters are set as follows. The maximum threshold U is set to 100 and the defined critical time C is set to 7 days. The smoothing window size is 1 h and re-sampling time resolution was set to 5 min. Lastly, the minimum R-squared score $R2_{min}$ for a line to be recognized as a good fit is set to 0.75. 65% of data was used for training and the rest for testing. However, we also show experiments on parameter sensitivity in this section.

4.1   Q1. Memory Leak Detection Accuracy

To demonstrate the effectiveness of the developed algorithm, we initially synthetically generated the timeseries. Table 2 shows the F1 score corresponding to each memory leak pattern and also the overall F1 score. Table 2 shows that Precog is able to reach an overall accuracy of 90%.

In addition, to demonstrate the effectiveness of the developed algorithm on the real cloud workloads, we evaluated Precog on the real Cloud dataset provided by Huawei Munich which consists of manually labeled memory leak data from 60 VMs spanned over 5 days and each time series consists of an observation every

---

Algorithm 1: Precog Algorithm

Input: input_Train_Ts,R2_score_min, input_Test_Ts, critical_time
Output: anomalous list a

1 **Function** CPD($x = input\_Ts, threshold = 3$):
2     $absDiffTs$ = first order absolute difference of x
3     $zScores$ = calculate z-scores of $absDiffTs$
4     $cpdIndexes$ = indexes of *(zScores > threshold)*
5   **return** $cpdIndexes$ `// return the change-points indexes`

6 **Function** TRAINING($x = input\_Train\_Ts, R2\_score\_min, C = critical\_time$):
    `// Train on` *input_Train_Ts*
7   P = **CPD**(x)   `// get Change-points`
8   p1 = 0
9   **while** *p1 <= length(P)* **do**
10   p2 = p1
11        $D_b, S_b, T_b = 0$ `// best local trend's duration, slope, exit time`
12   **while** *p2 <= length(P)* **do**
13           $exit\_time, r2, dur, slope \leftarrow$ ***LinearRegression****(ts)*   `// fitted`
              `line's exit time, R2 score, duration, slope`
14   **if** $r2 \geq R2\_score\_min$ *and* $dur \geq D_b$ *and* $slope \geq S_b$ **then**
15   **Update**(D_b, S_b, T_b)   `// update best local values`
16   p2 = p2 + 1
17   **if** $T_b \leq C$ **then**
18   **if** $D_b \geq D_{max}$ *and* $S_b \geq S_{max}$ **then**
19   **Update**(D_{max}, S_{max})   `// update global trend values`
20   **saveTrend**(D_b, S_b), **save**(D_{max}, S_{max})   `// save values`
21   p1 = p1 + 1

22 **Function** TEST($x = input\_Test\_Ts, C = critical\_time$):
    `// Test on the new data to find anomalous memory leak window`
23     $a = [0]$ `// anomalous empty array of size` *input_Test_Ts*
24   P = **CPD**(x)   `// get Change-points`
25   len = length(P)   `// length of change point indexes`
26   **while** $i \leq len$ **do**
27        $ts = x[P[len - i] : P[len]]$ `// i is a loop variable`
28        $exit\_time, r2, dur, slope$ = ***LinearRegression****(ts)*
29        $D_{max}, S_{max}, Trends$ = get saved values
30   **if** $exit\_time, \leq C$ *and* $r2 \geq R_{min}$ **then**
31   **if** $slope \geq S_{max}$ *and* $dur \geq D_{max}$ **then**
32              $a[P[len - i] : P[len]] = 1$ `// current trend greater than`
             `global saved so mark anomalous`
33   **else**
34   **For Each** t in $Trends$ **if** $slope \geq S_t$ *and* $dur \geq D_t$ **then**
35                 $a[P[len - i] : P[len]] = 1$ `// current trend greater than`
                `one of the saved trend so mark anomalous`
36        $i = i + 1$
37   **return** a `// list with 0s and anomalous indexes represented by 1`

Table 2. Synthetically generated timeseries corresponding to each memory leak pattern and their accuracy score.

Memory leak pattern	+ve cases	−ve cases	F1 score	Recall	Precision
Linearly increasing	30	30	0.933	0.933	0.933
Linearly increasing (with noise)	30	30	0.895	1.0	0.810
Sawtooth	30	30	0.830	0.73	0.956
Overall	90	90	0.9	0.9	0.91

minute. Out of these 60 VMs, 20 VMs had a memory leak. Such high number of VMs having memory leaks is due to the fact that applications with memory leak were deliberately run on the infrastructure. The algorithm achieved the F1-Score of 0.857, recall equals to 0.75 and precision as 1.0. Average prediction time per test data containing approximately 500 points is 0.32 s.

Furthermore, we present the detailed results of the algorithm on the selected 4 cases shown in the Fig. 3: simple linearly increasing memory utilization, sawtooth linearly increasing pattern, linearly increasing pattern with no trends detected in training data, and linearly increasing with similar trend as training data. The figure also shows the change points, training trends and the detected anomalous memory leak window for each of the cases.

For the first case shown in Fig. 3a, memory utilization is being used normally until it suddenly starts to increase linearly. The algorithm detected one training trend and reported the complete test set as anomalous. The test set trend is having similar slope as training trend but with a longer duration and higher memory usage hence it is reported as anomalous.

In the second case (Fig. 3b), the trend represents commonly memory leak sawtooth pattern where the memory utilization increases upto a certain point and then decreases (but not completely zero) and then again it start to increase in the similar manner. The algorithm detected three training trends and reported most of the test set as anomalous. The test set follows a similar trend as captured during the training but with the higher memory utilization, hence it is reported.

In the third case (Fig. 3c), no appropriate training trend was detected in the complete training data but, the algorithm is able to detect an increasing memory utilization trend in the test dataset.

In Fig. 3d, the VM does not have a memory leak but its memory utilization was steadily increasing which if observed without the historic data seems to be a memory leak pattern. However, in the historic data, the same trend is already observed and therefore it is a normal memory utilization pattern. Precog using the historic data for detecting the training trends and then comparing them with the test data correctly reports that trend as normal and hence does not flag the window as anomalous. It is also to be noted that, if the new data's maximum goes beyond the maximum in the training data with the similar trend then it will be regarded as a memory leak.

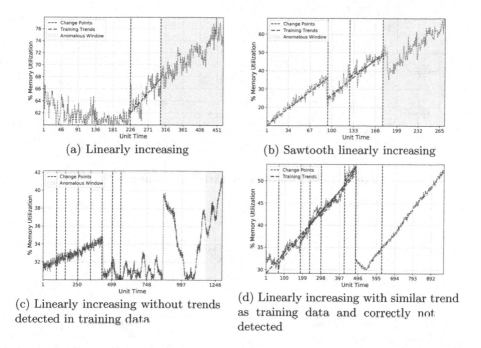

(a) Linearly increasing

(b) Sawtooth linearly increasing

(c) Linearly increasing without trends detected in training data

(d) Linearly increasing with similar trend as training data and correctly not detected

Fig. 3. Algorithm result on 3 difficult cases having memory leak (a–c) and one case not having a memory leak (d).

4.2 Q2. Scalability

Next, we verify that our prediction method scale linearly. We repeatedly dupli-cate our dataset in time ticks, add Gaussian noise. Figure 4b shows that Precog' predict method scale linearly in time ticks. Precog does provide the prediction results under 1s for the data with 100,000 time ticks. However, the training method shown in Fig. 4a is quadratic in nature but training needs to conducted once a week or a month and it can be done offline as well.

4.3 Q3. Parameter Sensitivity

Precog requires tuning of certain hyper-parameters like R2 score, and critical time, which currently are set manually based on the experts knowledge. Figure 5 compares performance for different parameter values, on synthetically generated dataset. Our algorithm perform consistently well across values. Setting minimum R2 score above 0.8 corresponds to stricter fitting of the line and that is why the accuracy drops. On the other hand, our data mostly contains trend lines which would reach threshold withing 3 to 4 days, therefore setting minimum critical time too less (less than 3 days) would mean the trend line never reaching thresh-old within the time frame and hence decreasing the accuracy. These experiments shows that these parameters does play a role in the overall accuracy of the algo-rithm but at most of the values algorithm is insensitive to them. Furthermore,

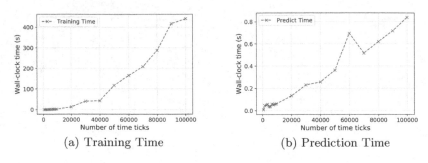

(a) Training Time   (b) Prediction Time

Fig. 4. Precog's prediction method scale linearly.

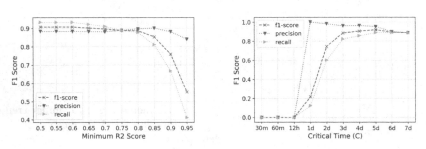

Fig. 5. Insensitive to parameters: Precog performs consistently across parameter values.

to determine these automatically based on the historic data is under progress and is out of the scope of this paper.

5   Conclusion

Memory leak detection has been a research topic for more than a decade. Many approaches have been proposed to detect memory leaks, with most of them looking at the internals of the application or the object's allocation and deallocation. The Precog algorithm for memory leak detection presented in the current work is most relevant for the cloud-based infrastructure where cloud administrator does not have access to the source code or know about the internals of the deployed applications. The performance evaluation results showed that the Precog is able to achieve a F1-Score of 0.85 with less than half a second prediction time on the real workloads. This algorithm can also be useful in the Serverless Computing where if a function is leaking a memory then its successive function invocations will add on to that and resulting in a bigger memory leak on the underneath system. Precog running on the underneath system can detect such a case.

Prospective directions of future work include developing online learning-based approaches for detection and as well using other metrics like CPU, network and storage utilization for further enhancing the accuracy of the algorithms and providing higher confidence in the detection results.

Acknowledgements. This work was supported by the funding of the German Federal Ministry of Education and Research (BMBF) in the scope of the Software Campus program. The authors also thank the anonymous reviewers whose comments helped in improving this paper.

References

1. Ataallah, S.M.A., Nassar, S.M., Hemayed, E.E.: Fault tolerance in cloud computing - survey. In: 2015 11th International Computer Engineering Conference (ICENCO), pp. 241–245, December 2015. https://doi.org/10.1109/ICENCO.2015.7416355
2. Chen, K., Chen, J.: Aspect-based instrumentation for locating memory leaks in java programs. In: 31st Annual International Computer Software and Applications Conference (COMPSAC 2007), vol. 2, pp. 23–28, July 2007). https://doi.org/10.1109/COMPSAC.2007.79
3. Clause, J., Orso, A.: LeakPoint: pinpointing the causes of memory leaks. In: 2010 ACM/IEEE 32nd International Conference on Software Engineering, vol. 1, pp. 515–524, May 2010. https://doi.org/10.1145/1806799.1806874
4. Gokhroo, M.K., Govil, M.C., Pilli, E.S.: Detecting and mitigating faults in cloud computing environment. In: 2017 3rd International Conference on Computational Intelligence Communication Technology (CICT), pp. 1–9, February 2017. https://doi.org/10.1109/CIACT.2017.7977362
5. Jain, N., Choudhary, S.: Overview of virtualization in cloud computing. In: 2016 Symposium on Colossal Data Analysis and Networking (CDAN), pp. 1–4, March 2016. https://doi.org/10.1109/CDAN.2016.7570950
6. Jump, M., McKinley, K.S.: Cork: dynamic memory leak detection for garbage-collected languages. In: Proceedings of the 34th Annual ACM SIGPLAN-SIGACT Symposium on Principles of Programming Languages, POPL 2007, New York, NY, USA, pp. 31–38. ACM (2007). https://doi.org/10.1145/1190216.1190224. http://doi.acm.org/10.1145/1190216.1190224
7. Mitchell, N., Sevitsky, G.: LeakBot: an automated and lightweight tool for diagnosing memory leaks in large java applications. In: Cardelli, L. (ed.) ECOOP 2003. LNCS, vol. 2743, pp. 351–377. Springer, Heidelberg (2003). https://doi.org/10.1007/978-3-540-45070-2_16
8. Pooja, Pandey, A.: Impact of memory intensive applications on performance of cloud virtual machine. In: 2014 Recent Advances in Engineering and Computational Sciences (RAECS), pp. 1–6, March 2014. https://doi.org/10.1109/RAECS.2014.6799629
9. Rudafshani, M., Ward, P.A.S.: Leakspot: detection and diagnosis of memory leaks in javascript applications. Softw. Pract. Exper. **47**(1), 97–123 (2017). https://doi.org/10.1002/spe.2406
10. Sor, V., Srirama, S.N.: A statistical approach for identifying memory leaks in cloud applications. In: CLOSER (2011)
11. Sor, V., Srirama, S.N.: Memory leak detection in java: taxonomy and classification of approaches. J. Syst. Softw. **96**, 139–151 (2014)
12. Sor, V., Srirama, S.N., Salnikov-Tarnovski, N.: Memory leak detection in plumbr. Softw. Pract. Exper. **45**, 1307–1330 (2015)

13. Vilk, J., Berger, E.D.: Bleak: automatically debugging memory leaks in web applications. In: Proceedings of the 39th ACM SIGPLAN Conference on Programming Language Design and Implementation, PLDI 2018, New York, NY, USA, pp. 15–29. ACM (2018). https://doi.org/10.1145/3192366.3192376. http://doi.acm.org/10.1145/3192366.3192376
14. Xie, Y., Aiken, A.: Context- and path-sensitive memory leak detection. SIG-SOFT Softw. Eng. Notes **30**(5), 115–125 (2005). https://doi.org/10.1145/1095430.1081728. http://doi.acm.org/10.1145/1095430.1081728

Multi-source Anomaly Detection in Distributed IT Systems

Jasmin Bogatinovski$^{(\boxtimes)}$ and Sasho Nedelkoski

Distributed Operating Systems, TU Berlin, Berlin, Germany
{jasmin.bogatinovski,nedelkoski}@tu-berlin.de

Abstract. The multi-source data generated by distributed systems, provide a holistic description of the system. Harnessing the joint distribution
of the different modalities by a learning model can be beneficial for critical applications for maintenance of the distributed systems. One such
important task is the task of anomaly detection where we are interested
in detecting the deviation of the current behaviour of the system from the
theoretically expected. In this work, we utilize the joint representation
from the distributed traces and system log data for the task of anomaly
detection in distributed systems. We demonstrate that the joint utilization of traces and logs produced better results compared to the single
modality anomaly detection methods. Furthermore, we formalize a learning task - next template prediction NTP, that is used as a generalization
for anomaly detection for both logs and distributed trace. Finally, we
demonstrate that this formalization allows for the learning of template
embedding for both the traces and logs. The joint embeddings can be
reused in other applications as good initialization for spans and logs.

Keywords: Multi-source anomaly detection · Multi-modal · Logs ·
Distributed traces

1 Introduction

The complexity of the multi-layered IT infrastructures such as the Internet of
Things, distributed processing frameworks, databases and operating systems, is
constantly increasing [10]. To meet the consumers' expectations of fluent service
with low response times guarantees and availability, the service providers highly
rely on the high volumes of monitoring data. The massive volumes of data lead to
maintenance overhead for the operators and require introducing of data-driven
tools to process the data.

A crucial task for such tools is to correctly identify the symptoms of deviation
of the current behaviour system from the expected one. Due to the large volumes of
data, the anomaly detector should produce a small number of false-positive alarms,

---

J. Bogatinovski and S. Nedelkoski—Equal contribution

© Springer Nature Switzerland AG 2021
H. Hacid et al. (Eds.): ICSOC 2020 Workshops, LNCS 12632, pp. 201–213, 2021.
https://doi.org/10.1007/978-3-030-76352-7_22

thus reducing the efforts of the operators, while at the same time producing a high detection rate. The benefit of timely detection allows prevention of potential failures and increases the opportunity window for conducting a successful reaction from the operator. This is especially important if urgent expertise and/or administration activity is required. The symptoms often are notified whenever there are performance problems or system failures and usually manifests as some fingerprints within the monitored data: logs, metrics or distributed traces.

The monitored system data represent the state of the system at any time point. They are grouped into three categories-modalities: metrics, application logs, and distributed traces [12]. The metrics are time-series data that represent the utilization of the available resources and the status of the infrastructure. Typically they involve measuring of the CPU, memory and disk utilization, as well as data as network throughput, and service call latency. Application logs are print statements appearing in code with semi-structured content. They represent interactions between data, files, services, or applications containing a rich representative structure on a service level. Service, microservices, and other systems generate logs which are composed of timestamped records. Distributed traces chains the service invocations as workflows of execution of HTTP or RPC requests. Each part of the chain in the trace is called an event or span. A property of this type of data is that it preserves the information for the execution graph on a (micro)service level. Thus, the information for the interplay between the components is preserved.

The log data can produce a richer description on a service level since they are fingerprints of the program execution within the service. On the other side, the traces do not have much information on system-level information but preserve the overall graph of request execution. Referring to the different aspects of the system, the logs and traces provide orthogonal information for the distributed systems behaviour. Building on this observation in this work, we introduce an anomaly detection multi-source approach that can consider the data from both the traces and logs, jointly. We demonstrate the usability of time-aligned log and tracing data to produce better results on the task of anomaly detection as compared to the single modalities as the main contribution to this work. The results show that the model build under the joint loss from both the logs and trace data can exploit some relationship between the modalities. The approach is trainable end-to-end and does not require the building of separate models for each of the modalities. As a second contribution, we consider the introduction of vector embeddings for the spans within the trace. The adopted approach allows the definition of the span vectors as a pooling over the words they are composed of. We refer to these vector embeddings as span2vec.

2   Related Work

The literature recognizes various approaches concerned with anomaly detection in distributed systems from single modalities. We review the single modalities approaches for both logs and traces. We also provide an overview of the existing multi-modal approaches, however, none of them jointly considers both traces and logs.

The most common approaches for anomaly detection from log data roughly follows a two-step composition - log parsing followed by a method for anomaly detection. The first step allows for an appropriate log representation. One challenge during this procedure is the reduction of the noise in the log data. This noise in a log message is present due to the various parameters parts of the log can take during execution. To this end, there are many proposed techniques for log parsing [3,7,14]. A detailed overview and comparison across benchmarks of these techniques are given in [18]. After the template extraction, there are two general approaches to represent the logs. The first one is based on word frequencies and metrics derived from the logs (e.g. TF-IDF) [2,5,15,17] or reusing word representation of the logs, based on corpora of words. The second approach aims at translating the templates into sequences of templates - most often represented as sequences of integers or sequences of vectors. Such representation allows modelling the sequential execution of a program workflow. One of the most commonly utilized approaches is RNN-based(e.g. LSTM, GRU) [1]. They often are coupled with an additional mechanism such as attention to allow for better preservation of the semantic information inside the logs [6]. Depending on the data representation, various methods are utilized from both the supervised and unsupervised domains of machine learning. However, due to easier practical adoption and the absence of labels, the unsupervised methods are preferred.

The available approaches for anomaly detection from tracing data are scarce. They usually model the normal execution of a workload, represented within the trace by utilizing history h of recent trace events as input. They decompose the trace in its building blocks, the events/spans, and predict the next span in the sequence. The anomaly detection is done with imposing thresholds on the number of errors the LSTM is making for the corresponding trace predicted [9, 10]. Further approaches aim to capture the execution of a complete workload into a finite state automata (FSA) [16]. However, the FSA approaches are dependent on specific tracing implementation systems. The unification of this approach with other types of modalities such as the log data due to the assumed homogeneous structure of the states building the FSA is harder.

Several works on multi-modal learning for anomaly detection demonstrate the feasibility of using different modalities of data for anomaly detection [11,13]. In the context of large scale ICT systems, the authors in [10] consider the joint exploitation of traces and the corresponding response times of the spans within the trace. More specifically, a multi-modal LSTM-based method, trained jointly on both modalities is introduced, showing the additional value added by the shared information, improves the anomaly detection scores. In [4] a Multimodal Variational Autoencoder approach is adopted for effectively learning the relationships among cross-domain data which provide good results for anomaly detection build on the logs and metrics as modalitites. However, they do not preserve the information for the overall microservice architecture.

To the best of our knowledge, the literature does not yet recognize methods for joint consideration of logs and traces as fundamentally complementary data sources describing the distributed IT systems. Hence in this work, we propose an approach on how to jointly consider the complement information within the logs and traces.

3 Multimodal Approach for Anomaly Detection from Heterogeneous Data

In this section, we describe the multi-source approach towards anomaly detection using logs and tracing data. First, we describe the logs and traces as generated by the system. We present their specifics that are exploited for the definition of the Next Template Prediction (NTP) pseudo-task. Second, we describe the NTP pseudo-task for anomaly detection. Thirdly, we describe one way to address the NTP task utilizing deep learning architecture on a single modality description of the system state. Next, we provide a solution that enables us to efficiently solve the NTP problem as a pseudo task for joint detection of anomalies from both logs and traces. Finally, we present an approach that uses the results from the NTP task and performs anomaly detection.

3.1 Data Representation

The raw logs and traces as generated by the system, contain various information about the specific operation being executed. Since some of the information is a sporadic description of the operations, proper filtering and representation should be done. Due to the specifics of the two modalities, we address them separately.

Logs. A log is a sequence of temporally ordered unstructured text messages $L = \{l_i : i = 1, 2, ...\}$. Each text message l_i is generated by a logging instruction (e.g. printf(), log.info()) within the software source code. Since the logging function is part of the body of the whole program, it can serve as a proxy for the program execution workflow. Hence one can infer the normal execution pattern within the program workflow.

The logs consist of a constant and a varying part, referred to as log template and log parameters. Due to the large variability of the parameters, they can introduce a lot of noise. To mitigate this problem common way to represent the logs is with the extraction of the constant part through a log parsing procedure. It allows for the creation of a dictionary of log templates from a given set of logs.

To unify the representations of the logs, the log templates are tokenized. A dictionary from the tokens, representing the vocabulary of all of the tokens in the logs - D_{logs_words} is created. Since the log templates can have a different number of tokens, for the uniform representation of the log templates a special <SPECLOG> token is added, such that each of the logs has an equal number of tokens. The maximal size of the log template is limited by a parameter called max_log_size.

$$L_i = \{W_0^i, W_1^i, \ldots, W_t^i\} \tag{1}$$

where each of the W_t is an extracted word mapped to index $t \in D_{logs\_word\_indecies}$.

Distrubted Traces. Distributed traces are a request-centred way to describe behaviour within the distributed system. It means that they follow the execution of the user issued a request through the distributed system in a record referred to as spans. The spans represent information (e.g. start time, end time, service name, HTTP path) about the operations performed when handling an external request in service. Formally, a trace is written as

$$T_i = \{S_0^i, S_1^i, \ldots, S_m^i\}, \tag{2}$$

where $i \in \{1, \ldots, N\}$ is a trace as part of an observation set of traces, and m is the length T_i or the number of spans in the trace.

One of the most characteristic properties of the spans is the function executed during the event and a corresponding endpoint. They usually represent either HTTP or RPC calls, denoting the interconnection between the spans within the trace. The HTTP calls are described with *path, scheme, method*. The RPC calls are represented with the functions they are executing. Since these features represent the intra-service communication in a trace, we assume that they are sufficient for structural analysis of possible anomalies. To provide a richer representation of the traces, further augmentation of the traces can be done. More specifically, two artificial spans (<START> and <END>) are added to the beginning and the end of the trace, accordingly. It preserves the knowledge for the length of the trace.

Represented in this form the spans have very similar representation as to the logs, with additional constraints that the spans are further bounded by the operation executed within the trace. It means that they also are facing the problem of the presence of noise into the representation induced by the varying parameters. Similar as for the logs, applying a template extraction technique produces a set of representative template spans. It allows for each of the trace to be represented as a sequence of template spans. Formally,

$$T_i = \{St_0^i, St_1^i, \ldots, St_k^i\} \tag{3}$$

where each of the St_k is an extracted template mapped to index $k \in D_{template\ indecies}$.

Observing that each function calls are sequences of characters, a dictionary of the sequences of characters appearing inside the given set of traces is constructed - $D_{span\ words}$. It provides a unique language for the description of all of the spans appearing in the observed traces. Formally a span is represented as

$$St_j = \{W_0^i, W_1^j, \ldots, W_q^j\} \tag{4}$$

where W_q is a sequence of characters as extracted from the dictionary of span words $D_{span\ words}$. Since there are spans with a different number of words, to provide spans in an appropriate representational format for later processing, each of the spans is augmented with a <SPECSPAN> token.

3.2 NTP: Pseudo-task for Anomaly Detection

Representation of both traces and logs in the previously described manner, allow us to take a unified approach towards their modelling. The appearance of the next log message is conditioned on the appearance of the history of the previous logs. Similarly, within a trace, the appearance of the next span is conditioned on the previous ones. Thus the modelling problem can be conceptualized formally as

$$P(A_{T_{win}:T}) = \prod_{t=T_{win}}^{T} P(A_t|A_{<t}) \tag{5}$$

where $A_{<t}$ denotes the templates traces or logs from A_{t-win} to A_t, with win denoting the size of the preserved history. Hence we refer to this task as the next template prediction (NTP).

3.3 Single Modality Anomaly Detection

Figure 1 depicts the proposed end to end architecture to solve the NTP task for single modalities. We use the same architecture for both the logs and the traces.

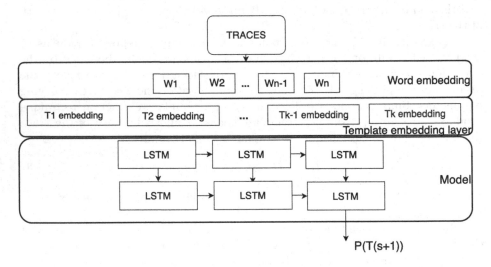

Fig. 1. Proposed architecture for single modality. The same approach can be utilized also for the logs data.

At the input, we provide the dictionary of the words as appearing in D_{logs_words} and D_{span_words}. We perform initialization with random vectors for each of the words with a specific size. This is a parameter of the method referred embedding size $N_{embedding}$. The template embedding layer uses the representations of the words to create the corresponding sequences of templates. These sequences are fed through an autoregressive deep learning LSTM method that is modelling the sequential dependence between the input samples represented with $f(x)$. Its output is used to calculate the softmax between the real next template and the output of the network. The softmax is calculated as

$$P(f(x)) = \frac{e^{f(x)}}{\sum\limits_{i=1}^{A} e^{f_i(x)}} \tag{6}$$

It calculates a distribution over the all possible templates. The one with the maximal probability is considered the most likely template to appear given the input sequence of templates.

LSTM architecture is a deep learning neural network method used for efficiently modelling sequential data. The representation of the system state is given via a single vector, refer to as a hidden state. The assumption the method is making, builds on top of the Markov property. It states that the state of the system at any particular point in time can be determined just from the previous state. To achieve this goal, it utilizes a selection mechanism build on abstractions of input, output and forget gates. This mechanism allows the network to selectively choose how much information from the previous inputs it should preserve and distribute towards the output. Hence it can model short and long term dependencies within a sequence and the structure appearing into the sequence of state events. Thus it is a handy solution for modelling our problem. Stacking of multiple LSTM cells provides greater representational power of the architecture.

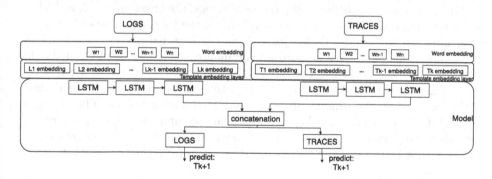

Fig. 2. Proposed architecture for joint analysis of logs and traces.

3.4 Multimodal LSTM

To account for both modalities and enable end to end learning system for anomaly detection, we propose the method as given on Fig. 2. It is composed of two models described in the previous section. On the inputs provided are the dictionary of logs and spans, simultaneously, to each of the two models. However, the output of both LSTMs is concatenated to one another and fed through an additional linear layer. It gives an advantage of including the information from both of the modalities, to improve the predictive performance. The shared information from the concatenation is then passed through two linear layers, one accounting for the traces and the other for the logs.

To account for both modalities the cost function is also changed. We calculate it as a joint cross-entropy loss of the most likely span and log to appear, given the joint information in a particular period. We calculated the joint loss as follows:

$$L((s,l), f(x,y)) = L(f(x), s) + L(f(y), l) \tag{7}$$

where $L(\cdot, \cdot)$ account for the categorical-cross entropy loss, and s and l for the ground truth span and log templates that should appear as the next relevant templates. Because the loss function includes the information from both modalities when the back-propagation step is done the gradients are calculated based on the information from both of the modalities.

One important detail for joint training the two modalities is providing the information from the same time intervals to the model from both of the modalities. The granularity representation of a log message is on a single time interval, on one side, and the spans span across multiple time stamps. To address this challenge we address block of logs of varying size. The size of a block of log messages is dependent on the corresponding spans within the trace appearing during the particular time interval. To create a block of log messages we stack multiple logs together to pair up with the corresponding time intervals determined by the spans. Such an approach requires the introduction of a maximal number of logs that are considered at once.

Given this coupling between the traces and logs, the question to ask is "What is the learning task for the joint method?". Since the time spanning of the spans determine the size of log blocks, just a *window_size* parameter on the traces imposed is. This parameter determines the number of spans the method should use to produce the next one. The block of log messages is created in a way that, the log messages that come from the start time of the first and the end time of the last span in the window of spans are joined into one block. The target is to predict the next expected log. An additional complication that can arise is the absence of logs in a particular time frame. To address this, we denote those windows that have a missing target and drop them from the learning set.

3.5   Anomaly Detection

NTP is utilized for anomaly detection for logs, however, the anomaly detection in the traces require additional anomaly detection procedure. We further provide a simple and effective method that acts on the output from the NTP solver to detect if there is an anomaly or not. The anomaly detection procedure for the single modality log model considers a log as normal if the prediction for the log is in the next *top_k_logs*. Otherwise, it is predicted as an anomaly.

For the detection of anomalous trace, the decision procedure should take into consideration the correct prediction among all of the spans in the trace subject to prediction. A span is correctly predicted if, for a given input sequence of spans, the true span is in the *top_k_span* ranked spans. For each trace, this procedure creates an accumulation of the correctly predicted spans. The ratio of incorrectly predicted spans (span error rate) $\frac{num\_err}{length(trace)}$ is considered as an anomaly score for the trace. Setting a threshold on this score can be used for anomaly detection. Finally, for the joint multimodal method, a combination of the previously described techniques is utilized.

4   Experiments and Results

In this section, we first describe the experimental design we used for evaluation. Second, we provide a detailed analysis of the results from the experiments to justify the improvements the joint information provides. Finally, we discuss the span2vec embedding as a consequence and further contribution of this work.

4.1   Experiments

Dataset Preprocessing Details. In the experiments we used the publicly available dataset[1] covering the trace and logs as monitoring components in overlapping time intervals. To the best of our knowledge, this is the only available dataset suited for multi-modal anomaly detection in distributed systems and as such it is utilized.

The experiments are generated from an OpenStack deployment testbed. We used the concurrent execution scenario, with 3 execution workloads: create an image, create a server, create a network, as described in [8]. As such we demonstrate the usefulness of our method in scenarios as close to real-world execution.

Train Test Split. The training dataset is composed of the traces appearing up to a particular time point, such that 70% of the normal traces are contained. The anomalous traces during this time-window are discarded. The logs that belong in the corresponding time intervals as generated by the trace are also preserved in the training set. We aim of modelling the normal behaviour of the system with preserving the normal traces and normal logs. To evaluate our model, the test set is composed of all of the remaining logs and traces appearing after the split time point.

Baselines. The main aim of this work is to demonstrate that the shared information between the logs and traces can improve anomaly detection in comparison to anomaly detection methods build from single modalities. As baselines we use the single modality LSTM method build separately for the traces and logs. The models are built on the same dataset as the multi-modal model and tested on the same test set to allow for a fair comparison.

---

[1] https://zenodo.org/record/3549604.

Table 1. Results from the experimental evaluation.

Score	Logs-joint	Trace-joint	Single logs	Single traces
Accuracy	0.976	0.990	0.974	0.955
Precision	0.904	0.992	0.897	0.992
Recall	0.996	0.984	0.996	0.909
f1	0.948	0.988	0.944	0.949

Implementation Details. The first step of the data preprocessing requires settings the values for the Drain parser. The values for the similarity and depth were set to 0.5, 0.4 and 4, 4, for the logs and traces accordingly. These values provide a concise template as evaluated by the domain expert. The $N\_embedding$ is set to 256. For the $window\_size$ parameter for the traces the value is set to 3. For optimization of the cost functions for the single and multiple modalities methods, we use SGD solver with standard values for the $learning\_rate = 0.001$ and $momentum = 0.9$. The $batch_size$ is set to 256 as a commonly chosen values. The number of $epochs$ is 100 for all of the tested methods.

For the anomaly detection procedure we further require the $logs\_top\_k$ and $trace\_top\_k$ parameters. They are set to 20 and 1 accordingly. For the error threshold on the anomaly score, the best value between 0.05 and 1 with a step of 0.05 chosen is.

4.2 Results

Table 1 summarize the results from the experiments. Firstly, one can observe that the results from the single modalities methods show that for the logs and traces, individually the approach can provide good results. It shows that the assumption made by the NTP task solver is sufficient for successful modelling of the normal state of the system.

Comparison of the results from the columns Trace-joint and Trace-single suggest that there is an improvement of the results for the traces for the multimodal method. More specifically, there can be observed improved value on the recall for the joint model for the traces in comparison to the single one. This suggests that the addition of the additional information from the logs can increase the number of correct predictions for the anomalous traces. The improvement is further depicted in the increased value for the F1 score on the joint traces. The results on the logs do not seem that change too much. One explanation of this behaviour is that the granularity of the information from the logs is truncated on the level of the data source with a lower frequency of generation - the trace is harder for the information in the trace to be transferred to the logs. The information that the multimodal method is receiving from the logs when it is aiming to predict the next relevant span complements the information as obtained just from the sequence of spans individually.

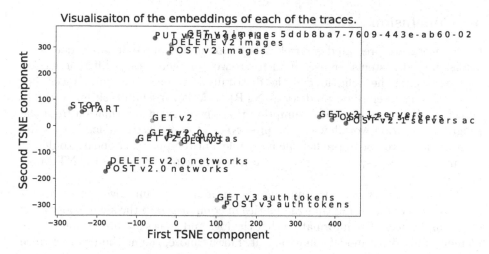

Fig. 3. Span2Vec embedding of the events in the tracing data from the whole vocabulary of spans for the three different workloads.

4.3   Span2Vec and Log2Vec

One element of the method is the ability to learn to embed both the logs and spans. The logs and spans are composed of words represented as vectors. The vectors are learned during the optimization procedure. Hence are optimized for the specific NTP task. Since the logs and spans are linear combinations from these words, pooling over the words belonging to the same span/log can be used to provide a unique vector mapping for them.

Figure 3 depicts a two-dimensional representation of the vector space of the spans embeddings. Three operations are executed. Close observation reviles that spans that are specific for a workload occur close to one another, while the ones that are shared co-occur in groups of their owns. For example, the spans *GET /v2.0/images/, PUT /v2.0/image/, GET /v2/images/* and *POST /v2.0/networks/* are unique for *create delete image* workload. As it can be observed, these spans are very close to one another in comparison to the other spans like the pair *POST /v2.0/networks* and *DELETE /v2.0/network/*. On the other side, the artificially added spans like *START* and *STOP* or the authentication span each of the workloads is utilizing are grouped, separated from the workload-specific spans. Close inspection of the Euclidean distance between the spans confirms the observations from the TSNE vector representation. The importance of these embeddings is the most emphasised in their future reuse for warm starting the methods. This can reduce the adoption time and the difficulty when a new machine model is deployed in production.

5 Conclusion

In this work, we presented a novel method for multi-source anomaly detection in distributed systems. It uses data from two complementary different modalities describing the behaviour of the distributed system - logs and traces. We utilize the next template prediction (NTP) task as a pseudo task for anomaly detection. It is based on the assumption that the relevant information from the program execution workflow can be preserved into one vector. Then it uses the corresponding vector to predict the most relevant template to appear. To detect the anomaly, a post-processing step that acts on the predictions of the NTP task is used.

The results show that the multimodal approach can improve the scores for anomaly detection for multiple modalities in comparison to the single modalities of logs and traces. The information that the logs and traces are preserving is complementary and the model can exploit it. Furthermore, the method can produce vector representation for both the logs and traces. These vector embeddings are used as a good bias for transferring and reusing the accumulated knowledge for faster training and adaptation.

In future work, we would investigate how adding additional information from the metric data can be incorporated into the model. It will allow for the creation of a unified model of the whole system behaviour, making the further processes of AIOps life-cycle easier. Additionally, we would investigate transfer learning approaches based on the generated embeddings. Specifically, we are interested in investigating how the learned embeddings can be reused for other types of workloads with a final aim to reduce the deploy time of the machine learning model in production.

References

1. Du, M., Li, F., Zheng, G., Srikumar, V.: DeepLog. In: Proceedings of the 2017 ACM SIGSAC Conference on Computer and Communications Security. Association for Computing Machinery, New York, NY, United States, pp. 1285–1298 (2017)
2. He, P., Zhu, J., He, S., Li, J., Lyu, M.R.: Towards automated log parsing for large-scale log data analysis. IEEE Trans. Dependable Secure Comput. **15**, 931–944 (2018)
3. He, P., Zhu, J., Zheng, Z., Lyu, M.: Drain: An online log parsing approach with fixed depth tree. In: IEEE International Conference on Web Services (ICWS). Curran Associates, Red Hook, NY, USA, pp. 33–40 (2017)
4. Ikeda, Y., Ishibashi, K., Nakano, Y., Watanabe, K., Kawahara, R.: Anomaly detection and interpretation using multimodal autoencoder and sparse optimization. arXiv preprint arXiv:1812.07136 (2018)
5. Lou, J.G., Fu, Q., Yang, S., Xu, Y., Li, J.: Mining invariants from console logs for system problem detection. In: Proceedings of the 2010 USENIX Conference on USENIX Annual Technical Conference. USENIX Association, USA, p. 24 (2010)
6. Meng, W., et al.: Loganomaly: unsupervised detection of sequential and quantitative anomalies in unstructured logs. In: Proceedings of the Twenty-Eighth International Joint Conference on Artificial Intelligence, IJCAI-19. International Joint Conferences on Artificial Intelligence Organization, pp. 4739–4745 (2019)

7. Nedelkoski, S., Bogatinovski, J., Acker, A., Cardoso, J., Kao, O.: Self-supervised log parsing. arXiv preprint arXiv:2003.07905 (2020)
8. Nedelkoski, S., Bogatinovski, J., Mandapati, A.K., Becker, S., Cardoso, J., Kao, O.: Multi-source distributed system data for AI-powered analytics. In: Brogi, A., Zimmermann, W., Kritikos, K. (eds.) ESOCC 2020. LNCS, vol. 12054, pp. 161–176. Springer, Cham (2020). https://doi.org/10.1007/978-3-030-44769-4_13
9. Nedelkoski, S., Cardoso, J., Kao, O.: Anomaly detection and classification using distributed tracing and deep learning. In: 19th IEEE/ACM International Symposium on Cluster. Cloud and Grid Computing (CCGRID), IEEE Computer Society, Los Alamitos, CA, USA, pp. 241–250 (2019)
10. Nedelkoski, S., Cardoso, J., Kao, O.: Anomaly detection from system tracing data using multimodal deep learning. In: 2019 IEEE 12th International Conference on Cloud Computing (CLOUD). IEEE Computer Society, Los Alamitos, CA, USA, pp. 179–186 (2019)
11. Park, D., Erickson, Z., Bhattacharjee, T., Kemp, C.C.: Multimodal execution monitoring for anomaly detection during robot manipulation. In: IEEE International Conference on Robotics and Automation (ICRA). Curran Associates, Red Hook, NY, USA, pp. 407–414 (2016)
12. Sridharan, C.: Distributed Systems Observability: A Guide to Building Robust Systems. O'Reilly Media (2018)
13. Srivastava, N., Salakhutdinov, R.: Multimodal learning with deep boltzmann machines. J. Mach. Learn. Res. **15**, 2949–2980 (2014)
14. Tang, L., Li, T., Perng, C.S.: Logsig: generating system events from raw textual logs. In: Proceedings of the 20th ACM International Conference on Information and Knowledge Management. Association for Computing Machinery, New York, NY, USA, pp. 785–794 (2011)
15. Xu, W., Huang, L., Fox, A., Patterson, D., Jordan, M.I.: Detecting large-scale system problems by mining console logs. In: Proceedings of the ACM SIGOPS 22nd Symposium on Operating Systems Principles. Association for Computing Machinery, New York, NY, USA, p. 117–132 (2009)
16. Yang, Y., Wang, L., Gu, J., Li, Y.: Transparently capturing request execution path for anomaly detection. arXiv preprint arXiv:2001.07276 (2020)
17. Zhang, Y., Sivasubramaniam, A.: Failure prediction in ibm bluegene/l event logs. In: Seventh IEEE International Conference on Data Mining (ICDM 2007), pp. 583–588 (2007)
18. Zhu, J., He, S., Liu, J., He, P., Xie, Q., Zheng, Z., Lyu, M.R.: Tools and benchmarks for automated log parsing. http://arxiv.org/abs/1811.03509 (2018)

TELESTO: A Graph Neural Network Model for Anomaly Classification in Cloud Services

Dominik Scheinert[✉] and Alexander Acker

Distributed and Operating Systems Group, TU Berlin, Berlin, Germany
{Dominik.Scheinert,Alexander.Acker}@tu-berlin.de

Abstract. Deployment, operation and maintenance of large IT systems becomes increasingly complex and puts human experts under extreme stress when problems occur. Therefore, utilization of machine learning (ML) and artificial intelligence (AI) is applied on IT system operation and maintenance - summarized in the term AIOps. One specific direction aims at the recognition of re-occurring anomaly types to enable remediation automation. However, due to IT system specific properties, especially their frequent changes (e.g. software updates, reconfiguration or hardware modernization), recognition of reoccurring anomaly types is challenging. Current methods mainly assume a static dimensionality of provided data. We propose a method that is invariant to dimensionality changes of given data. Resource metric data such as CPU utilization, allocated memory and others are modelled as multivariate time series. The extraction of temporal and spatial features together with the subsequent anomaly classification is realized by utilizing TELESTO, our novel graph convolutional neural network (GCNN) architecture. The experimental evaluation is conducted in a real-word cloud testbed deployment that is hosting two applications. Classification results of injected anomalies on a cassandra database node show that TELESTO outperforms the alternative GCNNs and achieves an overall classification accuracy of 85.1%. Classification results for the other nodes show accuracy values between 85% and 60%.

Keywords: Anomaly classification · Cloud computing · Cloud services · Time series classification · Graph neural network

1 Introduction

The rapid evolution of IT systems enables the development of novel applications and services in a variety of fields like medicine, autonomous transportation or manufacturing. Requirements of high availability and minimal latency together with general growth in distribution, size and complexity of these systems aggravate their operation and maintenance. Human experts require additional support to maintain control and ensure compliance with defined service level agreements (SLAs).

D. Scheinert and D. Acker—Equal contribution.

© Springer Nature Switzerland AG 2021
H. Hacid et al. (Eds.): ICSOC 2020 Workshops, LNCS 12632, pp. 214–227, 2021.
https://doi.org/10.1007/978-3-030-76352-7_23

Therefore, monitoring systems are employed to collect key performance indicators (KPIs) like network latency and throughput or system resource utilization from relevant IT system components. They provide detailed information about the overall system state, which can be used to identify imminent SLA violations.

One specific research direction in that area utilizes methods from machine learning (ML) and artificial intelligence (AI) for operation and maintenance of IT systems (AIOps) [4,9]. It includes methods for anomaly detection to identify problems ideally before SLAs are violated, anomaly localization to determine the origin of an ongoing anomaly, as well as recommendation and auto-remediation methods to execute actions and transfer anomalous system components back to a normal operation state. Significant research work is done on methods for anomaly detection [17,21] and root cause analysis [22]. However, existing solutions mostly fail to propose a holistic approach for an automated remediation execution. This is essential to autonomously transfer anomalous system components back into a normal operation state. The main reason is the focus on unsupervised methods that are usually trained on one class - the normal state. During the detection phase deviations from the learned normal class are labeled as anomalies. Although this is important to enable the detection of previously unseen anomalies, it imposes suboptimal implications. A generic anomaly class that summarizes all types of anomalies either implies actions that are able to remediate all anomaly types or pushes the responsibility for selecting an appropriate remediation to a subsequent instance - usually a human expert. Therefore, we propose a method to recognize reoccurring anomalies by training a classification model. Utilizing system metric data like CPU utilization, allocated memory or disk I/O statistics and model those as multivariate time series, our model is able to identify anomaly type specific patterns and to assign respective anomaly labels to those. Currently proposed time series classification methods assume a static dimensionality of input data, which is usually not the case for IT systems, which undergo frequent changes due to software updates, hardware modernization, etc.

To enable the automation of anomaly remediation, we propose a novel anomaly type classification solution, which is utilized to detect reoccurring anomaly types. To this end, our proposed model architecture TELESTO utilizes a novel graph neural network architecture to exploit multivariate time series modeled as graphs both in the spatial and temporal dimension. It is invariant to changing dimensionality and outperforms two other commonly used graph neural network methods.

The rest of the paper is structured as followed. In Sect. 2, we describe the preliminaries for our approach and present TELESTO in detail. A consolidation of the conducted evaluation is given in Sect. 3, encompassing the hyperparametrization and training setup, the testbed and experiment design as well as the results of the anomaly classification and their discussion. An excerpt of related approaches is presented in Sect. 4 capturing the state of the art of time series classification in the domain of anomalies. Lastly, Sect. 5 concludes this paper and gives an outlook for future work.

2 Anomaly Classification on Time Series Graphs

Our proposed model architecture operates on graphs and utilizes graph convolution to exploit both the spatial and temporal dimension of KPIs modelled as multivariate time series.

2.1 Preliminaries

AIOps systems require monitoring data, which is typically retrieved in form of tracing, logging and resource monitoring metrics. Latter are usually referred to as key performance indicators (KPIs). These can be formally expressed as time series, i.e. a temporally ordered sequence of vectors $X = (X_t(\cdot) \in \mathbb{R}^d : t = 1, 2, \ldots, T)$, where d is the dimensionality of each vector and T defines the last time stamp, at which a sample was observed. For $X_b^a(\cdot) = (X_a(\cdot), X_{a+1}(\cdot), \ldots, X_b(\cdot))$, we denote indices a and b with $a \leq b$ and $0 \leq a, b \leq T$ as time series boundaries in order to slice a given series $X_T^0(\cdot)$ and acquire a sub-series $X_b^a(\cdot)$. Additionally, we use the notion $X(i)$ to refer to a certain dimension i, with $1 \leq i \leq d$.

Our proposed method for anomaly classification relies on modelling time series as graphs. A Graph $G = (V, E)$ with n nodes consists of a set of vertices $V(G) = \{v_1, \ldots, v_n\}$ and a set of edges $E(G) \subseteq \{\{v_i, v_j\} | v_i, v_j \in V(G)\}$. An edge $\{v_i, v_j\} \in E(G)$ is a connection, i.e. an unordered pair, between vertex i and j, thus v_j is called a neighbor of v_i written as $v_i \sim v_j$. The adjacency matrix A of a graph G is an $n \times n$ matrix with entries a_{ij} such that $a_{ij} = 1$ if a connection $v_i \sim v_j$ exists, otherwise 0.

To represent time series as graphs, a sliding window of size w with a configurable stride is moved along the temporal dimension, extracting slices of time series data. This is also illustrated in Fig. 1, whereby each red rectangle is transformed into a graph with one node per series. Formally, the set of vertices for a graph G at time t is defined as

$$V^{(t)}(G^{(t)}) = \{v_i = \mathcal{F}(X_t^{t-w}(i)) \mid i = 1, 2, \ldots, d\}. \tag{1}$$

Thereby, $\mathcal{F}$ is a filter, extracting features from time series $X_t^{t-w}(i)$. Edges are used to express the relationship between time series feature vectors and can be either inferred from available data or set manually. We assume that KPIs where collected during known system states, i.e. either normal or one of a set of known anomaly types C. Therefore, we assign a label $c \in C$ to each graph $G^{(t)}$, defining them as tuples $(G^{(t)}, c)$.

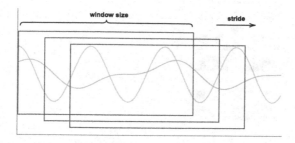

Fig. 1. A window is moved along the temporal dimension with a configurable stride while slices of time series data are extracted. Each slice is transformed into a graph.

2.2   The Architecture of TELESTO

Anomaly classification is done based on multivariate time series modelled as graphs, thus graph classification is required. Therefore, we employ a class of neural networks which incorporates concepts from graph theory. Graph convolutional neural networks (GCNNs) aim to generalize the convolution operation to be applied in non Euclidean domains. We utilize this to model the spatial domain of multivariate time series. Each node of a graph can have an arbitrary number of neighbors, thus making the method invariant to changing dimensionality. The convolution operation is applied on the neighborhood of each graph node. GCNN methods can be roughly clustered into spectral and spatial methods. Spectral methods are establishing frequency filtering by levering the fourier domain and the graph Laplacian. Spacial methods are essentially defining the graph convolution in the vertex domain by leveraging the graph structure and aggregating node information from the neighborhoods in a convolutional fashion. A comprehensive survey of existing methods was conducted in [25].

We propose TELESTO, a novel model architecture for graph classification consisting of multiple spatial methods. Our architecture is illustrated in Fig. 2. Building upon the definition in Eq. 1, $\mathcal{F}$ consists of a positional encoding and a subsequent 1D convolution layer. As argued and evaluated in [19], positional encoding allows for the injection of information about the relative or absolute position of values in a sequence. The convolution layer extracts N features from each time series $X_t^{t-w}(i)$. Thereby a constant filter size of 3 is used together with a suitable zero-padding to ensure the equivalence of input and output dimensions. Batch normalization is applied on the features to stabilize the training [11]. Initially, a fully connected graph is used resulting in an all-one adjacency matrix. A dropout layer randomly sets adjacency matrix entries to zero in order to achieve better generalization [18]. The dropout operation is followed by a single linear layer, which allows the model to adjust the dimensionality of the feature vector for the *graph transformation* module, i.e. by mapping from the node feature dimension to the dimensionality of the graph transformation module.

The graph transformation module is a core component of our approach and is composed of multiple levels/blocks. On each of its levels, a sublayer (or residual) connection (SLC) is used. This eases the training of deep neural network

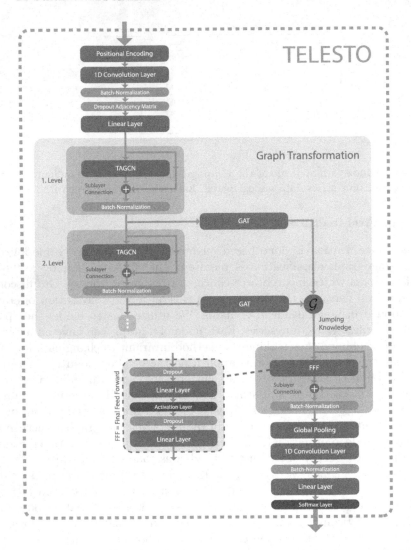

Fig. 2. The architecture of TELESTO. Green components indicate methods for generalization and red blocks indicate activation functions. An example of a graph transformation module with two blocks is depicted but can be arbitrarily increased. (Color figure online)

architectures [10]. For the graph transformation operation, topology adaptive graph convolutional networks (TAGCN) [5] and graph attention networks (GAT) [20] are used. Both aim to generate meaningful node feature representations for the task of anomaly classification. TAGCNs realize graph convolution on the vertex domain by applying a set of fixed-size learnable filters which are adaptive to the topology of the respective target graph. The TAGCN result is further used as input for a GAT layer which enables nodes to attend to generate features dependant on the vertex neighborhood features. Since we employ GAT layers with multi-head attention, the choice of the number of heads has a direct impact on the output dimensionality of the graph transformation module as the outputs of all parallel attention mechanisms are concatenated. At each GCNN level, the TAGCN layer exploits the spatial structure of its input graph and respectively updates the node features. Those are forwarded to the GAT layer as well as to the next TAGCN layer. TELESTO allows this combination of TAGCN and GAT layers to be arbitrarily stacked. Finally, the node feature matrices collected from the respective GAT layers across all levels are aggregated using Jumping Knowledge (JK) [24]. JK combines the GAT layer outputs by a configurable aggregator function $\mathcal{G}$. We choose the long short-term memory (LSTM) aggregator, due to its flexibility to learn weight coefficients and thus a weighted combination of the given matrices. This allows the model to prioritize certain level outputs.

The graph transformation module is followed by a final feed forward (FFF) block and a sublayer connection. The FFF block consists of two dropout layers and two linear layers in alternating order with a single activation layer in between. Next, graph embeddings are computed by utilizing the global pooling method *Global Attention* [16]. It employs a neural network to learn attention coefficients based on node features, which are then used to aggregate nodes graph-wise, resulting in an embedding for each graph. A final 1D convolution is applied on the graph embedding and the result is batch-normalized afterwards. The convolution layer is configured to use ten filters, a filter size of nine and suitable zero-padding to ensure the equivalence of input and output dimensions. The resulting ten feature maps are averaged. Assuming a classification problem with C classes, an additional linear layer needs to be added to the bottom of the architecture to transfer the graph embedding size to an output vector of size C. A subsequent softmax layer calculates a probability distribution over all classes.

3   Evaluation

We investigate a case study based on anomalous services deployed in an infrastructure as a service (IaaS) cloud environment to evaluate our model. Experiments are conducted to assess its goodness by classification of synthetically injected anomalies.

3.1   Testbed and Experiment Design

To evaluate the capabilities of our model, we deploy a cloud infrastructure and use a IaaS policy to run two applications. OpenStack 11.0.0 Stein[1] and Ceph 12.2.5 luminous[2] are installed on a commodity cluster of 21 nodes, each possessing an Intel Xeon X3450 CPU (4 cores @ 2.66 GHz), 16 GB RAM, 3×1 TB HDD and

---

[1] https://www.openstack.org/software/stein/.
[2] https://docs.ceph.com/docs/master/releases/luminous/.

2 × 1 GBit Ethernet connection. Twelve nodes are used as hypervisors, eight as storage nodes and one as a network and controller node. Furthermore, three hypervisor host groups (1/1/10 split) are created to separate the application load generation from the application deployment itself. A virtual IMS[3] and a content streaming service (CS) are used as example applications hosted within the cloud. Varying load is generated against both simulating user access. All VMs operate on Ubuntu 16.04.3 LTS under Linux kernel version 4.4.0-128-generic. All deployment scripts related to both services are available at github[4].

The proposed anomaly classification method requires labeled data to be evaluated. Therefore, different anomaly types are synthetically injected into the VMs and hypervisors at runtime. An injector agent was deployed on each hypervisor and application host VM. All injected anomaly types are listed in Table 1. A group-based injection policy was used throughout the experiments, means that all VMs of one service group (e.g. bono, hypervisor, etc.) are regarded as one entity. Note that all ten hypervisors are put into one component group. An injection into any component counts as an injection for the whole group. During the experiment, an initial period of six hours without anomaly injections is defined. Next, each anomaly is injected five times into each group for four to five minutes. After that, one minute of grace time is waited until the next injection is performed, i.e. there are no overlapping injections. Start and stop times are logged and used as the ground truth. Agents are deployed on the hypervisors for monitoring that sample KPIs such as CPU utilization or network I/O statistics at a frequency 2 Hz.

Table 1. List of injected anomalies and their abbreviations together with their respective description and parametrization.

Anomaly	Description
CPU overutilization (CPU)	Utilize 90% of available CPU
Abnormal disk utilization (ADU)	Constant disk read and write operations
Memory leak (MEL)	Incremental allocation of x MB main memory every y seconds x = 1, y = 3 for vIMS and CS VMs x = 2, y=3 for Hypervisors
Abnormal memory allocation (AMA)	Allocate x MB of memory x=450 for vIMS VMs x=900 for CS VMs x=2000 for Hypervisors
Network overload (NOL)	Start to download large files

[3] https://www.projectclearwater.org.
[4] https://github.com/IncrementalRemediation/testbed_deployment.

3.2   Hyper-Parametrization and Training Setup

Monitored KPIs of each node are modelled as multivariate time series X. Each sample of every single series is $X_t(i)$ is preprocessed by rescaling to the range $(0, 1)$ with min-max normalization. The limits of most KPIs are well-known (e.g. number of CPU cores and their frequency etc.). For some KPIs like latencies between network endpoints, context switches or cache misses, it is challenging to set upper or lower boundaries. Those are determined within the training dataset and used throughout testing.

Table 2. Hyper-Parametrization and Training Setup

Aspect	Configuration
Hardware	GeForce RTX 2080 Ti GPU
Implementation	PyTorch, PyTorch Geometric [7]
Graph Construction	Window size = 20, stride = 1
Optimizer	Adam [14] with learning rate = 10^{-3}, $\beta_1 = 0.9$, $\beta_2 = 0.999$
Regularization	Weight decay = 10^{-5}, dropout probability = 50%
Loss	Cross Entropy
Diverse	Epochs = 15, Batch size = 128, Xavier [8] weight initialization

Most specifications related to the training of TELESTO are listed in Table 2. We employ leave one group out (LOGO) cross-validation for data splitting, i.e. the five injections of each anomaly and each service component are split as 3/1/1 as a training/validation/test split. For TELESTO itself, we choose a graph node feature dimensionality of 64 and set the number of graph transformation levels to 5. For the TAGCN layers, we choose $k = 3$ fixed-size learnable filters as recommended in [5]. For the GAT layers, we choose $K = 8$ parallel attention mechanisms to produce rich node features with multi-head attention. Lastly, the JK LSTM-aggregator is equipped with seven layers in order to learn a reasonable node weighting based on node features. We choose ELU [3] as activation function for the FFF block. The final softmax calculates a distribution over anomaly classes, whereof the highest is used as the prediction target.

3.3   Anomaly Classification

In this section, the proposed model architecture will be evaluated on the data described in subsubsection 3.1. TELESTO and its default configuration is compared against a GCN architecture [15] and a GIN architecture [23]. GCN is a reasonable choice as it is a common benchmark, whereas GIN is selected due to it achieving state-of-the-art results both for node classification and graph classification on several benchmark data sets [23]. All models are trained on each service node individually. Moreover, each experiment is run 10 times and

Table 3. The results of anomaly classification on the cassandra data set.

Model	Metric	Split 1	Split 2	Split 3	Split 4	Split 5	∅
GCN	Accuracy	0.389	0.463	0.360	0.398	0.385	0.399
	Recall	0.389	0.462	0.358	0.396	0.385	0.398
	Precision	0.272	0.428	0.256	0.222	0.288	0.293
	F1-Score	0.310	0.436	0.295	0.269	0.320	0.326
GIN	Accuracy	0.447	0.455	0.562	0.452	0.400	0.463
	Recall	0.446	0.455	0.562	0.452	0.399	0.463
	Precision	0.408	0.422	0.569	0.363	0.289	0.410
	F1-Score	0.421	0.436	0.564	0.402	0.335	0.432
TELESTO	Accuracy	0.796	0.804	0.894	0.822	0.939	0.851
	Recall	0.796	0.803	0.894	0.822	0.939	0.851
	Precision	0.825	0.732	0.920	0.870	0.956	0.861
	F1-Score	0.810	0.764	0.906	0.844	0.948	0.854

the results are averaged in order to cancel out the effects of unfavorable weight initialization. Both the GCN architecture and the GIN architecture utilize two of their respective layers. For graph classification, the node features are added across the node dimension for each graph, followed by two linear layers with dropout in between and a final softmax layer. Specific to the GCN architecture is the hidden layer size of 32 and the row-normalization of input feature vectors. For the GIN architecture, the hidden layer size is set to 64 while for each GIN layer, the input is batch-normalized, the initial value of ϵ is set to 0 and a multilayer perceptron (MLP) is internally used for mapping the node features from the input dimension to the hidden dimension. If not specified otherwise, we use the values from the setup summarized in Table 2.

A detailed breakdown of the results is given in Table 3 with a focus on the cassandra service node. It can be seen that the proposed model outperforms both comparative models. Therefore we conclude that the proposed model is suitable for anomaly classification based on multivariate time series data. Note that for TELESTO, the average F1-score of 0.854 is close to its reported average accuracy. In general, this is an indication for a good model as the balanced recall and precision are not strongly different from the accuracy. In contrast to that, both comparative models appear to encounter difficulties during training. The GCN architecture achieves an average F1-score of 0.326 whereas the GIN architecture performs comparably better with an average F1-score of 0.432. It can be observed that the reported F1-scores differ from the achieved average accuracy of these architectures. Table 3 also shows a high variability between splits. For instance, TELESTO achieves an average F1-score of 0.906 on split 3 but an average F1-score of 0.764 on split 2. Moreover, an investigation of the corresponding confusion matrices shows that confusion exists between CPU anomalies and MEL anomalies as well as AMA anomalies and MEL anomalies. Similar observations

Table 4. The results of anomaly classification with TELESTO on all service nodes. The table shows the achieved accuracy scores for each split and in average across all splits.

Data set	Split 1	Split 2	Split 3	Split 4	Split 5	∅
Cassandra	0.796	0.804	0.894	0.822	0.939	0.851
Bono	0.865	0.853	0.825	0.566	0.729	0.768
Sprout	0.818	0.797	0.660	0.567	0.597	0.688
Backend	0.655	0.792	0.714	0.912	0.731	0.761
Chronos	0.806	0.741	0.984	0.898	0.620	0.810
Homer	0.214	0.529	0.725	0.846	0.677	0.598
Astaire	0.630	0.716	0.931	0.702	0.744	0.744
Load-balancer	0.801	0.880	0.989	0.767	0.680	0.823
Homestead	0.539	0.599	0.855	0.730	0.744	0.694

regarding diverse F1-scores can be made for all models between multiple splits. One possible explanation might be the high variability in simulated user load during our experiments. This resulted in a broad range of system states, from almost idle to almost over utilized. With the overall available training data being limited, such high load variability might lead to significant differences in classification performances between splits. For TELESTO, we observed that although the validation accuracy proportionally increases with the training accuracy, the losses on both data sets structurally diverge after a few epochs already, leading to an overfitting of the model which is intensified by the models complexity.

For completeness, we report the results of anomaly classification with TELESTO on all other service nodes in Table 4 while reporting only accuracy scores to omit redundancy. It can be seen that the performance of TELESTO varies across different service nodes and splits. While an average accuracy of 0.851 can be achieved on cassandra, the average accuracy on homer is 0.598. The reported scores also strongly vary between splits on homer, between 0.214 average accuracy on split 1 and 0.846 on split 4. While not listed in Table 4, the comparative models exhibit high variance in accuracy over splits and remain on average significantly inferior to TELESTO.

3.4   Limitations

The main problem encountered during our extensive experimentation is the contradiction between the expected amount of labeled anomaly data and required data to train a reliable model. Labeling anomaly data by human experts is costly, anomalies occur sparsely and IT environments undergo constant changes so once labeled data deprecates over time. Therefore, the generalization ability of our model represented by the prediction scores reveal a significant variance in dependence of a specific split. We plan to investigate methods and heuristics to generate additional training data from few anomaly examples and thus, synthetically increase the available training data size.

Another aspect is the lack of expression regarding the temporal dependence between consecutive graphs. Although sequential information of time series within a graph are encoded via positional encoding, consecutive graphs constructed via a moving window over the time series are regarded as independent. Possibilities to encode temporal information from preceding graphs for the classification of subsequent graphs is subject to future work.

4   Related Work

Concrete methods of identifying reoccurring anomalies within IT systems is sparsely covered on public research. We formulate it as a time series classification problem. Therefore, we analyse related work in both areas, IT system anomaly classification and time series classification in general.

Bodik et al. [1] referring to the classification of different data center crises as data center fingerprinting. Thereby, system resource metrics from all data center components are aggregated via quantile discretization and feature selection methods are applied to choose relevant metrics for distinguishing different data center crisis types. The reported results were achieved by an aggregation of system resource metrics collected over 30 min. Kajó and Nováczki [12] provide a comparison of different machine learning algorithms together with a genetic algorithm approach. Given monitoring data from a System Architecture Evolution (SAE) core network they select an optimized metric subset that is used as input for different classification algorithms. The focus lies on metric selection and classification models are trained on 850 anomaly observations. Having sparse occurrences of anomaly situations the expectation of many training data is a major limit. Cheng et al. [2] applies a multi-scale long short-term memory model to classify four different anomaly types based on update messages of the border gateway protocol (BGP). The approach expects the classification models to be trained with several hours of anomaly data. However, anomaly situations usually do not persist for an extended period of time in production systems.

A variety of time series classification exists. We focus on most recent published approaches. InceptionTime [6] is an ensemble of deep CNNs for time series classification. Each CNN consist of multiple inception modules, whereas every module utilizes bottleneck layers for regularization and applies both a sliding max-pooling operation and multiple sliding filters of different lengths for feature extraction. In InceptionTime, multiple architecturally equivalent networks with different initial weight values are utilized and their prediction outputs are evenly weighted to obtain a final prediction result. Another approach which incorporates the advantages of LSTM networks is named LSTM-FCN [13]. It consists of two parallel processing streams. In the first stream, the temporal structure of the input data is exploited by an LSTM module. The second stream leverages alternating convolution layers, batch normalization layers and activation layers and a final global average pooling (GAP) layer. In the end, the concatenation of both stream outputs is used for classification. Another method named T-GCN is presented in [26]. The model combines graph convolutional networks together with

GRU and thus aims at capturing the spatial and temporal dependencies simultaneously. The T-GCN model can be seen as an improvement of LSTM-FCN, since a convolution is applied that is not bound to the euclidean domain. After that, the convolution result is processed by a recurrent neural network (RNN). Although a recent publication, T-GCN utilizes a graph convolution method that is outperformed by other GCNN models.

5   Conclusion

In this paper we presented TELESTO, a novel time series classification model to identify reoccurring anomalies in services deployed in a IaaS cloud environment. Therefore, we model KPIs of hypervisors and virtual machines that are hosting applications as multivariate time series. A method to transform multivariate time series into graphs is presented. The proposed model is based on GCNNs and thus, invariant to changes of the input dimensionality. We apply convolution on both the spatial and temporal dimension to extract a set of features that are used for classifying anomalies via graph classification. To evaluate the method, a cloud system together with two applications hosted within an IaaS service model were deployed. Synthetic injections of anomalies provided the required ground truth for evaluation. TELESTO was able to outperform two state of the art GCNNs, revealed promising results for anomaly classification and thus, is able to detect reoccurring anomalies in services deployed in cloud environments.

For future work we want to examine ways to encode temporal information from preceding graphs for the classification of subsequent graphs. Further, different time series augmentation methods can be tested to synthetically increase the amount of data.

References

1. Bodik, P., Goldszmidt, M., Fox, A., Woodard, D.B., Andersen, H.: Fingerprinting the datacenter: automated classification of performance crises. In: Proceedings of the 5th European conference on Computer systems (2010)
2. Cheng, M., Li, Q., Lv, J., Liu, W., Wang, J.: Multi-scale lstm model for bgp anomaly classification. IEEE Trans. Serv. Comput. (2018)
3. Clevert, D., Unterthiner, T., Hochreiter, S.: Fast and accurate deep network learning by exponential linear units (elus). In: 4th International Conference on Learning Representations, ICLR 2016, Conference Track Proceedings (2016)
4. Dang, Y., Lin, Q., Huang, P.: Aiops: real-world challenges and research innovations. In: IEEE/ACM 41st International Conference on Software Engineering: Companion Proceedings (ICSE-Companion). IEEE (2019)
5. Du, J., Zhang, S., Wu, G., Moura, J.M.F., Kar, S.: Topology adaptive graph convolutional networks. arXiv preprint arXiv:1710.10370 (2017)
6. Fawaz, H.I., et al.: Inceptiontime: finding alexnet for time series classification. Data Min. Knowl. Disc. **34**, 1936–1962 (2020). https://doi.org/10.1007/s10618-020-00710-y

7. Fey, M., Lenssen, J.E.: Fast graph representation learning with pytorch geometric. arXiv preprint arXiv:1903.02428 (2019)
8. Glorot, X., Bengio, Y.: Understanding the difficulty of training deep feedforward neural networks. In: Proceedings of the Thirteenth International Conference on Artificial Intelligence and Statistics (2010)
9. Gulenko, A., Wallschläger, M., Schmidt, F., Kao, O., Liu, F.: A system architecture for real-time anomaly detection in large-scale nfv systems. Procedia Comput. Sci. **94**, 491–496 (2016)
10. He, K., Zhang, X., Ren, S., Sun, J.: Deep residual learning for image recognition. In: Proceedings of the IEEE Conference on Computer Vision and Pattern Recognition (2016)
11. Ioffe, S., Szegedy, C.: Batch normalization: accelerating deep network training by reducing internal covariate shift. In: International Conference on Machine Learning (2015)
12. Kajó, M., Nováczki, S.: A genetic feature selection algorithm for anomaly classification in mobile networks. In: 19th International ICIN conference-Innovations in Clouds, Internet and Networks (2016)
13. Karim, F., Majumdar, S., Darabi, H., Chen, S.: Lstm fully convolutional networks for time series classification. IEEE access **6**, 1662–1669 (2017)
14. Kingma, D.P., Ba, J.: Adam: a method for stochastic optimization. In: 3rd International Conference on Learning Representations, ICLR 2015, Conference Track Proceedings (2015)
15. Kipf, T.N., Welling, M.: Semi-supervised classification with graph convolutional networks. In: 5th International Conference on Learning Representations, ICLR 2017, Conference Track Proceedings (2017)
16. Li, Y., Tarlow, D., Brockschmidt, M., Zemel, R.S.: Gated graph sequence neural networks. In: 4th International Conference on Learning Representations, ICLR 2016, Conference Track Proceedings (2016)
17. Nedelkoski, S., Cardoso, J., Kao, O.: Anomaly detection from system tracing data using multimodal deep learning. In: 2019 IEEE 12th International Conference on Cloud Computing (CLOUD). IEEE (2019)
18. Srivastava, N., Hinton, G., Krizhevsky, A., Sutskever, I., Salakhutdinov, R.: Dropout: a simple way to prevent neural networks from overfitting. J. Mach. Learn. Res. **15**(1), 1929–1958 (2014)
19. Vaswani, A., et al.: Attention is all you need. In: Advances in Neural Information Processing Systems (2017)
20. Veličković, P., Cucurull, G., Casanova, A., Romero, A., Liò, P., Bengio, Y.: Graph attention networks. In: International Conference on Learning Representations (2018)
21. Wetzig, R., Gulenko, A., Schmidt, F.: Unsupervised anomaly alerting for iot-gateway monitoring using adaptive thresholds and half-space trees. In: 2019 Sixth International Conference on Internet of Things: Systems, Management and Security (IOTSMS). IEEE (2019)
22. Wu, L., Tordsson, J., Elmroth, E., Kao, O.: Microrca: root cause localization of performance issues in microservices. In: IEEE/IFIP Network Operations and Management Symposium (NOMS) (2020)
23. Xu, K., Hu, W., Leskovec, J., Jegelka, S.: How powerful are graph neural networks? In: 7th International Conference on Learning Representations, ICLR 2019 (2019)

24. Xu, K., Li, C., Tian, Y., Sonobe, T., Kawarabayashi, K., Jegelka, S.: Representation learning on graphs with jumping knowledge networks. In: Proceedings of the 35th International Conference on Machine Learning, ICML 2018. Proceedings of Machine Learning Research, vol. 80. PMLR (2018)
25. Zhang, S., Tong, H., Xu, J., Maciejewski, R.: Graph convolutional networks: a comprehensive review. Comput. Soc. Netw. **6**(1), 1–23 (2019)
26. Zhao, L., Song, Y., Zhang, C., Liu, Y., Wang, P., Lin, T., Deng, M., Li, H.: T-gcn: a temporal graph convolutional network for traffic prediction. IEEE Trans. Intell. Transp. Syst. **21**(9), 3848–3858 (2019)

Discovering Alarm Correlation Rules for Network Fault Management

Philippe Fournier-Viger[1]([✉])(iD), Ganghuan He[1], Min Zhou[2],
Mourad Nouioua[1,3], and Jiahong Liu[1]

[1] Harbin Institute of Technology (Shenzhen), Shenzhen, China
[2] Huawei Noah's Ark Lab, Shenzhen, China
`philfv@hit.cn`
[3] University of Bordj Bou Arreridj, El Anceur, Algeria

Abstract. Fault management is critical to telecommunication networks. It consists of detecting, diagnosing, isolating and fixing network problems, a task that is time-consuming. A promising approach to improve fault management is to find patterns revealing the relationships between network alarms, to then only show the most important alarms to network operators. However, a limitation of current algorithms of this type is that they ignore the network topology. But the network topology is important to understand how alarms propagate on a network. This paper addresses this issue by modeling a real-life telecommunication network as a dynamic attributed graph and then extracting correlation patterns between network alarms called Alarm Correlation Rules. Experiments on a large telecommunication network show that interesting patterns are found that can greatly compress the number of alarms presented to network operators, which can reduce network maintenance costs.

Keywords: Fault management · Dynamic graph · Correlation patterns

1 Introduction

In today's society, telecommunication networks are key to support personal communications as well as those of businesses and other organizations. To ensure the proper operation of large telecommunication networks, a crucial task is fault management, which consists of detecting, diagnosing, isolating and fixing network problems. The purpose of fault management is to preserve network availability, security, reliability and optimize its performance [1]. However, a key issue with fault management for large and heterogeneous telecommunication networks (e.g. covering cities) is that millions of alarms may be generated by network devices, and that the number of technicians or budget for maintaining a network is limited [3]. Thus, it is easy for technicians to be overloaded with thousands of alarms and being unable to investigate all of them. For example, the telecommunication network of a medium-sized city typically contains multiple device types where some devices may produce more than 300 different alarms. The alarms

H. Hacid et al. (Eds.): ICSOC 2020 Workshops, LNCS 12632, pp. 228–239, 2021.
https://doi.org/10.1007/978-3-030-76352-7_24

are recorded by each network device and can be stored centrally and analyzed to support fault management. Moreover, data is also collected about QPIs (Quality Performance Indicators) of each network device over time. For network experts, understanding the relationships between alarms is not easy because faults are often caused by complex interactions between network devices.

To improve fault management, some expert systems were designed that rely on a knowledge base created by hand to find the causes of network problems [2]. But this approach is costly, time consuming, prone to errors and cannot adapt to changes. As an alternative, an emerging approach is to rely on pattern mining techniques to automatically discover relationships between alarms in alarm logs and then to hide (compress) alarms that are correlated with previous alarms [3–7]. It was shown that this can greatly reduce the number of alarms presented to network operators and thus reduce maintenance costs. But such approaches generally represent alarm log data as a sequence of alarms and the network topology is ignored [3–7]. But the topology is important to understand how alarms propagate on a network.

A promising research direction is thus to consider the network topology as a dynamic graph and to extract richer and more complex patterns from it to reveal complex temporal relationships between alarms. Though, several algorithms have been proposed to mine patterns in dynamic graphs, none is specifically designs for alarm analysis [16–19]. To find more complex relationships between alarms based on the network topology, this paper models alarms data as a network (i.e. a dynamic graph) where vertices are devices and edges are communication links. Moreover, alarms are viewed as spreading following the information flow (which depends on the topology) and where QPIs are represented as attributes of network devices. From this representation, this paper proposes to extract a novel type of patterns called Alarm Correlation Rules using a novel correlation measure named ACOR (Alarm CORrelation). An experimental evaluation with real data from a large telecommunication network shows that the proposed rules can provide greater alarm compression than the state-of-the-art AABD system [3].

The paper is organized as follows. Section 2 reviews related work. Section 3 presents the proposed framework. Then, Sect. 4 describes results obtained for a large scale telecommunication network. Finally, Sect. 5 draws a conclusion.

2   Related Work

To discover relationships between alarms in telecommunication networks, several studies have applied pattern mining techniques [3–7] such as association rule mining [11] episode mining [8–10] and sequential pattern mining (SPM) [12,13].

The first system to discover alarm patterns is TASA (Telecommunication Alarm Sequence Analyzer) [5,6]. It takes as input a sequence of alarms with timestamps and applies an episode mining algorithm to find alarms that frequently appear together within a sliding window. Moreover, TASA offers a separate module that applies association rule mining to find sets of properties that are common to alarm occurrences (while ignoring time). TASA was applied to data from several telecommunication service providers.

Lozonavu et al. [7] proposed a system for mining alarm patterns that first partitions the input alarm sequence into a set of sequences such that alarms having close timestamps are grouped together. Then, a SPM algorithm is applied to find all subsequences of alarms appearing in many of those sequences. Patterns are then used to generate a graph indicating relationships between alarms, where the confidence (conditional probability) that an alarm precedes another is calculated. This visualization can help network operators to understand the relationships between alarms. The system was applied to a 3G mobile network.

Wang et al. [3] proposed a system called AABD (Automatic Alarm Behavior Discovery). This system first filters out invalid alarms (e.g. with missing timestamps) and transient alarms (that appear only for a short time) from the input alarm sequence. Then, the most frequent alarms are identified and the input sequence is partitioned based on these alarms. Then, a SPM [12,13] algorithm is applied to find frequent alarm sequences. These patterns are then used to generate rules indicating that an alarm may be caused by another alarm to perform alarm compression (reduce the number of alarms presented to network operators). AABD achieved good compression for alarms of a real telecommunication network where it was shown that this approach based on transient alarm detection can reduce the number of alarms presented to operators by more than 84%. But the rule generation process of AABD relies on a knowledge base provided by domain expert, which is time-consuming to create and maintain.

An alarm management system adopting a similar approach was designed by Raúl et al. [4]. It takes as input an alarm sequence with time where alarms have attributes. A modified SPM algorithm was applied to extract sequences of alarms frequently appearing in a sliding-window. Patterns are selected based on three measures that are the support, confidence and lift. The system was applied to data from a large Portugese telecommunication company and patterns were used to reduce the number of alarms presented to the user by up to 70%.

The above pattern mining approaches to study network alarms are useful but handle simple data types, that is mostly discrete sequences where alarms are viewed as events that have some attribute values and timestamps. To extract patterns that consider the network topology and provide different insights, this paper considers a more complex data representation by adding the spatial dimension (the network topology) to the pattern mining process. The network is viewed as a dynamic graph where alarms are spreading along edges (communication links) between vertices (network devices) to find spatio-temporal patterns.

3   The Proposed Framework

This section presents the proposed framework for discovering alarm correlation rules and performing alarm compression. This framework is illustrated in Fig. 1. It consists of three main steps: (1) obtaining and pre-processing alarm and network topology data, (2) extracting alarm correlation rules from it, and (3) utilizing the rules to select alarms to be presented to the user. These three steps are described in details in the next paragraphs.

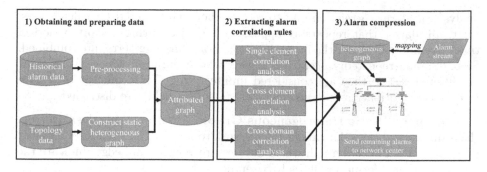

Fig. 1. The proposed alarm discovery and compression framework

Step 1. Obtaining and Preparing the Data. In previous studies, historical alarm logs were analyzed to find patterns involving multiple alarms. But most studies represent alarms log data as a sequence of alarms ordered by time. Because these studies ignore the network topology, it may lead to obtaining imprecise results or ignoring some important underlying patterns. In this work, we make the observation that telecommunication alarm data can be naturally modeled as a network (a dynamic graph) in which alarms can spread following the information flow. Thus, we not only consider the historical alarm log but also the network topology. The following paragraphs describes how these two types of data are obtained, pre-processed and then combined.

a) Preparing the Historical Alarm Log. The alarm log format considered in this study is presented in Table 1, and is more or less the same as in prior studies [3]. Each alarm has a name, a source (the device where the alarm was triggered), the domain of the device, an occurrence time and a clear time. For this study, five days of data was obtained from a large telecommunication network in Indonesia, from the 12th to 16th April, 2019. This dataset contains more than six million alarms, categorized into 300 types, triggered by different devices. To ensure privacy, alarm names and sources are not shown in Table 1.

Table 1. Part of an alarm log from an Indonesian telecommunication network

Alarm name	Domain	Alarm source	Occurrence time	Clear time
Alarm 1	ran-4g	Source 1	2019-04-12 10:40:23	2019-04-12 10:40:29
Alarm 2	Microwave	Source 2	2019-04-12 10:40:24	2019-04-12 11:30:44
Alarm 3	ran-2g	Source 3	2019-04-12 10:40:26	2019-04-12 10:40:36
...	...	...	...	...

After obtaining the alarm log, the proposed framework pre-processes the data to filter out some spurious alarms. This is done based on the recommendation of telecommunication network experts and allows to perform a more precise analysis of alarm correlation and to reduce the time required for calculations. First, all alarms that repeatedly appear in a device during a short period of time (five minutes as per the recommendation of domain experts) are combined. Second, some repeatedly occurring alarms whose duration time is very short are filtered out as they are considered uninteresting. Third, alarms that have incomplete information (e.g. an empty alarm source field) are discarded.

b) **Building a Static Heterogeneous Graph.** After preparing the alarm log, the framework obtains data about the network topology. This data is represented as a connected directed graph where devices are vertices and edges indicate the directions that information flows between devices.

In this study, the network topology was unavailable. Hence, a procedure was designed to extract the topology from logs indicating how information transited through the network. Table 2 depicts part of such log, where the basic component is paths. A path is an ordered list of devices through which some messages have transited. Note that a device may appear in multiple paths.

Table 2. Part of an information flow log

Path Id	Device name	Device type	Path Hop
1	Device 1	Router	0
1	Device 2	Microwave	1
1	Device 3	RAN	2
2	Device 1	Router	0
2	Device 4	Microwave	1
2	Device 5	RAN	2
...	...	...	...

By combining paths, a static heterogeneous graph is obtained representing the network topology such as the one shown in Fig. 2 (left). The constructed graph is hierarchical where each node represents a network device. Three types of devices are considered, namely routers, microwave devices and RAN (Radio Access Network) devices (also called NodeB). The information generally flows from routers to microwave devices, and then to RAN devices. The graph generated using the collected data contains 41,143 distinct vertices (devices) and nodes appears to be hierarchically organized into three layers (called domains) as a tree-like structure. However, it should be noted that some nodes are interconnected with others in the Microwave layer. Hence, there are some cycles between microwave nodes and the network must be represented as a graph rather than a tree.

c) **Mapping Alarms to the Network.** After obtaining the graph representing the network topology, the proposed framework maps each alarm from the alarm log to devices of the graph. This is done by matching values for the Device

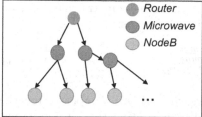

Node attribute dictionary:

Node type : Router/Microwave/NodeB

Historical Alarm data:
(Alarm type: appear time)
Alarm A: [t1, t2, t3...]
Alarm B: [t3, t4, t5...]
...

Fig. 2. The recovered network topology (left) with alarm attributes (right)

Name field from information flow paths (as in Table 2) to values in the Alarm
Source field of the alarm log (as in Table 1). The result is a graph-based data
representation where all triggered alarms are encoded in vertex attributes on
the topology and where the topology remains fixed. Such structure is depicted in
Fig. 2. For a given node, each attribute represents an alarm type and contains the
list of alarm occurrences of that type, sorted by time. This graph-based structure
indicates how alarms change over time and is a type of dynamic attributed
graph. Note that during the mapping process, alarms that are not mapped to
any device of the topology are discarded. The next paragraphs explains how
interesting correlation patterns are extracted from this data representation.

Step 2. Extracting Alarm Correlation Rules. After the data has been
prepared, the proposed framework extracts patterns indicating strong temporal
relationships between pairs of alarm types. A natural representation for such
relationships is rules of the form $A \rightarrow B$ indicating that if some alarm of type
A appears, an alarm of type B is also likely to appear. But finding interesting
rules requires to define a measure of the correlation of A and B.

In association rule mining [11,14], several measures have been proposed to
find strong rules such as the *support* (occurrence count of A with B) and the
confidence (occurrence count of A with B divided by the occurrence count of
A). But the support measure is not very suitable for alarm correlation analysis
because very frequent alarms are generally unimportant and may even be considered as noise. The *confidence* measure has the drawback that it is very sensitive
to the frequency of a rule's consequent (B) in the database. Another traditional
measure is the *Lift* measure [14], which is less influenced by the presence of rare
items but it is symmetric. In this study, we want an asymmetric measure to help
us judge how an alarm influences the other. The lift does not allow to distinguish
between the correlation of A with B and that of B with A.

To address the above limitations of the above measures, this paper presents
a novel correlation measure named ACOR (Alarm CORrelation) specifically
designed for evaluating the correlation between two alarms A and B. Some
advantages are that it consider the occurrence frequencies of A and B so
that it can minimize the impact of noisy data (some alarms always appear or
only appear once or twice). And ACOR amplifies the difference between the

associated values. It is worth noticing that it is an asymmetric measure, i.e., $acor_{A2B}$ does not equal $acor_{B2A}$. The measure is given as:

$$acor_{A2B} = \left(\frac{\frac{ABcount}{Acount}}{2 - \frac{ABcount}{Bcount}}\right), \tag{1}$$

where $ABcount$ is the number of time windows where A and B appeared together (e.g. within 5 min) which can be interpreted as indicating that A and B may have the same cause and $Acount$ (resp. $Bcount$) is the number of occurrences of alarm A (resp. alarm B) in the log data. The closer a $acor_{A2B}$ value is to the maximum of 1, the higher the correlation between the two alarms is.

Besides, it can be observed that the correlation measure is designed to not be strict about the order of occurrences between two alarms A and B, as long as they occur together closely enough (whithin a time window). The reason for not requiring a strict order between A and B is that clocks of network devices are not perfectly synchronized. As a result, some event may appear before another event in the alarm log although it actually appeared after.

Another contribution of this work is to not only find rules about alarms within a single device (single device rules) but also between devices from the same domain (cross device rules) and between devices of different domains (cross domain rules). This is useful because a telecommunication network is typically hierarchical, and devices within each layer (domain) behave quite differently. Nodes from different domains also have completely different types of alarms and communication link between nodes are also determined by the domains containing these nodes. Discovering cross device and cross domain rules allows to go beyond simple correlations occurring within a single device to find patterns applicable in other scenarios. This was not done in previous studies as the network topology was ignored.

To find correlation rules between alarms, a data mining algorithm is applied to the previous graph structure. As mentioned, this paper considers three scenarios for alarm correlation analysis: single device rules, cross device rules and cross domain rules. To find these rules, the correlation between all pairs of alarms is calculated according to the $acor_{A2B}$ formula. Algorithm 1 shows the pseudo code for calculating the correlation of a single device rule. It takes as input the graph data structure previously built, two alarm types A and B, and returns the correlation of $A \rightarrow B$. In the pseudocode, the notation $len(node.alarm\_A)$ represents the number of alarms of type A that have occurred in a given device called $node$, and $node.alarm\_A[i]$ refers to the i-th alarm occurrence of type A occurring in the device $node$. The algorithm can be easily extended to identify strongly correlated cross device and cross domain rules. The only difference between these different scenarios is that alarms must be in different positions in the network when calculating the correlation. Finally, the rules are ranked by decreasing order of correlation. The assumption is that rules having a high correlation are more interesting. The rules can be analyzed by an expert or used for alarm compression as it will be explained in the next subsection.

Algorithm 1: Calculating the correlation of a single device rule

---

   input : a dynamic attributed graph $\mathcal{G}$,
          two alarm types A and B
   output: Correlation value of A to B

1 Initialize $A\_count \leftarrow 0, B\_count \leftarrow 0, AB\_count \leftarrow 0$
2 **foreach** $node \in \mathcal{G}$ **do**
3     $A\_count \leftarrow A\_count + len(node.alarm\_A)$
4     $B\_count \leftarrow B\_count + len(node.alarm\_B)$
5   Initialize $i \leftarrow 0, j \leftarrow 0$
6   **while** $i < len(node.alarm\_A)$ and $j < len(node.alarm\_B)$ **do**
7   **if** $node.alarm\_A[i], node.alarm\_B[j]$ *appear together* **then**
8            $AB\_count \leftarrow AB\_count + 1$
9            $i \leftarrow i + 1$
10           $j \leftarrow j + 1$
11   **end**
12   **if** $node.alarm\_A[i]$ *appears before* $node.alarm\_B[j]$ **then**
13          $i \leftarrow i + 1$
14   **end**
15   **if** $node.alarm\_B[j]$ *appears before* $node.alarm\_A[i]$ **then**
16          $j \leftarrow j + 1$
17   **end**
18   **end**
19 **end**
20 $acor \leftarrow (AB\_count/A\_count)/(2 - (AB\_count/B\_count))$
21 return cor

Step 3. Compressing Alarms Using the Alarm Correlation Rules. After extracting alarm correlation rules, the framework utilizes the discovered alarm correlation rules for alarm compression. This is done in two steps.

a) Aggregating Rules and Inferring the Cause of an Alarm. First, the top-k alarm correlation rules are selected where k is a parameter that is set by the user. This is to avoid having to process a very large number of rules in the subsequent step.

For single device correlation analysis, if a more compact representation is required, an inference graph can be created between alarms of different domains. There will be a small connected subgraph in the inference graph where vertices are alarms and edges are the relations between alarms. Then we can get a number of connected subgraphs that is independent alarm sets. Note that the proposed method uses the property that the value of the correlation is not symmetric to delete edges in the inference graph, and simply obtain each P alarm through the inference graph. For cross device and domain correlation analysis, we directly infer the direction of the information flow in the network to get P alarms.

b) Filtering Alarms in Real-Time. Then, the framework applies alarm correlation rules selected in the previous step to filter alarms. But a challenge is that the network center in charge of the telecommunication network receive a constant flow of alarms. To be able to filter alarms in real-time, the proposed framework is adapted to uses a sliding window. For each window, alarms of that window are mapped to the graph representing the network topology to create an attributed graph. Then, the framework respectively performs pre-processing filtering, cross domain compression, cross device compression and single device compression using the rules obtained by the knowledge discovery process. Lastly, the remaining alarms from the network are reported to the network management center and some technicians will be dispatched to check and fix the nodes (devices) having alarms.

4   Experimental Evaluation

To evaluate the proposed framework, two experiments were done using real alarm data collected from an Indonesian telecommunication network (described in Sect. 3). Results where compared with rules found by the state-of-the-art AABD [3] system, obtained from its authors.

Rule Quality. The first experiment was carried out to verify the quality of the alarm correlation rules extracted by the proposed framework. For this purpose a comparison was made with the 135 rules found by the AABD system [3] to see if rules found by AABD could be rediscovered and if many other rules with a similar or higher correlation could be found. The rules found by AABD are used as baseline as they have been verified as valid by domain experts. Both approaches were applied using the same time window of 5 min, suggested by domain experts. Let A and B denote the sets of rules found by AABD and the proposed framework, respectively. The coverage ratio was calculated, which is defined as $coverage = |A \cap B|/|B|$. A high coverage ratio indicates that many rules of the proposed framework are valid (as A was validated by experts). However, it should be noted that this measure does not give a full picture as there may exist valid rules not found by AABD.

To select good rules, a minimum correlation threshold was applied in the proposed framework. As this parameter is set lower, more rules may be found, and then the coverage ratio may increase but the accuracy of rules may decrease. It is thus important to choose a suitable value for this parameter that is not too low to avoid finding many spurious rules. To choose a suitable value, we applied the empirical "elbow method" approach for setting a parameter, which is commonly used in data mining and machine learning [15]. It consists of drawing a chart representing the impact of a parameter on a measure to find the point where further increasing or decreasing the parameter would result in a huge change for that measure. In this study, we varied the minimum correlation threshold and noted the number of rules found for each value to draw a chart representing the frequency distribution of rules w.r.t correlation (shown in Fig. 3). We then observed that a large increase in the number of rules occurs for correlation values below 0.135. Assuming that those could be spurious rules, we set the minimum correlation to 0.135. For this parameter value, about 500 alarm correlation rules were discovered including 113 found by AABD. Thus, in this case, the coverage ratio is 113/135 = 84%. It was observed that many rules not found by the AABD system were discovered that have similar or higher correlation values than rules found by AABD. This is interesting as they are new rules exclusively discovered by the proposed framework that may be valid rules. Rules were presented to a domain expert who found that the majority of the new rules are interesting (Fig. 3).

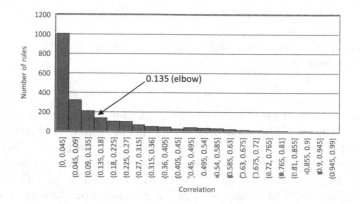

Fig. 3. The correlation distribution of rules

Alarm Compression Rate. The second experiment aimed at evaluating the number of alarms that could be compressed (removed) using the discovered alarm correlation rules. The original number of alarms and the number of remaining alarms after applying each compression procedure is shown in Table 3. After keeping only the alarms triggered by devices from the reconstructed topology, 4,481,273 alarms were kept from the original 6,199,650 ones. Then, preprocessing was applied, which further reduced that number to 992,966. Then, alarms were compressed using cross domain, cross device, and single device alarm correlation rules, respectively. In the end, 590,307 alarms remained, that is 9.5% of the original alarms. The **final compression rate** obtained by the system is thus $\frac{590,307}{4,481,273} = 87.9\%$. This is a big reduction that can greatly reduce the work of network operators.

To put this into perspective, the compression obtained using the alarm correlation rules was compared with that obtained using rules found by AABD. While the proposed framework can find three main types of rules (cross domain, cross device and single device), AABD only finds single devices rules in the RAN domain. Thus, a comparison of the obtained compression was made using only this type of rules. Using rules found by AABD, 39,603 alarms were removed, while 44,548 were removed using the proposed framework. Thus, the proposed method allowed to remove 12.5% more rules than AABD for the scenario of single device compression. If the other types of rules found by the proposed framework

Table 3. Remaining alarms count after applying each compression procedure

Compression procedure	Remaining alarms count
Original alarms count	6,199,650 (100%)
Available alarms on topology	4,481,273 (72.2%)
After pre-processing	992,966 (16.0%)
After cross domain compression	874,770 (14.1%)
After cross device compression	756,316 (12.2%)
After single device compression of microwave domain	634,855 (10.2%)
After single device compression of ran domain compression	590,307 (9.5 %)

are also used, a much greater compression can be obtained. For example, if cross domain rules are also used, 162,744 alarms are removed by the proposed framework, that is 123,141 more than AABD. And if both cross domain and cross device rules are utilized as well as single device rules from all domains (not just RAN), 284,463 more alarms are removed compared to AABD. Note that AABD could also be used to find rules in other domains but such rules were not provided.

5   Conclusion

To find interesting correlations between triggered alarms in telecommunication networks, we modeled a network as a dynamic attributed graph where alarms are viewed as device (node) attributes. A framework was designed to extract correlation rules for a single device, between different devices and across different domains. For this, a novel correlation measure named ACOR (Alarm CORrelation) was designed. By considering the network topology, the rules can reveal interesting relationships between alarms not found by prior approaches. The solution was applied to data from a large telecommunication network. Interesting patterns were discovered and it was found that the patterns can provide greater alarm compression than the state-of-the-art AABD system [3]. This reduces the number of alarms to be analysed by network operators and thus the costs of network maintenance. In future work, we plan to extract more complex graph-based patterns to reveal other types of interesting information from network alarm logs and designing distributed algorithms for processing very large alarm logs.

References

1. Dusia, A., Sethi, A.S.: Recent advances in fault localization in computer networks. IEEE Commun. Surv. Tutor. **18**(4), 3030–3051 (2016)

2. Ding, J., Kramer, B., Xu, S., Chen, H., Bai, Y.: Predictive fault management in the dynamic environment of IP networks. In: 2004 IEEE International Workshop on IP Operations and Management, pp. 233–239 (2004)
3. Wang, J., et al.: Efficient alarm behavior analytics for telecom networks. Inf. Sci. **402**, 1–14 (2017)
4. Costa, R., Cachulo, N., Cortez, P.: An intelligent alarm management system for large-scale telecommunication companies. In: Lopes, L.S., Lau, N., Mariano, P., Rocha, L.M. (eds.) EPIA 2009. LNCS (LNAI), vol. 5816, pp. 386–399. Springer, Heidelberg (2009). https://doi.org/10.1007/978-3-642-04686-5_32
5. Klemettinen, M., Mannila, H., Toivonen, H.: Rule discovery in telecommunication alarm data. J. Netw. Syst. Manage. **7**(4), 395–423 (1999)
6. Hatonen, K., Klemettinen, M., Mannila, H., Ronkainen, P., Toivonen, H.: TASA: telecommunication alarm sequence analyzer or how to enjoy faults in your network. In: Proceedings of NOMS 1996-IEEE Network Operations and Management Symposium, vol. 2, pp. 520–529. IEEE (1996)
7. Lozonavu, M., Vlachou-Konchylaki, M., Huang, V.: Relation discovery of mobile network alarms with sequential pattern mining. In: 2017 International Conference on Computing, Networking and Communications (ICNC), pp. 363–367. IEEE (2017)
8. Mannila, H., Toivonen, H., Verkamo, A.I.: Discovering frequent episodes in sequences. In: Proceedings of 1st International Conference on Knowledge Discovery and Data Mining (1995)
9. Ao, X., Shi, H., Wang, J., Zuo, L., Li, H., He, Q.: Large-scale frequent episode mining from complex event sequences with hierarchies. ACM Trans. Intell. Syst. Technol. (TIST) **10**(4), 1–26 (2019)
10. Fournier-Viger, P., Wang, Y., Yang, P., Lin, J. C.-W., Yun, U.: TKE: mining top-K frequent episodes. In: Proceedings of 33rd International Conference on Industrial, Engineering and Other Applications of Applied Intelligent Systems, pp. 832–845 (2020)
11. Luna, J.M., Fournier-Viger, P., Ventura, S.: Frequent itemset mining: a 25 years review. Wiley Interdisc. Rev. Data Min. Knowl. Disc. **9**(6), e1329 (2019)
12. Pei, J., et al.: Mining sequential patterns by pattern-growth: the PrefixSpan approach. IEEE Trans. Knowl. Data Eng. **16**(11), 1424–1440 (2004)
13. Truong, T., Duong, H., Le, B., Fournier-Viger, P.: FMaxCloHUSM: an efficient algorithm for mining frequent closed and maximal high utility sequences. Eng. Appl. Artif. Intell. **85**, 1–20 (2019)
14. Tsang, S., Koh, Y.S., Dobbie, G.: Finding interesting rare association rules using rare pattern tree. Trans. Large-Scale Data Knowl. Centered Syst. **8**, 157–173 (2013)
15. Liu, F., Deng, Y.: Determine the number of unknown targets in Open World based on Elbow method. IEEE Trans. Fuzzy Syst. (2020)
16. Kaytoue, M., Pitarch, Y., Plantevit, M., Robardet, C.: Triggering patterns of topology changes in dynamic graphs. In: IEEE/ACM International Conference on Advances in Social Networks Analysis and Mining (ASONAM 2014) (2014)
17. Fournier-Viger, P., et al.: A survey of pattern mining in dynamic graphs. Wiley Interdisc. Rev. Data Min. Knowl. Disc. **10**, e1372 (2020)
18. Fournier-Viger, P., Cheng, C., Cheng, Z., Lin, J.C.W., Selmaoui-Folcher, N.: Finding strongly correlated trends in dynamic attributed graphs. In: International Conference on Big Data Analytics and Knowledge Discovery, pp. 250–265 (2019)
19. Desmier, É., Plantevit, M., Robardet, C., Boulicaut, J.F.: Granularity of co-evolution patterns in dynamic attributed graphs. In: International Symposium on Intelligent Data Analysis, pp. 84–95 (2014)

Resource Sharing in Public Cloud System with Evolutionary Multi-agent Artificial Swarm Intelligence

Beiran Chen[1]([✉]), Yi Zhang[2], and George Iosifidis[1]

[1] Trinity College Dublin, Dublin, Ireland
{chenbe,iosifidg}@tcd.ie
[2] Huawei Technologies, Dublin, Ireland

Abstract. Artificial Intelligence for IT operations (AIOps) is an emerging research area for public cloud systems. The research topics of AIOps have been expanding from robust and reliable systems to cloud resource allocation in general. In this paper we propose a resource sharing scheme between cloud users, to minimize the resource utilization while guaranteeing Quality of Experience (QoE) of the users. We utilise the concept of recently emerged Artificial Swarm Intelligence (ASI) for resource sharing between users, by using Artificial-Intelligence-based agents to mimic human user behaviours. In addition, with the variation of real-time resource utilisation, the swarm of agents share their spare resource with each other according to their needs and their Personality Traits (PT). In this paper, we first propose and implement an Evolutionary Multi-robots Personality (EMP) model, which considers the constraints from the environment (resource usage states of the agents) and the evolution of two agents' PT at each sharing step. We then implement a Single Evolution Multi-robots Personality (SEMP) model, which only considers to evolve agent's PT and neglects the resource usage states. For benchmarking we also implement a Nash Bargaining Solution Sharing (NBSS) model which uses game theory but does not involve PT or risks of usage states. The objective of our proposed models is to make all the agents get sufficient resources while reducing the total amount of excessive resources. The results show that our EMP model performs the best, with least iteration steps leading to the convergence and best resource savings.

Keywords: Artificial Swarm Intelligence · AIOps · Resource sharing · Reinforcement learning

1 Introduction

The Artificial Intelligence for IT Operations (AIOps) has become one of the most industrial-favorite research areas in recent years, especially in public cloud systems. The scope of AIOps has been extending from conventional fault monitoring and recovery problems to the optimization of cloud resources, i.e. avoiding sub-optimal resource utilization and preventing potential failures related to

© Springer Nature Switzerland AG 2021
H. Hacid et al. (Eds.): ICSOC 2020 Workshops, LNCS 12632, pp. 240–251, 2021.
https://doi.org/10.1007/978-3-030-76352-7_25

it. In public cloud services, cloud hardware resources, such as CPU, memory, disk, networking, etc., are virtualised before being provided to the customers. These virtualised resources are abstracted as one layer above the physical infrastructure layer and exposed to the customers. With this resource virtualisation, customers all over the world are designated to share the same pool of resources in the cloud. In public cloud, customer usage patterns are highly dynamic and asynchronous because the customers locate at different time zones. Therefore, there exist opportunities for cloud operators (IT operation teams) to re-allocate resources between users dynamically to save the overall resource usage. Recent published literature suggest resource sharing to save resources in the cloud, e.g., [6, 15]. From the service providers perspective, a business model that encourages sharing excessive resource between customers helps to save the overall cloud resources for the service providers. In addition, since these shared resources are excessive, the QoE of users is not compromised. This concept of resource sharing provides a new option for IT operation team to optimize the resource utilization of the cloud.

A successful sharing scheme includes the following key points: 1) to give enough reward to the users who share; 2) to make sure that the users do not experience service disruptions while sharing; and 3) to give resources back to users when they need extra resources in some circumstances. In public cloud systems, the major challenges for designing an optimal sharing scheme are the following: 1) the real-time user demands for cloud resources, e.g. CPU, memory, and network bandwidth are asynchronous, dynamic and hard to predict, which makes the amount of excessive resources quite uncertain; 2) the users have different personalities, either conservative or generous, which affect their choices of sharing or requesting resources; 3) traditional centralized resource allocation algorithms are inefficient to solve this problem due to the difficulty of tracking and grouping large scale of users. Decentralized user-to-user sharing is more effective.

To cope with the resource sharing problem of large scale of cloud users, we bring in the concept of Artificial Swarm Intelligence (ASI) [9], which is a recently emerged research field under the umbrella of reinforcement learning. The goal of ASI is to provide solutions with AI agents to imitate human/animal behaviors/personalities in a swarm and make optimal decisions in a decentralized way. ASI is a multi-agent based reinforcement learning solution that achieves collaborative goals with a swarm of agents. With the concept of ASI applying in the cloud resource allocation system, AIOps team of public cloud system is capable to achieve optimal resource utilization without compromising the QoE of the customers.

To implement the ASI, in this paper, we design an EMP model for our sharing system, which is an evolutionary multi-agent model, originated from ASI. The goal of this sharing procedure is to dynamically allocate spare resources from one agent to another who needs that resource in the meantime. Besides, we define that each agent has Personality Traits (PT) which leads them to have different preferences in different situations. In addition, the evolution of their PT during the sharing procedure causes them to make different actions at each iteration steps to achieve the goal of sharing, i.e. all agents have sufficient resources and QoE is guaranteed. During the sharing procedure, we use Nash Bargaining method in game theory to obtain optimal policies.

Our contributions in this paper are the following: 1) building a multi-agent EMP model based on Artificial Swarm Intelligence for user resource sharing in the public cloud system; 2) apply PT and game theory for the EMP model to optimize the sharing policy; and 3) the comprehensive analysis of the performance of our algorithm by comparing our EMP model (including EMP with Asymmetric Nash Bargaining strategy (EMP-A) and EMP with Fixed step strategy (EMP-F)) with 2 other baseline models, SEMP and NBSS). This paper provides a new scope of optimal resource allocation for the AIOps research field in public cloud system.

2   Background and Related Work

Swarm Intelligence is a concept of designing de-centralized and self-organized systems for collective behaviours, first introduced in 1989 in the field of cellular robotics [1]. This concept was firstly derived from natural animal behaviors in swarms, and later on, has been applied in the Artificial Intelligence field to establish a research area of ASI since 2015 [9]. The goal of ASI is to design robotics to optimize the performance of a self-organized system that involves a group of AI agents. Since it emerged, the ASI system has been firstly used in many research areas. For instance, in intelligence transportation and logistics systems, the authors in [13] presented a model to track and chase a target, by designing an algorithm to control a swarm of autonomous robots to move in a 2-D space. Besides, other research areas using ASI include "human swarming" for financial market forecasting [11] and disease diagnosing [8], etc.

In the meantime, the evolution of personalities and the corresponding effects on collaboration of agents have also been brought to the swarm intelligence research. For example, the authors in [2] used personality evolution to build a self-organized algorithm to research on a problem of multiple robots to leave a room in a self-organized way. The authors in [10] proposed a game-theory based approach to swarm robots to collaborate with each other to chase a target. The authors in [3] deal with the evolved control of a swarm of robots to decide the signalling and connectivity of communications with each other. The authors in [14] uses ASI to build a SWARM AI platform to measure the personalities of human groups.

In this paper, we bring the concept of the evolution of the personality of swarm agents to solve our resource sharing problem in cloud systems. In addition, we use game theory when defining rewards in the system for the agents to take optimal actions. Our paper use ASI to deal with the resource sharing problem in cloud system, which brings a novel scope for AIOps researchers to optimize resource allocation when operating the public cloud system.

3   System Model

As mentioned in previous sections, our system encourages public cloud users to share spare resources to each other, while agents with Artificial Intelligence act as the advisors of the users to perform the sharing. In this section, we design the EMP model, which is based on ASI that considers the personalities of the agents. In the EMP model, we design two algorithms EMP-A and EMP-F, considering both PT and risks of running short of resources. After that we design 2 algorithms as control groups for benchmarking: one is SEMP, in which every agent only cares about its own PT at each iteration step without considering the risk of shortage of resources; the other one is NBSS that uses Nash Bargaining Solution (NBS) for agents without involving the agents' PT in the sharing.

3.1   The EMP Model

In this EMP model, we focus on the cooperation between the agents without considering the competitions between them, which aligns with the service scenario in Infrastructure as a Service (IaaS) of public cloud where users should not realize any competition between them when sharing resources. To quantify the personality of agents in EMP, we introduce PT for agents, which describe agents' intention of selecting their sharing behaviours (i.e. actions).

We set all agents in our system as homogeneous agents with the only difference between them on the parameter settings of the agents based on the agents' preferences. To describe the agents' different preferences, we use one pair of numerical value named PT [4], which is also called the orientation of the agents [2]. After that, when calculating the probability of sharing conducted by the agents, we adopt Game Theory [10] and Asymmetric Nash Bargaining Solution (ANBS) [17] to derive mathmatical formulation.

The System Definition. We use a multi-agent robotic system imitating the cloud users to give sharing suggestions. Each agent i in the system has the same resource value R_{fixed} at beginning and has a flexible extra resource quota $Re(t)$, which denotes the resources it would get from other agents at time t, with the upper boundary Re_{max} restricting the greediness of each agent. The resource usage set is denoted as $U = \{U_i | U_i \in [0, R_{fixed} + Re_{max}], i = 1, 2, 3....n\}$, and the real-time resource at time t holds $Rr_i(t) = R_{fixed} + Re_i(t)$.

When agents get a real-time resource usage information (e.g. the system measures agents' usage amount in a time interval), the agents start to evaluate whether they have enough resources for their current usage. Then the EMP model will guide the agents to adjust their resource allocation in the system. Besides, the whole resource in the system holds conservation. For each agent i, the spare resource is denoted as $Rs_i(t) = Rr_i(t) - U_i(t)$. According to the real-time resource, we defined all agents in three groups: 1) needy agents ($Rs_i(t) < 0$); 2) rich agents ($Rs_i(t) > 0$); 3) self-sufficient agents($Rs_i(t) = 0$).

The Action and Payoff Definition. In every sharing step, we pick up two agents: one rich agent and one needy agent. In principle both agents have two action choices, giving out resource, denoted as a_0, or getting resources, denoted as a_1, the action space can be defined as $A = [a_0, a_1]^T$. The payoff matrix for agents is designed in Table 1. Nash Equilibrium point will be reached when the rich agent selects 'give' and the needy agent selects 'get'.

Table 1. Payoff for individual agent

Ma		Payoff for needy agent	
		give a_0	get a_1
Payoff of	give a_0	0, −1	3, 3
rich agent	get a_1	−5, −5	−1, 0

$M\theta$	State risk	
	high risk θ_0	low risk θ_1
Payoff of agent	−1	5
for state	4	−2

The Agent PT and State Definition. Our EMP model uses the Theory of Evolution to define the agents' PT, which makes the agents imitating the human emotions. These emotions act as an internal motivation to their behaviours. In our case, the vector PT, $\beta = [\beta_0, \beta_1]^T$ ($\beta_0 \geq 0, \beta_1 \geq 0$), describes the agents' normalized characteristics, where the β_0 represents 'generous' trait and β_1 represents 'eager' trait, and $\beta_0 + \beta_1 = 1$.

Other than the PT, the resource availability state s is another factor to influence the agents' actions that we have to take into account together with the PT. The agent in safe resource state means an agent has enough resource for its usage and more likely to share. Otherwise it is more likely to ask help from the other agents. We describe the state risk at time t by $P\theta_i(t) = [P\theta_{i0}(t), P\theta_{i1}(t)]^T$, where $P\theta_{i0}(t)$ denotes the probability of the agent i in risky state θ_0, and $P\theta_{i1}(t)$ denotes the probability of the agent in safe state θ_1, derived by the following Eq. (1):

$$P\theta_{i0}(t) = \frac{U_i(t)}{Rr_i(t)} + (\beta_{i0}(t) - \beta_{i1}(t)); \quad P\theta_{i1}(t) = 1 - P\theta_{i0}(t) \tag{1}$$

Besides, we describe the value function for agents when they are in sharing step as the following:

$$V_i(s, a_m) = E\{J_i(s, a_m)\} = M\theta_i \cdot P\theta_i^T = \begin{bmatrix} M\theta_{11}, M\theta_{12} \\ M\theta_{21}, M\theta_{22} \end{bmatrix} \cdot \begin{bmatrix} P\theta_i(\theta_0) \\ P\theta_i(\theta_1) \end{bmatrix} \cdot J_i(s, a_m)$$

$$(i, j = 1, 2....n \quad i \neq j, \quad m = 0, 1) \tag{2}$$

where, $s = [\theta_0, \theta_1]^T$ and

$$J_i(s, a_m) = Ma \cdot Pa_j^T = \begin{bmatrix} Ma_{11}, Ma_{12} \\ Ma_{21}, Ma_{22} \end{bmatrix} \cdot \begin{bmatrix} Pa_j(a_0) \\ Pa_j(a_1) \end{bmatrix}$$

$$(i, j = 1, 2....n \quad i \neq j, \quad m = 0, 1) \tag{3}$$

The payoff matrices $M\theta$ and Ma are defined in Table 1. Once the agent recognizes its state situation and its value function, its action selection strategy uses randomized strategy [5], which considers the agent is 'exploring' new action as well as 'exploiting' learnt action. The probability of action selection holds:

$$Pa_i\left(a_0|s\right) = \frac{k^{V_i(s,a_0)}}{k^{V_i(s,a_0)} + k^{V_i(s,a_1)}}; \quad Pa_i\left(a_0|s\right) + Pa_i\left(a_1|s\right) = 1 \qquad (4)$$

where the coefficient k represents how often the agent would like to 'explore' rather than 'exploit'. In our application, we set it to be $e = 2.718$.

The Sharing Strategy. At each sharing step, assuming both agents are rational and intend to maximize their spare resource utility in the bargain, EMP model considers the two-agent cooperation as a bargain problem. The set of spare resource utility function can be described as $\Gamma = \{\gamma_i | i = 1, 2\}\}$, where $\gamma_i = \{(Rr_i - U_i)|i = 1, 2\}$, which is a nonempty compact convex set with boundary ([16], [7]). For each agent, status quo point is where their Rs equals 0. When the real resource equivalent to their usage will get the status point for each agent. In this case the sharing problem can be described as [17]:

$$\Gamma^* = \underset{Rr_i}{argmax} \prod_i (Rr_i - U_i)^{\lambda_i},$$

$$s.t. \qquad \sum_{i=1}^{2} Rr_i = \Phi$$

$$\sum_{i=1}^{2} \lambda_i = 1 \qquad (5)$$

$$Rr_i \geq U_i, \quad i = 1, 2$$

$$Rr_i \geq 0, \quad i = 1, 2$$

$$\lambda_i = \frac{\beta_{i1}}{\beta_{i1} + \beta_{j1}} \qquad (i = 1, 2, 3...n) \qquad (6)$$

Where Φ is the total real-time resource of those two agents. λ_i denotes the bargaining power of agent. At each allocation step, the agent gets their real-time resource as $Rr_i = U_i + \lambda_i \cdot Rs$.

The Evolutionary Strategy. After the sharing, the real-time resource hold by every agents involved in the sharing step has been changed, leading their PT to be evolving. This will affect their action decisions in next sharing steps. We define the updating rule as the following [12]:

$$\beta_m\left(t\right) = \beta_m\left(t - 1\right) + \alpha\Delta\beta_m(t), \quad (m = 0, 1), \quad \alpha \in (0, 1) \qquad (7)$$

where, α is the learning rate and the $\Delta\beta_m(t)$ holds as :

$$\Delta\beta_m(t) = \frac{\Delta J_m(s, a_m; t)}{\Delta J_m(s, a_m; t) + \Delta J_l(s, a_l; t)}, \quad (m, l = 0, 1 \quad m \neq l) \qquad (8)$$

where, $\Delta J_i(s, a_i; t) = J_i(s, a_i; t) - J_i(s, a_i; t-1), (i = m, l)$. Then we implement the aforementioned EMP model by Algorithm 1 with two different sharing policies, i.e. EMP-A using ANBS, and EMP-F using fixed-value sharing strategy during the sharing (i.e., all the agents share a fixed value of resources during a step of sharing).

Algorithm 1. for EMP, including EMP-A and EMP-F

1: Initialisation:
2: Initial$\beta = [0.5, 0.5]^T$,$Pa = [0.5, 0.5]^T$
3: Split agents into three groups according to $Rs_i(t)$
4: Sharing step:
5: **while** there are needy agents in the system **do**
6:     **for** pick one rich agent **do**
7:         **for** pick one needy agent **do**
8:             1) evaluate state risk by equation (1)
9:             2) calculate expected value by equations (2-3)
10:             3) calculate Pa by equation (4)
11:             4) select and execute actions by Pa
12:             5) share spare resource:
13:             **if** sharing by ANBS strategy(EMP-A) **then**
14:                 sharing value by equation (5)
15:             **if** sharing by fixed value strategy(EMP-F) **then**
16:                 sharing with fixed value (e.g.5 units).
17:             6) update PT by equation (7-8)
18:             7) update the groups
19:             **if** no needy agent in the system **then**
20:                 **break**

3.2   Other Models

The SEMP Model. We build an SEMP model to be one of the baseline models to compare with our EMP model in Sect. 3.1. Unlike the EMP model that involves both agents in state evaluation before sharing and updates PT after sharing, the SEMP model, implemented as Algorithm 2, only considers the agents' PT when making actions and neglects the evaluation of the state risk probabilities. During the sharing steps, the SEMP algorithm also adopts the ANBS strategy and updates the rich agent's PT at the end of sharing. Without considering the state risk probability, the PT updating functions are simplified as:

$$Pa_i(a_0) = \frac{e^{\beta_{i0}}}{e^{\beta_{i0}} + e^{\beta_{i1}}}; \quad \beta_i(t) = \beta_i(t-1) + \alpha \cdot Ma, \quad \alpha \in (0,1) \qquad (9)$$

Algorithm 2. for SEMP

1: Initialisation:
2: Initialize $\beta = [0.5, 0.5]^T, Pa = [0.5, 0.5]^T$
3: Split agents into three groups according to $Rs_i(t)$
4: Sharing step:
5: **while** there are needy agents in the system **do**
6:     **for** pick one rich agent **do**
7:         **for** pick one needy agent **do**
8:           · 1) calculate Pa by equation(9)
9:              2) select and execute actions by Pa
10:             3) share spare resource by ANBS strategy
11:             4) update PT by equation(9)
12:             5) update the groups:
13:             **if** no needy agent in the system **then**
14:                 **break**

The NBSS Model. We build an NBSS model as another baseline algorithm to compare with our EMP model and SEMP model. In the NBSS model, without considering the state influence, we assume all agents have the same personality and the needy-rich agents pairs share their total spare resource by NBS as $Rr_i = U_i + 1/2 \cdot Rs$.

4   Experimental Results

4.1   Experimental Setup

In this section, we present experimental results of our sharing models. We simulate a 100-agent system for sharing resources in a public cloud. To initiate our experiment, we assume that each agent has fixed resource of 70 units at the beginning, and the maximum extra resource it could get from the other agents is 30 units. The real-time resource usage for each agent is a set of numbers in the range of $[0, 100]$ units. We define one "resource usage measuring round" as the time window between two consecutive instances of measuring on the agents' resource usage, when the resource sharing between all agents should start and finish. During each "resource usage measuring round", each agent conducts multiple "sharing steps" with other agents to achieve the goal of eliminating all needy agents. We investigate two use-case scenarios: "independent sharing" and "continuous sharing". "Independent sharing" means the real-time resource usage in each usage measuring round is randomly generated, and the remaining resource from the previous round is not rolled over to the next round. However, "continuous sharing" means the real-time resource usage follows a sinusoidal pattern during time of the day and the remaining resource is rolled over to the next round. We investigate the performance of our four aforementioned models: 1) The EMP-A model; 2) The EMP-F model; 3) The SEMP model; 4) The NBSS model.

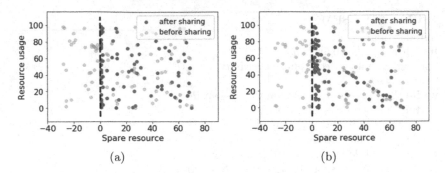

Fig. 1. 100-agent distribution of independent sharing in one resource usage measuring round by different algorithms (orange dots show the agents' distribution before sharing; blue dots shows the agents' distribution after sharing): (a) EMP-A; (b)EMP-F. (Color figure online)

4.2 Experimental Results for Different Models

Independent Sharing with Random Resource Usage Pattern. In this subsection, we investigate the independent sharing case with usage pattern randomly generated in the rage of $[0, 100]$. Figure 1 shows the experimental results for algorithms EMP-A and EMP-F in terms of the amount of spare resource before and after sharing in one usage measuring round. Each dot on the figure corresponds to one agent. The negative value of spare resource of an agent on x axis means the agent's real-time resource is not sufficient for its usage, while the positive value means the agent has some spare resource. The results show that both algorithms help every agent in the system to get enough resource for their current need as all the agents place at non-negative space on the right side of the figures after sharing (the blue dots).

Figure 2(a) shows the process of reducing the number of the needy agents during the sharing steps within one single resource usage measuring round for the four algorithms. The fewer sharing steps it takes to reach 0, the better the performance of the algorithm is. Figure 2(b) show the sharing steps the algorithms take during 100 independent usage measuring rounds. Both figures show that the algorithm EMP-A performances most effectively as it uses the least sharing steps to make the number of needy agents to reach 0. The performance of the other three algorithms, i.e., EMP-F, SEMP and NBSS are close to each other.

Continuous Sharing with Sinusoidal Usage Pattern. In this section, we consider continuous sharing in sinusoidal usage pattern (the resource usage follows a sinusoidal function over time of the day). As mentioned before in this case the remaining resource in the previous round is rolled over to the next round. Besides, we assume every agent has different phase of sinusoidal daily usage pattern, i.e., every agent reaches its peak usage value at different time of the day.

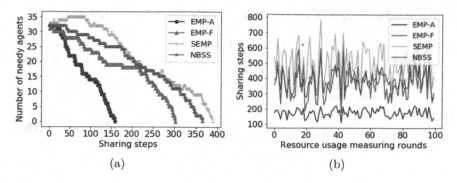

Fig. 2. (a) Number of needy agents during one independent usage measuring round vs. sharing steps; (b) Number of Sharing steps in 100 independent rounds with random usage pattern.

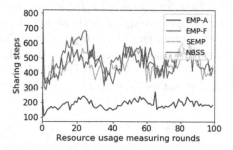

Fig. 3. Number of sharing steps in 100 continuous resource usage measuring rounds with sinusoidal usage pattern.

This setup emulate users from different time zone of the world that are sharing the same pool of cloud computing resources.

In this case, the performances of the four algorithms are shown in Fig. 3. Similar to the previous scenario with random usage, the EMP-A needs fewer sharing steps than the other three algorithms to make the number of needy agents to 0. The performance of the other three algorithms are also close to each other. However, in this continuous sharing case with sinusoidal usage pattern, the advantage of EMP-A over the other three algorithms is more obvious than the independent sharing case. It is reasonable because this continuous sharing scenario is more stable and closer to the reality (Table 2).

4.3 Average Percentage of Satisfaction Time of Users

We assume that, in a public cloud system, the system gets real-time usage information measured every 15 min slot and schedule one sharing step every 1.5 s, i.e., the system can schedule up to 600 sharing steps in 15 min. As shown in previous Fig. 2 and 3, most of the time all 4 algorithms can make number of needy users to 0 (all users reaching satisfaction) within 600 sharing steps. However, for

Table 2. Average percentage of satisfaction time of users

Usage pattern	Without sharing	With sharing			
		EMP-A	EMP-F	SEMP	NBSS
100 rounds of independent sharing with random usage pattern	69%	95%	87%	87%	87%
100 rounds of continuous sharing with sinusoidal usage pattern	63%	95%	85%	88%	84%

different algorithms, the needy users spend different sharing steps to get sufficient resources before reaching satisfaction. We calculate and show the average percentage of satisfaction time for all users in Table reftab:4modelsspssufficient, The EMP-A has the highest percentage of satisfaction time as 95% during the sharing. However without sharing the percentage is only 69% in random usage pattern and 63% in sinusoidal usage pattern. The other three algorithms have similar performance of around 85%-88% during the sharing, which is higher than the case without sharing but not as good as the EMP-A.

5   Conclusion

With the trend of AIOps research fields in public cloud extending to resource optimization, in this paper, we integrate the concept of Artificial Swarm Intelligence and Personality Traits to design a multi-agent system on the resource sharing of the cloud system. We have designed and implemented our main algorithm, i.e., EMP-A, as well as three other algorithms, EMP-F, SEMP and NBSS for comparison. All algorithms are capable to re-allocate spare resources to the needy agents through the sharing procedure between the agents, without adding external resource.

To evaluate the performance of the 4 algorithms, we simulated a 100-agent system, and executed 100 usage measuring rounds with two different use-case scenarios. The results showed that EMP-A performance much better than the other three algorithms (SEMP, EMP-F and NBSS) in terms of fewer sharing steps as well as higher user satisfaction rate.

References

1. Beni, G., Wang, J.: Swarm intelligence in cellular robotic systems. In: Proceedings of NATO Advanced Workshop on Robots and Biological Systems (1989)
2. Ding, Y., He, Y., Jiang, J.P.: Self-organizing multi-robot system based on personality evolution. In: IEEE International Conference on Systems, Man and Cybernetics, Systems, Man and Cybernetics, 2002 IEEE International Conference on, Systems, man and cybernetics 5 (2002)

3. Dorigo, M., et al.: Evolving self-organizing behaviors for a swarm-Bot. Auton. Robot. **17**(2–3), 223–245 (2004)
4. Givigi Jr., S.N., Schwartz, H.M.: Swarm robot systems based on the evolution of personality traits. Turk. J. Electr. Eng. Comput. Sci. **15**(2), 257–282 (2007)
5. Kaelbling, L., Littman, M., Moore, A.: Reinforcement learning: a survey. J. Artif. Intell. Res. **4**, 237–285 (1996)
6. Li, C., Yang, C.: A novice group sharing method for public cloud. In: 2018 IEEE 11th International Conference on Cloud Computing (CLOUD), pp. 966–969 (2018)
7. Ma, X.P., Dong, H.H., Li, P., Jia, L.M., Liu, X.: A multi service train-to-ground bandwidth allocation strategy based on game theory and particle swarm optimization. IEEE Intell. Transp. Syst. Mag. **10**(3), 68–79 (2018)
8. Rosenberg, L., Lungren, M., Halabi, S., Willcox, G., Baltaxe, D., Lyons, M.: Artificial swarm intelligence employed to amplify diagnostic accuracy in radiology. In: 2018 IEEE 9th Annual Information Technology, Electronics and Mobile Communication Conference (IEMCON), pp. 1186–1191 (2018)
9. Rosenberg, L.: Artificial Swarm Intelligence, a human-in-the-loop approach to A.I. In: AAAI 2016: Proceedings of the Thirtieth AAAI Conference on Artificial Intelligence (2016)
10. Givigi Jr., S.N., Schwartz, H.M.: A game theoretic approach to swarm robotics. Appl. Bionics Biomech. **3**(3), 131–142 (2006)
11. Schumann, H., Willcox, G., Rosenberg, L., Pescetelli, N.: "Human swarming" amplifies accuracy and ROI when forecasting financial markets. In: 2019 IEEE International Conference on Humanized Computing and Communication (HCC), pp. 77–82 (2019)
12. Schwartz, H.M.: Multi-agent Machine Learning: A Reinforcement Approach. Wiley, Hoboken (2014)
13. Van Le, D., Tham, C.: A deep reinforcement learning based offloading scheme in ad-hoc mobile clouds. In: IEEE INFOCOM 2018 - IEEE Conference on Computer Communications Workshops (INFOCOM WKSHPS), pp. 760–765, April 2018. https://doi.org/10.1109/INFCOMW.2018.8406881
14. Willcox, G., Askay, D., Rosenberg, L., Metcalf, L., Kwong, B., Liu, R.: Measuring group personality with swarm AI. In: 2019 First International Conference on Transdisciplinary AI (TransAI), pp. 10–17 (2019)
15. Xu, J., Palanisamy, B.: Cost-aware resource management for federated clouds using resource sharing contracts. In: 2017 IEEE 10th International Conference on Cloud Computing (CLOUD), pp. 238–245 (2017)
16. Yaiche, H., Mazumdar, R., Rosenberg, C.: A game theoretic framework for bandwidth allocation and pricing in broadband networks. IEEE/ACM Trans. Netw. **8**(5), 667–678 (2000)
17. Yin, T., Hong-hui, D., Li-min, J., Si-yu, L.: A bandwidth allocation strategy for train-to-ground communication networks. 2014 IEEE 25th Annual International Symposium on Personal, Indoor and Mobile Radio Communication (PIMRC), p. 1432 (2014)

SLMAD: Statistical Learning-Based Metric Anomaly Detection

Arsalan Shahid[1,2]([⊠]), Gary White[2]([⊠]), Jaroslaw Diuwe[2]([⊠]),
Alexandros Agapitos[2]([⊠]), and Owen O'Brien[2]([⊠])

[1] School of Computer Science, University College Dublin, Belfield, Dublin 4, Ireland
arsalan.shahid@ucd.ie
[2] Huawei Ireland Research Centre, Townsend Street, Dublin 2 D02 R156, Ireland
{gary.white,jaroslaw.diuwe,alexandros.agapitos,owen.obrien}@huawei.com

Abstract. Technology companies have become increasingly data-driven, collecting and monitoring a growing list of metrics, such as response time, throughput, page views, and user engagement. With hundreds of metrics in a production environment, an automated approach is needed to detect anomalies and alert potential incidents in real-time. In this paper, we develop a time series anomaly detection framework called Statistical Learning-Based Metric Anomaly Detection (SLMAD) that allows for the detection of anomalies from key performance indicators (KPIs) in streaming time series data. We demonstrate the integrated workflow and algorithms of our anomaly detection framework, which is designed to be accurate, efficient, unsupervised, online, robust, and generalisable. Our approach consists of a three-stage pipeline including analysis of time series, dynamic grouping, and model training and evaluation. The experimental results show that the SLMAD can accurately detect anomalies on a number of benchmark data sets and Huawei production data while maintaining efficient use of resources.

Keywords: Anomaly detection · Unsupervised learning · Online machine learning · Streaming time series · Cloud computing

1 Introduction

Artificial Intelligence for IT Operations (AIOps) is an emerging field arising in the intersection between the research areas of machine learning, big data, and the management of IT operations [4]. The main aim is to analyze system information of different kinds (metrics, logs, customer input, etc.) and support administrators by optimizing various objectives like prevention of Service Level Agreements (SLA) violation, early anomaly detection, auto-remediation, energy-efficient system operation, providing optimal Quality of Experience (QoS) for customers, predictive maintenance and many more [13,22]. Over the years, a constantly growing interest can be observed in this field, which has led to the development of practical tools from academia and industry.

© Springer Nature Switzerland AG 2021
H. Hacid et al. (Eds.): ICSOC 2020 Workshops, LNCS 12632, pp. 252–263, 2021.
https://doi.org/10.1007/978-3-030-76352-7_26

With the large amount of data being collected, there is a need to identify rare events and possible failures. Anomalies in time series data can potentially result in losses to the business in terms of revenue and market reputation. Accurate anomaly detection can be used to trigger prompt troubleshooting and help to avoid downtime or SLA violations for a company. Once the anomalies are detected, alerts are either sent to the operators to make timely decisions related to the incidents or automatically handled by self-healing mechanisms [7]. We now summarize the challenges posed in the construction of a desired industrial-grade time series anomaly detection framework:

- *Lack of labels:* In production-level business scenarios, the systems often process millions of metrics. There is no easy way to label data on this scale manually. Moreover, if the time series is in a dynamic environment where the data distribution is constantly changing, then the model will need to be retrained frequently on new data. Labelling this data can introduce a significant delay and cost with a need to continuously update the models [3]. This makes supervised models insufficient for the industrial scenario.
- *Generalization:* There are a large number of different metrics and scenarios that time series anomaly detection can be applied to. In a typical production environment, time series data can exhibit a range of different patterns and variability. Industrial anomaly detection must work well on all kinds of time series patterns. However, some existing approaches assume the shape of data and are not designed to generalize and adapt to different patterns. For example, Holt-Winters shows good performance in seasonal data but does not generalise well to unstable data [14].
- *Efficiency and accuracy:* In production scenarios, a monitoring system must process millions of time series in near real-time. Especially for the sub-minute-level time series, the anomaly detection procedure needs to be finished within a limited time, i.e., before the next data point. Furthermore, if the monitoring and anomaly detection system is running on a production node, it must not use a lot of computing resources to respect the quota assigned for customers. Therefore, even though models with large time complexity may achieve good accuracy, they are often of little use in a production scenario.
- *Online training and update:* Once trained, the model should be able to update online as new data is collected. Many existing anomaly detection approaches are designed for settings that send data in batches rather than run-time streams. This can lead to an old model being used until the new model has been updated with new data. In a dynamic environment where the data can change suddenly, using an old model can lead to a decrease in anomaly detection accuracy.
- *Parameter tuning:* The existing anomaly detection algorithms require a lot of parameter tuning to generate meaningful results. For example, a neural network or deep learning-based method requires tuning of optimal hyper-parameters and window-size for each metric. This can become complex and time-consuming in production environments with a large number of metrics. An approach that can automatically set parameters or requires little parameter tuning can be more easily deployed in a production environment.

In this paper, we focus on the framework for our anomaly detection service for time series data. We propose a Statistical Learning-Based Metric Anomaly Detection Framework (SLMAD) to spot anomalies at run-time for streaming data while maintaining high efficiency and accuracy. SLMAD is based on an unsupervised approach for anomaly detection that does not assume the shape of time series data, requires very little parameter tuning, detects anomalies, and updates online. SLMAD uses statistical-learning and employs a robust box-plot algorithm and Matrix Profile (MP) [23] to detect anomalies.

The framework is based on a three-stage pipeline. In the first stage, the framework analyses the time series in terms of its characteristics such as stationary or non-stationary, discrete or continuous, seasonal or non-seasonal, trend, or flat. In the second stage, based on the presence of seasonal components in the continuous time series data, we calculate the period using a statistical methodology and perform dynamic grouping for model training. If there is a lack of seasonal components and/or time series is discrete, we make use of matrix profiling to find the discords and identify them using robust box-plot in streaming data. In the third stage, we evaluate the models and analyse statistically and visually the spotted anomalies. We evaluate the efficiency and accuracy of SLMAD using publicly available data sets as well as production data from Huawei products.

The rest of the paper is organised as follows: Sect. 2 outlines the related work. Section 3 presents our proposed SLMAD framework. Section 4 and 5 describes the experimental approach used to evaluate our framework and the results of the experiments. Finally, Sect. 6 concludes the paper.

2   Related Work

Existing approaches in anomaly detection can be categorised into statistical, supervised, semi-supervised, and unsupervised approaches.

Statistical Anomaly Detection Methods: In the past years, several models have been proposed in the statistical literature, including hypothesis testing [15], wavelet analysis [11], SVD [12] and auto-regressive integrated moving average (ARIMA) [24]. Statistical methods have become popular in recent years as they consume much fewer resources than deep learning models, making them suitable for production deployment. However, there can be some limitations in terms of anomaly detection accuracy.

Supervised Anomaly Detection Methods: Supervised models have shown impressive accuracy in a range of cases. Opprentice has outperformed other traditional detectors by using statistical feature extractors and leveraged a random forest classifier to detect anomalies [10]. EGADS developed by Yahoo utilises a collection of anomaly detection and forecasting methods with an anomaly filtering layer for scalable anomaly detection on time series data. Google has recently leveraged deep learning models to detect anomalies in their data sets achieving promising results [17]. However, continuous labels cannot be generated to retrain these models in an industrial environment, making these algorithms insufficient in online applications. To mitigate the lack of labelled training data, Microsoft injected synthetic anomalies into the time series and trained different types of deep networks as binary classifiers [14]. In cases where the time series is stationary and exhibits no significant concept drift, Active Learning can reduce the data labelling requirements by selectively requesting user-feedback on the most informative examples. Instance-based Transfer Learning has been combined with domain adaptation to further reduce labelling requirements with similar source and target domains [19]. However, even with the reduction in labelled data, some feedback or labelled data need to be used, which may not be available in a production environment.

Semi-supervised Anomaly Detection Methods: Semi-supervised anomaly detection techniques do not require the entire data-set to be labelled. The model is instead trained only on the normal data. Recent approaches have used neural network techniques, such as generative adversarial networks (GANs) [1]. In this method, an encoder-decoder network is used to create a generator, which enables the model to map the data to a lower dimension vector, which is then used to reconstruct the original data. A distance metric from the learned data is then used to indicate that a new data point is an outlier from the distribution. Semi-supervised approaches share some of the problems as supervised approaches as we need to know that there are no anomalies in the data.

Unsupervised Anomaly Detection Methods: Due to the limitations of supervised and semi-supervised approaches in real industrial applications, unsupervised approaches have been proposed. DONUT is an unsupervised anomaly detection method based on Variational Auto Encoder (VAE) [5]. A large number of technology companies have also developed anomaly detection methods due to deal with the large streams of data. LinkedIn developed a method based on time series bitmaps, which allows for assumption-free detection [21]. A number of the approaches have used forecasting, such as Alibaba RobustTAD [6], Facebook Prophet [18], Amazon DeepAR+ [16], Uber RNN [9] and Microsoft SR-CNN [14]. Deep-learning techniques have also become popular in recent years, but they can be expensive to deploy in production. Twitter SH-ESD [8] employs a lightweight approach to detect anomalies using the robust measures of scale.

Table 1 summarizes the state-of-the-art unsupervised anomaly detection methods developed by industry along with their features such as the approach used (supervised, semi-supervised, or unsupervised), forecasting or statistical learning-based, evaluation data, online implementation, and computational complexity.

3   Statistical Learning-Based Metric Anomaly Detection (SLMAD)

3.1   Terminology, Problem Formulation, and Employed Statistical Methods

This section presents the terminology used in this paper for different types of anomalies, the problem statement, and the multiple implementations of box-plot

Table 1. Current State-of-the-art in Industry (Open Source and Commercial). *Abbreviations: US (Unsupervised), F (Forecasting), SL (Statistical Learning), OSB (Open-Source Benchmarks), and PD (Production Data)*

AD solutions	US	F/SL	Generalized	Evaluation OSB/PD	Online	Complexity
Alibaba RobustTAD [6]	✓	F	✓	OSB	✓	High
Facebook Prophet [18]	✓	F	✓	OSB		Medium
Amazon DeepAR+ [16]	✓	F	✓	OSB		High
Uber RNN [9]	✓	F	✓	OSB		High
Microsoft SR-CNN [14]	✓	F	✓	PD	✓	Medium
Twitter SH-ESD [8]	✓	SL		PD		Low
Proposed SLMAD	✓	SL	✓	OSB and PD	✓	Low

based statistical outlier detection rules used in SLMAD. The anomalies in time series data can be categorised into two types: point anomalies and contextual anomalies. A data point is considered a point anomaly if its value is far outside the range of data set in which it is found. A contextual anomaly, however, is an instance in a data-set that is an outlier based on the context [3].

Problem Statement: The problem of anomaly detection in time series data has been explored in different scientific fields, such as computer science, biology, and astronomy using a variety of techniques (Sect. 2). The drive towards autonomous networks, in a cloud-based production environment, yields a specialized problem of detection for anomalies in metrics (produced in a run-time environment) by maintaining the high performance and keeping resource cost low. The goal of achieving a higher performance translates into low computational complexity, online decision-making, higher accuracy and efficiency, and the ability to detect anomalies for streaming data. The low resource cost of the algorithm means low utilization of computing resources, such as CPU and memory, as well as a reduction in the storage costs. Finally, the anomaly detection approach must be able to detect all types of anomalies (point and contextual) without assuming the shape of the time series.

Box-plot Implementations in SLMAD: We now present three box-plot methods for anomaly detection implemented in SLMAD. We used the box-plot rules in combination with custom build methodologies using dynamic grouping and matrix profile (discussed in Sect. 3.2). The first Turkey's box-plot can be constructed using three quantities: first quantile (Q1), second quantile (Q2), and third quantile (Q3). The Tukey's lower and upper boundaries (LB_t and UB_t) are defined using the concept of interquartile range or IQR (where, $IQR = Q3 - Q1$) as shown in Eq. 1.

$$LB_t = Q1 - 1.5 \times IQR$$
$$UB_t = Q3 + 1.5 \times IQR \tag{1}$$

The second implementation is based on the robust measure of scale, i.e., median absolute deviation (MAD), and known as robust box-plot. The lower and upper boundaries for robust box-plot (LB_r and UB_r) are calculated using Eq. 2.

In terms of accuracy, the robust box-plot performs better than Tukey's box-plot [2]. By default, SLMAD trains the models using robust box-plots because of our objective to achieve higher performance.

$$LB_r = Q1 - 1.44 \times MAD$$
$$UB_r = Q3 + 1.44 \times MAD \tag{2}$$

The third box-plot implementation is based on Bowley's coefficient that adjusts the fences and overcomes some statistical limitations of Tukey's box-plot [20]. The lower and upper boundaries of Bowley's box-plot (LB_b and UB_b) can be calculated using Eq. 3.

$$LB_b = Q1 - 1.5 \times (\frac{SIQR_l}{SIQR_u})$$
$$UB_b = Q3 + 1.5 \times (\frac{SIQR_l}{SIQR_u}) \tag{3}$$

Where $SIQR_l = Q2 - Q1$ and $SIQR_u = Q3 - Q2$.

3.2 Construction of SLMAD Framework

Figure 1 gives an overview of the three-stage process of SLMAD: 1) analysis, 2) dynamic grouping, 3) modelling and evaluation. Figure 2 presents the stages of the SLMAD algorithm in more detail. Since one of our objectives is to have an anomaly detection framework with low time-complexity, we designed and tested each stage of SLMAD to have a worst-case complexity of O(n).

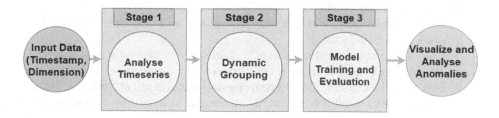

Fig. 1. Overview of the SLMAD framework

We now briefly explain each stage of SLMAD:

1. Analyze time series
   - We first use the statistical methods to evaluate a small subset of time series data in terms of its characteristics. We check whether the time series is stationary using the augmented Dickey-Fuller (ADF) and Kwiatkowski Phillips Schmidt Shin (KPSS) test. If both tests conclude that the series is stationary, it is labeled as stationary. If only the ADF test confirms that the series is stationary then it is labelled difference stationary and if only the KPSS test confirms the data is stationary, we label it as trend stationary.

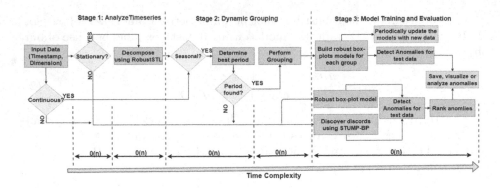

Fig. 2. Workflow of the SLMAD framework

- We then check if the time series is discrete or continuous. If there are more than twenty distinct points in the data set we label it as continuous, otherwise, it is discrete.
- We make sure that the time series is not flat, i.e., it does not possess a constant dimension for all the timestamps.
- We check if the time series is seasonal using auto-correlation and if seasonality is detected, we find the best period. The seasonality analysis and period determination methodology are given in Fig. 3. If the autocorrelation is over 0.9, we label the data as seasonal. We then determine the best period using peaks in the time series and calculate the number of points between the peaks as the period.

2. Dynamic Grouping
   - Based on the output of the analysis stage, we build the following two cases, case 1 and case 2:
   - *Case 1:* Provided the time series is stationary, continuous, and seasonal, we use a subset of data as a training set and group them based on the best period using a dynamic grouping approach. Since at least five points are needed to build a box-plot, we use equal to five or more periods and group them.
   - *Case 2:* If the time series does not possess all the aforementioned characteristics in case 1, we use the Matrix Profile for streaming data (STUMP Incremental or STUMPI) to determine the discords in time series. The top discords are the anomalies. We select Matrix Profile as it can work without a need for parameter tuning and does not assume the shape of time series.

3. Model Training and Evaluation
   - Finally in Stage 3, we train the models and evaluate them with new data. We build models for cases 1 and 2 in the dynamic grouping stage separately as explained below.
   - *Case 1:* For each group, we build one box-plot. For example, if the best period is 10, we use at least 50 points to build the models in the training set. In other words, 10 box-plot thresholds are constructed.

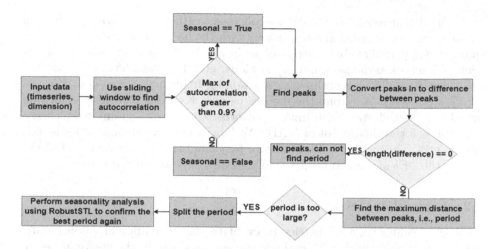

Fig. 3. Methodology for seasonality and period analysis

- *Case 2:* The discords in the Matrix Profile given by STUMPI are directed upwards. Therefore, we use the upper bound of a robust box-plot (UP_r) to detect the top discords. We call the custom Matrix Profile and box-plot approach as *STUMPI-BP* and employ it for all types of time series that are non-stationary, non-seasonal, or discrete. For case 2, we found that STUMPI-BP successfully detects the contextual anomalies in complex time series but misses some prominent point anomalies. Therefore, we use a robust box-plot constructed from the points equal to the window-size (default=100) and rank the anomalies reported by both methods.
- The models are also updated at the run-time. For case 1, we update the box-plot threshold for each group as soon as we collect the data equal to one period. For case 2, we update the Matrix Profile for each streaming data point. The results can then be saved to a file, visualised, or used as the start of an automated remediation process.

4 Experimental Setup

4.1 Data Set Characteristics

We use multiple data sets including open-source anomaly detection benchmarks and in-house Huawei network production data to evaluate our model. The open-source benchmarks include labelled data-sets from Numenta Benchmarking Suite and Yahoo lab7. Due to the high complexity of labelling the data from the production environment, the in-house collected metrics at Huawei are labelled using a simple statistical-based labelling methodology that is build upon box-plot thresholds.

We find that a total of 96 metrics collected in Huawei network's production environment are seasonal and continuous with the best period equal to one hour (found using period analysis methodology in Fig. 3). These metrics include KPIs such as network response times, utilizations, etc. The data has been collected for a duration of four months with a sampling frequency of one hour (i.e., 24 samples per day). We group the metrics using a K-means clustering algorithm into 14 distinct clusters with unique patterns. Out of each group, we pick one time series for evaluation of SLMAD. We label the selected time series as ts1, ts2, ..., ts14. The metrics are collected for four months. To evaluate the SLMAD, we use the 70% of data to train the models and 30% as test data sets.

4.2   Evaluation Metrics

We use a combination of metrics to evaluate that our approach is accurate, efficient, and generalizable. To evaluate the accuracy, we use the precision, recall, and F1 score of the actual and identified anomalies. To evaluate the resource cost, we measure the CPU and memory utilization as well as the training and testing time. We evaluate the generalizability of our approach by testing the anomaly detection accuracy on a number of different data sets. We compare our approach against a lightweight persistence anomaly detection model that uses the previous data point as a forecast for the next point and classifies the point as an anomaly if the actual point is outside the standard deviation expected.

5   Results

In Table 2, we evaluated a range of continuous and discrete-time series data from the art_daily data set in the Numenta benchmark suite. Four of the time series (Art_daily_no_noise, Art_noisy, Art_flatline, and Art_daily_perfect_square_wave) have no points labelled as anomalies and SLMAD does not report any false anomalies for them. We also evaluated the non-stationary and complex data sets (cpu_utilization_asg_misconfiguration and ec2_cpu_utilization_5f5533) where no best period could be found by using *STUMPI-BP*. Our proposed SLMAD approach achieves high precision across the range of data sets. When we combine the precision and recall scores in the F1 score and compare it against the persistence baseline approach, we can see that SLMAD has achieved better results for each of the data sets.

Table 3 shows the results for the real data in the Yahoo benchmark. The recall values are slightly lower than precision given the unbalanced amount of anomalies in the data set. We can see that the SLMAD approach achieved an improved F1 score compared to the persistence approach again for all of the data sets. Finally, we evaluate our approach in a production environment using network data from Huawei, as shown in Table 4. We can see that the SLMAD approach achieved higher precision scores than recall, due to the unbalanced nature of production anomaly data. Overall, the SLMAD approach shows an improved F1 score for all of the production data sets compared to the persistence model.

Table 2. Numenta metrics accuracy

Numenta metrics	Precision	Recall	SLMAD F1	Persistence F1
Art_daily_Flat_middle	0.65	1	0.78	0.75
Art_daily_Jumps_down	0.936	0.377	0.537	0.50
Art_daily_Jumps_up	0.932	0.4	0.56	0.53
Art_daily_Spike_density	1	0.75	0.85	0.77
Art_daily_No_jump	0.92	0.39	0.55	0.54
cpu_utilization_asg_misconfig	0.83	0.48	0.60	0.55
ec2_cpu_utilization_5f5533	0.72	0.56	0.63	0.61

Table 3. Yahoo metrics accuracy

Data set	Precision	Recall	SLMAD F1	Persist F1
real_7	0.966	0.9	0.92	0.90
real_9	0.99	0.959	0.974	0.827
real_15	0.988	0.7	0.821	0.74
real_17	0.973	0.974	0.974	0.84
real_20	0.959	0.873	0.91	0.9
real_26	0.867	0.825	0.847	0.48
real_31	0.982	0.843	0.9	0.66
real_34	0.994	0.956	0.973	0.6
real_46	0.9	0.788	0.82	0.72
real_47	0.98	0.95	0.972	0.84
real_51	0.99	0.93	0.961	0.84
real_55	0.99	0.72	0.83	0.89
real_58	0.98	0.962	0.968	0.831
real_65	0.98	0.61	0.744	0.72

Table 4. Huawei metrics accuracy

Data	Precision	Recall	SLMAD F1	Persist F1
ts1	0.83	0.53	0.64	0.57
ts2	0.96	0.49	0.64	0.62
ts3	0.66	0.73	0.69	0.63
ts4	0.88	0.57	0.69	0.68
ts5	0.76	0.56	0.64	0.62
ts6	0.82	0.66	0.73	0.72
ts7	0.69	0.65	0.67	0.63
ts8	0.82	0.73	0.77	0.75
ts9	0.86	0.53	0.65	0.64
ts10	0.91	0.57	0.70	0.67
ts11	0.98	0.63	0.76	0.75
ts12	0.77	0.75	0.76	0.74
ts13	0.78	0.65	0.70	0.63
ts14	0.97	0.56	0.71	0.70

Table 5 shows the results for SLMAD when evaluated in terms of efficiency and resource consumption. The results are obtained on a virtual machine that is representative of a modern multicore platform with 4 CPU cores and 32 GB of memory. We implemented the box-plot models with a dynamic grouping approach in Python3 and Golang to analyse their resource costs, training, and testing times. SLMAD trains on 10K data points in 0.072 s and 100K data points in 0.5 s. This allows our framework to update online for a large number of points. The Golang implementation is much more efficient than the Python implementation, dramatically reducing the package size (up to $\approx 60\times$), CPU and memory utilization. The Golang implementation yields substantial improvements of 98%, 36%, 95%, and 99% in terms of package size, CPU consumption, memory consumption, and execution times, respectively. This improvement is due to Golang being better able to handle concurrency problems.

Table 5. Efficiency and Resource Costs for SLMAD

Parameters	Go	Python
Package size	7.5 MB	447 MB
CPU utilization	65%	102%
Memory utilization	4320KB	102056 KB
Training time (with 1K data points)	0.014 s	15 s
Training time (with 10K data points)	0.072 s	70 s
Training time (with 100K data points)	0.501 s	N/A
Testing time (With 1K data points)	0.5 s	0.74 s

6   Conclusion and Future Work

Time series anomaly detection is a critical module for operations and management in a production environment. A resource-efficient, general, and accurate anomaly detection system is indispensable in real applications and can be used to trigger root cause analysis (RCA) and automated remediation. This paper has introduced a lightweight time series anomaly detection framework that has shown improved F1 scores on a number of anomaly detection benchmarks when compared against a persistence-based anomaly detection approach. We also evaluated our approach using real production data from Huawei, where it showed impressive recall and precision scores while maintaining low training and testing times as well as low CPU and memory utilization.

For future work, we will evaluate our framework against a number of other anomaly detection approaches using the current anomaly detection benchmarks as well as additional production data. Once we have evaluated the best anomaly detection approach for our production environment, we will use it to trigger RCA and automated remediation.

References

1. Akcay, S., Atapour-Abarghouei, A., Breckon, T.P.: GANomaly: semi-supervised anomaly detection via adversarial training. In: Jawahar, C.V., Li, H., Mori, G., Schindler, K. (eds.) ACCV 2018. LNCS, vol. 11363, pp. 622–637. Springer, Cham (2019). https://doi.org/10.1007/978-3-030-20893-6_39
2. Andrea, K., Shevlyakov, G., Smirnov, P.: Detection of outliers with boxplots (2013)
3. Chandola, V., Banerjee, A., Kumar, V.: Anomaly detection: a survey. ACM Comput. Surv. (CSUR) **41**(3), 1–58 (2009)
4. Dang, Y., Lin, Q., Huang, P.: AIOps: real-world challenges and research innovations. In: 2019 IEEE/ACM 41st International Conference on Software Engineering: Companion Proceedings (ICSE-Companion), pp. 4–5. IEEE (2019)
5. Doersch, C.: Tutorial on variational autoencoders. arXiv preprint arXiv:1606.05908 (2016)

6. Gao, J., Song, X., Wen, Q., Wang, P., Sun, L., Xu, H.: RobustTAD: robust time series anomaly detection via decomposition and convolutional neural networks. arXiv preprint arXiv:2002.09545 (2020)
7. Gulenko, A.: Autonomic self-healing in cloud computing platforms (2020)
8. Hochenbaum, J., Vallis, O.S., Kejariwal, A.: Automatic anomaly detection in the cloud via statistical learning. arXiv preprint arXiv:1704.07706 (2017)
9. Laptev, N., Yosinski, J., Li, L.E., Smyl, S.: Time-series extreme event forecasting with neural networks at Uber. In: International Conference on Machine Learning, vol. 34, pp. 1–5 (2017)
10. Liu, D., Zhao, Y., Xu, H., Sun, Y., Pei, D., Luo, J., Jing, X., Feng, M.: Opprentice: towards practical and automatic anomaly detection through machine learning. In: Proceedings of the 2015 Internet Measurement Conference, pp. 211–224 (2015)
11. Lu, W., Ghorbani, A.A.: Network anomaly detection based on wavelet analysis. EURASIP J. Adv. Signal Process. **2009**, 1–16 (2008)
12. Mahimkar, A., et al.: Rapid detection of maintenance induced changes in service performance. In: Proceedings of the Seventh Conference on Emerging Networking Experiments and Technologies, pp. 1–12 (2011)
13. Masood, A., Hashmi, A.: AIOps: predictive analytics & machine learning in operations. Cognitive Computing Recipes, pp. 359–382. Apress, Berkeley, CA (2019). https://doi.org/10.1007/978-1-4842-4106-6_7
14. Ren, H., et al.: Time-series anomaly detection service at microsoft. In: Proceedings of the 25th ACM SIGKDD International Conference on Knowledge Discovery & Data Mining, pp. 3009–3017 (2019)
15. Rosner, B.: Percentage points for a generalized ESD many-outlier procedure. Technometrics **25**(2), 165–172 (1983)
16. Salinas, D., Flunkert, V., Gasthaus, J., Januschowski, T.: Deepar: probabilistic forecasting with autoregressive recurrent networks. Int. J. Forecast. **36**(3), 1181–1191 (2020)
17. Shipmon, D.T., Gurevitch, J.M., Piselli, P.M., Edwards, S.T.: Time series anomaly detection; detection of anomalous drops with limited features and sparse examples in noisy highly periodic data. arXiv preprint arXiv:1708.03665 (2017)
18. Taylor, S.J., Letham, B.: Forecasting at scale. Am. Stat. **72**(1), 37–45 (2018)
19. Vercruyssen, V., Meert, W., Davis, J.: Transfer learning for anomaly detection through localized and unsupervised instance selection. In: The Thirty-Fourth AAAI Conference on Artificial Intelligence, AAAI 2020, pp. 6054–6061. AAAI Press (2020)
20. Walker, M., Dovoedo, Y., Chakraborti, S., Hilton, C.: An improved boxplot for univariate data. Am. Stat. **72**(4), 348–353 (2018)
21. Wei, L., Kumar, N., Lolla, V.N., Keogh, E.J., Lonardi, S., Ratanamahatana, C.A.: Assumption-free anomaly detection in time series. SSDBM **5**, 237–242 (2005)
22. White, G., Clarke, S.: Short-term qos forecasting at the edge for reliable service applications. IEEE Transactions on Services Computing pp. 1–1 (2020)
23. Yeh, C.C.M., Zhu, Y., Ulanova, L., Begum, N., Ding, Y., Dau, H.A., Silva, D.F., Mueen, A., Keogh, E.: Matrix profile i: all pairs similarity joins for time series: a unifying view that includes motifs, discords and shapelets. In: 2016 IEEE 16th international conference on data mining (ICDM). pp. 1317–1322. Ieee (2016)
24. Zhang, Y., Ge, Z., Greenberg, A., Roughan, M.: Network anomography. In: Proceedings of the 5th ACM SIGCOMM conference on Internet Measurement. pp. 30–30 (2005)

Software Reliability Engineering for Resilient Cloud Operations

Michael R. Lyu$^{(\boxtimes)}$ and Yuxin Su

The Chinese University of Hong Kong, Hong Kong, China
{lyu,yxsu}@cse.cuhk.edu.hk

Abstract. In the last decade, cloud environments become the most sophisticated software systems. Due to the inevitable occurrences of failures, software reliability engineering is top priority for cloud developers and maintainers. In this essay, we introduce several frameworks to provide resilient cloud operations from different development phases, ranging from fault prevention before deployment and fault removal at run-time.

Keywords: Software reliability engineering · Resilient cloud operation · Fault prevention · Fault removal

1 Introduction

In the recent years, IT enterprises have drastically increased development of their applications and services on cloud computing platforms, such as search engine, instant messaging app and online shopping. As cloud systems continue to grow in terms of complexity and volume, cloud failures become inevitable and critical, which may lead to service interruptions or performance degradation. Whether cloud failures can be properly managed will greatly impact company revenue and customer satisfaction. For example, in 2017, a downtime in Amazon led to a loss of 150+ million US dollars. Thus, the reliability of modern software is of paramount importance. Consequently, we identified several critical challenges commonly seen in industrial cloud systems and provide a general road-map from fault prevention and fault removal to improve the cloud reliability by resilient operations. First, as cloud systems are actively undergoing continuous feature upgrade and system evolution, the statistical properties of system monitoring data may change from time to time. Hence, to impede the deployment of the buggy cloud service, a fast and effective fault prevention mechanism for the source code and cloud services interface is a crucial task to address. In practice, however, fault prevention is hard to offer perfect cloud services without any runtime bugs or errors. Fault removal mechanisms can come to rescue after cloud deployment.

© Springer Nature Switzerland AG 2021
H. Hacid et al. (Eds.): ICSOC 2020 Workshops, LNCS 12632, pp. 264–268, 2021.
https://doi.org/10.1007/978-3-030-76352-7_27

2   Fault Prevention for Cloud Services

In this section, we introduce fault prevention before the deployment of cloud services. We attempt to detect buggy code while the service is under development and discuss the testing approaches to verify the correctness of cloud service before actual deployment.

2.1   RESTful API Testing

Most industrial scale cloud services are programmatically accessed through Representational State Transfer (REST) APIs, which are a clear trend as composable paradigm to create and integrate cloud software. One of the key benefits of involving RESTful APIs is a systematic approach to software logic modeling leveraged by a growing usage of standardized cloud software stack. In the last few years, the OpenAPI Specification (OAS) has gradually become the de-facto standard to describe RESTful APIs from a functional perspective. The adequate testing of stateful cloud services via OpenAPI is difficult and costly. Failures generated by complex stateful interactions can be of high impact to customer, but they are hard to replicate.

To address the testing problem in an automatic way would certainly increase the reliability of cloud services. Fuzzing is a widely adopted approach to find bugs in software by feeding variety of test input. RESTler [1] first performs a lightweight static analysis on the API specification of a target cloud service and detects dependencies among test input. However, the automatically-generated fake paths will limit the combinatorial explosion of the fuzzing space due to the lack of feedback about grammar. To effectively induce the fuzzers to focus on fake paths (or branches), we consider the following design aspects. We maintain a resource pool which stores a sufficient number of fake paths to affect the fuzzing policy. Typically, as the fuzzer generates various mutations from one startup input, fake paths should provide different request coverage and be directly affected by the input so that the fuzzer will keep uncovering the trap. Various mechanisms for RESTful API testing based on this direction will be investigated and evaluated.

2.2   Software Defect Prediction

To improve software reliability, software defect prediction is utilized to assist developers in finding potential bugs and allocating their testing efforts. Traditional defect prediction studies mainly focus on designing hand-crafted features, which are input into machine learning classifiers to identify defective code. However, these hand-crafted features often fail to capture the semantic and structural information of programs. Such information is important in modeling program functionality and can lead to more accurate defect prediction. Software defect prediction is a process of building classifiers to predict code areas that potentially contain defects, using information such as code complexity and change history. The prediction results (i.e., buggy code areas) can place warnings for code reviewers and allocate their efforts. The code areas could be files, changes or methods.

In this essay, we introduce a framework called Defect Prediction via Convolutional Neural Network (DP-CNN) [4], which leverages deep learning for effective feature generation. We evaluate our method on seven open source projects in terms of F-measure in defect prediction. The experimental results show that in average, DP-CNN improves the state-of-the-art method by 12%.

3   Fault Removal after Deployment

In this section, we introduce several fault removal approaches from different perspectives, e.g. log analysis, emerging incident detection and fault localization.

3.1   Automated Log Mining for Fault Management

Logs are semi-structured text generated by logging statements in software source code. In recent decades, logs generated from cloud service have become imperative in the reliability assurance mechanism of cloud systems because they are often the only data available that traces cloud runtime information.

This essay presents a general overview of log mining techniques including how to automate and assist the writing of logging statements and how to employ logs to detect anomalies, predict failures, and facilitate diagnosis [3]. Traditional log analysis that is mainly based on ad-hoc domain knowledge or manually constructed and maintained rules is inefficient and ineffective for cloud systems due to its large scale and high complex in structure. This brings three major challenges to modern log analysis for cloud services. (1) Quality of the logging statements varies to a large extent because developers from different groups usually write logging statements based on their own domain knowledge and ad-hoc designs. (2) log mining based on manual rules is prohibited due to large volume of logs generated in a short time. (3) Due to the wide adoption of the DevOps software development concept, a new software version often appears in a short-term manner. Thus, corresponding logging statements update frequently as well. To address these challenges, we introduce several work about automated rule construction and critical information extraction.

3.2   Automatic Emerging Incident Mining from Discussion

When a high-damaging incident happens in cloud system, developers and maintainers generate an incident ticket or establishes a war-room to discuss the potential reasons and possible solution to fix the incident. Detecting emerging bugs or errors timely and precisely is crucial for developers and maintainers to provide resilient cloud services. However, the tremendous quantities of discussion comments, together with their imprecise and noisy descriptions increase the difficulties in accurately identifying newly-appearing issues. In this essay, we introduce an automated framework IDEA [2] to identify any new issues based on maintainers' discussions. IDEA takes the discussion of different incident tickets or war-room about the same target as input. To track the topic variations over discussion, AOLDA (Adaptively Online Latent Dirichlet Allocation) is employed for generating discussion-sensitive topic distributions. The emerging topics are then identified based on the typical anomaly detection method. Finally, IDEA visualizes the variations of different issues along with discussions, and highlights the emerging ones for better understanding.

3.3   Fault Localization from Structural Information

A critical research direction in cloud computing has been the defense against inevitable cloud failures and their prevention from causing service interruption or service degradation. We have consequently identified two critical challenges commonly seen in industrial cloud systems. First, when diagnosing failures for large-scale cloud systems, there is currently a lack of means to incorporating expert knowledge into the training of automated detection models. Second, although the dependencies of cloud service/resource can provide rich information for tracking the cascading effect of cloud failures, they have not been explicitly considered in existing methods of root cause analysis.

To address the above challenges, we introduce a resilient cloud systems framework by incorporating structural information and knowledge about the cloud systems. Our goal is to comprehensively improve the reliability of cloud systems and services. Particularly, the framework consists of an end-to-end pipeline of software reliability engineering, i.e., *anomaly detection, failure diagnosis*, and *fault localization*. Anomaly detection looks for systems' anomalous patterns that do not conform to normal behaviors, such as high CPU usage, low throughput. We propose a log-based anomaly detection model which can quickly learn unprecedented log patterns in an online manner and dynamically adapt to concept drift caused by system evolution. Moreover, failure diagnosis attempts to find the most significant problems directly induced by the failures; for example, abnormal Key Performance Indicator (KPIs). We therefore introduce an adaptive failure diagnosis algorithm with a human-in-the-loop strategy for efficient model training. Lastly, fault localization locates the root cause of a failure, such as a failed microservice or device. We also develop a novel fault localization technique for microservice architecture using dependency-aware collaborative filtering. Experimental evaluations will be conducted on this end-to-end framework regarding its effectiveness in providing resilient cloud operations.

References

1. Atlidakis, V., Godefroid, P., Polishchuk, M.: RESTler: stateful REST API fuzzing. In: Proceedings of the 41st International Conference on Software Engineering, ICSE 2019 (2019)
2. Gao, C., Zeng, J., Lyu, M.R., King, I.: Online app review analysis for identifying emerging issues. In: Proceedings of the 40th International Conference on Software Engineering, ICSE (2018)

3. He, S., Zhu, J., He, P., Lyu, M.R.: Experience report: system log analysis for anomaly detection. In: Proceedings of the 27th IEEE International Symposium on Software Reliability Engineering (ISSRE) (2016)
4. Li, J., He, P., Zhu, J., Lyu, M.R.: Software defect prediction via convolutional neural network. In: Proceedings of IEEE International Conference on Software Quality, Reliability and Security (QRS)

AI-Enabled Process Automation
(AI-PA 2020)

Introduction to the 1st International Workshop on AI-enabled Process Automation (AI-PA 2020)

The 1st International Workshop on AI-enabled Process Automation (AI-PA 2020) was held as one of the workshops of the 18th International Conference on Service Oriented Computing (ICSOC 2020). The AI-PA workshop aims at providing a forum for researchers and professionals interested in Artificial Intelligence (AI) enabled Business Processes and Services, and in understanding, envisioning, and discussing the opportunities and challenges of intelligent Process Automation, Process Data Analytics, and providing Cognitive Assistants for knowledge workers. Recognizing the broad scope of the potential areas of interest, the workshop was organized into four themes as follows:

- Theme 1: Artificial Intelligence (AI), Services and Processes
- Theme 2: BigData, Services and Processes
- Theme 3: Smart Entities, Services and Processes
- Theme 4: Industry Applications

The papers selected for presentation and publication in this volume showcase fresh ideas from exciting and emerging topics in service-oriented computing and case studies in Artificial Intelligence (AI) enabled Business Processes and Services.

We have selected 11 high-quality papers from the AI-PA 2020 submissions, keeping the acceptance rate at around 50%. We have also included a short invited paper from our keynote speaker, Prof. Aditya Ghose, from the University of Wollongong, Australia, providing insights into the future of "Robotic Process Automation (RPA)". Each paper was reviewed by a team comprising a senior Program Committee member and at least two regular Program Committee members who engaged in a discussion phase after the initial reviews were prepared. AI-PA 2020 paid the workshop registration fees for all accepted papers, and offered $500.00 for the best paper award, thanks to the AI-PA workshop **sponsors**:

- ITIC Pty Ltd (https://www.itic.com.au/)
- dAIta Pty Ltd (https://www.daita.com.au/)
- AIP Research Centre (https://aip-research-center.github.io/)

We are grateful for the support of our authors, sponsors, Program Committee members, and the ICSOC 2020 Organizing Committee. We very much hope you enjoy reading the papers in this volume.

December 2020   AI-PA 2020 Workshop Organizers

Organization

Workshop Organizers

Amin Beheshti (Co-chair)	Macquarie University, Australia
Boualem Benatallah (Co-chair)	UNSW Sydney, Australia
Hamid Motahari (Co-chair)	EY, USA
Ladjel Bellatreche	ISAE-ENSMA, France
Mohamed Adel Serhani	UAE University, UAE
Francois Charoy	Inria/Université de Lorraine, France
Li Qing	Hong Kong Polytechnic University, Hong Kong

Program Committee

Schahram Dustdar	Vienna University of Technology, Austria
Fabio Casati	Servicenow, USA
Aditya Ghose	University of Wollongong, Australia
Ramana Reddy	West Virginia University, USA
Anup Kalia	IBM Research, USA
Michael Sheng	Macquarie University, Australia
Mark Burgin	University of California, Los Angeles, USA
Farouk Toumani	Blaise Pascal University, France
Hakim Hacid	Zayed University, UAE
Gordana Dodig Crnkovic	Chalmers University of Technology, Sweden
Mehdi Elahi	University of Bergen, Norway
Daniela Grigori	Paris-Dauphine University, France
Enayat Rajabi	Dalhousie University, Canada
Sajib Mistry	Curtin University, Australia
Fabrizio Messina	University of Catania, Italy
Qiang Qu	Shenzhen Institutes of Advanced Technology, China
Azadeh Ghari Neiat	Deakin University, Australia
Rama Akkiraju	IBM Watson, USA
Shayan Zamanirad	UNSW Sydney, Australia
Marcos Baez	University of Trento, Italy
Fariborz Sobhanmanesh	Macquarie University, Australia
Adrian Mos	NAVER LABS Europe, France

Best Paper Award

Ghodratnama S., Zakershahrak M., and Sobhanmanesh F.: "Am I Rare? An Intelligent Summarization Approach for Identifying Hidden Anomalies".

Workshop Website

https://aip-research-center.github.io/AIPA_workshop/2020/.

The Future of Robotic Process Automation (RPA)

Aditya Ghose[✉], Geeta Mahala, Simon Pulawski, and Hoa Dam

Decision Systems Lab, School of Computing and IT, University of Wollongong,
Wollongong, Australia
{aditya,hoa}@uow.edu.au, {gm168,spp701}@uowmail.edu.au

Abstract. While there has been considerable industry interest in the deployment and uptake of Robotic Process Automation (RPA) technology, very little has been done by way of generating technological foresight into the manner in which RPA systems might evolve in the short- to medium-term. This paper seeks to fill that gap.

1 Introduction

The idea that business process execution can be automated via the deployment of robotic components or *bots*, leading to the notion of Robotic Process Automation (or RPA) [1,13,27], has gained significant traction in industry. This technology trend started with cost-saving as the driving value proposition (e.g., automating many of the human-mediated tasks often outsourced to call centres could lead to signficant cost reductions). There is, however, a growing realisation that RPA can deliver a significant improvement in the quality of enterprise functionality.

A key trend is the shift in emphasis from *process automation* to *enterprise automation*. In the current thinking (and product offerings) around RPA, the unit of analysis is the business process. We argue that agent technology affords the opportunity to automate all aspects of enterprise functionality (traditionally conceived as a collection of disparate business processes) within a single unifying conceptual (and programming) framework. In the most general sense, an enterprise can be viewed as a collection of agents. This does not imply that we must view every distinct enterprise actor (people, roles, machines) as a distinct agent. Instead, we create distinct agents to accommodate situations where the separation of knowledge or capabilities (driven by business competition constraints, or compliance requirements) needs to be maintained. In the rest of this paper, we will base our arguments mainly around the Belief-Desire-Intention (BDI) agent architecture [21] which is arguably the most expressive and most comprehensive conception of agent design and implementation [20] on offer in the literature.

Much of the discourse around business process management (BPM) assumes a clear distinction between the coordination machinery (the process engine, driven by a coordination model) and the machinery that actually executes the required functionality. We have come to a point where we have to acknowledge

H. Hacid et al. (Eds.): ICSOC 2020 Workshops, LNCS 12632, pp. 273–280, 2021.
https://doi.org/10.1007/978-3-030-76352-7_28

the need for *ubiquitous, hierarchic coordination*. In other words, what is an activity or task at a given level abstraction is in fact a complex sub-process requiring its own coordination model at a lower level of abstraction. Similarly, an action in an abstract BDI agent plan is a goal requiring its own set of plans at a lower level of abstraction. This sets the stage for a compare-and-contrast exercise between the coordination capabilities of BDI agents and those of current business process engines.

In the following, when we refer to RPA, we mean the RPA technology of the future, built around sophisticated agent technology (such as BDI agent systems).

2   Related Work

Agent technology has been used in RPA in a variety of settings. The following is a brief survey of existing work in using agents or bots in RPA implementations in various contexts. Agents have been used in RPA to enable automation in multiple industries and enterprise functions including Telecommunication [18], Human Resources (HR) & Recruitment [6,8], and Banking [11]. Rule-based and monotonous tasks of a business process can be automated with the help of rule-based agents whereas agents utilizing AI algorithms can be used to automate more complex business processes involving decision-making [23]. Agents in RPA combine AI technologies such as machine learning, computer vision, image and natural language processing with internet of things (IoT) and block chain technologies to perform automation tasks. [16] represents an orchestration of agents with different functionalities into a single bot to automate a business function.

Autonomous agents have been used in an innovative approach to debt collection [28]. This approach uses optical character recognition (OCR) to identify the main objects from a document and employs deep learning methods, such as temporal convolutional networks and convolutional neural networks to enhance the quality. Lin et al. [14] demonstrates RPA agents in semiconductor smart manufacturing without the help of an RPA vendor, where an OCR technique is implemented for reading handwritten or printed text and an artificial neural network (ANN) is implemented for identifying the fuzzy or clear pictures. Kobayashi et al. [9] illustrates requirements for utilizing RPA agents in consumer services, especially using robots to assist elderly people by communicating with their families and placing grocery orders for them. Desirable technologies for implementation with RPA agents include the internet of things (IoT), artificial intelligence (especially the Watson Natural Language Classifier for learning communication), the Google Cloud Platform for voice recognition, and the NeoFace Cloud for facial recognition. Zhang et al. [30] has provided an RPA implementation framework for audit engagements which involves audit workflow analysis, audit tasks are automated using either RPA agents or combination of RPA agents and AI or cognitive computing based on the complexity of the audit task (structured, semi-structured or unstructured respectively). Rizk et al. [22] explores interactive automation in the form of a conversational digital assistant. It allows the business users to interact with customers and customize their automation solutions through natural language. The framework, responsible for creating such assistants, relies on a multi-agent orchestration model and conversational wrappers to create autonomous agents including RPA agents. These agents are equipped with skills that help them understand and generate natural language alongside task automation. The orchestrator expects agents to adhere to a specific contract to determine which agents respond to a user's utterance.

Little research has been conducted to improve RPA technology using Belief-Desire-Intention (BDI) agents. Mahala [15] explores using BDI agents as a promising alternative to the current approaches in the implementation and design of complicated software systems and uses BDI agent architecture to implement BDI agents in RPA. BDI agents can handle multiple competing agendas which is not possible in process-oriented approaches. BDI agents provide additional flexibility by using an option selection function and allowing high priority events to interrupt lower priority goals.

3   Agent-Based Process Execution

It is useful to compare and contrast agent-based process execution with current approaches to process execution based on BPM technology.

3.1   Where BPM Outperforms Agent-Based Execution

- Process models are easier to understand than agent models
- Diagrammatic executable specification languages are in widespread use (e.g., BPMN)
- BPM technology in general enjoys widespread use and adoption
- There is a vast amount of associated tooling on offer
- Mature technology for data-driven extraction of process models is available and in widespread use (process mining)

3.2   Where Agent-Based Execution Outperforms BPM Technology

- BDI agent plans have pre-conditions
- In some extensions, they also have post-conditions
- There are sophisticated mechanisms for hierarchic structuring of workflows (via sub-goals)
- Sophisticated mechanisms exist for the interruption, re-consideration and re-deliberation of agent plans (again, leveraging sub-goals)
- Sophisticated event listeners and event management exist (e.g., the event set and the event selection function)
- Sophisticated XOR-gateways, which might leverage run-time non-functional assessments are on offer (via the option selection function)
- Sophisticated pre-emptive scheduling mechanisms exist (e.g., intention stacks and intention selection)

– Sophisticated agent communication languages exist
– A large body of other AI results can be brought to bear (learning, uncertainty handling, knowledge representation and reasoning, optimisation, game-tree search...)

4   The Evolution of Enterprise Functionality

4.1   Personalization

Personalization has received considerable attention in the recent past. The question of business process personalization has been considered in the literature [29] (although tool support is not in widespread use in industry). The adoption of RPA makes the prospect of *extreme personalization* or *enterprise-wide personalization* - where all business processes personalize to cater to the individual's unique characteristics or needs every time an external client (or internal user) engages with the enterprise - feasible. For instance, the simple act of an employee engaging with the corporate leave booking process can lead to multiple downstream consequences. It can trigger a process to search for temporary replacement from the internal pool of employees, or a process to externally recruit a temporary replacement (which process is actually executed might be determined by the preferences of the employee seeking leave). Also, in a manner driven by employee preferences, a leave notification email might be sent to a set of internal or external stakeholders, a temporary hold might be placed on the employee's corporate gym and childcare centre membership, and a similar temporary hold might be placed on room cleaning services for the employee's office. Depending on the employee's preferences, a request for a travel quote might be generated from the firm's corporate travel agency and an alert might be generated for the firm's internal security service seeking additional patrolling of the employee's office area for the duration of the leave. In terms of RPA implementation, the personalization exercise would be driven by a multi-dimensional characterization of the individual (which might be explicitly coded or learnt over a series of interactions). An agent planning approach, where the variations in the preconditions satisfied might lead to a wide variety of plans or sub-plans invoked, would be far more amenable to achieving these behaviour variations than a more traditional approach based on existing BPM technology.

4.2   Variation

While personalization is one significant driver for process variability [2], a range of other factors require the generation, selection and enactment of business process variants. These include the local needs of various industry segments, local compliance requirements and so on. An insurance claim handling process, for instance, must vary according to the legislative and regulatory frameworks that apply in various jurisdictions and according to the specific line of business, while adhering, where possible, to the intent of an enterprise-wide reference process model [19]. An agent planning approach that would be feasible in an RPA context would enable us to support a very large space of process variants, for very similar reasons to those listed above.

4.3   Adaptation

The notion of business process adaptation generalizes many of the questions around process personalization and variation management [7]. Specifically, adaptation of enterprise functionality involves responding to changes in the operating context, ideally in real-time. Reactivity is difficult to achieve in a BPM context, but is often a natural consequence of the adoption of agent technology (specially, BDI agent systems). BDI agent systems provide for event listeners which feed into an *event set*. The event set can be viewed as a FIFO queue, or can be provided as input to an *event selection function* which can support sophisticated machinery for identifying the event most deserving of immediate attention. A combination of event listeners and event-driven, context-sensitive agent planning will enable future RPA systems to achieve enterprise functionality adaptation on a scale and at a level of ubiquity not achievable with current BPM technology.

4.4   Distribution

Distributed business process execution has received limited attention in the literature [4], but represents an important challenge. Future RPA systems will likely still require a mix of human and machine functionality. In general, the locus of execution of these functionalities might be different (even for fully automated processes). Distributed process execution may be necessary for a variety of reasons. The knowledge required for process execution might be distributed and not easily shared on account of business competition or compliance constraints. Sometimes connectivity or network latency issues might impede knowledge sharing and hence centralized process execution. Distributed process execution is not uncommon in the current context. A field service technician might execute some parts of an equipment maintenance or repair process on-site, while the automated components might be executed at the corporate headquarters. Future RPA systems will enable seamless distribution of process execution (something eminently achievable with current agent technology). This will also likely involve elements of edge computing (driven by similar considerations underpinning current deployments of edge computing).

4.5   Distributed Optimization

While business process optimization has received considerable research attention [26], and some modicum of tool support, very little attention has been paid to the problem of distributed optimization. The notion of distribution here includes geographic distribution, but also distribution over multiple loci for decision making and multiple actuators (or effectors or business process execution engines).

The optimization exercise can apply to design-time artefacts (plans, coordination models, business process designs etc.) or to run-time artefacts (executing agents, business process instances and such). For a simple example, consider a situation where seeking agreement from a customer to wait for a few hours can lead to better alignment with the trucking providers schedule (leading to discounts in trucking costs) thus leading to an overall reduction in costs incurred by both the producer and customer. This is a clear instance of run-time optimization, but if this is a repeated pattern, it can also play out at design time. Future RPA systems will be able to exploit such opportunities by leveraging enterprise business process architectures [10] and techniques for managing the designs of business process ecosystems [12]. For run-time optimization, techniques for solving Distributed Constraint Optimization Problems (DCOP) such as [3].

5 Novel Forms of Governance

In this section, we summarize future prospects for RPA systems governance.

Normative Governance: Future RPA systems will need to be governed as a socio-technical system based on norm-driven agents. In this conception [24] the socio-technical system will be populated by agents (humans or machines/RPA systems) each of which retain individual autonomy but are incentivized to comply with a set of norms that govern inter-agent interactions (obligations, prohibitions and such). This approach offers the opportunity to ensure that the behaviour of the overall multi-agent RPA system meets the required objectives while not removing the opportunities for autonomous behaviour on the part of the constituent agents.

6 Conclusion

This paper reports on a preliminary exercise in technology foresight in relation to RPA. It is by no means a detailed research roadmap, which remains an important item for future work. The formation of optimal coalition structures [25] can play an important role in delivering enterprise functionality via RPA, but remains the subject of future study. The role of agent belief dynamics (revision, merging etc.) [5,17] in RPA also requires further study.

References

1. van der Aalst, W.M.P., Bichler, M., Heinzl, A.: Robotic process automation. Bus. Inf. Syst. Eng. **60**(4), 269–272 (2018). https://doi.org/10.1007/s12599-018-0542-4
2. Bendoly, E., Cotteleer, M.J.: Understanding behavioral sources of process variation following enterprise system deployment. J. Oper. Manag. **26**(1), 23–44 (2008)
3. Billiau, G., Chang, C.F., Ghose, A.: SBDO: a new robust approach to dynamic distributed constraint optimisation. In: Desai, N., Liu, A., Winikoff, M. (eds.) PRIMA 2010. LNCS (LNAI), vol. 7057, pp. 11–26. Springer, Heidelberg (2012). https://doi.org/10.1007/978-3-642-25920-3_2

4. Cheng, E.: An execution model for process mapping and process automation in a distributed business environment. In: The Second Pacific Asia Conference on Information Systems, PACIS 1995, Singapore, 29 June–2 July 1995, p. 32. AISeL (1995). http://aisel.aisnet.org/pacis1995/32
5. Ghose, A., Goebel, R.: Belief states as default theories: studies in non-prioritized belief change. ECAI **98**, 8–12 (1998)
6. Gupta, P., Fernandes, S.F., Jain, M.: Automation in recruitment: a new frontier. J. Inf. Technol. Teach. Cases **8**(2),118–125 (2018). https://doi.org/10.1057/s41266-018-0042-x. http://link.springer.com/10.1057/s41266-018-0042-x
7. Hermosillo, G., Seinturier, L., Duchien, L.: Using complex event processing for dynamic business process adaptation. In: 2010 IEEE International Conference on Services Computing, pp. 466–473. IEEE (2010)
8. Šimek, D., Šperka, R.: How robot/human orchestration can help in an hr department: a case study from a pilot implementation. Organizacija 52(3) (2019). http://organizacija.fov.uni-mb.si/index.php/organizacija/article/view/1026
9. Kobayashi, T., Arai, K., Imai, T., Tanimoto, S., Sato, H., Kanai, A.: Communication robot for elderly based on robotic process automation. In: 2019 IEEE 43rd Annual Computer Software and Applications Conference (COMPSAC), vol. 2, pp. 251–256, July 2019. https://doi.org/10.1109/COMPSAC.2019.10215
10. Koliadis, G., Ghose, A.K., Padmanabhuni, S.: Towards an enterprise business process architecture standard. In: 2008 IEEE Congress on Services-Part I, pp. 239–246. IEEE (2008)
11. Kukreja, M.: Study of robotic process automation (RPA). Int. J. Recent Innov. Trends Comput. Commun. **4**(6), 434–437 (2016)
12. Kurniawan, T.A., Ghose, A.K., Lê, L.-S., Dam, H.K.: On formalizing inter-process relationships. In: Daniel, F., Barkaoui, K., Dustdar, S. (eds.) BPM 2011. LNBIP, vol. 100, pp. 75–86. Springer, Heidelberg (2012). https://doi.org/10.1007/978-3-642-28115-0_8
13. Lacity, M.C., Willcocks, L.P.: A new approach to automating services. MIT Sloan Manag. Rev. **58**(1), 41–49 (2016)
14. Lin, S.C., Shih, L.H., Yang, D., Lin, J., Kung, J.F.: Apply RPA (robotic process automation) in semiconductor smart manufacturing. In: 2018 e-Manufacturing Design Collaboration Symposium (eMDC), pp. 1–3, September 2018
15. Mahala, G.: Improving RPA Technology. Master of research (computer science) thesis, School of Computing and Information Technology, University of Wollongong (2020)
16. Mendling, J., Decker, G., Hull, R., Reijers, H.A., Weber, I.: How do machine learning, robotic process automation, and blockchains affect the human factor in business process management? Commun. Assoc. Inf. Syst. **43**(19), 297–320 (2018)
17. Meyer, T., Ghose, A., Chopra, S.: Social choice, merging, and elections. In: Benferhat, S., Besnard, P. (eds.) ECSQARU 2001. LNCS (LNAI), vol. 2143, pp. 466–477. Springer, Heidelberg (2001). https://doi.org/10.1007/3-540-44652-4_41
18. Penttinen, E., Kasslin, H., Asatiani, A.: How to choose between robotic process automation and back-end system automation? In: 26th European Conference on Information Systems: Beyond Digitization - Facets of Socio-Technical Change, ECIS 2018, Portsmouth, UK, 23–28 June 2018, p. 66 (2018). https://aisel.aisnet.org/ecis2018_rp/66
19. Ponnalagu, K., Ghose, A., Narendra, N.C., Dam, H.K.: Goal-aligned categorization of instance variants in knowledge-intensive processes. In: Motahari-Nezhad, H.R., Recker, J., Weidlich, M. (eds.) BPM 2015. LNCS, vol. 9253, pp. 350–364. Springer, Cham (2015). https://doi.org/10.1007/978-3-319-23063-4_24

20. Rao, A.S.: AgentSpeak(L): BDI agents speak out in a logical computable language. In: Van de Velde, W., Perram, J.W. (eds.) MAAMAW 1996. LNCS, vol. 1038, pp. 42–55. Springer, Heidelberg (1996). https://doi.org/10.1007/BFb0031845

21. Rao, A.S., Georgeff, M.P.: Modeling rational agents within a BDI-architecture. KR **91**, 473–484 (1991)

22. Rizk, Y., et al.: A conversational digital assistant for intelligent process automation. In: Asatiani, A., et al. (eds.) BPM 2020. LNBIP, vol. 393, pp. 85–100. Springer, Cham (2020). https://doi.org/10.1007/978-3-030-58779-6_6

23. Roy, P., Dickinson, P.: How robotic process automation and artificial intelligence will change outsourcing. Mayer Brow, 65–70 (2016. https://m.mayerbrown.com/files/Event/7b819bc0-d4e5-4042-9aa5-f38885e1dbee/Presentation/EventAttachment/a71c0fcc-14a6-40a6-8e60-ab700e8eb357/160607-CHI-SEMINAR-BTS-Best-Practices-Robotic.pdf. Accessed 17 Jan 2019

24. Singh, M.P.: Norms as a basis for governing sociotechnical systems. ACM Trans. Intell. Syst. Technol. (TIST) **5**(1), 1–23 (2014)

25. Sombattheera, C., Ghose, A.: A best-first anytime algorithm for computing optimal coalition structures (short paper)

26. Vergidis, K., Tiwari, A., Majeed, B.: Business process analysis and optimization: beyond reengineering. IEEE Trans. Syst. Man Cybern. Part C (Appl. Rev.) **38**(1), 69–82 (2008). https://doi.org/10.1109/TSMCC.2007.905812

27. Willcocks, L., Lacity, M., Craig, A.: Robotic process automation: strategic transformation lever for global business services? J. Inf. Technol. Teach. Cases **7**(1), 17–28 (2017)

28. Wroblewska, A., Stanislawek, T., Prus-Zajaczkowski, B., Garncarek, L.: Robotic process automation of unstructured data with machine learning. Annals Comput. Sci. Inf. Syst. **16**, 9–16 (2018)

29. Xue, S., Wu, B., Chen, J.: An end-user oriented approach for business process personalization from multiple sources. In: Ghose, A., et al. (eds.) ICSOC 2012. LNCS, vol. 7759, pp. 87–98. Springer, Heidelberg (2013). https://doi.org/10.1007/978-3-642-37804-1_10

30. Zhang, A.C.: Intelligent process automation in audit. J. Emerging Technol. Account. 1–38, July 2019. https://papers.ssrn.com/abstract=3448091

Adaptive Summaries: A Personalized Concept-Based Summarization Approach by Learning from Users' Feedback

Samira Ghodratnama[1(✉)], Mehrdad Zakershahrak[2], and Fariborz Sobhanmanesh[1]

[1] Macquarie University, Sydney, Australia
{samira.ghodratnama,fariborz.sobhanmanesh}@mq.edu.au
[2] Arizona State University, Arizona, USA
mehrdad@asu.edu

Abstract. Exploring the tremendous amount of data efficiently to make a decision, similar to answering a complicated question, is challenging with many real-world application scenarios. In this context, automatic summarization has substantial importance as it will provide the foundation for big data analytic. Traditional summarization approaches optimize the system to produce a short static summary that fits all users that do not consider the subjectivity aspect of summarization, i.e., what is deemed valuable for different users, making these approaches impractical in real-world use cases. This paper proposes an interactive concept-based summarization model, called *Adaptive Summaries*, that helps users make their desired summary instead of producing a single inflexible summary. The system learns from users' provided information gradually while interacting with the system by giving feedback in an iterative loop. Users can choose either reject or accept action for selecting a concept being included in the summary with the importance of that concept from users' perspectives and confidence level of their feedback. The proposed approach can guarantee interactive speed to keep the user engaged in the process. Furthermore, it eliminates the need for reference summaries, which is a challenging issue for summarization tasks. Evaluations show that *Adaptive Summaries* helps users make high-quality summaries based on their preferences by maximizing the user-desired content in the generated summaries.

Keywords: Multi-document summarization · Interactive summarization · Adaptive summaries · Personalized summaries · Preference-based summaries

1 Introduction

The expansion of Internet and Web applications, followed by the growing influence of smartphones on every aspect of our lives, has induced an everyday growth of textual information. As a result, data summaries as a solution are becoming of paramount importance. Therefore, carefully constructed summaries make the data analytic possible by improving scalability and efficiency. Summarization has been widely used in

ⓒ Springer Nature Switzerland AG 2021
H. Hacid et al. (Eds.): ICSOC 2020 Workshops, LNCS 12632, pp. 281–293, 2021.
https://doi.org/10.1007/978-3-030-76352-7_29

many applications and domains, using a variety of techniques [2,4,31]. A good summary should keep the main content while helping users understand large volumes of information in a small amount of time. However, the summarization problem is subjective because different users have different attitudes toward what is considered valuable. Consequently, producing a generic summary that can satisfy everyone makes the problem challenging. Therefore, despite much research in this area, it is still a significant challenge to produce summaries that can satisfy all users.

Traditional state-of-the-art approaches produce only a single, globally short summary for all users [15]. They optimize a system towards one single best summary without considering users' interests and needs in seeking their desired information [3]. However, this is not useful in real-world scenarios where different users may explore diverse interests in the same corpus, thus need a distinct summary. Furthermore, these high-level interests vary over time. To be more specific, a person might be interested in a different area based on *background knowledge*, and *context* due to their cognitive bias. For instance, there is various information available on the Internet about COVID-19. While one might be interested in symptoms, the other could be looking for the outbreak locations, while others are searching about the death toll. An example of background knowledge is when a researcher works on a research topic, for instance, "Summarization". She could be eager to know *what is* the definition of summarization. Then her interest may turn to different categories of summarization, such as extractive or abstractive approaches. Therefore, a good summary should change correspondingly based on the interest and preference of its reader.

A recent definition of summarization is given by Radev et al. [30] as "a text that is produced from one or more texts, which conveys the important information in the original text, and usually significantly less than that." However, the *importance* interpretation in this definition is different even for one person according to the need, time, knowledge. Besides, humans quickly assess the importance of concepts from their side. Previous approaches mainly select the most informative sentences as the summary and try to employ users' feedback in selecting sentences, not content, which makes the summaries vague. Therefore, it would be advantageous if users can interact with the system to incorporate their desired information into the summary. While there exist many automatic summarization approaches, only a few methods focused on the needs of individuals. Among them, a few considered the notion of the importance of a concept included in a summary, where this notion does not refer to users' attitudes and is statistical properties of the content, such as the frequency of occurrence of a word in a body of text [32]. Therefore, they fail to heed what is deemed to be valuable from individuals' perspectives. One way to achieve personalized summary is thus by integrating the advantages of personal feedback in defining what is considered as important.

We put the human in the loop and create a personalized summary that better captures the users' needs and their different notions of importance. Besides, the notion of having the human in the loop is very popular in different aspects of Explainable AI (XAI) [37–39]. In this setting, users can give feedback in an iterative loop in selecting or rejecting a concept, defining the level of importance or being unrelated, and giving the confidence level in their feedback. By doing this, we allow even novice users to interactively explore, manipulate, and analyze sizeable unstructured text document

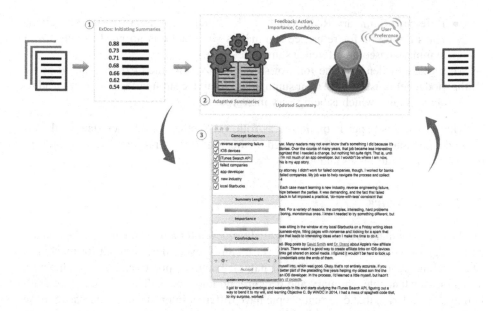

Fig. 1. An overview of the proposed approach (Adaptive Summaries). 1) Summaries are initiated with ExDos [12]. 2) Users integrate their preferences in making summaries by giving feedback in an iterative loop. 3) An example of user interaction.

collections to find their desired information and integrate their user-specific notion of importance. Our model employs an integer linear programming (ILP) optimization function to maximize user-desired content selection. Besides, most existing document summarization techniques require access to reference summaries to train their systems. However, obtaining reference summaries is very expensive. Lin in [24] explains that 3,000 h of human effort is required for a simple evaluation of the summaries for the Document Understanding Conference (DUC). *Adaptive Summaries* does not require reference summaries since it optimizes the summaries based on user-specific needs and not the goals standard summaries. An overview of the proposed approach is illustrated in Fig. 1. *Adaptive Summaries* can be used in multiple application scenarios where there are numerous documents, and the users seek information for getting personal insights. The main contributions can be summarized as follows:

– We have proposed an algorithm, called *Adaptive Summaries*, to include the user's needs and knowledge in making summaries. Adaptive summaries help users select the content of summaries based on their perspective, defining the degree of importance, and confidence in their feedback which benefits users in various ways including:
  - Customized Summary Length: The user can choose the length of the summary.
  - Interaction: Users interact with the summary, and it provides a better understanding of the topic. Besides, interacting with summary hint users to understand what is important related to the documents.

- Reference Summary Requirements: The summaries' dynamic structure eliminates the needs for reference summaries since there is no need to optimize a summary based on reference summaries.
- We provide evidence in the form of simulated user-oriented experiments to prove the model helps users make their summaries. Adaptive summaries also have an additional advantage, which is being very interpretable.

The rest of this paper is organized as follows. Section 2 discusses state-of-the-art methods. Section 3 presents the proposed method and Sect. 4 presents the experimental results. Section 5 discusses and justifies the obtained results.

2   Related Work

Producing a summary is quite a complicated task even for a person who has the domain knowledge of words and concepts, and yet it can be even more difficult for machines. The machine should have the ability of natural language processing and producing a human-understandable summary and background knowledge. This is even more challenging, considering that different people have different interests and concerns to make their summaries, the subjectivity problem of summarization. There exist different categorization for document summarization problem. There exist different categories for document summarization. For instance, one is based on the goal of the summarization task, which includes generic, domain-based (topic-focused) [10], or query-based summarization algorithm [34]. We also have other categories for document summarization which is based on the application of summarization such as article summarization [35], review summarization [18], news summarization [36], and also summarization for anomaly detection [1]. In this paper, we consider the problem of summarization from a traditional perspective as well as recent personalized and interactive approaches, as discussed below.

2.1   Traditional Approaches

Traditional state-of-the-art approaches produce only a single, globally short summary for all users. There are different perspectives to categorize traditional summarization approaches. The main aspect is considering the process and the output type of the summarization algorithms, which include *extractive* and *abstractive approaches*. The problem in both tasks is defined as summarizing a set of related articles and producing a short (e.g., 3–6 sentences) single summary, which conveys the most informative information. Abstractive approaches generate summaries by interpreting the main concepts of a document and then stating those contents in another format. Therefore, abstraction techniques are a substitute for the original documents rather than a part of it. Consequently, abstractive approaches require deep natural language processing, such as semantic representation and inference. However, they are challenging to produce and yet have not arrived at a mature stage [15]. On the other hand, the extractive text summarization approach selects some sentences as representative of the original documents. These sentences are then concatenated into a shorter text to produce a meaningful and coherent summary [17]. Early extractive approaches focused on shallow features, employing graph structure, or extracting the semantically related words. Different machine learning approaches are also used for this purpose, such as naive-Bayes, decision trees, log-linear, and hidden Markov models [13–15]. Recently, the focus for both extractive and abstractive approaches is mainly on neural network-based and deep reinforcement learning methods, which could demonstrate promising results. They employ word embedding [29] to represent words at the input level. Then, feed this information to the network to gain the output summary. These models mainly use a convolutional neural network [8], a recurrent neural network [9,25] or combination of these two [26,33]. The problem is that these approaches do not consider the users' opinions and are not interactive. Consequently, the summaries are not well-tailored from the users' perspective.

2.2 Personalized and Interactive Approaches

While most state-of-the-art approaches produce a single general summary for all users, a few attempts to take a user's preferences into account are defined as personalized or interactive summarization techniques.

Interactive summarization approaches include approaches which require human to interact with the system to make summaries. Unlike non-interactive systems that only present the system output to the end-user, Interactive NLP algorithms ask the user to provide certain feedback forms to refine the model and generate higher-quality outputs tailored to the user. Most approaches in this category create a summary and then require humans to cut, paste, and reorganize the critical elements to make the final summary [27,28]. Multiple forms of feedback also have been studied including mouse-clicks for information retrieval [6], post-edits and ratings for machine translation [11], error markings for semantic parsing [21], and preferences for translation [20]. Other interactive summarization systems include the iNeATS [22] and IDS [19] systems that allow users to tune several parameters (e.g., size, redundancy, focus) to customize the produced summaries.

The closest work to ours is proposed by Avinesh and Meyer [3], an interactive summarization approach that asks users to label important bigrams within candidate summaries. Then they used integer linear programming (ILP) to extract sentences, covering as many important bigrams a possible. However, importance is a binary value in this system, important and unimportant. The work by Orasan and Hasler [28] is also closely related to ours since they assist users in creating summaries for a source document based on the output of a given automatic summarization system. However, their system is neither interactive nor considers the user's feedback in any way. Instead, they suggest the output of the state-of-the-art (single-document) summarization method as a summary draft and ask the user to construct the summary without further interaction. The problem of concept-based ILP summarization framework was first introduced by [7]. However, they used bigrams as concepts [5,23] and either use document frequency (i.e. the number of source documents containing the concept) as weights [5,32]. As our interactive approach we allows for any combination of words, even sentence as concepts and also the weights are user defined parameters.

Our models also employ an optimization function to maximize user-desired content selection. *Adaptive Summaries* creates a personalized summary that better captures the users' needs and their different notions of importance by keeping the human in the loop. Instead of binary labeling of concepts as important and unimportant, users can give feedback to either select or reject a concept, define the level of importance or being unrelated, and the user's level of confidence in providing an iterative feedback loop. In the following section, we formalize the proposed approach.

3   Adaptive Summaries

The goal of *Adaptive Summary* is to interact with users to maximize the user-desired content in generating personalized summaries for users by interactions between system and user. In this problem, the input is a set of documents where the output is a human-readable summary consisting of a set of sentences with the user's preferred size. The novelty of this paper is that the user can select the desired content in making personalized summaries. In this setting, users can choose either reject or accept action for selecting a concept being included in the summary, the importance of that concept from users' perspectives, and the confidence level of users' feedback. This is modeled as an objective function to maximize the score of sentences based on the user-selected budget. Besides, to guarantee interactive speed to keep the user engaged, we propose a heuristic approach for selecting users' queries. In the following, we formally define the summarization tasks considered in this paper.

3.1   Problem Definition

The input is a set of documents $\{D_1, D_2, ..., D_N\}$ while each document consists of a sequence of sentences $S = [s_1, s_2, ..., s_n]$. Each sentence s_i is a set of concepts $\{c_1, c_2, .., c_k\}$ where a concept can be a word (unigram) or a sequence of words (Name entity or bigram). This framework optimizes the summarization outcome for a specific user. Therefore, the user interacts with the system and gives feedback to make summaries. This feedback is in the form of: i) Action A which perform on a concept where the values can be *accept (A=1)* or $reject(A = -1)$, ii) concept weight, W, corresponding to concepts' importance according to the user's opinion, and iii) the level of confidence for the chosen action, $conf$. The output is a set of sentences S define as the summary according to the budget limit (B) defined by the user.

3.2   Methodology

The goal of *Adaptive Summaries* is to incorporate the user preference in iteratively making summaries. Therefore, a continuous objective function is defined for analytically optimizing the user preference. In the first iteration, a summary is generated using our previous work, ExDos [12], that ranks sentences based on a general notion of importance using dynamic local feature weighting. It also demonstrates sentences in groups based on similarity defined in [12] to help users in selecting content. The user then can select an action A, which performs on a concept where the values can be *accept (A=1)* or $reject(A = -1)$. Next, for each concept user can define a weight, W, corresponding to the concepts' importance based on the user's opinion. Next, the user defines the level of confidence, $conf$, for the chosen action. When the action is accepted, this weight represents the importance of the concept, and when the action is rejection, the weights are the value for being unrelated. The logic behind this is that not all concepts have an equal level of importance. For instance, when users search for an illness's symptoms, a *headache* may not be as important as *sneezing* from users' perspectives. On the other hand, a *fever* may not be as unrelated as *acne*. The overall objective function, which is an Integer Linear Programming(ILP), is defined as:

$$maximize \sum_{s_i \in D} \sum_{c_j \in s_i} A \times conf(A) \times W_{c_j}$$

$$s.t. \sum_{s \in Summary} length(s) < B \tag{1}$$

Where A is the action, c_j is the concept in a given sentence (s_i), D the source documents, W_{c_j} is the corresponding user-preference weight for the conceptc_j and B is the summary length given by user. The objective function 1 maximizes the occurrence of concepts with maximum weights and confident level. The sudo-code of the proposed algorithm is reported in Algorithm 1. The following is the high-level description of our approach:

- To accelerate the process of making a summary, in first iterations, the sentences are ranked by our previous approach, ExDos. Then these weights are updated based on users' feedback.
- In order to prevent users from being overwhelmed, the similar sentences using our previous approach, ExDos, are grouped and shown to the user simultaneously.
- If weights of a concept gets updated in an iteration, the weights are updated for every occurrence of that concept.
- If the user rejects a sentence ($A_{s_i} = -1$), then the weight of the sentence is set to zero ($W_{s_i} = 0$). However, the system does not update the weights of concepts included in the sentence as there may be different reasons for rejection of a sentence such as redundancy or not being important.
- A concept is only selected if it is present in at least one of the selected sentences.
- The number of sentences is a user parameter define in each iteration and the confidence in feedback is set 1 by default.
- If there are no more concepts to query, the process terminated. To optimize the summary creation based on user feedback, concept weights iteratively change the in the objective function.

4   Experiment

In this section, we present the experimental setup for implementing and assessing our summarization model's performance. We discuss the datasets, give implementation details, and explain how system output was evaluated.

Algorithm 1 Adaptive Summaries

Input: Document Cluster D.
Output: Optimal Summary Generated by user (S).

procedure ADAPTIVE SUMMARIES.
   Ranked Sentences ← ExDos(D)
   while user is not satisfied **do**
      Concepts ← ExtractNewConcepts(Ranked Sentences)
      if Concepts ≠ ∅ **then**
         Ask user for action (A), importance(W), and confidence($Conf$).
         Select sentences to maximize Equation 1.
      end if
      return Summary(S)
   end while
   return Summary(S)

4.1 Data

To compare the performance of *Adaptive Summaries* with the existing leading approaches, experiments on two benchmark datasets, DUC2002[1] and CNN/Daily Mail [16][2] are performed. The documents are all from the news domain and are grouped into various topic clusters. We analyze our system based on different criteria, including selecting different units of concepts, number of iterations, and the ROUGE score.

4.2 Evaluation

We evaluate the quality of summaries using ROUGE (Recall-Oriented Understudy for Gisting Evaluation) measure [24][3] defined below. It compares produced summaries against a set of reference summaries. The three variants of ROUGE (ROUGE-1, ROUGE-2, and ROUGE-L) are used. ROUGE-1 and ROUGE-2 are used to evaluate informativeness, and ROUGE-L (longest common subsequence) is used to assess the fluency. We used the limited length ROUGE recall-only evaluation (75 words) for comparison of DUC to avoid being biased. Besides, the full-length F1 score is used for the evaluation of the CNN/DailyMail dataset.

$$ROUGE_n = \frac{\sum_{S \in \{ReferenceSummaries\}} \sum_{gram_n \in S} Count_{match}(gram_n)}{\sum_{S \in \{ReferenceSummaries\}} \sum_{gram_n \in S} Count(gram_n)} \quad (2)$$

In traditional approaches, to evaluate a summarization system, the mean ROUGE scores across clusters using all the reference summaries are reported. *Adaptive Summaries* is

---

[1] Produced by the National Institute of Standards and Technology (https://duc.nist.gov/).
[2] https://github.com/abisee/cnn-dailymail.
[3] We run ROUGE 1.5.5: http://www.berouge.com/Pages/default.aspx with parameters -n 2 -m -u -c 95 -r 1000 -f A -p 0.5 -t 0.

Table 1. ROUGE score comparison on CNN/DailyMail using F1 variant of ROUGE.

Model	Rouge-1 Score	Rouge-2 Score	Rouge-L Score
LEAD-3	39.2	15.7	35.5
NN-SE	35.4	13.3	32.6
SummaRuNNer	39.9	16.3	35.1
HSSAS	42.3	17.8	37.6
BANDITSUM	41.5	18.7	37.6
Adaptive dictionary	42.9	20.1	38.2
Adaptive reference	41.4	19.7	32.1

Table 2. ROUGE score (%) comparison on DUC-2002 dataset.

Model	Rouge-1 Score	Rouge-2 Score	Rouge-L Score
LEAD-3	43.6	21.0	N/A
NN-SE	47.4	23.0	N/A
SummaRuNNer	46.6	23.1	N/A
HSSAS	52.1	24.5	N/A
Upper Bound	47.4	21.6	18.7
Avinesh-Al	44.8	18.8	16.8
Avinesh-Joint	44.4	18.2	16.5
Adaptive dictionary	50.4	22.1	18.4
Adaptive reference	46.5	20.1	18.8

evaluated based on the mean ROUGE scores across clusters per reference summary in personalized summarization approaches. It is worth mentioning that this approach aims at facilitating making summaries for individual users, not improving the general accuracy of summaries. Since this approach is interactive, it requires humans to interact with the system for a user study based evaluation. However, collecting data for different settings from different humans is too expensive. Thus we simulate the users' behavior by generating feedback. To simulate users' behaviors, we analyze two variations of the proposed approach. In the first approach (AdaptiveDictionary), to simulate the users' behavior, we define a dictionary for ten clusters of topics, including the essential concepts and weights with defined actions for each concept. In the second one (AdaptiveReference), the reference summaries are considered as the users' feedback. The concepts are essential if they are presented in the reference summary. Therefore, we assign the maximum weight for the presented concepts. We compare our approach with both traditional and personalized approaches. The results are reported in Tables 1 and 2 for both datasets.   From the results, it can be seen that the proposed approach nearly reaches the upper bound for both datasets. Besides, the ROUGE analysis with real users does not show any pattern of increasing or decreasing. However, it is an expected result since this approach aims to optimize the summary for individual users, not the gold standard summary.

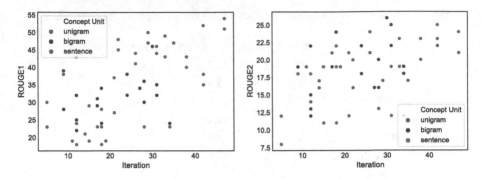

Fig. 2. Th left image shows the ROUGE1 based on iteration number and the right image shows the ROUGE2 based on iteration number. The green samples are when the permitted concept unit are unigram, blue bigram and red the sentences. (Color figure online)

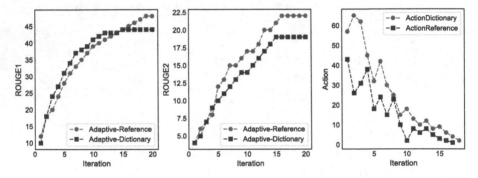

Fig. 3. 1) Number of iteration and ROUGE1 for DUC-2002. 2) Number of iterations versus ROUGE2 values. 3) Number of actions versus iterations for DUC2002.

To compare the concepts' unit's effect, we evaluate our approach based on three-unit measures, including uni-gram, bi-gram, and sentences. Although our model reaches the upper bound when using unigram-based feedback, they require significantly more iterations and much feedback to converge, as shown in Fig. 2. We analyze the speed (iterations) and the accuracy (ROUGE1 and ROUGE2) for different concepts units for DUC2002. CNN/DailyMail dataset follows the same trend. From the image, we see that when the permitted selection unit is unigram, the ROUGE1 score is higher. However, it takes more iterations to converge. For ROUGE2, both bigram and unigram have higher scores, however, when the unit is bigram, it converges sooner.

Another experiment is considering the ROUGE scores versus the number of iterations. In Fig. 3, the results for the DUC-2002 data set for two versions of adaptive, using a dictionary as feedback and reference summary as feedback is depicted. In the third image, we evaluated the models based on the number of actions (A) taken by the oracles to converge to the upper bound within ten iterations.

5   Conclusion and Future Work

We propose an interactive and personalized multi-document summarization approach using users' feedback. The selection or rejection of concepts, defining the importance of a concept, and the level of confidence engage users in making their desired summary. We empirically checked the validity of our approach on standard datasets using simulated user feedback. We observed that our framework shows promising results in terms of ROUGE score and also human evaluation. Results show that users' feedback can help them to find their desired information. As future work, we plan to include the reasons behind any action to optimize the system's performance.

Acknowledgement. We acknowledge the AI-enabled Processes (AIP) Research Centre (https:// aip-research-center.github.io/) for funding this research. We also acknowledge Macquarie University for funding this project through IMQRES scholarship.

References

1. Ahmed, M.: Intelligent big data summarization for rare anomaly detection. IEEE Access **7**, 68669–68677 (2019)
2. Amouzgar, F., Beheshti, A., Ghodratnama, S., Benatallah, B., Yang, J., Sheng, Q.Z.: iSheets: a spreadsheet-based machine learning development platform for data-driven process Analytics. In: Liu, X., et al. (eds.) ICSOC 2018. LNCS, vol. 11434, pp. 453–457. Springer, Cham (2019). https://doi.org/10.1007/978-3-030-17642-6_43
3. Avinesh, P., Meyer, C.M.: Joint optimization of user-desired content in multi-document summaries by learning from user feedback. In: Proceedings of the 55th Annual Meeting of the Association for Computational Linguistics (Volume 1: Long Papers), pp. 1353–1363 (2017)
4. Beheshti, A., et al.: iProcess: enabling IoT platforms in data-driven knowledge-intensive processes. In: Weske, M., Montali, M., Weber, I., vom Brocke, J. (eds.) BPM 2018. LNBIP, vol. 329, pp. 108–126. Springer, Cham (2018). https://doi.org/10.1007/978-3-319-98651-7_7
5. Berg-Kirkpatrick, T., Gillick, D., Klein, D.: Jointly learning to extract and compress. In: Proceedings of the 49th Annual Meeting of the Association for Computational Linguistics: Human Language Technologies, pp. 481–490 (2011)
6. Borisov, A., Wardenaar, M., Markov, I., de Rijke, M.: A click sequence model for web search. In: The 41st International ACM SIGIR Conference on Research & Development in Information Retrieval, pp. 45–54 (2018)
7. Boudin, F., Mougard, H., Favre, B.: Concept-based summarization using integer linear programming: from concept pruning to multiple optimal solutions. In: Conference on Empirical Methods in Natural Language Processing (EMNLP) 2015 (2015)
8. Cao, Z., Wei, F., Li, S., Li, W., Zhou, M., Houfeng, W.: Learning summary prior representation for extractive summarization. ACL **2**, 829–833 (2015)
9. Cheng, J., Lapata, M.: Neural summarization by extracting sentences and words. In: 54th Annual Meeting of the Association for Computational Linguistics, pp. 484–494 (2016)
10. Codina-Filbà, J., et al.: Using genre-specific features for patent summaries. Inf. Process. Manage. **53**(1), 151–174 (2017)
11. Denkowski, M., Dyer, C., Lavie, A.: Learning from post-editing: Online model adaptation for statistical machine translation. In: Proceedings of the 14th Conference of the European Chapter of the Association for Computational Linguistic, pp. 395–404 (2014)

12. Ghodratnama, S., Beheshti, A., Zakershahrak, M., Sobhanmanesh, F.: Extractive document summarization based on dynamic feature space mapping. IEEE Access **8**, 139084–139095 (2020)
13. Ghodratnama, S., SadrAldini, M.: An innovative sampling method for massive data reduction in data mining. In: The 3rd Iran Data Mining Conference, Tehran (2009)
14. Ghodratnama, S., Boostani, R.: An efficient strategy to handle complex datasets having multimodal distribution. In: Sanayei, A., E. Rössler, O., Zelinka, I. (eds.) ISCS 2014: Interdisciplinary Symposium on Complex Systems. ECC, vol. 14, pp. 153–163. Springer, Cham (2015). https://doi.org/10.1007/978-3-319-10759-2_17
15. Gupta, V., Lehal, G.S.: A survey of text summarization extractive techniques. J. Emerg. Technol. Web Intell. **2**(3), 258–268 (2010)
16. Hermann, K.M., Kocisky, T., Grefenstette, M., Blunsom, P.: Teaching machines to read and comprehend. In: Advances in Neural Information Processing Systems, pp. 1693–1701 (2015)
17. Heu, J.U., Qasim, I., Lee, D.H.: Fodosu: multi-document summarization exploiting semantic analysis based on social folksonomy. Inf. Process. Manage. **51**(1), 212–225 (2015)
18. Hu, Y.H., Chen, Y.L., Chou, H.L.: Opinion mining from online hotel reviews-a text summarization approach. Inf. Process. Manage. **53**(2), 436–449 (2017)
19. Jones, S., Lundy, S., Paynter, G.W.: Interactive document summarisation using automatically extracted keyphrases. In: Proceedings of the 35th Annual Hawaii International Conference on System Sciences, pp. 1160–1169. IEEE (2002)
20. Kingma, D.P., Ba, J.: Adam: a method for stochastic optimization. arXiv preprint arXiv:1412.6980 (2014)
21. Lawrence, C., Riezler, S.: Counterfactual learning from human proofreading feedback for semantic parsing. arXiv preprint arXiv:1811.12239 (2018)
22. Leuski, A., Lin, C.Y., Hovy, E.: ineats: interactive multi-document summarization. In: The Companion Volume to the Proceedings of 41st Annual Meeting of the Association for Computational Linguistics, pp. 125–128 (2003)
23. Li, C., Qian, X., Liu, Y.: Using supervised bigram-based ilp for extractive summarization. In: Proceedings of the 51st Annual Meeting of the Association for Computational Linguistics (Volume 1: Long Papers), pp. 1004–1013 (2013)
24. Lin, C.Y.: Rouge: A package for automatic evaluation of summaries. In: Text Summarization Branches Out (2004)
25. Nallapati, R., Zhai, F., Zhou, B.: Summarunner: A recurrent neural network based sequence model for extractive summarization of documents. In: AAAI (2017)
26. Narayan, S., Cohen, S.B., Lapata, M.: Ranking sentences for extractive summarization with reinforcement learning. In: Association for Computational Linguistics (2018)
27. Narita, M., Kurokawa, K., Utsuro, T.: A web-based English abstract writing tool using a tagged ej parallel corpus. In: LREC (2002)
28. Orasan, C., Hasler, L.: Computer-aided summarisation-what the user really wants. In: LREC, pp. 1548–1551 (2006)
29. Pennington, J.E.A.: Glove: global vectors for word representation. In: EMNLP (2014)
30. Radev, D.R., Hovy, E., McKeown, K.: Introduction to the special issue on summarization. Comput. Linguist. **28**(4), 399–408 (2002)
31. Schiliro, F., et al.: iCOP: IoT-enabled policing processes. In: Liu, X., et al. (eds.) ICSOC 2018. LNCS, vol. 11434, pp. 447–452. Springer, Cham (2019). https://doi.org/10.1007/978-3-030-17642-6_42
32. Woodsend, K., Lapata, M.: Multiple aspect summarization using integer linear programming. In: Proceedings of the 2012 Joint Conference on Empirical Methods in Natural Language Processing and Computational Natural Language Learning, pp. 233–243. Association for Computational Linguistics (2012)

33. Wu, Y., Hu, B.: Learning to extract coherent summary via deep reinforcement learning. In: Thirty-Second AAAI Conference on Artificial Intelligence (2018)
34. Xiong, S., Ji, D.: Query-focused multi-document summarization using hypergraph-based ranking. Inf. Process. Manage. **52**(4), 670–681 (2016)
35. Xu, H., Wang, Z., Weng, X.: Scientific literature summarization using document structure and hierarchical attention model. IEEE Access **7**, 185290–185300 (2019)
36. Yang, P., Li, W., Zhao, G.: Language model-driven topic clustering and summarization for news articles. IEEE Access **7**, 185506–185519 (2019)
37. Zakershahrak, M., Sonawane, A., Gong, Z., Zhang, Y.: Interactive plan explicability in human-robot teaming. In: RO-MAN, pp. 1012–1017. IEEE (2018)
38. Zakershahrak, M., Gong, Z., Sadassivam, N., Zhang, Y.: Online explanation generation for human-robot teaming. arXiv preprint arXiv:1903.06418 (2019)
39. Zakershahrak, M., Marpally, S.R., Sharma, A., Gong, Z., Zhang, Y.: Order matters: generating progressive explanations for planning tasks in human-robot teaming. arXiv preprint arXiv:2004.07822 (2020)

TAP: A Two-Level Trust and Personality-Aware Recommender System

Shahpar Yakhchi$^{(\boxtimes)}$, Seyed Mohssen Ghafari, and Mehmet Orgun

Department of Computing, Macquarie University, Sydney, Australia
{shahpar.yakhchi,seyed-mohssen.ghafari}@hdr.mq.edu.au,
mehmet.orgun@mq.edu.au

Abstract. Recommender systems (RSs) have been adopted in a variety set of web services to provide a list of items which a user may interact with in near future. Collaborative filtering (CF) is one of the most widely used mechanism in RSs that focuses on preferences of neighbours of similar users. Therefore, it is a critical challenge for CF models to discover a set of appropriate neighbors for a particular user. Most of the current approaches exploit users' ratings information to find similar users by comparing their rating patterns. However, this may be a simple idea and over-tested by the current studies, which may fail under data sparsity problem. Recommender system as an intelligent system needs to help users with their decision making process, and facilitate them with personalized suggestions. In real world, people are willing to share similar interest with those who have the same personality type; and then among all similar personality users pope may only take advice and recommendation from the trustworthy ones. Therefore, in this paper we propose a two-level model, TAP, which analyzes users' behaviours to first detect their personality types, and then incorporate trust information to provide more customized recommendations. We mathematically model our approach based on the matrix factorization to consider personality and trust information simultaneously. Experimental results on a real-world dataset demonstrate the effectiveness of our model.

Keywords: Recommendation system · Personality information · Trust relation

1 Introduction

In the last two decades, we have witnessed the emerging growth of the generated information by people's daily activities (e.g., browsing, clicking, listening to music, and purchasing items). Due to this information explosion, people are surrounded by too many options and services. In this regard, RSs can help customers with their daily living activities from eating, and deciding on clothes to housing and traveling, and to alleviate the information overload problem. Traditionally, Recommender Systems have been recognized as playlist generators for

© Springer Nature Switzerland AG 2021
H. Hacid et al. (Eds.): ICSOC 2020 Workshops, LNCS 12632, pp. 294–308, 2021.
https://doi.org/10.1007/978-3-030-76352-7_30

video/music services (e.g., Netflix[1] and Spotify[2]), e-commerce product recommenders (e.g., Amazon[3] and eBay[4]), or social content recommenders (e.g., Facebook[5] and Twitter[6]). Today, almost every organization leverage recommender systems for better building their users' profile and make a personalized suggestion, accordingly.

Models in RSs can be mainly classified in to three major classes of approaches: (i) *content-based RS:* which uses the past preferred items' descriptions for recommendation process; ii) *collaborative Filtering:* they look in like-minded people to predict users' preferences; and iii) *Hybrid:* they combine some of the previous methods in a single model. Among all techniques in recommender systems, Collaborative Filtering (CF) has been taken lots of attention from research community [50]. The main idea behind CF is that people with similar preference will share similar items, and usually most of the CF approaches exploit user-item interaction matrix to discover similar users [49]. The rating matrix records previously observed items in which users explicitly express their interests by providing ratings to different items. However, there may be too simplistic assumption that if two users have the same ratings pattern, they are similar to each other. More importantly, these models may fail under data sparsity problem (i.e., when there is a lack of available information).

Personality as a domain-independent factor which a person tends to show regardless of her situation has inspired researchers to use this factor into their model. Personality has a strong correlation with individuals' interests and people with similar personality types tend to share similar interests. Due to these benefit of personality, researchers start to exploit personality into their model. Incorporating personality characteristics into RSs to not only help users with a diverse set of items [37], but also provide a better group recommendation [42] and improve the accuracy of RSs in Music, Movies, e-learning and web searches [24,40]. Although incorporating personality factor may help at finding a set of similar neighbours for a target user, not all of similar personality type neighbours are trustworthy. In real-world scenarios people usually ask their friends when they are looking for an item (e.g., movie, music, and products), because they have similar preference. But, a particular friend may not be reliable for suggesting items, and thus user's trustworthiness plays an important role in recommendations. Based on this observation, a good recommender system not only should take personality as an important individuals' feature, but also their trustworthiness is another critical parameter which needs to be taken in into account. To do so, we propose a novel recommender system in which personality and trust are considered as the main parameters for neighbours selection. We detect users' personality type implicitly by analyzing their written

---

[1] https://www.netflix.com/.

[2] https://www.spotify.com/.

[3] https://www.amazon.com/.

[4] https://www.ebay.com/.

[5] https://www.facebook.com/.

[6] https://twitter.com/.

comments/reviews with no extra burden on users. Then, the trustworthy users will be found by looking into their social context factors supported by theories from social psychology. At the end, we propose a novel mathematical model which integrates personality and trust information. The unique contributions of this paper are:

- We propose a novel Recommender System called TAP, integrating two main factors which may affect a user's decision making process, users' personality types, and the trust information.
- We detect users' personality type implicitly with no need to users' efforts and through analyzing their online-generated contents.
- We calculate trust values by using social context factors inspired by psychological theories such as Social Penetration Theory.
- We conduct a comprehensive experiment on a real-world dataset to demonstrate the superiority of TAP compared to the state-of-the-art approaches.

2   Related Work

In this section, first we give an introduction of personality and trust, next we discuss personality-aware RSs, and trust-aware RSs.

2.1   What is Personality?

Personality is described as "consistent behavior pattern and interpersonal processes originating within the individual" [10], which can explain the wide variety of human behavior. From the psychology point of view, personality is an important factor which is able to explain "patterns of thought, emotion, and behavior" [12]. Personality as a stable behavioral pattern plays an important role in people's decisions making process [39], which similar personality type people tend to share similar taste. According to the Rentfrow et al., [41] "reflective" personality type people with high openness really like to listen to the jazz, blues and classical music, while "energetic" people with high degree of extraversion and agreeableness usually prefer rap, hip-hop, funk and electronic music. There are several personality traits models which can explain human behaviors, and among them, Five Factor Model(FFM) or Big Five Model has drawn more attention both in psychology and computer science research. FFM model is "the dominant paradigm in personality research, and one of the most influential models in all of the psychology" [47]. Based on the FFM, people's personality types can be categorized into five main traits which is briefly called OCEAN [36]: Openness to Experience, Conscientiousness, Extroversion, Agreeableness, and Neuroticism.

2.2   Personality Acquisition

The main issue that personality-based RSs are confronted with is how to identify user's personality type. Basically, there are two different ways which can be can be grouped into: Explicit techniques (filling a questionnaire questionnaires according to the chosen model); and Implicit techniques (observing users' behavioral patterns and then proposed a regression/classification model). There are several questionnaire types based on FFM model, NEO-Personality-Inventory Revised (NEOPI-R, 240 items) for instance, in which the participants' personality types are revealed after they answer several questions [11,55]. Except explicit ways of identifying users' personality types, user's personality types can be extracted by analyzing posts and activities like written review texts from online social media [44]. For instance, in the work presented by the Kosinski et al. [29], users' facebook activities are recorded and then logistic regression is applied to predict the people's personality type. Although personality detection with questionnaires might reveal a better understanding of a user's personality, it is a tedious and time-consuming task and thus users may be unwilling to attend to it. In contrast, in implicit personality detection models, user's digital footprints, and their behaviors and actions could be analyzed with no extra burden on users [4].

2.3 Personality-Aware RS

In contrast to traditional CF approaches, users' personality types, which can explain the wide variety of human behavior, have inspired some recommender systems [5,52,54]. For instance, Hu and Pu [25] detect user personality types by providing a questionnaire for users and ask them to answer the questions. Then, the results show a better performance compared to the purely collaborative filtering methods. TWIN is another example of integrating personality in recommender system which calculates the user personality types based on the NEO-Personality Inventory-Revised classification scheme (also known as the Big Five model) [43]. Adopting users' personality types not only can provide users with a diverse set of items [37], and make an intelligent recommender system [6], but also generate a better group recommendation [42], and improve the accuracy of RSs in Music, Movies, e-learning and web searches [24,40]. Another advantage of taking users' personality is to better deal with the data sparsity and cold-start problems [8,52]. Except all mentioned above benefits of incorporating personality features into a recommender system, it can offer a divers set of items for user to discover unexpected items [2,27,53]. Brynjolfsson et al. [9] investigate the effect of different personality type people in selecting divers items and realize that "'reactive, excited and nervous persons (high in Neuroticism)" like to select movies form a diverse directors, while "suspicious/antagonistic users (low in Agreeableness)" prefer diverse movie countries.

2.4 What is Trust?

In early human societies people tend to interact with each other to fulfil their needs. Soon after, they realized that not all surrounded people that they have interacted are trustworthy and they need to filter those who they can trust. Trust can be defined as the 'willingness of a party to be vulnerable to actions of another party based on the expectation that the other will perform a particular action important to the trustor, irrespective of the ability to monitor or control that other party' [35]. 'Trust is necessary in order to face the unknown, whether that unknown is another human being, or simply the future and its contingent events'[7]. Sociologically speaking, 'a complete absence of trust would prevent [one] even getting up in the morning' [31].

Trust may be considered as a key property in users' behaviors in Online Social Networks (OSNs). In computer science, a definition of trust is defined by Tang et al. [45] as 'Trust provides information about with whom we should share information, from whom we should accept information and what considerations to give to information from people when aggregating or filtering data'. Trust can be used in too many applications, including fake news detection [14], retweet behaviour detection [1,7] and recommender systems [32,57]. The main step in all of these applications is trust prediction between users, and of course there are many different views of how to measure and use trust.

2.5   Trust Acquisition

Trust is a context-dependent concept, a trust relationship is a connection between a source user (trustor) and a target user (trustee) that indicates that the trustor trusts the trustee. With the help of trust, the trustor may seek information from the trustee, to avoid being confused by the huge amount of available data (i.e., mitigated information overload) and to be confident about the credibility of the received information (i.e., increased information credibility) [46]. Trust can be acquired explicitly or implicitly [15]. The explicit trust prediction model try to use the pre-established relationships which is manly based on the analyzing the web of trust for a particular user (i.e., if user A trusts user B, and user B trusts user C, then it is more likely that user A trusts user C) [17,18,21]. While implicit based trust prediction model try to use the basis of the item ratings for inferring trust among users [22], in which how users rate the items is a main basis for building a trust network.

2.6   Trust-Aware RSs

Due to the strength property of trust in discovering a set of trustworthy neighbours around a target user, trust-aware recommender systems have been gaining an increasing amount of attention in a number of research communities [13]. It has been shown by the existing studies that incorporating trust information in the recommender systems leads to increase the quality of the recommendations [16,20,30,34]. FilmTrust [19] is an example of trust-based RS which first integrates web-based social networking information, and then analyzes their features to generate movie recommendations. In another study presented by Moradi and Ahmadian [38], the accuracy of the predicted ratings is considered as one of the main challenges of collaborative filtering models. To overcome this problem,

---

[7] https://reviews.history.ac.uk/review/287a.

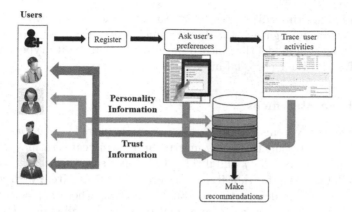

Fig. 1. A framework of our model consists of two-level; at first level we integrate personality information, and next, at the second level trust information is exploited to the TAP.

the authors have proposed a novel model to first construct a trust network for an active user, then the initial rate is predicted to compare it with actual one in order to calculate the reliability measure. Next, based on the results of the previous steps another trust network is constructed for the final rate predication task. similarity problem. Yao et al. [56] have proposed a trust inference approach based on matrix factorization (MF). They have treated a trust prediction problem as a recommendation task and model 'characterizes multiple latent factors for each trustor and trustee from the locally-generated trust relationships'. The authors also have adopted prior knowledge (e.g., trust bias and trust propagation) in order to improve the accuracy of their approach. In the work presented by Zhang et al. [58], it is discussed that in a trust relationship the source user may accept the recommendations from similar neighbour nodes (i.e., other users directly connected to the target user).

3   Overview and Framework

Figure 1 depicts our novel framework called TAP. For the first phase of this framework, we analyze the contents and contexts of users' generated information and contextualize these raw data to discover users' main characteristics, like their personality types. For the next phase, we monitor a user's activities and those of her neighbors' activities in a particular domain to figure out how other users are influenced by her opinions/comments to ascertain her level of knowledge. In the following sections, we give more details about each main phase of TAP. Note that, our framework aims to capture users' interests in order to make recommendations more personalized, improve users' satisfaction, and boost business profits. To do that, we first construct a user-item interactions matrix with the real values of ratings in order to preserve the degree of users' interests and their personal interests. For example while it is true that between two users who gave 2 and 5 stars (in a Likert scale 0–5) to a specific item, they both may like this item, but one who gave 5 stars likes it much more. Then, we propose a novel matrix factorization model that incorporates users' personality type, users' level of knowledge and their personal interests.

3.1 Problem Statement

Suppose there are N items $V = \{v_1, v_2, \cdots, v_N\}$, M users $U = \{u_1, u_2, \cdots, u_M\}$. Let $R \in R^{M \times N}$ represents the rating matrix, and R_{ij} indicates ratings which have been given to item i by user j, and if the user has not seen this item, thus there is not any available rating for that item, we show this unrated item with unr in this paper. Let $L \in R^{N \times N}$ defines the personality matrix, where $L_{ij} = \{0, 1\}$, there is a direct connection between u_i and u_j if they have similar personality type. In the other word, $L_{ij} = 0$ means that u_i and u_j do not have similar personality type. Most of the existing approaches consider that all the rated items are the same and equal to 1, and thus there is no any difference between all rated items, while ratings value can indicate the level of users' preferences on an item [23]. However, in this paper in order to capture users' interests completely and express the level of preferences, we construct this matrix with the Eq. 1 [51]:

$$w_{ij} = \begin{cases} 0, & \text{if } R_{ij} = unr \\ R_{ij}, & \text{otherwise} \end{cases} \tag{1}$$

where $W \in R^{M \times N}$, and $w_{ij} > 0$ represents the interest of user i on item j, and $w_{ij} = 0$

3.2 Detecting Users' Personality Types

Unlike most of the current methods which detect users' personality types by asking them to fill a questionnaire, in our work, the users' personality types are measured implicitly with no need to their' effort. To do so, we collect all written users' reviews to categorize them according to the Linguistic Inquiry and Word Count (LIWC) tool to understand how many words of users' reviews are related to each its 88 categories (such as positive emotions, cognitive process, and social processes). Inspired by Roschina et al. [43], we employ a linear regression model to measure a user's personality traits as follows:

$$E = w_1 X_1 + w_2 X_2 + w_3 X_3 + \cdots + w_i X_i \tag{2}$$

where X_i and w_i denote a category of LIWC and its corresponding weight, respectively, $i \in \{1, 2, ..., n\}$ and $n = 88$, which is the total number of LIWC categorises. A final score $E =$ can show the level of each of five personality traits, if we only place their correlated categorizes form LIWC and their corresponding weights which can be extracted by Mairesse et al. [33]. In addition to the users' personality types, the level of an individual's knowledge can be one of the main factors to determine the acceptance rate of her/his recommendations, which we term it as users' level of knowledge. In real-world, individuals may have different level of knowledge about various domains, but they may be an expert in one or some of them. We mark the level of knowledge of u_i in domain d, as kl_i^d, and can be computed as follows:

$$kl_i^d = \frac{1}{n_i^d} \sum_{p=1}^{n_i^d} h_p^{i,d} \tag{3}$$

where n_i^d is the total number of reviews left by user u_i in that domain, and $h_p^{i,d}$ represents the ratings that are given to each review p by other users in this domain.

3.3 Trust Acquisition

As we discussed in Sect. 2, there are a wide range of trust prediction models in the literature. However, in the context of recommender system we have a limited source of available information for measuring the trust among users. Therefore, to calculate the trust, we leverage our previous work [14], which considers social context factors in trust prediction task. Our proposed model is based on the consideration of the level of expertise, interest, number of followers, frequency and quality of previous interactions, and self-disclosure for capturing the trust relations. We take the level of expertise because a recommendation from an expert person in a certain domain is more acceptable compared to the less knowledgeable person. Next, the reason behind incorporating interest is that it could be conceived of as an individual's attitude towards a set of objects. Then, the higher number of followers may be a good indicator of being an expert in a particular domain. We consider the frequency and quality of previous interactions between two users since it may show a high potential of establishing a trust relation between them. Finally, according to the Social Penetration Theory (SPT): "as relationships develop, interpersonal communication moves from relatively shallow, non-intimate levels to deeper, more intimate ones" [3]. Based on this theory, self-disclosure which means revealing personal information (e.g., personal motives or thoughts, feelings and etc.), which can be a reason behind constructing a relationship. In this section we avoid going into too much details due to the space limitation, and we encourage a motivated reader to have a look at our previous work [14]. Moreover, since this is an ongoing work, we do not train and test our proposed model, TAP, with the extracted trust relation. While our mathematical model presented in Sect. 3.4 take both personality and trust information into account, we only test our model when it considers personality feature to evaluate the results and leave evaluating a model with trust information as our future work.

3.4 Our Framework

Let $U = \{u_1, u_2, ..., u_{|i|}\}$ denotes the user set and $V = \{v_1, v_2, ..., v_{|j|}\}$ indicates the item set, where $|i|$ and $|j|$ are the total number of users and items, respectively. In Eq. 4, $u_i^{(d)}$ represents the latent feature vector of user i in domain d, and the latent feature vector of item j in domain d is shown by $v_j^{(d)}$. Matrix L contains personality information, and $\varphi_i^{+(d)}$ is the set of users who have the same personality type with u_k, where $l_{ik} = 1$ means that u_i and u_k have the same personality type. Matrix T records the trust values, where T_{ik} is the trust relation between u_i and u_k, and when there is a trust relation between these two users $T_{ik} = 1$. For the sake of simplicity, we denote $\gamma_i^d = \beta + kl_i^d$, where β is the controlling parameter which controls the weight of users preferences, and kl_i^d is the level of knowledge of u_i in domain d which we already discuss in Sect. 3.2. Finally, the ratings score for unobserved items can be computed as follows:

$$R_{ij}^{(d)} = \gamma_i^d u_i^{(d)^T} v_j^{(d)} + (1 - \gamma_i^d) \sum_{k \in \varphi_i^{+(d)}} L_{ik} T_{ik} u_k^{(d)^T} v_j^{(d)} \tag{4}$$

where $R_{ij}^{(d)}$ predicts the ratings values for unobserved items, which are based on the combination of users' personality type, and their trust information.

$$min \frac{1}{2} \sum_{i=1}^{N} \sum_{j=1}^{M} I_{ij}^{(d)} \left(R_{ij}^d - \left(\gamma_i^d u_i^{(d)^T} v_j^{(d)} + \left((1 - \gamma_i^d) \sum_{k \in \varphi_i^{+(d)}} L_{ik} T_{ik} u_k^{(d)^T} v_j^{(d)} \right) \right) \right)^2 \tag{5}$$
$$+ \alpha_1 \|U^{(d)}\|_F^2 + \alpha_2 \|V^{(d)}\|_F^2$$

where, $I_{ij}^{(d)} = 1$, if user i has rated item j, otherwise $I_{ij}^{(d)} = 0$. In order to prevent overfitting, we introduce $\eta(i, i')$ as the personality coefficient between u_i and $u_{i'}$ with some features (1) $\eta(i, i') \in \{0, 1\}$, (2) $\eta(i, i') = \eta(i', i)$ and (3) if $\eta(i, i') = 1$, means that u_i and $u_{i'}$ are more likely to have in common interests. Then, we have personality regularization as follows,

$$min \sum_{i=1}^{n} \sum_{j=1}^{m} \eta(i, i') \|U(i, :) - V(j, :)\|_2^2 \tag{6}$$

where U, and V are the user and item latent matrices, respectively. After some derivations for a particular u_i, we have the following regularization:

$$\frac{1}{2} \sum_{i=1}^{n} \sum_{j=1}^{m} \eta(i, i') \|U(i, :) - V(j, :)\|_2^2 = \frac{1}{2} \sum_{i=1}^{n} \sum_{j=1}^{m} \sum_{k=1}^{d} \eta(i, i') \left(U(i, :) - V(j, :) \right)^2$$
$$= \frac{1}{2} \sum_{i=1}^{n} \sum_{j=1}^{m} \sum_{k=1}^{d} \eta(i, i') U^2(i, k) + \frac{1}{2} \sum_{i=1}^{n} \sum_{j=1}^{m} \sum_{k=1}^{d} \eta(i, i') V^2(j, k) \tag{7}$$
$$- \sum_{i=1}^{n} \sum_{j=1}^{m} \sum_{k=1}^{d} \eta(i, i') U(i, k) - V(j, k) = \sum_{k=1}^{d} U^T(:, k)(D - Z)V(:, k) = Tr(U^T Y V)$$

Next, we have the updating rule as follows:

$$U(i, j) \leftarrow U(i, j) \sqrt{\frac{A(i, j)}{B(i, j)}}, \qquad V(i, j) \leftarrow V(i, j) \sqrt{\frac{C(i, j)}{D(i, j)}} \tag{8}$$

where A, B, C, and D are represented below:

$$A = v^{(d)^T} R^{(d)} \gamma^{(d)} + (1 - \gamma_i^{(d)}) L^T T^T R^{(d)^T} v^{(d)^T} + \gamma^{(d)} R^{(d)} v^{(d)^T} \tag{9}$$

$$\begin{aligned}
B = &\gamma^{(d)} v^{(d)} u^{(d)} v^{(d)^T} + v^{(d)^T} \gamma^{(d)} u^{(d)} v^{(d)} + \gamma^{(d)} u^{(d)} L^T T^T v^{(d)^T} \\
&+ (1 - \gamma_i^{(d)}) L^T T^T v^{(d)^T} u^{(d)^T} + (1 + \gamma_i^{(d)}) L^T T^T R^{(d)} v^{(d)^T} + (1 - \gamma_i^{(d)}) L^T T^T v^{(d)} u^{(d)} v^{(d)^T} \\
&+ (1 - \gamma_i^{(d)}) v^{(d)^T} L u^{(d)} T u^{(d)} v^{(d)} + (1 - \gamma_i^{(d)}) L^T T^T u^{(d)^T} v^{(d)^T} + (1 - \gamma_i^{(d)}) L^T T^T v^{(d)^T} u^{(d)^T} \\
&+ YU + Y^T U + \alpha_1 2u
\end{aligned} \tag{10}$$

$$C = R^{(d)} \gamma^{(d)} u^{(d)^T} + (1 - \gamma_i^{(d)}) u^{(d)^T} L^T T^T R^{(d)^T} + \gamma^{(d)} u^{(d)^T} R^{(d)} \tag{11}$$

$$\begin{aligned}
D = &\gamma^{(d)} u^{(d)^T} v^{(d)} u^{(d)} + \gamma^{(d)} u^{(d)} v^{(d)} u^{(d)^T} + (1 - \gamma_i^{(d)}) u^{(d)^T} v^{(d)^T} u^{(d)^T} L^T T^T \\
&+ 1 - \gamma_i^{(d)}) u^{(d)^T} L^T T^T v^{(d)^T} + (1 + \gamma_i^{(d)}) u^{(d)^T} L^T T^T R^{(d)} + (1 - \gamma_i^{(d)}) u^{(d)^T} L^T T^T v^{(d)} u^{(d)} \\
&+ (1 - \gamma_i^{(d)}) L u^{(d)} T u^{(d)} v^{(d)} u^{(d)^T} + (1 - \gamma_i^{(d)}) u^{(d)^T} L^T T^T v^{(d)^T} u^{(d)^T} L^T T^T \\
&+ (1 - \gamma_i^{(d)}) u^{(d)^T} L^T T^T v^{(d)^T} u^{(d)^T} L^T T^T + \alpha_2 2v
\end{aligned} \tag{12}$$

4 Experimental Evaluation of TAP and Analysis

4.1 Experimental Setup

Dataset. In order to evaluate our proposed model and compare the performance of our model with other state-of-the-art methods, we use Amazon dataset, which has been widely used in RSs, and consists a rich source of reviews. Amazon dataset provides a wide range of useful information (e.g., ratings, reviews), which consists of 2000 of users, 1500 items, 86690 reviews, 7219 number ratings, 3.6113 average number of rates per user, 0.2166 average number of rates per item and user ratings density is 0.0024. In this paper, we select Amazon Instant Video which is a subset of Amazon dataset due to the strong correlation between users' preferences on video and their personality types and leave evaluation on the other domains for our future work. We select users who have more than three reviews in this dataset in order to have a better contextual analysis. We use five-fold cross-validation and set parameters as $\gamma = 0.5$, $d = 100$, $\alpha_1 = 0.1$, and $\alpha_2 = 0.1$ in order to have the best recommendation accuracy.

Evaluation Metrics. We select two well-known evaluation metrics: Mean Absolute Error (MAE) and Root Mean Squared Error (RMSE) for evaluating the performance of our model compared to the baselines. The smaller MAE and RMSE demonstrate a better recommendation accuracy;

$$MAE = \frac{\sum_{(i,j) \in R_{test}} |\overline{R_{ij}} - R_{ij}|}{|R_{test}|} \qquad RMSE = \sqrt{\frac{\sum_{(i,j) \in R_{test}} |\overline{R_{ij}} - R_{ij}|}{|R_{test}|}}, \tag{13}$$

where R_{ij} and $\overline{R_{ij}}$ are the real and estimated ratings values, respectively, and R_{test} represents the total number of ratings in the test dataset.

Baselines. We compare our proposed model, TAP, with the following methods: TWIN [43] and Hu [26] as the personality-based recommender systems. CTR as the state-of-the-art model which using topic molding for recommendation [48] and SVD++ which only utilizes user-item ratings matrix [28]. We do not compare our model with the trust-based RSs, since we do not test our model with trust information which we leave it for our future work.

4.2   Performance Comparison and Analysis

As it is clear from the Table 1 the accuracy of all models improve when we increase the volume of the training data. However, among all compared methods, our proposed model, TAP, achieves the best performance in terms of both RMSE and MAE. It may be because of paying more attention to the real values of ratings, which can explain the level of interests of a user for an item. Additionally, TAP considers users' personality types and their level of knowledge in addition to their personal interests, which results in a making more personalized suggestions for a particular user. TWIN is the second best model, which can explain the benefit of integrating users' personality type compared to the purely rating-based CF models. Between personality-based recommender systems, TWIN also performs better than Hu, because Hu integrates both ratings and personality information and it may be more prone to suffer from data sparsity problem. The recommendation performance of CTR is higher than that of SVD++ by around 11% and 14% in terms of MAE and RMSE, respectively, since SVD++ only takes users' ratings into account. Finally, unlike the majority of existing approaches, TAP which constructs a matrix with actual ratings score which can help to better understanding users' interests. It achieves the recommendation performance of CTR by 39%, 42% and TWIN by 11%, 31% in terms of MAE and RMSE, respectively.

Table 1. Performance analysis on the Amazon dataset

Training Data	Metrics	Hu	SVD++	CTR	TWIN	TAP
60%	MAE	2.983	2.314	1.887	1.353	**1.165**
60%	RMSE	3.32	1.887	2.129	2.074	**1.554**
70%	MAE	2.763	1.864	1.69	1.215	**1.005**
70%	RMSE	3.152	2.29	2.101	1.898	**1.295**
80%	MAE	2.581	1.719	1.52	1.132	**0.936**
80%	RMSE	2.978	2.011	1.986	1.620	**1.058**
90%	MAE	2.426	1.547	1.391	0. 95	**0.850**
90%	RMSE	2.649	1.99	1.718	1.428	**0.995**

5   Work Plan and Implications

In this paper, we have proposed a novel personality and trust-based RS, TAP. We first construct a rating matrix with the real value of ratings in order to preserve the actual level of user' interest. Then, we analyze user generated contents to detect their personality type implicitly. Next, we mathematically model our approach based on the MF to employ personality and trust information. We have tested the impact of incorporating the users' personality types into our recommender system, and we observe a significant improvement compared to the existing studies. While we propose a novel technique and algorithm to exploit personality and trust information, we leave out the testing of the impact of trustworthy users as our future work.

References

1. Abdullah, N.A., Nishioka, D., Tanaka, Y., Murayama, Y.: Why i retweet? exploring user's perspective on decision-making of information spreading during disasters. In: HICSS, pp. 1–10 (2017)
2. Adomavicius, G., Kwon, Y.: Improving aggregate recommendation diversity using ranking-based techniques. IEEE Trans. Knowl. Data Eng. **24**(5), 896–911 (2012)
3. Altman, I., Taylor, D.: Social Penetration: The Development of Interpersonal Relationships. Holt, Rinehart & Winston (1973)
4. Azaria, A., Hong, J.: Recommender systems with personality. In: Sen, S., Geyer, W., Freyne, J., Castells, P. (eds.) Proceedings of the 10th ACM Conference on Recommender Systems, Boston, MA, USA, September 15–19, 2016, pp. 207–210. ACM (2016)
5. Beheshti, A., Hashemi, V.M., Yakhchi, S., Motahari-Nezhad, H.R., Ghafari, S.M., Yang, J.: personality2vec: enabling the analysis of behavioral disorders in social networks. In: Caverlee, J., Hu, X.B., Lalmas, M., Wang, W. (eds.) Conf. on WSDM, USA, pp. 825–828. ACM (2020)
6. Beheshti, A., Yakhchi, S., Mousaeirad, S., Ghafari, S.M., Goluguri, S.R., Edrisi, M.A.: Towards cognitive recommender systems. Algorithms **13**(8), 176 (2020)
7. Bild, D.R., Liu, Y., Dick, R.P., Mao, Z.M., Wallach, D.S.: Aggregate characterization of user behavior in twitter and analysis of the retweet graph. ACM Trans. Internet Technol. **15**(1), 1–24 (2015)
8. Braunhofer, M., Elahi, M., Ricci, F.: User personality and the new user problem in a context-aware point of interest recommender System. In: Tussyadiah, I., Inversini, A. (eds.) Inf. Commun. Technol. Tourism 2015, pp. 537–549. Springer, Cham (2015). https://doi.org/10.1007/978-3-319-14343-9_39
9. Brynjolfsson, E., Hu, Y.J., Simester, D.: Goodbye pareto principle, hello long tail: the effect of search costs on the concentration of product sales. Manag. Sci. **57**(8), 1373–1386 (2011)
10. Burger, J.: Introduction to Personality. Cengage Learning (2011)
11. Costa, P. T.,M.R.R.: Domains and facets: hierarchical personality assessment using the revised neo personality inventory. J. Pers. Assess. **64**(1), 21–50 (1995)
12. Funder., D.: Personality. Ann. Rev. Psychol. **52**, 197–221 (2001)
13. Ghafari, S.M., et al.: A survey on trust prediction in online social networks. IEEE Access **8**, 144292–144309 (2020)

14. Ghafari, S.M., Beheshti, A., Yakhchi, S., Orgun, M.: Social context-aware trust prediction: a method for identifying fake news. In: Conference on WISE, pp. 161–177 (2018)

15. Ghafari, S.M.: Towards time-aware context-aware deep trust prediction in online social networks. arXiv preprint arXiv:2003.09543 (2020)

16. Ghafari, S.M., Joshi, A., Beheshti, A., Paris, C., Yakhchi, S., Orgun, M.A.: DCAT: a deep context-aware trust prediction approach for online social networks. In: Haghighi, P.D., Salvadori, I.L., Steinbauer, M., Khalil, I., Anderst-Kotsis, G. (eds.) Conference on MoMM, pp. 20–27. ACM (2019)

17. Ghafari, S.M., Yakhchi, S., Beheshti, A., Orgun, M.: SETTRUST: social exchange theory based context-aware trust prediction in online social networks. In: Hacid, H., Sheng, Q.Z., Yoshida, T., Sarkheyli, A., Zhou, R. (eds.) QUAT 2018. LNCS, vol. 11235, pp. 46–61. Springer, Cham (2019). https://doi.org/10.1007/978-3-030-19143-6_4

18. Ghafari, S.M., Yakhchi, S., Beheshti, A., Orgun, M.: Social context-aware trust prediction: methods for identifying fake news. In: Hacid, H., Cellary, W., Wang, H., Paik, H.-Y., Zhou, R. (eds.) WISE 2018. LNCS, vol. 11233, pp. 161–177. Springer, Cham (2018). https://doi.org/10.1007/978-3-030-02922-7_11

19. Golbeck, J.: Generating predictive movie recommendations from trust in social networks. In: Trust Management, 4th International Conference, iTrust, Italy, pp. 93–104 (2006)

20. Guo, G., Zhang, J., Thalmann, D.: Merging trust in collaborative filtering to alleviate data sparsity and cold start. Knowl. Based Syst. **57**, 57–68 (2014)

21. Guo, G., Zhang, J., Yorke-Smith, N.: Leveraging multiviews of trust and similarity to enhance clustering-based recommender systems. Knowl. Based Syst. **74**, 14–27 (2015)

22. Guo, G., Zhang, J., Yorke-Smith, N.: Trustsvd: collaborative filtering with both the explicit and implicit influence of user trust and of item ratings. In: Bonet, B., Koenig, S. (eds.) Conference on AAAI, USA. pp. 123–129. AAAI Press (2015)

23. He, X., Zhang, H., Kan, M., Chua, T.: Fast matrix factorization for online recommendation with implicit feedback. In: Conference on SIGIR, Italy, pp. 549–558 (2016)

24. Hu, R., Pu, P.: A study on user perception of personality-based recommender systems. In: De Bra, P., Kobsa, A., Chin, D. (eds.) UMAP 2010. LNCS, vol. 6075, pp. 291–302. Springer, Heidelberg (2010). https://doi.org/10.1007/978-3-642-13470-8_27

25. Hu, R., Pu, P.: Enhancing collaborative filtering systems with personality information. In: Mobasher, B., Burke, R.D., Jannach, D., Adomavicius, G. (eds.) Proceedings of the 2011 ACM Conference on Recommender Systems, RecSys 2011, Chicago, IL, USA, October 23–27, 2011, pp. 197–204. ACM (2011)

26. Hu, Y., Lee, P., Chen, K., Tarn, J.M., Dang, D.: Hotel recommendation system based on review and context information: a collaborative filtering appro. In: Conference on PACIS, Taiwan, p. 221 (2016)

27. Hurley, N., Zhang, M.: Novelty and diversity in top-n recommendation-analysis and evaluation. ACM Trans. Internet Techn. **10**(4), 14:1–14:30 (2011)

28. Koren, Y.: Factorization meets the neighborhood: a multifaceted collaborative filtering model. In: Conference on SIGKDD, USA, pp. 426–434 (2008)

29. Kosinski, M., Stillwell, D., Graepel, T.: Private traits and attributes are predictable from digital records of human behavior. Proc. National Acad. Sci. **110**, 5802–5805 (2013)

30. Lika, B., Kolomvatsos, K., Hadjiefthymiades, S.: Facing the cold start problem in recommender systems. Expert Syst. Appl. **41**, 2065–2073 (2014)
31. Luhmann, N.: Trust and power. Wiley, Chichester (1979)
32. Ma, X., Lu, H., Gan, Z.: Implicit trust and distrust prediction for recommender systems. In: Conference on WISE, USA, pp. 185–199 (2015)
33. Mairesse, F., Walker, M.A., Mehl, M.R., Moore, R.K.: Using linguistic cues for the automatic recognition of personality in conversation and text. J. Artif. Intell. Res. **30**, 457–500 (2007)
34. Martínez-Cruz, C., Porcel, C., Bernabé-Moreno, J., Herrera-Viedma, E.: A model to represent users trust in recommender systems using ontologies and fuzzy linguistic modeling. Inf. Sci. **311**, 102–118 (2015)
35. Mayer, R.C.: An integrative model of organizational trust. Acad. Manag. Rev. **20**(3), 709–734 (1995)
36. McCrae, R.: The five-factor model of personality traits: consensus and controversy. In: The Cambridge Handbook of Personality Psychology, pp. 148–161
37. McNee, S.M., Riedl, J., Konstan, J.A.: Being accurate is not enough: how accuracy metrics have hurt recommender systems. In: Olson, G.M., Jeffries, R. (eds.) Con. on CHI, Canada, pp. 1097–1101. ACM (2006)
38. Moradi, P., Ahmadian, S.: A reliability-based recommendation method to improve trust-aware recommender systems. Expert Syst. Appl. **42**(21), 7386–7398 (2015)
39. Nunes, M.A.S.N., Hu, R.: Personality-based recommender systems: an overview. In: Cunningham, P., Hurley, N.J., Guy, I., Anand, S.S. (eds.) Conference on RecSys, Ireland, pp. 5–6. ACM (2012)
40. Paiva, F.A.P., Costa, J.A.F., Silva, C.R.M.: A personality-based recommender system for semantic searches in vehicles sales portals. In: Martínez de Pisón, F.J., Urraca, R., Quintián, H., Corchado, E. (eds.) HAIS 2017. LNCS (LNAI), vol. 10334, pp. 600–612. Springer, Cham (2017). https://doi.org/10.1007/978-3-319-59650-1_51
41. Peter J. Rentfrow, Samuel D. Gosling, e.a.: The do re mi's of everyday life: the structure and personality correlates of music preferences. J. Personal. Soc. Psychol. **84**(6), 1236 (2003)
42. Recio-García, J.A., Jiménez-Díaz, G., Sánchez-Ruiz-Granados, A.A., Díaz-Agudo, B.: Personality aware recommendations to groups. In: Conference on RecSys, USA, pp. 325–328. ACM (2009)
43. Roshchina, A., Cardiff, J., Rosso, P.: TWIN: personality-based intelligent recommender system. J. Intell. Fuzzy Syst. **28**(5), 2059–2071 (2015)
44. Schwartz, H.A., et al.: Toward personality insights from language exploration in social media. In: Analyzing Microtext, Papers from the 2013 AAAI Spring Symposium, Palo Alto, California, USA, March 25–27, 2013. AAAI Technical Report, vol. SS-13-01. AAAI (2013)
45. Tang, J., Gao, H., Hu, X., Liu, H.: Exploiting homophily effect for trust prediction. In: Conference on WSDM, Italy, pp. 53–62 (2013)
46. Tang, J., Liu, H.: Trust in social media. Morgan and Claypool Publishers, San Rafael, California (2015)
47. Vinciarelli, A., Mohammadi, G.: A survey of personality computing. IEEE Trans. Affective Comput. **5**(3), 273–291 (2014)
48. Wang, C., Blei, D.M.: Collaborative topic modeling for recommending scientific articles. In: Confereence on SIGKDD, USA, pp. 448–456 (2011)

49. Wang, J., de Vries, A.P., Reinders, M.J.T.: Unifying user-based and item-based collaborative filtering approaches by similarity fusion. In: SIGIR 2006: Proceedings of the 29th Annual International ACM SIGIR Conference on Research and Development in Information Retrieval, Seattle, Washington, USA, August 6–11, 2006, pp. 501–508 (2006)

50. Xie, F., Chen, Z., Shang, J., Huang, W., Li, J.: Item similarity learning methods for collaborative filtering recommender systems. In: 29th IEEE International Conference on Advanced Information Networking and Applications, AINA 2015, Gwangju, South Korea, March 24–27, 2015, pp. 896–903 (2015)

51. Xue, H., Dai, X., Zhang, J., Huang, S., Chen, J.: Deep matrix factorization models for recommender systems. In: Conference on IJCAI, Australia, pp. 3203–3209 (2017)

52. Yakhchi, S., Beheshti, A., Ghafari, S.M., Orgun, M.A., Liu, G.: Towards a deep attention-based sequential recommender system. IEEE Access **8**, 178073–178084 (2020)

53. Yakhchi, S., Beheshti, A., Ghafari, S.M., Orgun, M.A.: Enabling the analysis of personality aspects in recommender systems. In: Wei, K.K., Huang, W.W., Lee, J.K., Xu, D., Jiang, J.J., Kim, H. (eds.) Conference on PACIS, China, p. 143 (2019)

54. Yakhchi, S., Ghafari, S.M., Beheshti, A.: CNR: cross-network recommendation embedding user's personality. In: Hacid, H., Sheng, Q.Z., Yoshida, T., Sarkheyli, A., Zhou, R. (eds.) Data Quality and Trust in Big Data - 5th International Workshop, QUAT 2018, Held in Conjunction with WISE 2018, Dubai, UAE, November 12–15, 2018, Revised Selected Papers. Lecture Notes in Computer Science, vol. 11235, pp. 62–77. Springer (2018)

55. Yakhchi, S., Ghafari, S.M., Tjortjis, C., Fazeli, M.: Armica-improved: a new approach for association rule mining. In: Li, G., Ge, Y., Zhang, Z., Jin, Z., Blumenstein, M. (eds.) Conference on KSEM, Australia. Lecture Notes in Computer Science, vol. 10412, pp. 296–306. Springer (2017)

56. Yao, Y., Tong, H., Yan, X., Xu, F., Lu, J.: Matri: a multi-aspect and transitive trust inference model. In: Proceedings of the 22nd International Conference on World Wide Web, pp. 1467–1476 (2013)

57. Yu, Y., Gao, Y., Wang, H., Wang, R.: Joint user knowledge and matrix factorization for recommender systems. In: Conference on WISE, China, pp. 77–91 (2016)

58. Zhang, Y., Chen, H., Wu, Z.: A social network-based trust model for the semantic web. In: ATC, pp. 183–192 (2006)

Am I Rare? an Intelligent Summarization Approach for Identifying Hidden Anomalies

Samira Ghodratnama[1(✉)], Mehrdad Zakershahrak[2],
and Fariborz Sobhanmanesh[1]

[1] Macquarie University, Sydney, Australia
{samira.ghodratnama,fariborz.sobhanmanesh}@mq.edu.au
[2] Arizona State University, Arizona, USA
mehrdad@asu.edu

Abstract. Monitoring network traffic data to detect any hidden patterns of anomalies is a challenging and time-consuming task which requires high computing resources. To this end, an appropriate summarization technique is of great importance, where it can be a substitute for the original data. However, the summarized data is under the threat of removing anomalies. Therefore, it is vital to create a summary that can reflect the same pattern as the original data. Therefore, in this paper, we propose an INtelligent Summarization approach for IDENTifying hidden anomalies, called *INSIDENT*. The proposed approach guarantees to keep the original data distribution in summarized data. Our approach is a clustering-based algorithm that dynamically maps original feature space to a new feature space by locally weighting features in each cluster. Therefore, in new feature space, similar samples are closer, and consequently, outliers are more detectable. Besides, selecting representatives based on cluster size keeps the same distribution as the original data in summarized data. *INSIDENT* can be used both as the preprocess approach before performing anomaly detection algorithms and anomaly detection algorithm. The experimental results on benchmark datasets prove a summary of the data can be a substitute for original data in the anomaly detection task.

Keywords: Anomaly detection · Summarization · Network data · Clustering · Classification

1 Introduction

Monitoring the fast and large volume of Internet traffic data that is being generated is paramount since they may have instances of anomalous network traffic, which makes the system vulnerable. However, detecting anomalies when we face big data is computationally expensive and still an open challenge. To this end, summarization is a practical approach that produces a condensed version of the

© Springer Nature Switzerland AG 2021
H. Hacid et al. (Eds.): ICSOC 2020 Workshops, LNCS 12632, pp. 309–323, 2021.
https://doi.org/10.1007/978-3-030-76352-7_31

original data. Therefore, a summary of the network traffic data helps network managers quickly assess what is happening in the network. For instance, the summary should still give insight into most visited websites, frequently used applications, and incoming traffic patterns. In [23] authors defined three scenarios in which summarization can help in traffic data, including: Summarizing network traffic can give an overview of what is going on in the network to the administrator. Summarized network traffic can be used as input to anomaly detection algorithms to reduce the cost. A summary of intrusion detection alarms facilitates the administrator's duty. In all mentioned scenarios, a concise representation of the data helps both the administrator and the analysis algorithms.

Different data summarization techniques are designed for other applications such as transactional data or stream data [1], which can be applied to traffic data. However, they have some drawbacks to be used for anomaly detection purposes, including:

- Clustering is the most used approach for summarization, where centers are considered as the summarized data. The problem is that the centroids may not be a part of the original data.
- Detecting frequent itemsets is another approach which only captures frequent items in the summaries. Therefore, they ignore or leave out anomalies that may be infrequent. Consequently, anomaly detection techniques do not perform well on summaries as they do not contain any anomalies.
- Semantics-based techniques do not keep the same samples in the summarized data.
- Statistical based techniques such as sampling do not guarantee the representation of anomalies in summary since they use a sampling-based summarization technique.

Therefore, not all summarization approaches are proper for anomaly detection purposes. Consequently, there is a need for an efficient network traffic summarization technique so that the summary more closely resembles the original network traffic In this context, summarization aims to create a summary from original data that includes interesting patterns, especially anomalies, and normal data for further analysis.

This paper proposes an intelligent summarization approach suitable for anomaly detection on network traffic datasets, which guarantees the preservation of original data distribution. We investigate the adaptation of clustering and KNN algorithms to create a summary. The proposed algorithm is used in two scenarios: i) as the preprocess approach for performing anomaly detection, ii) to detect anomalies in supervised problems as it reveals the hidden structure of data. The proposed summarization technique can also be adapted to other domains where big data requires being minded for interesting and relevant information. The rest of this paper is organized as follows. Section 2 discusses the state-of-the-art methods. Section 3 presents the proposed method, and Sect. 4 explains the experimental results and justifies the obtained results. Finally, Sect. 5 concludes the paper and discusses future work.

2 Related Work

Summarization has been widely explored in many domains and applications, using a variety of techniques [8, 12, 33]. When data size increases, the anomaly detection techniques perform poorly due to increasing false alarms and computational cost. Detecting anomalies from a summary could address these issues. However, existing summarization techniques cannot accurately represent the rare anomalies present in the dataset. In this section, we will present related work on traffic data summarization, along with anomaly detection techniques. It is worth mentioning although the general goal is to represent an input dataset in a condensed version, there is no definition of a good summary since each application requires a unique technique. For anomaly detection purposes, a good summary should be representative of all samples in the original dataset.

2.1 Network Analysis Tools

Different network analysis tools summarize network traffic data, such as Traffic Flow Analysis Tool, Flow-tools, Network Visualization Tools, and Network Monitoring Tools [2]. They produce a graphical report using different measurements, such as network bandwidth or latency. However, they only characterize and aggregate traffic instances based on a single attribute, such as the source/destination address or protocol. As a result, they are suitable to extract insights, not for further processing tasks such as anomaly detection. Besides, the objective of a summary is to provide an accurate report of the network's traffic patterns. Consequently, the summarization technique should identify traffic patterns based on arbitrary combinations of attributes efficiently.

2.2 Statistical Approaches

Statistical approaches aim to estimate the statistical distribution of data that could approximate the data set pattern. Sampling is a common technique in this category where a sample is a subset of the dataset. There are different kinds of sampling in practice, including i) simple random sampling, ii) stratified random sampling, iii) systematic sampling, iv) cluster random sampling, and v) multi-stage random sampling [15, 17]. However, summarized data using sampling is under the threat of removing anomalies. To solve this problem, in a recent work [2], the author proposed a sampling-based summarization technique, called SUCh, which integrated the concept of sampling using the modified Chernoff bound to include anomalous instances in summary. SUCh is computationally effective than the existing techniques and also performs better in identifying rare anomalies. However, an essential aspect of the summarization is representing all different types of traffic behavior. Although SUCh ensures the presence of anomalies, it ignores other types of traffic as they focus only on anomalous data.

2.3   Machine Leaning Approaches

Supervised and unsupervised learning techniques are two widely used knowledge discovery techniques. Two common machine learning algorithms used in summarizing network traffic data are *frequent itemsets* and *clustering*. Frequent itemsets are a set of items that appears more frequently than the rest of the samples. Different algorithms are used to detect frequent itemsets [14]. However, they are proper for detecting frequent items, not rare anomalies. Two main clustering-based algorithms for network traffic data summarization include centroid-based and feature-wise intersectin clustering algorithms. In a centroid-based summarization, after clustering samples, centroids are used to form the summary. Different variations of the k-means algorithm are widely used due to its simplicity, which can handle high-dimensional data [20,37]. In a feature-wise intersection-based summarization, the summary is created from each cluster using the feature-wise intersection of the data instances after clustering [14,23]. Consequently, summaries from all the clusters are combined to produce the final summary. This approach is best fitted for datasets with identical attribute values and, therefore, not suitable for detecting rare anomalies.

2.4   Semantic-Based Approaches

Semantic-based approaches are not suitable for anomaly detection since they do not produce a summary, which is part of the original data. Examples are linguistic summaries, which are based on the fuzzy. These approaches produce natural language expressions that describe important facts about the given data to enhance the human understanding of the network traffic summaries [31]. Attribute Oriented Induction (AOI) is another semantic-based approach aims to describe data in a concise and general manner [21]. AOI is a generalization process that abstracts a large dataset from a low conceptual level to a relatively higher conceptual level. Other semantic-based approaches include Fascicles [24], which relies on an extended form of association rules and perform lossy semantic compression. SPARTAN is another semantic-based summarization technique [10], which generalizes the fascicles approach.

2.5   Anomaly Detection Techniques

Anomaly detection is an important data analysis task that detects anomalous or abnormal data from a given dataset. Anomalies are patterns in data that do not follow the well-defined characteristic of typical patterns. Anomalies are important because they indicate significant but rare events that may have a detrimental impact on the system. Therefore, they require prompt critical actions to be taken in a wide range of application domains. An anomaly can be categorized in the following ways [3].

– Point anomaly: When a data instance deviates from the normal pattern of the dataset, it can be considered a point anomaly.

Table 1. Example of network traffic samples.

Source IP	Source port	Destination IP	Destination port	Protocol
192.168.5.10	1234	192.168.1.1	80	TCP
192.168.5.12	4565	192.168.1.2	20	TCP
192.168.5.10	20	192.168.28.80	119	HTTP
192.168.5.10	70	192.168.1.1	50	TCP
211.204.12.10	31	192.168.28.80	119	HTTP
192.168.5.1	3214	192.168.1.2	86	TCP

- Contextual anomaly: When a data instance behaves anomalously in a particular context, it is called a contextual or conditional anomaly.
- Collective anomaly: When a collection of similar data instances behave anomalously compared to the entire dataset, the group of data instances is called a collective anomaly.

Different supervised, unsupervised, and semi-supervised approaches have been proposed for this purpose. These techniques, including classification based network anomaly detection such as support vector machine [11], Bayesian network [27], neural network [30], and rule-based approaches [38]. Statistical anomaly detection techniques, including mixture model [16], signal processing technique [36], and principal component analysis (PCA) [34]. Other category includes information theory-based and clustering-based [1]. The proposed summarization approach is a general approach used in two scenarios: i) as the preprocessing approach where results are used as the input for anomaly detection algorithm, and ii) as an anomaly detection technique in a supervised setting discussed in the next section.

3   The Proposed Approach (INSIDENT)

This section discusses our proposed methodology. At first, we define the problem and then discuss our algorithm.

3.1   Problem Definition

In this paper, x_i is a sample vector and $X = [x_1, x_2, ..., x_N]$ is traffic data consists of N sample where $x_i \in R^d$ which d denotes the number of features. K is the number of clusters, and cluster centroids are denoted by c. $x_=$ is the closest similar sample to x, and $x_{\neq}$ is the closest different sample. An example of network traffic data with few attributes is reported in Table 1. The goal is to find a cluster of similar samples and find representatives for each cluster as the summary S where they keep the same distribution but less in size.

3.2 Methodology

Previous approaches used different clustering or sampling algorithms to summarize data. However, there is no guarantee that the summarized data has the same distribution as the original data, and therefore as the substitute for the original data. In this paper, we investigate the adaptation of clustering and the KNN algorithm to understand the data's underlying structure. In our previous work, this structure was used in the context of multi-document summarization [19] and image retrieval [18], demonstrating promising results. For this reason, the error rate of the nearest neighbor classifier in each cluster is minimized by locally weighting features in each cluster. INSIDENT transforms the feature space into a new feature space by weighting features separately in each cluster, where outliers are recognized easier in the new feature space. To this end, the weighted Euclidean distance is used. In our problem, these weights are arranged in a $d \times K$ weight matrix $W = \{w_{ij}, 1 \leq i \leq d, 1 \leq j \leq K\}$ where d is the number of features, and is K the number of clusters. To be more specific, for each cluster we have a vector of weights corresponding to each feature which are representative of the importance of each feature in each cluster. Our objective function is designed to minimize the error of 1NN in each cluster by regulating weights of each feature, and consequently cluster centers. To estimate the error of 1NN the following approximation function defined in [29] is used:

$$J(\mathbf{W}) = \frac{1}{N} \sum_{s \in XS} S_\beta \left(\frac{d_w(x, x_=)}{d_w(x, x_\neq)} \right) \tag{1}$$

where the sample $x_=$ is the nearest similar sample, and the sample $x_\neq$ is the closest different sample to the input sample x. Respectively d_w is the weighted Euclidean distance, and S_β is the sigmoid function, defined as:

$$S_\beta(z) = \left(\frac{1}{1 + e^{\beta(1-z)}} \right) \tag{2}$$

The objective function of K-means, which aims to minimize the errors of each cluster, is defined as:

$$J(\mathbf{W}, \mathbf{C}) = \sum_{k=1}^{K} \sum_{i=1}^{|N_K|} d_{W_K}^2 (x_i, c_K) \tag{3}$$

Thus, the overall objective function is defined as:

$$J(\mathbf{W}, \mathbf{C}) = \left(\sum_{k=1}^{K} \sum_{i=1}^{|N_K|} d_W^2 (s_i, c_K) + \frac{1}{N} \sum_{k=1}^{K} \sum_{i=1}^{|N_K|} S_\beta \left(\frac{d_w(x, x_=)}{d_w(x, x_\neq)} \right) \right) \tag{4}$$

where the first term is the objective function of K-means, and the second term is the summation of the classification errors over the K clusters.

Two parameters are optimized in this objective function. The first is the weights matrix. The feature-dependent weights associated with the sample are trained to make it closer to x, while making the sample $x_{\neq}$ further from x. Then, the cluster centroid update is based on the learned weighted distance. Since this function is differentiable, we can analytically use gradient descent for estimating the matrix W, guaranteeing convergence. The iterative optimization of a learning parameter like w is given below.

$$W^{t+1} = W^t - \alpha(\frac{J(\mathbf{W}, \mathbf{C})}{\delta(W)})$$
(5)

To simplify the formula, the function $R(x)$ is defined [29] as:

$$R_w(x_i) = (\frac{d_w(x_i, x_{i,=})}{d_w(x_i, x_{i,\neq})})$$
(6)

The partial derivative of $J(W, C)$ with respect to W is calculated by:

$$\frac{\delta J(\mathbf{W}, \mathbf{K})}{\delta W_K} \cong \sum_{i=1}^{|N_K|} 2W_K \odot (x_i - C_K)^2 + \frac{1}{N} \sum_{i=1}^{|N_K|} S'_\beta(R(x_i))\frac{\delta R(x_i)}{\delta W_k}$$
(7)

where $\odot$ is the inner product and $\frac{\delta R(x_i)}{\delta W_K}$ is :

$$\frac{\delta R(s_i)}{\delta W_K} = \frac{1}{d^2_{W_K}}(x_i, x_{i,\neq})(\frac{1}{R(x_i)}W_K \odot (x_i - x_{i,=})^2 - R(x_i)W_K \odot (x_i - x_{i,\neq})^2)$$
(8)

The derivative of $S_\beta(z)$ is defined as:

$$S_\beta(z)' = \frac{\delta S_\beta(z)}{\delta z}$$
$$= \frac{\beta e^{\beta(1-z)}}{(1 + e^{\beta(1-z)})^2}$$
(9)

The partial derivative of $J(\mathbf{W}, \mathbf{C})$ with respect to C is calculated as:

$$\frac{J(\mathbf{W}, \mathbf{C})}{\delta C_k} \cong \sum_{i=1}^{|N_k|} -2W_k^2 \odot (x_i - C_k)$$
(10)

Since we need to optimize the weight of features for each cluster's samples, along with the center of clusters, we first update W in each cluster, and then we update C (center of clusters). The INSIDENT algorithm is depicted in Algorithm 1 for more clarification. Since the algorithm performs in an iterative process using gradient descent, the simplest clustering (k-means) and (KNN) algorithms are used for efficiency. However, K-means is one of the most reliable and most widely used clustering algorithms. Besides, the K-nearest neighbor (NN) has been successfully used in many pattern-recognition applications [9]. Similar samples are close to each other in new feature space, making a point, and contextual type anomalies easily detectable. In the case of collective anomalies, we select the number of each cluster's representative based on its size to keep the distribution the same as the original data.

Algorithm 1. INSIDENT

Input: Traffic Data X, learning rate γ and α.
Output: Summary (S).

procedure INSIDENT.
  while $iter < MaxIterations$ **do**
    Clusters (C) $\leftarrow$ K-means(X)
    for each clusters c in C **do**
      for each sample x in c **do**
        $x_= \leftarrow findSimilarCloseSample()$
        $x_{\neq} \leftarrow findDifferentCloseSample()$
        $W^{iter+1} = W^{iter} - \gamma \frac{\delta J(W)}{W}$
      end for
    end for
    Update Clusters
  end while
  return Summary(S)

---

4 Experiments and Evaluation

In this section, the dataset, the evaluation method, and the performance of INSI-DENT are explained and compared with existing state-of-the-art approaches.

4.1 Data Set

Experiments on six benchmark datasets are performed. The details of this dataset and the distribution of normal and anomalous samples in each dataset are reported in Table 2. KDD1999 contains collective anomalies were the other five datasets contain only rare anomalies. These rare anomalous datasets are from SCADA network, including real SCADA (WTP), simulated anomalies (Sim1 and Sim2), and injected anomalies (MI and MO).

4.2 Evaluation Metrics

To evaluate network traffic summary, we explain two widely used summary evaluation metrics including *conciseness*, and *information loss* [5].

- Conciseness: The size of the summary influences the quality of the summary. At the same time, it is important to create a summary that can reflect the underlying data patterns. Conciseness is defined as the ratio of input dataset size (N) and the summarized dataset size (S) defined as:

$$Conciseness = \frac{N}{S} \tag{11}$$

– Information Loss: A general metric used to describe the amount of information lost from the original dataset due to the summarization. Loss is defined as the ratio of the number of samples not present by samples present in summary defines as:

$$InformationLoss = \frac{L}{T} \tag{12}$$

where T is the number of unique samples represented by the summary, and L defines the number of samples not present in the summary.

Table 2. Dataset sescription.

Dataset	Sample number	Normal Percentage	Anomalies percentage
KDD1999	494020	19.69	80.310
WTP	527	97.34	2.66
MI	4690	97.86	2.14
MO	4690	98.76	1.24
Sim1	10501	99.02	0.98
Sim2	10501	99.04	0.96

Besides, to evaluate the performance of the anomaly detection algorithms used in supervised approaches, three measures, including accuracy, recall, and F1 discussed below, are used. Before we define these measure, four values included in the confusion needs to be discussed [3].

– True Positive (TP): Number of anomalies correctly identified as anomalous.
– False Positive (FP): Number of normal data incorrectly identified anomaly.
– True Negative (TN): Number of normal data correctly identified as normal.
– False Negative (FN): Number of anomalies incorrectly identified as normal.

Based on the above definitions, we define the evaluation metrics.

$$Accuracy = \frac{TP + TN}{TP + TN + FP + FN} \tag{13}$$

$$Recall = \frac{TP}{TP + FN} \tag{14}$$

$$F1 = \frac{2TP}{2TP + FP + FN} \tag{15}$$

4.3 Result Analysis

In this section, we discuss the performance evaluation of the existing summarization methods compared to INSIDENT, along with the anomaly detection result.

Table 3. Real SCADA dataset (WTP) result.

Model	WTP-Recall	WTP-Accuracy	WTP-F1
KNN	85.71	97.39	85.71
LOF	78.57	97.38	78.57
COF	57.14	97.35	57.14
LOCI	85.71	97.39	85.71
LoOP	42.85	97.33	42.85
INFLO	57.14	97.35	57.14
CBLOF	92.85	97.40	92.85
LDCOF	85.71	97.39	85.71
CMGOS	57.14	97.35	57.14
HBOS	28.57	97.32	28.57
LIBSVM	85.71	97.39	85.71
INSIDENT	94.87	97.91	94.87

Anomaly Detection Evaluation. This section contains the performance analysis of anomaly detection techniques. The baseline algorithms include Nearest Neighbor-based algorithms (K-NN [32], LOF [13], COF [35], LOCI [28], LoOP [26], INFLO [25]), clustering-based approach(CBLOF [22], LDCOF [6], CMGOS [6]), and statistical appraoches (HBOS and LIBSVM [7]). These approaches are compared with INSIDENT on different variations of the SCADA dataset, including WTP, MI, MO, Sim1, and Sim2, where their values are reported by [3]. Results are reported respectively in Table 3, Table 4, and Table 5.

From Table 3, it can be seen that for the real SCADA dataset(WTP), INSIDENT has higher values. Then the clustering-based anomaly detection technique, CBLOF, performs best, and third, the nearest-neighbor-based approach attains the best performance. It is an expected result showing the combination of clustering and KNN can perform better. Statistical based approach HBOS dis not perform well. Table 4 displays the results on simulated datasets (Sim1 and Sim2). LIBSVM has better recall than others, and INCIDENT performs as the second best. Clustering-based approaches are not well suited for the simulated datasets. For the datasets with injected anomalies (MI, MO), INCIDENT, along with clustering-based approaches, are the best considering the evaluation measures. Nearest neighbor-based approaches are the next best. It is interesting to observe that the Recall and F1 values are identical for all the anomaly detection techniques. The reason is that since the top N anomalies detected by the techniques match the actual N number of anomalies in the dataset, the Recall, and F1 scores are always the same.

Network Traffic Summarization Evaluation. For summarization evaluation, the KDD dataset is used. Summarization size, which defines conciseness, is

Table 4. Simulated SCADA datasets result(Sim1 and Sim2).

Model	Sim1Recall	Sim1Accuracy	Sim1F1	Sim2Recall	Sim2Accuracy	Sim2F1
KNN	64.7	99.03	64.7	63	99.05	63
LOF	0	99.01	0	0	99.03	0
COF	0	99.01	0	2	99.03	2
LOCI	0	99.01	0	0	99.03	0
LoOP	0.98	99.01	0.98	0	99.03	0
INFLO	0	99.01	0	0	99.03	0
CBLOF	0	99.01	0	0	99.03	0
LDCOF	0	99.01	0	0	99.03	0
CMGOS	18.62	99.02	18.62	97	99.05	97
HBOS	30.39	99.02	30.39	27	99.04	6
LIBSVM	74.50	99.03	74.50	68	99.05	68
INSIDENT	72.13	99.07	72.13	78.21	99.05	78.21

Table 5. Simulated SCADA datasets with Injected Anomalies result (MI and MO).

Model	MI-Recall	MI-Accuracy	MI-F1	MO-Recall	MO-Accuracy	MO-F1
KNN	96	97.09	96	91.37	98.77	91.37
LOF	38.33	97.43	38.33	55.17	98.76	55.17
COF	9	97.82	9	25.86	98.75	25.86
LOCI	91	97.9	91	84.48	98.77	84.48
LoOP	10	97.83	10	27.58	98.75	27.58
INFLO	12	97.83	12	43.1 0	98.76	43.10
CBLOF	24	97.84	24	63.79	98.76	63.79
LDCOF	100	97.91	100	63.79	98.76	63.79
CMGOS	100	97.91	100	50	98.76	50
HBOS	98	97.91	98	65.51	98.76	65.51
LIBSVM	86	97.9	86	91.37	98.77	91.37
INSIDENT	100	98.76	100	94.21	99.04	94.21

considered as a constraint in summarization algorithms. When the summary is small, it has maximum information loss. On the other hand, when conciseness is small, the summary contains the whole dataset has no information loss. Therefore, information loss and conciseness are orthogonal parameters. Our experiments used five different summary sizes, and then information loss was measured for each summary size. In practice, the network manager/analyst decides the summary size based on the network. The results are compared with NTS and FIB approaches [4]. Since our algorithm is based on k-means, we test three times with different initial points for each summary size. Results are depicted in Fig. 1. Besides, the percentage of anomalies compared with SUCh [2] is reported in Table 6 proving that INSIDENT well-preserved the percentage of anomalies in generated summaries.

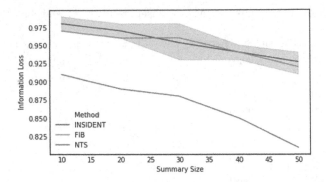

Fig. 1. The result of comparing information loss based on different summary size.

Table 6. Comparing the distribution of anomalies in summaries and original data.

Dataset	Original data	SUCh Alg	INSIDENT
WTP	2.66	N/A	2.33
MI	2.14	2.61	2.76
MO	1.24	1.46	1.52
Sim1	0.98	1.04	1.11
Sim2	0.96	0.94	1.01

5   Conclusion and Future Work

Monitoring network traffic data to detect any hidden patterns of anomalies is a challenging and time-consuming task which requires high computing resources. Therefore, in this paper, we proposed an INtelligent Summarization approach for IDENTifying hidden anomalies, called *INSIDENT*. In data summarization, it is always a dilemma to claim the best summary. The proposed approach claim is to guarantee to keep the original data distribution in summarized data. The INSIDENT's backbone is the clustering and KNN algorithm that dynamically maps original feature space to a new feature space by locally weighting features in each cluster. The experimental results proved that the proposed approach helps keep the distribution the same as the original data, consequently making anomaly detection easier. In future work, we aim to focus on real-time network traffic summarization.

Acknowledgement. We acknowledge the AI-enabled Processes (AIP) Research Centre (https://aip-research-center.github.io/) for funding this research. We also acknowledge Macquarie University for funding this project through IMQRES scholarship.

References

1. Ahmed, M.: Data summarization: a survey. Knowl. Inf. Syst. **58**(2), 249–273 (2019)
2. Ahmed, M.: Intelligent big data summarization for rare anomaly detection. IEEE Access **7**, 68669–68677 (2019)
3. Ahmed, M., Anwar, A., Mahmood, A.N., Shah, Z., Maher, M.J.: An investigation of performance analysis of anomaly detection techniques for big data in scada systems. EAI Endorsed Trans. Indust. Netw. Intell. Syst. **2**(3), e5 (2015)
4. Ahmed, M., Mahmood, A.N., Maher, M.J.: A novel approach for network traffic summarization. In: Jung, J.J., Badica, C., Kiss, A. (eds.) INFOSCALE 2014. LNICST, vol. 139, pp. 51–60. Springer, Cham (2015). https://doi.org/10.1007/978-3-319-16868-5_5
5. Ahmed, M., Mahmood, A.N., Maher, M.J.: An efficient technique for network traffic summarization using multiview clustering and statistical sampling. EAI Endorsed Trans. Scalable Inf. Syst. **2**(5), (2015)
6. Amer, M., Goldstein, M.: Nearest-neighbor and clustering based anomaly detection algorithms for rapidminer. In: Proceedings of the 3rd RapidMiner Community Meeting and Conference (RCOMM 2012), pp. 1–12 (2012)
7. Amer, M., Goldstein, M., Abdennadher, S.: Enhancing one-class support vector machines for unsupervised anomaly detection. In: Proceedings of the ACM SIGKDD Workshop on Outlier Detection and Description, pp. 8–15 (2013)
8. Amouzgar, F., Beheshti, A., Ghodratnama, S., Benatallah, B., Yang, J., Sheng, Q.Z.: iSheets: a spreadsheet-based machine learning development platform for data-driven process analytics. In: Liu, X., et al. (eds.) ICSOC 2018. LNCS, vol. 11434, pp. 453–457. Springer, Cham (2019). https://doi.org/10.1007/978-3-030-17642-6_43
9. Anava, O., Levy, K.: k*-nearest neighbors: From global to local. In: Advances in Neural Information Processing Systems, pp. 4916–4924 (2016)
10. Babu, S., Garofalakis, M., Rastogi, R.: Spartan: a model-based semantic compression system for massive data tables. ACM SIGMOD Rec. **30**(2), 283–294 (2001)
11. Balabine, I., Velednitsky, A.: Method and system for confident anomaly detection in computer network traffic, uS Patent 9,843,488, 12 December 2017
12. Beheshti, A., et al.: iProcess: enabling IoT platforms in data-driven knowledge-intensive processes. In: Weske, M., Montali, M., Weber, I., vom Brocke, J. (eds.) BPM 2018. LNBIP, vol. 329, pp. 108–126. Springer, Cham (2018). https://doi.org/10.1007/978-3-319-98651-7_7
13. Breunig, M.M., Kriegel, H.P., Ng, R.T., Sander, J.: Lof: identifying density-based local outliers. In: Proceedings of the 2000 ACM SIGMOD International Conference on Management of Data, pp. 93–104 (2000)
14. Chandola, V., Kumar, V.: Summarization-compressing data into an informative representation. Knowl. Inf. Syst. **12**(3), 355–378 (2007)
15. Cochran, W.G., William, G.: Sampling Techniques. Wiley, New York (1977)
16. Eskin, E.: Anomaly detection over noisy data using learned probability distributions (2000)
17. Ghodratnama, S., SadrAldini, M.: An innovative sampling method for massive data reduction in data mining. In: The 3rd Iran Data Mining Conference, Tehran (2009)
18. Ghodratnama, S., Abrishami Moghaddam, H.: Content-based image retrieval using feature weighting and C-means clustering in a multi-label classification framework. Pattern Anal. Appl. **24**(1), 1–10 (2020). https://doi.org/10.1007/s10044-020-00887-4

19. Ghodratnama, S., Beheshti, A., Zakershahrak, M., Sobhanmanesh, F.: Extractive document summarization based on dynamic feature space mapping. IEEE Access **8**, 139084–139095 (2020)
20. Ghodratnama, S., Boostani, R.: An efficient strategy to handle complex datasets having multimodal distribution. In: Sanayei, A., E. Rössler, O., Zelinka, I. (eds.) ISCS 2014: Interdisciplinary Symposium on Complex Systems. ECC, vol. 14, pp. 153–163. Springer, Cham (2015). https://doi.org/10.1007/978-3-319-10759-2_17
21. Han, J., Fu, Y.: 16 exploration of the power of attribute-oriented induction in data mining. In: Advances in Know Ledge Discover and Data Mining, pp. 399–421. AAAI/'&I1T Press, Cambridge (1996)
22. He, Z., Xu, X., Deng, S.: Discovering cluster-based local outliers. Pattern Recogn. Lett. **24**(9–10), 1641–1650 (2003)
23. Hoplaros, D., Tari, Z., Khalil, I.: Data summarization for network traffic monitoring. J. Netw. Comput. Appl. **37**, 194–205 (2014)
24. Jagadish, H., Madar, J., Ng, R.T.: Semantic compression and pattern extraction with fascicles. VLDB **99**, 186–97 (1999)
25. Jin, W., Tung, A.K.H., Han, J., Wang, W.: Ranking outliers using symmetric neighborhood relationship. In: Ng, W.-K., Kitsuregawa, M., Li, J., Chang, K. (eds.) PAKDD 2006. LNCS (LNAI), vol. 3918, pp. 577–593. Springer, Heidelberg (2006). https://doi.org/10.1007/11731139_68
26. Kriegel, H.P., Kröger, P., Schubert, E., Zimek, A.: Loop: local outlier probabilities. In: Proceedings of the 18th ACM Conference on Information and Knowledge Management, pp. 1649–1652 (2009)
27. Kruegel, C., Mutz, D., Robertson, W., Valeur, F.: Bayesian event classification for intrusion detection. In: 19th Annual Computer Security Applications Conference. Proceedings. pp. 14–23. IEEE (2003)
28. Papadimitriou, S., Kitagawa, H., Gibbons, P.B., Faloutsos, C.: Loci: fast outlier detection using the local correlation integral. In: Proceedings 19th International Conference on Data Engineering (Cat. No. 03CH37405), pp. 315–326. IEEE (2003)
29. Paredes, R., Vidal, E.: Learning weighted metrics to minimize nearest-neighbor classification error. IEEE Trans. Pattern Anal. Mach. Intell. **28**(7), 1100–1110 (2006)
30. Poojitha, G., Kumar, K.N., Reddy, P.J.: Intrusion detection using artificial neural network. In: 2010 Second International conference on Computing, Communication and Networking Technologies, pp. 1–7. IEEE (2010)
31. Pouzols, F.M., Lopez, D.R., Barros, A.B.: Summarization and analysis of network traffic flow records. In: Mining and Control of Network Traffic by Computational Intelligence, pp. 147–189. Springer, Heidelberg (2011). https://doi.org/10.1007/978-3-642-18084-2_4
32. Ramaswamy, S., Rastogi, R., Shim, K.: Efficient algorithms for mining outliers from large data sets. In: Proceedings of the 2000 ACM SIGMOD International Conference on Management of Data, pp. 427–438 (2000)
33. Schiliro, F., et al.: iCOP: IoT-enabled policing processes. In: Liu, X., et al. (eds.) ICSOC 2018. LNCS, vol. 11434, pp. 447–452. Springer, Cham (2019). https://doi.org/10.1007/978-3-030-17642-6_42
34. Shyu, M.L., Chen, S.C., Sarinnapakorn, K., Chang, L.: A novel anomaly detection scheme based on principal component classifier. Miami Univ Coral Gables FL Dept of Electrical and Computer Engineering, Technical report (2003)
35. Tang, J., Chen, Z., Fu, A.W., Cheung, D.W.: Capabilities of outlier detection schemes in large datasets, framework and methodologies. Knowl. Inf. Syst. **11**(1), 45–84 (2007)

36. Thottan, M., Ji, C.: Anomaly detection in IP networks. IEEE Trans. Signal Process. **51**(8), 2191–2204 (2003)
37. Wendel, P., Ghanem, M., Guo, Y.: Scalable clustering on the data grid. In: 5th IEEE International Symposium Cluster Computing and the Grid (ccGrid) (2005)
38. Yang, Y., McLaughlin, K., Littler, T., Sezer, S., Wang, H.: Rule-based intrusion detection system for scada networks (2013)

On How Cognitive Computing Will Plan Your Next Systematic Review

Maisie Badami[1(✉)], Marcos Baez[2], Shayan Zamanirad[1], and Wei Kang[3]

[1] University of New South Wales (UNSW), Kensington, Australia
{m.badami,shayan.zamanirad}@unsw.edu.au
[2] LIRIS – University of Claude Bernard Lyon 1, Villeurbanne, France
marcos.baez@liris.cnrs.fr
[3] Data61, CSIRO, Eveleigh, Australia
wei.kang@data61.csiro.au

Abstract. Systematic literature reviews (SLRs) are at the heart of evidence-based research, setting the foundation for future research and practice. However, producing good quality timely contributions is a challenging and highly cognitive endeavor, which has lately motivated the exploration of automation and support in the SLR process. In this paper we address an often overlooked phase in this process, that of planning literature reviews, and explore under the lenses of cognitive process augmentation how to overcome its most salient challenges. In doing so, we report on the insights from 24 SLR authors on planning practices, its challenges as well as feedback on support strategies inspired by recent advances in cognitive computing. We frame our findings under the cognitive augmentation framework, and report on a prototype implementation and evaluation focusing on further informing the technical feasibility.

Keywords: Systematic review · Cognitive process · Web services · Word embedding

1 Introduction

Systematic Literature Reviews (SLRs) are valuable research contributions that follow a well-known, comprehensive, and transparent research methodology. It is at the heart of evidence-based research, allowing researchers to systematically collect and integrate empirical evidence regarding research questions. Given their demonstrated value, SLRs are becoming an increasingly popular type of publication in different disciplines, from medicine to software engineering [11].

Despite the valuable contributions of systematic reviews to science, producing good quality timely reviews is a challenging endeavor. Studies have shown that SLRs might fail to provide a good and complete coverage of existing evidence, missing up to 40% of relevant papers [6], and even end up being outdated by the time of publication [6,22] – this without considering those never published.

The reasons behind these challenges have been documented in several studies [7,8,20], which attribute them to the demanding nature of the involved tasks,

© Springer Nature Switzerland AG 2021
H. Hacid et al. (Eds.): ICSOC 2020 Workshops, LNCS 12632, pp. 324–333, 2021.
https://doi.org/10.1007/978-3-030-76352-7_32

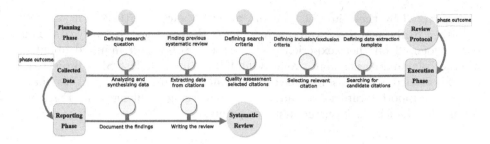

Fig. 1. The SLR process, defined by planning, execution and reporting activities

lack of expertise, limitations of support technology, and issues with primary studies. Recent advances in cognitive computing and collaborative technology offer an opportunity to address these challenges, and support researchers in planning, running and reporting SLRs (see Fig. 1 for an overview of the SLR process). We have seen new techniques and platforms enabling large-scale collaboration [13,25,26], and automation opportunities [9,16,19], offering promising results in different research activities relevant to the SLR process. Most of these efforts however are centered around the screening and identification of relevant scientific articles – and rightly so as it is one of the most time-consuming phases – but leaving other critical tasks largely unexplored.

In this paper we address a much less explored phase of the SLR process, that of planning the reviews. Guidelines and recommendations (e.g., [11]) define the main activities in this phase as i) identifying the need for undertaking the review, ii) defining the research questions (RQs), iii) defining the search and eligibility criteria, and iv) the data extraction template. These tasks are fundamental to guiding the SLR process and setting the foundation to having meaningful and original contributions, good coverage of the literature and a process free of bias. Yet, as we will see, they are often poorly performed, if at all.

In what follows we investigate how cognitive augmentation can support the planning phase of SLRs. We build on the insights and feedback from SLR authors to identify challenges and support strategies inspired by recent advancements in cognitive computing, framing the results under the framework for cognitive process augmentation. We also report on our early prototype and evaluation runs, showing the potential of augmentation in identifying the need for undertaking a review by leveraging word embeddings to find relevant SLRs from an input RQs.

2   Challenges in Planning SLRs

The challenges in running an SLR can be found throughout the entire process. These have been observed in the literature [7,8,20] as well as in our preliminary work, where we run an open-ended survey with more than 50 SLR authors tapping on their experience running SLRs. The results indicated that planning tasks are generally perceived as difficult to manage, requiring higher level of expertise and domain knowledge compared to more labor-intensive tasks.

Motivated by these insights, we run a second survey with 24 authors who published SLRs in top software engineering outlets in the last two years. This survey focused on their experience in planning SLRs, and inquired about i) whether planning tasks were properly addressed in their last SLR project, ii) the importance of addressing some salient challenges, and asked for iii) feedback on some support strategies to address the emerging challenges. In Fig. 2 we summarise the feedback on the first two points.

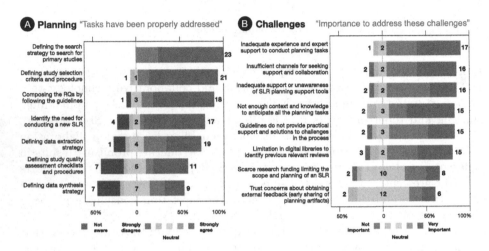

Fig. 2. Summary of feedback from SLR authors. A) Feedback on how planning tasks are addressed. B) Importance of addressing salient challenges in the planning.

While most authors reported positively to having addressed the tasks properly – not surprisingly given the quality of the outlets – there was still a significant number of researchers reporting neutral to negative, and in some cases even not being aware of certain tasks. More illuminating is to observe the challenges that we identified and the importance authors put in addressing them (Fig. 2), which we summarise and group below and address in the next section:

C1 Inadequate experience and support by current tools and guidelines.
C2 Insufficient context and knowledge to anticipate tasks in the planning phase.
C3 Limitations of digital libraries to identify relevant SLRs.
C4 Inadequate expert support and trust concerns in obtaining external feedback.

3   Cognitive Support in Planning SLRs

In this section, we present the conceptual design of a platform to support the planning of SLRs. In the following we describe the solutions and strategies addressing the main challenges, and the feedback obtained from SLR authors.

Strategies to Address Planning Challenges. The strategies were derived in brainstorming sessions among the co-authors, taking as input the insights from the first round of interviews with SLR authors on workarounds and strategies employed in the process, our own experience and prior work. The resulting strategies leverage techniques from machine learning and data-mining to address the main challenges. Below we describe the strategies for each of the challenges (C#).

- **Chatbot assistant** that allows authors to ask questions about the SLR process and best practices (C1). The chatbot provides a natural language interface to all information encoded in guidelines and recommendations.
- **Step-by-step guides** for each of the tasks (C1). The interface provides a conversational interface that assists authors in the preparation of each "artefact", (e.g., RQs), by providing step-by-step prompts based on guidelines.
- **Incremental and iterative process** that can be adapted as more information becomes available to the researcher (C2). For example, RQs and inclusion/exclusion criteria can be refined as we identify similar literature reviews and learn more about the topic.
- **Incorporating context and knowledge** (C2). The system leverages information available in seed papers and similar literature reviews by extracting SLR-specific metadata relevant to the protocol (e.g., RQs, search strategy) and making them available to the authors as a reference point at each step.
- **Search focused on similar SLRs** (C3). Instead of defining complex queries and terms to identify similar reviews, the search focuses only on SLRs. The search results provide SLR-specific information, including RQs, inclusion/exclusion criteria, search strategy, etc.
- **From RQs to similar SLRs** and papers (C3). The system allows users to go from their (partial) RQs directly to similar SLRs by enriching information with extra data from seed papers.
- **Expert networking** (C4). We presented strategies including artefact-specific evaluation criteria, improved discovery of experts, and leveraging groups of trust. However, we limit the discussion on human-human collaboration.

We illustrate how the above strategies can be combined in a the concept tool in Fig. 3, depicting the iteration over the first two steps of the planning.

Feedback from SLR Authors. We requested feedback on the concept tool and strategies in our second wave of survey with 24 SLR authors. Authors were shown a mockup and descriptions,[1], and asked to i) rate the strategies on whether they addressed the specific challenges, ii) provide feedback on potential barriers for adopting them, as well as iii) other alternatives not foreseen.

The strategies to address the inadequate experience and support (C1) received positive feedback by almost all participants (15/16). For example, a

---

[1] All materials related to the study can be found at https://github.com/maisieb01/Cognitive_SLR.git.

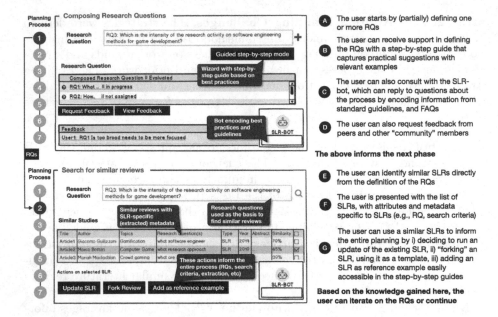

Fig. 3. Interaction illustrating the first two steps in the planning: defining the RQs and identifying the need for the SLR. The same strategies apply to the rest of the steps.

participant stated *"I had to search for hours to understand how to get tasks done, so I think a Bot that helps with the common question would be really helpful"*. Another one stated *"Support from peers is the most important help that I've been missing when conducting my study. [..] I didn't know much about the process and where to start, the step by step guideline would have helped me a lot"*.

Participants reported potential barriers being, i) mismatch between the assistance provided by the tool and practices of the target domain of the SLR, and ii) the quality of the underlying algorithms, including the actual bot recommendations.

Regarding the lack of context and knowledge (C2), all participants provided positive feedback (17/17). The only concern was about extracting information from multiple SLRs following different approaches and the tool mixing them up.

The strategies to address the limitation of digital libraries (C3) also received positive feedback (15/17). The argument against it came from an author not sure about the effectiveness *"I'm not sure if searching by [RQs] will help identify an area which needs more research or just other SLRs .. [it] just seems like a 'type' filter as e.g. in Scopus or adding 'review' in Title from Google Scholar"*.

Among the suggestions we can mention: i) adapt the builtin guidelines to the target domain of the SLR, and consider other frameworks such as PICO, ii) extend the SLR extraction capabilities to recommend highly reputable venues, infer the guidelines to follow and create a to-do list authors could follow, iii) expand the search to suggest research that is very relevant, and iv) include explanations for as to why papers are suggested.

4   Conceptual Architecture

In delivering on the vision of cognitive support in the planning of SLRs, we rely on the framework for cognitive process augmentation by Barukh et al. [2].

As we will see, the strategies we devised inform this architecture at different layers.

Foundation: Existing technology provides support for coordinating *data, tasks* and *collaboration* that we can leverage to build our vision of a cognitively-augmented planning process. Starting from the process itself, we have seen the planning of SLRs to be an incremental and iterative process leading to a review protocol. The process management in this context could rely on lightweight artifact-centric systems (e.g., Gelee [1]) where the researcher drives the process while the system advises on the steps to take based on community-specific guidelines. Along the process, some tasks are already supported by current online services, such the search and access to scientific articles. Digital libraries and search engines provide access to article data and metadata, but under the limitations pointed out in the previous section, requiring researchers to engage in significant manual effort. Thus, although the data and knowledge required to elaborate the research protocol and inform the process is available, identifying, curating and adopting such knowledge is a challenging endeavour.

Enablement: The next layer leverages existing data sources and services to apply domain-specific data extraction and enrichment that will enable cognitive augmentation. Components such as *article recommendation*, enabling search for similar SLRs and papers, *article augmentation*, enriching SLRs with domain-specific metadata, *activity recommendation*, recommending steps based on process definitions and progress, and *knowledge graph*, aggregating knowledge about the process in queryable format, are among the enabling components. In this context, SLRs, primary research articles, and guidelines on how to run the SLR process, are the main sources of information. Lower-level algorithms such as named-entity recognition, word-embeddings and similarity serve as building blocks for these higher level components.

Delivery: The researcher finally experiences the cognitive augmentation in the planning through Conversational AI as well as intelligent GUIs. Conversational AI helps in delivering assistance in the process, providing a natural language interface to query the vast knowledge encoded in guidelines, and receive practical assistance in each step of the process. In the form of more general conversational interfaces, guided prompts would provide step-by-step guidance to assist researchers in knowledge-intensive tasks (e.g., defining RQs). We also recognise the need for serving more traditional delivery systems such as GUIs, to dote complex tasks with intelligent features (e.g., domain-specific search).

5   Prototype Implementation and Evaluation

In this section we describe our ongoing exploration into the technical feasibility of our approach.

We started with *article recommendation* as it emerged as a promising component based on the feedback from SLR authors, but raised concerns in terms of feasibility. We note however that providing support for identifying related SLRs has all the potential benefits but fewer of the concerns (e.g., recall and accuracy) with respect to assisting the identification and screening of relevant literature [12]. In the planning, bringing up the most relevant SLRs and papers can provide the additional context in a task that some people are not even aware of.

Prototype. The prototype is our initial exercise into understanding the technical requirements of the system, as well as a working tool to serve evaluation purposes. On the surface, the current version of the prototype provides a set of REST APIs (and an accompanying user interface) that given a set of RQs in input, it returns the most relevant SLRs, along with the relevance score computed based on the available models. Figure 4 presents the pipeline of the implementation.

The source of information is currently a curated database of SLRs, where domain-specific information has been manually extracted so as to evaluate the recommendation component in isolation. The end goal is to have a data layer that can interface with existing services to access structured and unstructured data, which can then be processed to automatically extract relevant domain-specific information. While we currently use a MySQL database to store and retrieve raw and curated data, the concept of *Data Lake* [3] emerges as a promising direction to store and query structured and unstructured SLR data.

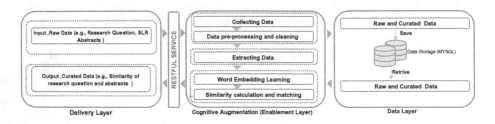

Fig. 4. Architecture pipeline

In entering the augmentation layer, the data is pre-processed in two steps: (a) normalizing text corpus (e.g., removing special characters and stop words, all to lowercase); and (b) lemmatizing and converting each word to its base form. We leverage Stanford CoreNLP toolkit[2] to perform this process.

Following the data cleaning, the next step is to extract meaningful information from the data. The inspiration behind our proposed approach is leveraging a similar approach to the word embedding's model [18] that represents words in a Vector Space Model (VSM). We have extended the idea of considering a "word"

---

[2] https://nlp.stanford.edu/software/tagger.html.

as a vector to represent the SLR-related corpus (e.g., RQs and SLR abstracts) in a vector space. To create the VSMs, we employ an N-gram selector component to extract all the keywords (nouns and verbs) from sentences of the given context. We leverage Stanford Part-Of-Speech (POS) Tagger [15] to achieve this. Then, we create a list of n-grams out of these keywords and transform them into vector representations for each corpus. After encoding the given corpora into vectors, these vectors are used to calculate the similarity between desired corpora. Information is then augmented based on a pool of word embedding models. Our work leverages state-of-the-art algorithms widely used in NLP communities. Several such algorithms (e.g. GloVe [21], Word2Vec [17], Numberbatch [24], WikiNewsFast [4], and GoogleNews [5]) come with efficient implementations that are readily available as libraries to use.

The REST APIs then expose the functionality of the article recommendation component for programmatic access. A front-end application takes these services and wraps them up in a user interface.

Planned Evaluation. The goal of the initial evaluation is to inform specific design decisions regarding the algorithmic support for recommending papers. Among the main design decisions we consider: i) *What models will better serve the specific task?* The idea is to identify among the embedding models and architectures the most promising candidates to build on, and understand whether investing in domain-specific embedding models is required. Then, ii) *What information should we leverage when assessing the relevance calculations?* (e.g., title, title-abstract, RQs, full-text). The aim is to understand what (combinations of) information to focus when assessing the relevance of SLRs from an input RQs, and therefore to consider in the extraction process.

The prototype supports these two dimensions, models and selective information, so as to serve the evaluations. The dataset of SLRs is being manually constructed to incorporate for each SLR a set of related SLRs (as reported in the reviews) and not relevant SLRs as judged by human experts. Armed with the human-annotated dataset, we evaluate the quality of word embedding models by assessing how well the similarity scores of the word vectors correlate with human judgment [23]. The similarity is calculated as the distance between the vectors representing RQs and SLRs, using cosine similarity as measure, which has been found suitable for SLRs in prior work [14].

We rely on Spearman's rank (r_s) correlation between the word embedding models similarity score and researchers annotations to evaluate how well the similarity of the given pairs (e.g., RQs and abstracts) agrees with human judgments [10]. We performed a preliminary test run to tune the experimental setup, and the early results are encouraging. Results under limited settings (e.g., dataset of 160 SLRs, only abstract-RQ comparisons) already show good level of agreement (rs = 0.67), for the best performing model, although at this point this is anecdotal since more comprehensive tests are required.

6   Conclusion and Future Work

We have seen that planning is a challenging endeavor, requiring resources, expertise or context that is often missing when undertaking a new topic, when performing it for the first time, or when resources are lacking – all typical scenarios in research. This paper shows that cognitive processes provide the ingredients to address these issues and support researchers in this often overlooked but impactful phase. As for ongoing and future work, we are in the process of refining the technical details of the first experiment, and planning an evaluation with end-users so as to assess the actual benefits of the approach when compared to standard tools. We also continue with our human-centered design approach to the development of the overall tool and algorithms, which will inform all components of the platform.

Acknowledgments. We acknowledge CSIRO Data61 for funding scholarship on this research.

References

1. Báez, M., et al.: Gelee: cooperative lifecycle management for (composite) artifacts. In: Baresi, L., Chi, C.-H., Suzuki, J. (eds.) ICSOC/ServiceWave -2009. LNCS, vol. 5900, pp. 645–646. Springer, Heidelberg (2009). https://doi.org/10.1007/978-3-642-10383-4_50
2. Barukh, M.C., et al.: Cognitive augmentation in processes. In: Aiello, M., Bouguettaya, A., Tamburri, D.A., van den Heuvel, W.-J. (eds.) Next-Gen Digital Services. A Retrospective and Roadmap for Service Computing of the Future. LNCS, vol. 12521, pp. 123–137. Springer, Cham (2021). https://doi.org/10.1007/978-3-030-73203-5_10
3. Beheshti, A., Benatallah, B., Nouri, R., Chhieng, V.M., Xiong, H.T., Zhao, X.: Coredb: a data lake service. In: Proceedings of the 2017 ACM on Conference on Information and Knowledge Management, pp. 2451–2454 (2017)
4. Bojanowski, P., Grave, E., Joulin, A., Mikolov, T.: Enriching word vectors with subword information. Trans. Assoc. Comput. Linguist. **5**, 135–146 (2017)
5. Kenneth Ward Church: Word2vec. Nat. Lang. Eng. **23**(1), 155–162 (2017)
6. Créquit, P., Trinquart, L., Yavchitz, A., Ravaud, P.: Wasted research when systematic reviews fail to provide a complete and up-to-date evidence synthesis: the example of lung cancer. BMC Med. **14**(1), 8 (2016)
7. Garousi, V., Felderer, M.: Experience-based guidelines for effective and efficient data extraction in systematic reviews in software engineering. In: Proceedings of the 21st International Conference on Evaluation and Assessment in Software Engineering, pp. 170–179 (2017)
8. Hassler, E., Carver, J.C., Hale, D., Al-Zubidy, A.: Identification of SLR tool needs-results of a community workshop. Inf. Software Technol. **70**, 122–129 (2016)
9. Howard, B.E., et al.: Swift-review: a text-mining workbench for systematic review. Syst. Rev. **5**(1), 87 (2016)
10. Huang, E.H., Socher, R., Manning, C.D., Ng, A.Y.: Improving word representations via global context and multiple word prototypes. In: Proceedings of the 50th Annual Meeting of the Association for Computational Linguistics (Volume 1: Long Papers), pp. 873–882 (2012)

11. Kitchenham, B.: Procedures for performing systematic reviews. Keele, UK, Keele University **33**(2004), 1–26 (2004)
12. Kontonatsios, G., et al.: A semi-supervised approach using label propagation to support citation screening. J. Biomed. Inform. **72**, 67–76 (2017)
13. Krivosheev, E., Casati, F., Baez, M., Benatallah, B.: Combining crowd and machines for multi-predicate item screening. Proc. ACM Hum.-Comput. Interact. **2**(CSCW), 1–18 (2018)
14. Lopes, A.A., Pinho, R., Paulovich, F.V., Minghim, R.: Visual text mining using association rules. Comput. Graph. **31**(3), 316–326 (2007)
15. Manning, C.D., Surdeanu, M., Bauer, J., Finkel, J.R., Bethard, S., McClosky, D.: The Stanford Corenlp natural language processing toolkit. In: Proceedings of 52nd Annual Meeting of the Association for Computational Linguistics: System Demonstrations, pp. 55–60 (2014)
16. Marshall, I.J., Kuiper, J., Wallace, B.C.: Robotreviewer: evaluation of a system for automatically assessing bias in clinical trials. J. Am. Med. Inform. Assoc **23**(1), 193–201 (2016)
17. Mikolov, T., Chen, K., Corrado, G., Dean, J.: Efficient estimation of word representations in vector space. arXiv preprint arXiv:1301.3781 (2013)
18. Mikolov, T., Sutskever, I., Chen, K., Corrado, G.S., Dean, J.: Distributed representations of words and phrases and their compositionality. In Advances in Neural Information Processing Systems, pp. 3111–3119 (2013)
19. Ouzzani, M., Hammady, H., Fedorowicz, Z., Elmagarmid, A.: Rayyan-a web and mobile app for systematic reviews. Syst. Control Found. Appl. **5**(1), 210 (2016)
20. Palomino, M., Dávila, A., Melendez, K.: Methodologies, methods, techniques and tools used on SLR elaboration: a mapping study. In: Mejia, J., Muñoz, M., Rocha, Á., Peña, A., Pérez-Cisneros, M. (eds.) CIMPS 2018. AISC, vol. 865, pp. 14–30. Springer, Cham (2019). https://doi.org/10.1007/978-3-030-01171-0_2
21. Pennington, J., Socher, R., Manning, C.D.: Glove: global vectors for word representation. In: Proceedings of the 2014 Conference on Empirical Methods in Natural Language Processing (EMNLP), pp. 1532–1543 (2014)
22. Sampson, M., Shojania, K.G., Garritty, C., Horsley, T., Ocampo, M., Moher, D.: Systematic reviews can be produced and published faster. J. Clin. Epidemiol. **61**(6), 531–536 (2008)
23. Shi, B., Lam, W., Jameel, S., Schockaert, S., Lai, K.P.: Jointly learning word embeddings and latent topics. In: Proceedings of the 40th International ACM SIGIR Conference on Research and Development in Information Retrieval, pp. 375–384 (2017)
24. Speer, R., Chin, J., Havasi, C.: Conceptnet 5.5: an open multilingual graph of general knowledge. arXiv preprint arXiv:1612.03975 (2016)
25. Sun, Y., et al.: Crowdsourcing information extraction for biomedical systematic reviews. arXiv preprint arXiv:1609.01017 (2016)
26. Vaish, R., et al.: Crowd research: open and scalable university laboratories. In: Proceedings of the 30th Annual ACM Symposium on User Interface Software and Technology, pp. 829–843 (2017)

Security Professional Skills Representation in Bug Bounty Programs and Processes

Sara Mumtaz[1]([✉]), Carlos Rodriguez[2]([✉]), and Shayan Zamanirad[1]

[1] School of Computer Science and Engineering, UNSW Sydney,
Kensington, NSW 2052, Australia
{s.mumtaz,shayan.zamanirad}@unsw.edu.au
[2] Universidad Católica Nuestra Señora de la Asunción, Asunción, Paraguay
carlos.rodriguez@uc.edu.py

Abstract. The ever-increasing amount of security vulnerabilities discovered and reported in recent years are significantly raising the concerns of organizations and businesses regarding the potential risks of data breaches and attacks that may affect their assets (e.g. the cases of Yahoo and Equifax). Consequently, organizations, particularly those suffering from these attacks are relying on the job of security professionals. Unfortunately, due to a wide range of cyber-attacks, the identification of such skilled security professional is a challenging task. One such reason is the "skill gap" problem, a mismatch between the security professionals' skills and the skills required for the job (vulnerability discovery in our case). In this work, we focus on platforms and processes for crowdsourced security vulnerability discovery (bug bounty programs) and present a framework for the representation of security professional skills. More specifically, we propose an embedding-based clustering approach that exploits multiple and rich information available across the web (e.g. job postings, vulnerability discovery reports) to translate the security professional skills into a set of relevant skills using clustering information in a semantic vector space. The effectiveness of this approach is demonstrated through experiments, and the results show that our approach works better than baseline solutions in selecting the appropriate security professionals.

Keywords: Bug bounty programs and processes · Skills representation · Embeddings models · Ethical hackers · Cyber security

1 Introduction

The advancement in the Web 2.0 technology and its widespread use in virtually all types of businesses has increasingly exposed us to security threats and cyber-attacks during the last years. These attacks result in several security breaches events targeting not only individuals but giant organizations including the US

© Springer Nature Switzerland AG 2021
H. Hacid et al. (Eds.): ICSOC 2020 Workshops, LNCS 12632, pp. 334–348, 2021.
https://doi.org/10.1007/978-3-030-76352-7_33

Department of Defence[1], JP Morgan[2] and many more. Perhaps, among these, the most notable is the Equifax data breach, which exposed the sensitive information of 147 million people, with an estimated settlement of ~650 million US dollars[3].

In response to these security breaches, organizations are increasingly relying on security professionals (SecPros) and investing in their services through security crowdsourcing platforms and processes (i.e. bug bounty programs) to find and address security vulnerabilities [2]. A bug bounty program offers rewards to external parties (through crowdsourcing) allowing them to perform a security assessment of their assets (e.g. software, hardware) [9].

These bug bounty programs are a useful complement to existing internal security programs and widely accepted by organizations [19]. Additionally, due to the nature of crowdsourcing, organizations are benefitting from its speed and the vast pool of available SecPros with diverse skills and expertise. For instance, one study [31] found that through these outsourced programs, a greater number of vulnerabilities can be found, and more quickly compared to the time required by in-house testers making the process more time- and cost-effective. However, despite the large pool of these SecPros, there is a lack of sufficiently skilled cybersecurity professionals [26]. For example, the MIT Technology Review[4] and Cybersecurity Venture[5] predicted that the demand for cybersecurity professionals is expected to increase by 350%, from one million in 2013 to 3.5 million in 2021.

There could be many reasons for skills shortages, one of the main ones being the "skills gap" problem [7], that is, a mismatch between the skills of security professionals and the skills required for a particular job (vulnerability discovery in our context). Secondly, the different types of vulnerabilities require different levels of skills and expertise [2]. For example, Web application vulnerabilities require knowledge about the software itself, networking protocols, Web frameworks, and vulnerabilities that target Web technologies.

To address the aforementioned challenges, it is essential that SecPros get selected for tasks based on their skills. In turn, having the right SecPros assigned to tasks contributes to making bug bounty programs and processes successful. In this context, we propose an embedding-based clustering technique, which translates the *SecPro skills* into a set of relevant skills using clustering information in the semantic space. Firstly, the data related to SecPros skills is collected from heterogeneous, multiple sources and grouped them as semantically correlated clusters in an embedding space using clustering algorithms [15]. Then, when a vulnerability discovery task is presented, it is placed (vectorized) in the same embedding space as the skills. Lastly, the cosine distance between clusters of

---

[1] https://thehill.com/policy/cybersecurity/483853-defense-department-agency-suffers-potential-data-breach.

[2] https://www.theguardian.com/business/2014/oct/02/jp-morgan-76m-households-affected-data-breach.

[3] https://www.nytimes.com/2019/07/22/business/equifax-settlement.html.

[4] https://www.technologyreview.com/s/612309/a-cyber-skills-shortage-means-students-are-being-recruited-to-fight-off-hackers/.

[5] https://cybersecurityventures.com/.

skills and a task vector is computed to either recommend a set of skills, or SecPros for the task. The core of our approach is the representation of a task and skills in the same embedding space, which helps to mitigate the "skill gap" problem.

The rest of the paper is structured as follows. Section 2 introduces our approach to representing security professionals' skills. The experiments and evaluations are presented in Sect. 3. Section 4 provides background information and related work on general and crowdsourced approaches for skills extraction and representation. Finally, Sect. 5 provides concluding remarks and future work.

2   Representing Security Professionals' Skills and Recommending to Vulnerability Discovery Task

This section presents our proposal for representing SecPros skills and recommending them to vulnerability discovery task. Figure 1 presents an outline of our framework.

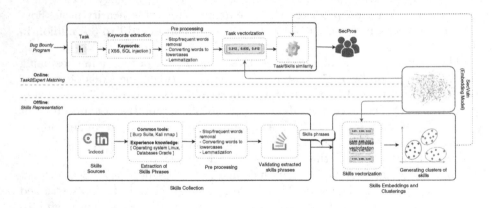

Fig. 1. Overall framework

Our proposed framework exploits the heterogeneous information available across the Web such as job postings, resumes from job search portals, and complements these with other notable sources such as the skills declared on SecPros profiles across different platforms (e.g. Cobalt). More precisely, our approach consists of two phases (see Fig. 1):

– **Skills Representation.** This phase collects the SecPros' skills related information scattered across the Web. Then, by leveraging the property of word embeddings [20], we represent them in a semantic space via clustering. The generated skills clusters are then stored offline for further use by the following phase.

– **Task-SecPro Matching.** This phase represents the vulnerability discovery task in the same space as the skills (built in the previous phase). Then, this representation is matched with either expertise of SecPros or skills to find appropriate SecPros for the task.

In the following sections, we discuss in more details each of these two phases.

2.1   Skills Representation

Skills Collection. Skills collection consists of four steps: (i) Identification of skills sources, (ii) extraction of skills from identified sources, (iii) pre-processing, and (iv) normalization and validation of the extracted skills.

(i) Identification of Skills Sources. Skills can be extracted from multiple sources, including job postings (e.g. job descriptions and requirements section) from job portals (e.g. CareerBuilder.com), technical/skills sections in online resumes (e.g. indeed.com), and self-declared skills list from various platforms (e.g. LinkedIn, Cobalt). In this work, we prefer to use "skill phrase" (also called n-grams [3]), considering that skills are often made up of multiple-words (e.g. "penetration testing", "source code review"). After the identification of skills sources, the next step involves the extraction of skill-related phrases.

(ii) Extraction of Skills Phrases. The literature offers several techniques for the extraction of skills phrases. For example, [13] used Term-Frequency Inverse-Document-Frequencey (TF-IDF) [5] to extract relevant and essential keywords from job descriptions and resumes. Likewise, LinkedIn argued in [3] that users on LinkedIn use a comma-separation technique to provide a skills list in the "skills and expertise" section (e.g. "Java", "SQL", "Reinforcement Learning"). They utilized the comma-separated technique for identification and extraction of skills phrases. Similarly, we use TF-IDF and topic modelling [4] techniques to extract the important keywords representing the skills set of SecPros from notable sources (i.e. vulnerability discovery report, job descriptions). Furthermore, we also utilize the comma-separation technique and Web scrapping methods when necessary (e.g. in case of the self-declared skills list).

(iii) Pre-Processing. We apply basic text pre-processing techniques to make our collected skill phrases available for further processing. These techniques include the removal of stop words, converting the whole dataset to lowercase and lemmatization. More importantly, frequently occurring words (e.g. knowledge, proficient, team-oriented in job requirements), are discarded as they can act as outliers and make the skills data noisy [14].

(iv) Normalizing and Validating the Extracted Skills Phrases. The goal of this step is to retain the valid skills phrases and discard any other keywords that are not valid skills phrases. As mentioned previously, the skills phrases are human-generated (job postings, resumes), and everyone has different ways of expressing them (i.e. different representations of the same concept/skill). For example, some may prefer to write a vulnerability type as "XSS", and others

may write it as "Cross-Site Scripting". As a result, there could be a great deal of redundancy in the users' skills set.

We apply a normalization technique to express them in a standard (base) form. An example of a base form would be *penetration testing, pen testing, and pen test* into *penetration testing*. However, the lemmatization is usually done through Wordnet [21], which is a general-purpose database and, as expected, does not have specialized terminology. Likewise, there are several skills knowledge bases available to validate skill phrases, such as O*Net (used by US public recruitment services) [6], and ESCO (a European skills taxonomy)[6]. Nevertheless, all these skills knowledge bases are for general purpose recruitment and do not necessarily contain terminology that is specific to cyber security domain. To tackle this problem, we utilize Wikipedia open search [14], and tags (Stack Overflow and Stack Exchange), and also rely on keywords from our previous work dataset [24] and other cybersecurity domain-specific sources (e.g. National Institute of Standard and Technology (NIST)[7]).

Skills Representation via Clustering. This step involves a semantic representation of skills phrases to reduce the skill gap problem. To do so, we present an embedding-based clustering method. Embedding models, more precisely, word embeddings, generate a dense, continuous, low-dimensional representation of words from the raw corpus in an unsupervised way [20]. The words (in our case, skills phrases) that have a similar context or semantics have close embeddings in the vector space. These vectors of skills phrases are further represented using clusters so that similar and semantically coherent skills should be in the same cluster. The assumption is that, since word embeddings span a semantic space, the clusters based on word embeddings would give a higher semantic space for the skills phrases [8].

Clustering. A cluster is a collection of items that are similar to each other and dissimilar to other clusters' items [15]. Clustering is essentially an unsupervised, machine learning method and is mainly used to classify unlabeled data. Examples of applications of clustering include text analysis, pattern recognition, segmentation (image processing) and collaborative filtering. Recently, it has been used successfully to represent taxonomies for topics based on academic papers [29] and experts finding [8].

Generating Clusters. Given m number of skill phrases $S = \{s_1, s_2, \ldots .s_m\}$, we utilize our cyber security vulnerability word embedding (SecVuln) [24] to generate a vector representation for each skill phrase. Then, we apply a clustering algorithm, specifically hierarchical clustering [8] to group them into k clusters, that is, $C = \{C_1, C_2...C_k\}$ (e.g. $C_1 = \{burpsuite, kalilinux, nmap, metasploit\}$) such that semantically correlated skill phrases belong to the same cluster.

SecPro Expertise Representation. This step represents SecPro expertise to match it with a vulnerability discovery task. Using statistical language modeling [23], the expertise and skills of SecPros can be inferred from their relevant

---

[6] https://ec.europa.eu/esco/portal/skill.
[7] https://csrc.nist.gov/glossary.

documents (e.g. email communications or answers in Q&A web sites). In our context, vulnerability discovery reports and self-declared skills of SecPros are an excellent illustration of their expertise. However, as previously mentioned, self-declared skills listed in profiles are human-generated and therefore prone to incompleteness or bias. Therefore, after the initial collection of skills phrases, we enriched it with the discovered vulnerabilities given in SecPros profiles.

Next, we leverage the clusters generated in the previous step to represent SecPros in a cluster form. The purpose of this step is to recognize the unspecified skills of SecPros. To do so, the skills phrases of each SecPro are matched with the clusters of skills phrases using a simple keyword-matching algorithm. It is worth mentioning that the matching takes place at a certain threshold (e.g. if 50% of skills phrases are matched, then a cluster is chosen, otherwise discarded). The clusters are further aggregated using vector averaging technique [23] to represent the cluster as a vector. Unlike the result of skills phrases' vector averaging, the vector average of a clusters gives more accurate result as shown later in our experiments reported in Sect. 3, having the advantage of being semantically similar to each other.

For instance, skills phrases extracted from a vulnerability report would consist of keywords with different semantics (e.g. "Persistent XSS via filename in projects", a title of a vulnerability discovery report on HackerOne[8]). However, the skills phrases within a cluster are already related to each other, and hence would be more useful in accurately matching SecPros with tasks. The representations of the selected cluster vectors $\overrightarrow{C} = \left\{ \overrightarrow{C_1}, \overrightarrow{C_2}...\overrightarrow{C_k} \right\}$ are then stored offline as a distribution over skills.

2.2 Task-SecPro Matching

The purpose of this phase is to recommend either SecPros to the given task or skill phrases to the given task. Upon the arrival of a task to the crowdsourcing platform (e.g. HackerOne[9]), we perform similar pre-processing and keywords extraction (from the description of task) as in the previous phase to obtain a list of keywords T. Then, we leverage the word embedding model to generate a vector representation $\overrightarrow{T}$ for the task based on the extracted keywords T.

After obtaining the vector $\overrightarrow{T}$, the task matching between $\overrightarrow{T}$ and the clusters of skills phrases $\overrightarrow{C}$ takes place using cosine similarity [5], which is defined as follows:

$$sim(T, C) = \frac{\overrightarrow{T}.\overrightarrow{C}}{|\overrightarrow{T}||\overrightarrow{C}|}$$

The similarity score ranges from $[-1, 1]$, where the closer the value to 1 the more relevant to the task is to the expertise of a SecPro.

---

[8] https://hackerone.com/reports/662204.
[9] https://www.hackerone.com/.

3   Experiments

In this section, we present the experimental results of our approach using the following evaluation techniques.

- *Validation of cluster quality:* To examine how closely the skills phrases are related to each other within the cluster.
- *Validation based on information retrieval:* To determine the effectiveness of our approach in selecting the appropriate SecPros for a given task.

3.1   Dataset

In this work, we collected data from popular job search portals such as *indeed.com*[10] and *monster.com*[11] with cyber security jobs related query (e.g. "penetration testing", "code reviews") [26]. The collected data is further enriched with vulnerability discovery reports from HackerOne. Specifically, we focused on the section where the required skills are listed. Moreover, we utilized SecPros profiles on Cobalt for collecting self-declared skills along with the vulnerabilities they had discovered. The intuition is, if a set of skills and discovered vulnerabilities appear in the same profile (co-occurred), then they are important for each other.

Test Data. For test purposes, we select the vulnerability discovery tasks (e.g. Sony Vulnerability Discovery Program[12]) that are available on the HackerOne platforms. It is worth mentioning here, that during cluster generation we did not consider these tasks as a source, so that test data and training do not overlap.

Ground Truth. To examine how well our technique can determine the right SecPro for a given task, we need to have a ground truth for comparison (between the actual SecPros and the SecPros returned by our technique). To do so, we collected the profiles of top 100 SecPros from Cobalt[13]. Cobalt rank these SecPros according to the vulnerabilities that have discovered along with the quality of reports they submitted to the platform.

3.2   Embedding Model

We utilized the embedding model (SecVuln) built for the cybersecurity domain in our previous work [24]. However, to cope with the new terminologies in the job advertisements, we enriched our previous model with information extracted from job descriptions and resumes. We followed the same parameter settings as reported in [24].

---

[10] https://au.indeed.com/.
[11] https://www.monster.com/.
[12] https://hackerone.com/sony.
[13] https://app.cobalt.io/pentesters.

3.3   Evaluation

Comparison Method. In order to demonstrate the effectiveness of our proposed approach, we compared it with the a baseline approach, that is, the vectors averaging technique [23].

Evaluation Metrics. To determine the effectiveness of our proposed approach in terms of quality of clusters and retrieving the appropriate SecPro, we used the (i) silhouette index [15], and (ii) information retrieval measure such as Precision at N (P@N) [5].

(i) *Cluster Quality.* Embedding-based clustering is expected to learn coherent and semantically correlated skills phrases within the clusters to facilitate the semantic understanding of these phrases. Hence, we evaluate the coherence of clusters using a silhouette index [15]. The silhouette index indicates the compactness and separation of clusters. For example, a set of skills clusters represented by $C = C_1, C_2 \ldots C_k$, consists of n number of vectors; then, the silhouette index is given below:

$$S(C) = \frac{\frac{1}{n} \sum_{i=1}^{n} (b_i - a_i)}{max(a_i, b_i)}$$

where a_i denotes the average distance of skill i to other skills in the cluster, whereas b_i is the minimum of average distance of a skill b_i to other skills of clusters. The value of the silhouette index ranges from –1 to 1. A higher value represents a better quality of clusters. In our case, the result amounts to approximately 0.75, which indicates the quality of our clusters.

(ii) *Precision.* Precision is one of the widely used information retrieval measures for expert finding [23], which measures the percentage of correct results (relevant SecPro found) out of total results (total number of SecPros returned) from the system. Formally, let R_c and R_w represent correct (true positives) and wrong results (false positives) respectively. Then, precision is defined as $P = \frac{R_c}{R_c + R_w}$. Instead, Precision at N (P@N) is the percentage of relevant SecPros found at the top N retrieved, ranked results (e.g. P@5 shows the total relevant SecPros until 5).

Table 1 shows that the proposed clustering-based technique perform better compared to the baseline technique. The clustering technique has an advantage over the keywords' vector-averaging technique. For instance, vector averaging technique, combines all available keywords extracted from multiple sources (which may consist of skill phrases and other words)

3.4   Discussions and Limitations

The use of keywords other than skill phrases may add noise and lead to an inaccurate vector representation. The clustering-based approach presented in this paper groups the semantically related skills phrases, which helps in overcoming

Table 1. Task-to-SecPros Matching

Technique	P@5	P@10
Vector Averaging	0.55	0.45
Clustering (our proposal)	0.60	0.57

this problem. Furthermore, our proposed approach offers the following advantages:

Skills Representation. Skills representation can help educational institutions to address the skills gap between industry and current curriculum offerings (as these skills come from the 'hands-on expertise' of SecPros (ethical hackers)). For instance, organizations like NIST have already initiated a program called NICE (National Initiative for Cybersecurity Education)[14] to fill the gap; they can further leverage our work for improvement. Secondly, the organization can also benefit from this pool of skills; for example, they can train their internal security (testers) on a specific type of vulnerability like Web API vulnerability.

SecPros Expertise Representation. Moreover, the representation of SecPros' expertise can help crowdsourcing platforms, after launching bug bounty programs, to directly contact SecPros (mapping between task and SecPro expertise) and invite them to participate.

Limitations. Despite its advantages, our approach has limitations. For instance, to represent SecPros expertise, we rely on textual contents only, and moreover, only one source (i.e. self-declared profiles on Cobalt) is taken into account. This approach can be further improved by incorporating SecPros' social activities and their interactions on social networks (e.g. Twitter) [25] through network embedding.

Regarding the computing of SecPro ranking in terms of their expertise, we consider only one expertise signal (i.e. report quality on Cobalt). However, as mentioned in [23], "expertise" is an umbrella term and comprises many signals (e.g. SecPros certifications, platforms ranking, badges, hall of fame). Moreover, [2] conducted a comprehensive study and found different indicators such as certifications and number of the vulnerabilities discovered as signals of SecPro expertise. Our work can leverage that study and add more signals for computing the expertise.

SecPros data is scattered across the Web and different platforms provide different information (expertise signals). For instance, HackerOne discloses the reports submitted to their platform following their bug bounty policy (not every organization discloses its reports). BugCrowd, on the other hand, provides information about the type and severity of vulnerabilities discovered by SecPros. The key challenge here is to combine all those signals and information about a specific SecPro from different platforms. However, the prevalence of social platforms (LinkedIn and Twitter) and the presence of SecPros on these platforms

---

[14] https://www.nist.gov/itl/applied-cybersecurity/nice.

can mitigate this problem by using SecPros' social identifiers to recognize them on different platforms.

Moreover, we observed from experiments that the proposed clustering technique is prone to the problem of over-representation of users' skills and expertise. On the one hand, clustering helps in identifying any unspecified skills. However, some clusters list skills which are not necessarily a substitution of skills. For instance, the cluster defining the skills phrases indicates that there are different techniques for finding vulnerabilities; they do not need to have knowledge of all of them. As mentioned in [18], sometimes they prefer low hanging fruit and finding vulnerabilities and utilize the tools they already have.

4 Related Work

Our work in this direction inherits a rich ecosystem of commercial job search platforms and general skills modeling techniques and draws on the insights offered by previous works in regard to the selection of workers in security crowdsourced platforms (bug bounty).

4.1 General Approaches for Skills Extraction and Representation

One of the most challenging tasks for any employer is the hiring of new people from a large pool of job applications. [16] developed a system, Elisit (Expertise Localization from Informal Sources and Information Technologies), that peruses data from Wikipedia and LinkedIn to extract skills from text documents. The authors claim that their approach could be easily integrated with any skills search engine or HR automation in any automatic meta-data extraction systems.

However, the self-declared skills (e.g. those explicitly given in the LinkedIn profile) may be incomplete or biased. To address this problem, [27] introduced approaches to analyze individuals' communication data (e.g. emails, discussion forums) to infer their skills. [28] also utilized personal skill information derived from social media platforms (e.g. Twitter) for skills inferences. They proposed a joint prediction factor graph model to infer user skills automatically from their connections on social networks.

Commercial Based Approaches. Several works address the skills representation in commercial job search portals for talent search using their built-in systems [11,14]. Some of the works from notable job search portals (e.g. CareerBuilder and LinkedIn) are described below.

CareerBuilder. To overcome the "skill gap" in the labor market, CareerBuilder (US most prominent human capital solution) [14,30] presented an in-house skill terms extraction system, SKILL, for the extraction of keywords (aka skills) from both job descriptions and users' resumes. More specifically, in this work [14], the authors assumed the contents of individuals' resumes (technical section) and job ads (descriptions) as indicative of specific skills. They utilized a well-known algorithm, Word2vec [20] with the assumption that related skills are likely to appear in the same documents (resumes and job ads). For instance, "Python" would always be a *programming language* in their system instead of a *snake*. This work has achieved almost 91% accuracy and 76% recall, and the system is successfully deployed in multiple business intelligence projects.

As an improvement on their previous work, the authors [32] quantified the relevance of the skills to the job titles. To do so, they used a simple yet effective technique, TF-IDF (term frequency and inverse document frequency) [5], to measure the skills-job title relationship, assuming that a particular skill is important if it constitutes part of the job title.

In further work, they proposed a representation learning [7] to jointly represent job titles and skills in the same vector space for skills to skill similarity via three networks/graphs (i.e. job skill graph, job transition graph, and skills co-occurrences graph). These graph are constructed using skills (nodes) from the same resume. For example, an edge is formed between skills (e.g. Data mining and Machine learning) if they both appear in the same resume. Likewise, they extended this work and proposed [17] a tripartite vector representation of job posting (i.e. job, skills and location) for a better job recommendation. The vector representation of job title and the skills required for that job are added to a personalized vector for a specific position in one vector representation. Then, this vector is further concatenated with the location vector, and is currently being used within CareerBuilder.

LinkedIn. LinkedIn is the world's largest professional online social network with 500 million users profiles, indicating their professional identity. Their talent search system is widely used by job seekers and employers and generates approximately 65% of company revenue [11]. LinkedIn presented [3] "Skills and Expertise" feature as a part of their current system. They built a folksonomy (often used for categorization of contents) using a data-driven approach.

To further improve their in-house system, LinkedIn introduced another technique [11] to address the problem of personalized expertise search. More specifically, this work utilized collaborative filtering and matrix factorization techniques to infer the member' skills and expertise from the existing set of skills. Next, they combined these skills with other personalized (e.g. location, social connections) and non-personalized (e.g. textual contents) features to rank members accordingly against the query.

4.2   Workers (SecPro) Selection in Bug Bounty

As previously mentioned, bug bounty programs inherit all the properties of crowdsourced platforms [19]. Hence, they have implemented the same strategies for crowd/SecPro selection as those used by general crowdsourced platforms, such as qualification tests [1] The qualification test is a pre-selection criterion used to screen potential workers. It is used to assess the level of expertise of SecPros before recruiting them for the real task of vulnerability discovery. Like general crowdsourced platforms (e.g. Amazon Mechanical Turk), the bug bounty platforms also ask SecPros to correctly answer the questions with already-known solutions. For instance, [10] developed a conceptual expertise tool that relies on a set of questions to distinguish a novice from an expert SecPros. However, it relies on the self-declared skills and assessment of the expertise of the SecPro. Similarly, Synack[15], a crowdsourced vulnerability discovery platform, evaluates the SecPros through written and practical tests to ensure that candidates are eligible to join the platform. Likewise, Upwork[16], an online freelancer market, assesses the competency of the freelancer using prior knowledge like certification and then determines the skills via online testing. Apart from the preliminary tests, some organizations may also impose specific predefined criteria (e.g. eligibility) for participation in the task. For instance, Mozilla bug bounty[17] do not allow their own employees to participate in any of their bug bounty programs.

Furthermore, some of the bug bounty platforms (e.g. HackerOne, BugCrowd) maintain the SecPro' profiles utilizing their details (e.g. certifications) and ongoing activities (e.g. number of vulnerabilities they have discovered, relative ranking, and any reward they received) on the platforms. After launching bug bounty programs, organizations may invite the top SecPros (the top 100, for example) to participate.

Several studies have been conducted for worker/people selection in general crowdsourcing platforms. However, to the best of our knowledge, we did not come across any such work for security crowdsourced platforms (bug bounty) other than empirical studies. For example, [31] performed an empirical study to determine the characteristics of SecPros. Their study focuses on the tools and methods used by SecPros for discovering vulnerabilities and the type of vulnerability is common in the community. They determined how SecPros approach vulnerability discovery task. However, their study did not explore the criteria for SecPros' expertise indicators to accomplish the task. On the other hand, [12] investigated the heterogeneity among the SecPros participating in crowdsourced vulnerability discovery tasks. The authors discovered that there are two different types of SecPros participating in crowdsourced vulnerability discovery. Most SecPros are non-project-specific (i.e. submit vulnerabilities to multiple tasks) and are different from traditional SecPros who work on specific projects (i.e. submit vulnerabilities only to tasks that they are interested in making the software secure). However, unlike the previous approaches, [1] conducted a comprehensive empirical study to determine SecPros expertise indicators to improve the quality of software vulnerability discovery.

Keeping the limitations of previous works in mind, our study aimed to propose computational techniques for skills representation and task matching for crowdsourced vulnerability discovery platforms and processes (bug bounty programs).

---

[15] https://www.synack.com/red-team/.

[16] https://www.upwork.com/.

[17] https://www.mozilla.org/en-US/security/bug-bounty/.

5   Conclusion

In this paper, we addressed the skills gap problem in the context of platforms and processes for crowdsourced vulnerability discovery by proposing a word embedding-based clustering method for skill representation. The key to our approach is the representation of skills phrases and task keywords in the same semantic space to minimize any differences and offer the best mapping between them. To this end, by combining different and multiple skills-related information, we create an embedding space that incorporates the syntactic and semantic relationship between skills, SecPros expertise and vulnerability discovery tasks. The clustering algorithm further grouped them in semantically correlated groups. Furthermore, we have conducted experiments that demonstrate the effectiveness of our approach in finding the promising SecPros for vulnerability discovery tasks. These encouraging results open up opportunities for improving people-to-task assignment in crowdsourced vulnerability discovery processes and programs. Directions for future work include the use of additional sources that can help improve our skills representation model as well as the integration of other indicators as identified in [1].

Acknolwedgement. This research was done in the context of the first author's Ph.D. thesis [22]. We thank Scientia Prof. Boualem Benatallah for the useful feedbacks provided on this work.

References

1. Al-Banna, M., Benatallah, B., Barukh, M.C.: Software security professionals: expertise indicators. In: 2016 IEEE 2nd International Conference on Collaboration and Internet Computing (CIC), pp. 139–148 (2016)
2. Al-Banna, M., Benatallah, B., Schlagwein, D., Bertino, E., Barukh, M.C.: Friendly hackers to the rescue: how organizations perceive crowdsourced vulnerability discovery. In: PACIS, p. 230 (2018)
3. Bastian, M., et al.: Linkedin skills: large-scale topic extraction and inference. In: Proceedings of the 8th ACM Conference on Recommender Systems, pp. 1–8 (2014)
4. Blei, D.M., Ng, A.Y., Jordan, M.I.: Latent Dirichlet allocation. J. Mach. Learn. Res. **3**, 993–1022 (2003)
5. Christopher, D.M., Prabhakar, R., Hinrich, S.: Introduction to information retrieval. Int. Inf. Retrieval **151**(177), 5 (2008)
6. Council, N.R., et al.: A database for a changing economy: review of the Occupational Information Network (O* NET). National Academies Press (2010)
7. Dave, V.S., Zhang, B., Al Hasan, M., AlJadda, K., Korayem, M.: A combined representation learning approach for better job and skill recommendation. In: Proceedings of the 27th ACM International Conference on Information and Knowledge Management, pp. 1997–2005. ACM (2018)
8. Dehghan, M., Abin, A.A.: Translations diversification for expert finding: a novel clustering-based approach. ACM Trans. Knowl. Discov. Data (TKDD) **13**(3), 1–20 (2019)

9. Finifter, M., Akhawe, D., Wagner, D.: An empirical study of vulnerability rewards programs. In: Proceedings of the 22nd USENIX conference on Security, pp. 273–288 (2013)

10. Giboney, J.S., Proudfoot, J.G., Goel, S., Valacich, J.S.: The security expertise assessment measure (seam): developing a scale for hacker expertise. Comput. Secur. **60**, 37–51 (2016)

11. Ha-Thuc, V., et al.: Search by ideal candidates: next generation of talent search at linkedin. In: Proceedings of the 25th International Conference Companion on World Wide Web, pp. 195–198 (2016)

12. Hata, H., Guo, M., Babar, M.A.: Understanding the heterogeneity of contributors in bug bounty programs. In: 2017 ACM/IEEE International Symposium on Empirical Software Engineering and Measurement (ESEM), pp. 223–228. IEEE (2017)

13. Hughes, S.: How we data-mine related tech skills (2015). https://insights.dice.com/2015/03/16/how-we-data-mine-related-tech-skills/?ads_kw=idf

14. Javed, F., Hoang, P., Mahoney, T., McNair, M.: Large-scale occupational skills normalization for online recruitment. In: Twenty-Ninth IAAI Conference (2017)

15. Kaufman, L., Rousseeuw, P.J.: Finding Groups in Data: An Introduction to Cluster Analysis, vol. 344. Wiley, New York (2009)

16. Kivimäki, I., et al.: A graph-based approach to skill extraction from text. In: Proceedings of TextGraphs-8 Graph-Based Methods for Natural Language Processing, pp. 79–87 (2013)

17. Liu, M., Wang, J., Abdelfatah, K., Korayem, M.: Tripartite vector representations for better job recommendation. arXiv preprint arXiv:1907.12379 (2019)

18. Maillart, T., Zhao, M., Grossklags, J., Chuang, J.: Given enough eyeballs, all bugs are shallow? revisiting eric raymond with bug bounty programs. J. Cybersecur. **3**(2), 81–90 (2017)

19. Malladi, S.S., Subramanian, H.C.: Bug bounty programs for cybersecurity: practices, issues, and recommendations. IEEE Software **37**(1), 31–39 (2019)

20. Mikolov, T., Chen, K., Corrado, G., Dean, J.: Efficient estimation of word representations in vector space. arXiv preprint arXiv:1301.3781 (2013)

21. Miller, G.A.: Wordnet: a lexical database for English. Commun. ACM **38**(11), 39–41 (1995)

22. Mumtaz, S.: People selection for crowdsourcing tasks: representational abstractions and matching techniques. Ph.D. thesis, School of Computer Science and Engineering, Faculty of Engineering, UNSW Sydney (2020)

23. Mumtaz, S., Rodriguez, C., Benatallah, B.: Expert2vec: experts representation in community question answering for question routing. In: International Conference on Advanced Information Systems Engineering, pp. 213–229 (2019)

24. Mumtaz, S., Rodriguez, C., Benatallah, B., Al-Banna, M., Zamanirad, S.: Learning word representation for the cyber security vulnerability domain. In: 2020 International Joint Conference on Neural Networks (IJCNN), pp. 1–8. IEEE (2020)

25. Mumtaz, S., Wang, X.: Identifying top-k influential nodes in networks. In: the 26th ACM International Conference on Information and Knowledge Management, pp. 2219–2222 (2017)

26. Potter, L.E., Vickers, G.: What skills do you need to work in cyber security?: a look at the australian market. In: Proceedings of the 2015 ACM SIGMIS Conference on Computers and People Research, pp. 67–72 (2015)

27. Shankaralingappa, D.M., De Fransicsi Morales, G., Gionis, A.: Extracting skill endorsements from personal communication data. In: Proceedings of the 25th ACM International on Conference on Information and Knowledge Management, pp. 1961–1964 (2016)
28. Wang, Z., Li, S., Shi, H., Zhou, G.: Skill inference with personal and skill connections. In: Proceedings of COLING 2014, the 25th International Conference on Computational Linguistics: Technical Papers, pp. 520–529 (2014)
29. Zhang, C., et al.: Taxogen: unsupervised topic taxonomy construction by adaptive term embedding and clustering. In: Proceedings of the 24th ACM SIGKDD International Conference on Knowledge Discovery & Data Mining, pp. 2701–2709 (2018)
30. Zhao, M., Javed, F., Jacob, F., McNair, M.: Skill: a system for skill identification and normalization. In: Twenty-Seventh IAAI Conference (2015)
31. Zhao, M., Grossklags, J., Liu, P.: An empirical study of web vulnerability discovery ecosystems. In: Proceedings of the 22nd ACM SIGSAC Conference on Computer and Communications Security, pp. 1105–1117 (2015)
32. Zhou, W., Zhu, Y., Javed, F., Rahman, M., Balaji, J., McNair, M.: Quantifying skill relevance to job titles. In: 2016 IEEE International Conference on Big Data (Big Data), pp. 1532–1541. IEEE (2016)

Stage-Based Process Performance Analysis

Chiao-Yun Li[1][✉], Sebastiaan J. van Zelst[1,2][✉],
and Wil M. P. van der Aalst[1,2][✉]

[1] Fraunhofer FIT, 53754 Sankt Augustin, Germany
chiao-yun.li,sebastiaan.van.zelst,wil.van.der.aalst}@fit.fraunhofer.de
[2] RWTH Aachen University, Templergraben 55, 52062 Aachen, Germany
{s.j.v.zelst,wvdaalst}@pads.rwth-aachen.de

Abstract. Process performance mining utilizes the event data generated and stored during the execution of business processes. For the successful application of process performance mining, one needs reliable performance statistics based on an understandable representation of the process. However, techniques developed for the automated analysis of event data typically solely focus on one aspect of the aforementioned requirements, i.e., the techniques either focus on increasing the analysis interpretability or on computing and visualizing the performance metrics. As such, obtaining performance statistics at the higher level of abstraction for analysis remains an open challenge. Hence, using the notion of process stages, i.e., high-level process steps, we propose an approach that supports human analysts to analyze the performance at the process-stage-level. An extensive set of experiments shows that our approach, without much effort from users, supports such analysis with reliable results.

Keywords: Process performance analysis · Event abstraction · Process stage · Performance visualization

1 Introduction

The goal of all the business is to maximize the return on investment, which can be realized through improving the efficiency of their business processes [10]. Analyzing process efficiency enables companies to locate and diagnose the causes of the bottlenecks in order to optimize workload scheduling and the distribution of resources. *Process mining* is a technology that empowers companies to analyze a process by exploiting an event log, i.e., records of events executed during the execution of a process [20]. *Process performance mining* is a subfield that focuses on the process performance, often referred to as the *time* dimension, to identify and diagnose the inefficiencies during the operation of a business process [4].

The effectiveness of the analysis of the process performance depends on two aspects, i.e., interpretability of the results as well as the reliability of the performance metrics provided. To achieve this, most process performance mining

H. Hacid et al. (Eds.): ICSOC 2020 Workshops, LNCS 12632, pp. 349–364, 2021.
https://doi.org/10.1007/978-3-030-76352-7_34

techniques project the performance information on a discovered or predefined process model [5,9,21]. However, as the behavior in a process becomes more complex, the results may be no longer reliable and interpretable for human analysts. For example, as shown in Fig. 1, each of the bottlenecks highlighted only occurs once and the graph is difficult to read. As such, diagnosis remains difficult.

Fig. 1. Performance projection at activity level using Disco [5].

Problem Statement. Consider the medical domain, where specific activities are performed *prior*, *during*, and *after* the surgery. In some cases, the exact scheduling of activities within such a *part* of the process might be arbitrary, leading to the complex behavior. Such complexity results from the original level of granularity of the process, i.e., activities. As shown in Fig. 1, the complexity leads to the unreliable diagnosis due to the noninterpretable results and the relatively low frequency of the bottlenecks.

Moreover, an effective diagnosis of the inefficiencies in a process requires the context of cases. Consider the same example of performing a surgery in the medical domain. Assume that the average duration of conducting a surgery (*during*) is 5 h and preparation for a surgery (*prior*) is 15 min. Suppose that the efficiency of conducting a surgery depends on how much the necessary materials or information are prepaid for the surgery in advance. Without the performance statistics presented in the context of a case, the bottleneck may be identified as performing the surgery. However, the cause of the bottleneck, i.e., how well the surgery is prepared in advance, may not be diagnosed. The paper aims to enable analysts to locate and diagnose the inefficiencies in complex processes by addressing the following questions:

Q1. What is the major *part* in a complex process that forms the bottleneck which *actually* causes the additional costs in the process?
Q2. What causes the bottleneck and how does the bottleneck impact the following *parts* of the process?
Q3. How reliable is the performance statistics given?

Solution Overview. In this paper, we formally define such *parts* as the concept of *stages* to elevate the analysis of process performance to the *stage* level. By only consider the activities that would determine the performance of a stage, we provide an overview of the stage performance with the throughput time of a stage and the time that the stage is *in active* in a process. The details of a stage, i.e., the executions within the stage, are not of concern considering such details may result in losing the focus of the big picture as shown in Fig. 1. Then, we visualize the performance metrics for the diagnosis at the stage level.

By doing so, the process can be easier understood and, thus, diagnosed at a higher level of abstraction. Moreover, the reliability of the results can be enhanced by including a sufficient number of performance measurements (compared with the bottleneck that occurs only once in the process). Meanwhile, to support the diagnosis, the visualizations emphasize on the performance statistics in the context of cases. To summarize, the proposed approach supports an analyst to identify the bottlenecks at the stage level and diagnose the root causes before drilling down the process at the original granularity level.

Our implementation and the datasets used for evaluation are available for replicating the experiments. We evaluate our solution by analyzing two event logs and compare the results with the existing techniques. The contributions are summarized as follows:

1. We develop an approach to extract the execution of stages defined by a user. According to our evaluation, the proposed approach reaches the balance between the usability and the reliability of the results compared with the existing techniques.
2. Provided the definition of stages, we introduce performance metrics that are straightforward for a user, allowing one to perform an unbiased analysis.
3. We provide a visualization that shows the evolution of the performance at the stage level and the interaction between stages. Meanwhile, the performance distribution of cases are presented together at a glance for diagnosis.

The remainder of the paper is organized as follows. Section 2 introduces related work in the field of process mining. The proposed approach is presented in Sect. 3. In Sect. 4, we perform analysis of stage performance by experimenting with various techniques and summarize the paper in Sect. 5.

2 Related Work

To the best of our knowledge, this is the first work that focuses on the performance analysis at a coarser granularity level by extracting the instances at the corresponding abstraction level and compute and visualize the duration accordingly. Using the existing techniques, such analysis may be performed by applying filtering or combining different techniques. Filtering based on per target coarse-granular instance may result in biased performance results. Alternatively, one may combine different techniques, which we classified into event abstraction and performance analysis techniques, to analyze the performance based on the instances extracted at a coarser granularity level.

Event Abstraction. Based on the output of the techniques, the event abstraction techniques can be further classified into model-based and non model-based.

In [15], the authors decompose a process model into groups of *activities*, i.e., well-defined process steps, by mimicking the intuition of a human analyst identifying stages according to the modularity of the graph. Mannhardt and Tax identify the coarse-granular instances based on user-defined patterns which are represented by process models [14]. However, both approaches do not guarantee the availability of the results as shown in the experiments.

Other approaches do not require a process model. A supervised learning technique predicts the stage of an event using a probabilistic model [18,19]. The prediction model is trained with an event log in which all the events are labeled with the target instances at the coarser granularity level. Assuming that there exist patterns of the occurrence of the activities within a user-defined time interval, de Leoni and Dündar cluster activities and annotate the corresponding clusters to the events [12]. To analyze the performance of the process at the abstracted level, one has to apply other performance analysis techniques on top of the results of using the event abstraction techniques. For example, one may discover a process model at a higher granularity level from an abstracted event log and project the performance statistics on the model [9,11,21].

Performance Analysis. Similarly, we categorize the techniques into model-based and non model-based. Model-based techniques project the performance statistics on a predefined or discovered process model [5,9,21]. The performance can be analyzed with the context that is presented with the model. Nevertheless, due to the modeling formalism, i.e., certain process models do not allow for expressing all possible control-flow behaviors, the resulting process behavior highly depends on the discovery technique or the modeling method applied. The performance metrics may, thus, present biased results. Also, the aggregated performance metrics on a model limits the depth of the diagnosis. One needs to look into the cases in order to identify the root causes.

Other performance analysis techniques do not require a process model. The objectives of such methods vary and, therefore, it is hard to point out a common technique applied. Generally speaking, they aim to present unbiased results by showing the raw performance measures. For example, the *dotted chart* is a simple, yet powerful, technique that allows batched executions to be observed [17]. However, additional calculation is required to quantify the observed behavior and the context, i.e., relationship between activities, is lost. Another work focuses on the process performance over time with a parallel plot showing the duration of each process step on a absolute timeline [2,3]. The batched executions, which are often the causes of the delays in a process, can be easily observed and some behaviors, e.g., the overtake of the activities, may be discovered. The visualization emphasizes on the performance of and the interaction between process steps. However, without the context of a case, some diagnosis may be limited. For example, the influence of the performance of a process step to another one which is not directly following the one may not be observed and compared at the case level. The work is extended by incorporating a process model such that the performance of a process step can be analyzed with the context and more advanced process behavior may be presented with the model [1]. However, using a model suffers from the modeling formalism mentioned. In [16], the authors visualize the metrics such as the number of cases that arrive at each phase of a process per day. The workload and the efficiency in a phase over a specific time frame can then be observed. However, every event in an event log must be assigned to a phase and the phases must occur in a specific order, i.e., no phase could be skipped and no parallel phases is possible.

The existing event abstraction techniques may suffer from the biased results due to the modeling formalism or the assumptions of a process which the techniques are developed based on. The current performance analysis techniques are either insufficient for analyzing the influence of bottlenecks in the context of cases, one of the objectives of our research, or restricted to the assumed process behavior. To conclude, simply combining the existing techniques is insufficient to analyze the process performance at a coarser granularity level.

3   Stage-Based Performance Mining

An overview of the approach is presented in Fig. 2. It is a two-fold approach, which extracts stage instances, i.e., the execution of stage classes, and visualizes the performance metrics. The approach consists of four core components: *Mine for Stage Instances, Compute Stage Performance Metrics Visualize Stage Performance Evolution* and *Visualize Stage Performance Summary*. Based on the stage classes specified by a user, *Mine for Stage Instances* extracts the stage instances. The performance metrics is computed and visualized with *Visualize Stage Performance Evolution* and *Visualize Stage Performance Summary*. This section formally defines the terms mentioned and explains the components shown in Fig. 2 after briefly introducing the basic concept used in our approach.

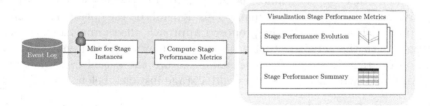

Fig. 2. Schematic overview of the proposed approach.

3.1   Preliminaries

Given an arbitrary set X, we write $\mathscr{P}(X)=\{X'|X'\subseteq X\}$ to denote its powerset. A sequence of length n over X is a function $\sigma\colon\{1,2,...n\}\to X$. Let X^* denote the set of all sequences over X. We write $\sigma=\langle x_1,x_2,...,x_n\rangle\in X^*$, where

$\sigma(1)=x_1, \sigma(2)=x_2, ..., \sigma(n)=x_n$. Given a sequence $\sigma \in X^*$, $|\sigma|$ denotes the length of the sequence. The empty sequence is written as $\langle \rangle$, i.e., $|\langle \rangle|=0$. We overload the set notation and write $x \in \sigma$ if and only if $\exists 1 \leq i \leq |\sigma| (\sigma(i)=x)$.

In a process, an execution of an *activity* is recorded as an *event* with the timestamp of the execution in the context of a process instance, i.e., a *case*. The events of a process are collected in an *event log*, the input for any process mining technique. In practice, many additional data attributes can be associated with an event. For example, event data typically captures the resource executing the activity, the cost of such an activity, etc. In this paper, we represent an event by a pair $e=(a,t)$ executed in the context of a case represented by a trace. The definitions of an event, trace, and event logs are as follows.

Definition 1 (Event, Trace & Event Log). *Let $\mathscr{A}$ denote the universe of process activities and $\mathscr{T}$ denote the universe of time. An event $e=(a,t) \in \mathscr{A} \times \mathscr{T}$ represents the execution of activity a at time t. We let $\mathscr{E}=\mathscr{A} \times \mathscr{T}$ denote the universe of events. Given $e=(a,t) \in \mathscr{E}$, we let $\pi_{act}(e)=a$ and $\pi_{ts}(e)=t$. A trace σ is a sequence of events, i.e., $\sigma \in \mathscr{E}^*$, such that $\forall 1 \leq i < j \leq |\sigma| (\pi_{ts}(\sigma(i)) \leq \pi_{ts}(\sigma(j)))$. An event log L is a collection of traces, i.e., $L \subseteq \mathscr{E}^*$.*[1]

3.2   Mine for Stage Instances

A natural way to analyze the performance of a complex process is to firstly elevate the process to the *stage* level. The performance of such a stage might not be impacted (significantly) by the arbitrary scheduling of activities inside a stage. Therefore, only the activities that a stage might start and end with are of interest when analyzing the performance at a higher level of abstraction, i.e., *at the stage level*.

Meanwhile, the level of abstraction of a process depends on the organization and the analysis objectives. For example, for a process operated by several companies, a stage can be defined as the part of the process within a specific company or a department of a company. This depends on the question of interest. To support such an analysis, the proposed approach allows analysts to specify the stages with the activities that a *stage class* may start and end with. The execution of stage classes, i.e., *stage instances*, are the actual operations in a case. We formally define a stage class and a stage instance below.

Definition 2 (Stage Class). *A stage class S is a pair of non-empty sets of disjoint activities, i.e., $S=(A_s, A_c) \in (\mathscr{P}(\mathscr{A}) \setminus \{\emptyset\}) \times (\mathscr{P}(\mathscr{A}) \setminus \{\emptyset\}) \wedge A_s \cap A_c = \emptyset$, where A_s (start activities) represents the activities that the stage class may start with and A_c (end activities) represents the activities that the stage class may end with. We let $\mathcal{S}=(\mathscr{P}(\mathscr{A}) \setminus \{\emptyset\}) \times (\mathscr{P}(\mathscr{A}) \setminus \{\emptyset\})$ denote the set of all stage classes.*

---

[1] We assume that an event can only appear once in a trace and that no two cases have the same trace in an event log. This can be enforced by adding more event attributes.

Definition 3 (Stage Instance). *Let $\sigma \in \mathscr{E}^*$ be a trace and $S=(A_s, A_c) \in \mathcal{S}$ a stage class. We define a function $\gamma:\mathscr{E}^* \times \mathcal{S} \to \mathscr{P}(\mathscr{E} \times \mathscr{E})$ that returns a set of pairs of events such that $\gamma(\sigma, S) \subseteq \{(\sigma(i), \sigma(j)) | 1 \leq i < j \leq |\sigma| \wedge \pi_{act}(\sigma(i)) \in A_s \wedge \pi_{act}(\sigma(j)) \in A_c\}$. Each pair of events $si=(e, e') \in \gamma(\sigma, S)$ is a stage instance. For simplicity, we write $\gamma(\sigma, S)$ as $\gamma_\sigma(S)$.*

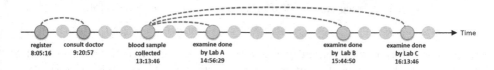

Fig. 3. A trace of a medical process of a patient in a hospital.

How to extract the stage instances depends on the business context of the analysis. Consider an example of a medical process. Suppose that we are interested in the duration of a patient being registered in the hospital until the doctor consultation and the time for the laboratories to examine the blood sampled from the patient. We define two stage classes, $S_1=(\{registered\}, \{consult\ doctor\})$ and $S_2=(\{sample\ collected\}, \{examine\ done\ by\ Lab\ *\})$, where $*$ denotes any string fitting the pattern. Figure 3 illustrate the trace of a case, where the dots denote the events in the trace. The events of interest are labeled and colored in blue and green for two stage classes. The grey dots are the activities, e.g., consulting nurses, that are not of concern for the analysis. The stage instances are the pairs of events connected with the dashed lines. The example shows that there are many possible ways to extract stage instances, depending on the process and the objectives of analysis.

The realization of extracting stage instances is by applying a generic technique that we developed in [13]. We extract the maximal number of stage instances in a trace. Considering the scenario mentioned, we allow one to flexibly define how the events of the start and the end activities should be mapped. We assume that the closer two events are, the more likely they form a stage instance. Such *distance* between events can be specified based on domain knowledge. Without the knowledge of how the events should be paired and the distance specified, we assume one-to-one mapping of events and use the order of the events in a trace by assuming that the closer two events are in a trace, the more possible that they belong to the same stage instance.

3.3   Compute Stage Performance Metrics

Given the stage instances extracted, the duration of a stage instance i, i.e., *cycle time* is naturally derived, which indicates the duration that a stage class is *in active*. To differentiate the duration that a stage class is actually being executed and the idle time of a stage class, we define *flow time* as the first occurrence of *any* start activity of the stage class until the last occurrence of *any* end activity of the

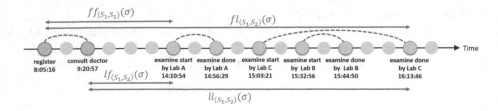

Fig. 4. A trace of a medical process of a patient in a hospital.

stage class. Moreover, to quantify the behavior among stage classes executed, we introduce the metrics for inter-stage performance. The metrics is formally defined as follows.

Definition 4 (Cycle Time). *Given an event log L and a stage class S, cycle time (ct) is the duration of a stage instance $si\in\gamma_{ts}(S)$ where $\sigma\in L$, i.e., $ct_S(si) = \pi_{ts}(si(2)) - \pi_{ts}(si(1))$. For each trace, we aggregate all the $ct_S(si)$, $\forall si\in\gamma_{ts}(S)$, into $ct_S^{stat}(\sigma)$, where stat stands for the target statistics, e.g., $ct_S^{avg}(\sigma)$ refers to the average duration of $ct_S(si)$, $\forall si\in\gamma_S(\sigma)$. For the process, we collect all the cycle time of S in L, i.e., $CT_S(L)=[ct_S(si)|\forall si\in\gamma_\sigma(S),\sigma\in L]$, and aggregate into $CT_S^{stat}(L)$.*

Definition 5 (Flow Time). *Given an event log L of a process and a stage class $S=(A_s,A_c)$, flow time (ft) is the duration that a stage class lasts in a case, i.e., $\forall 1\leq i<j\leq|\sigma|$ where $\sigma\in L$, $ft_S(\sigma)=\pi_{ts}(\sigma(max(\{j|\pi_{act}(\sigma(j))\in A_c\})))-\pi_{ts}(\sigma(min(\{i|\pi_{act}(\sigma(i))\in A_s\})))$ if and only $\gamma_\sigma(S)\neq\emptyset$. In other words, flow time only exists when S is executed, i.e., $\gamma_\sigma(S)\neq\emptyset$. For the process, we aggregate all the flow time of S in L into $FT_S^{stat}(L)$, where stat stands for the target statistics.*

Consider a process of applying for a mortgage. An application, i.e., a case with its trace σ, is finally approved after several rejections and re-submissions (S). The flow time $ft_S(\sigma)$ represents the duration from the first submission until it is finally approved and the duration that the application is under review is $ct_S^{sum}(\sigma)$.

Definition 6 (Inter-Stage Performance Metrics). *Given a trace σ and a stage class $S=(A_s,A_c)$ which $\gamma_\sigma(S)\neq\emptyset$, $\forall 1\leq i<j\leq|\sigma|$, we obtain two timestamps $t_S^{min}(\sigma)=\pi_{ts}(\sigma(min(\{i|\pi_{act}(\sigma(i))\in A_s\})))$ and $t_S^{max}(\sigma)=\pi_{ts}(\sigma(max(\{j|\pi_{act}(\sigma(j))\in A_c\})))$. Let S_1 and S_2 be two stage classes which $\gamma_(\sigma)(S_1)\neq\emptyset\wedge\gamma_(\sigma)(S_2)\neq\emptyset$. We compute the performance between S_1 and S_2 with the following metrics:*

$$- \textit{ff}_{(S_1,S_2)}(\sigma) = t_{S_2}^{min}(\sigma) - t_{S_1}^{min}(\sigma)$$
$$- \textit{fl}_{(S_1,S_2)}(\sigma) = t_{S_2}^{max}(\sigma) - t_{S_1}^{min}(\sigma)$$
$$- \textit{lf}_{(S_1,S_2)}(\sigma) = t_{S_2}^{min}(\sigma) - t_{S_1}^{max}(\sigma)$$
$$- \textit{ll}_{(S_1,S_2)}(\sigma) = t_{S_2}^{max}(\sigma) - t_{S_1}^{max}(\sigma)$$

Given the similar scenario in a hospital with a trace σ of a case, we define $S_1=(\{registered\}, \{consult\ doctor\})$ and $S_2=(\{examine\ start\ by\ Lab\ *,\ examine\ done\ by\ Lab\ *\}$. Figure 4 visualizes the four metrics, $ff_{(S_1,S_2)}(\sigma)$, $fl_{(S_1,S_2)}(\sigma)$, $lf_{(S_1,S_2)}(\sigma)$, and $ll_{(S_1,S_2)}(\sigma)$. Since $lf_{(S_1,S_2)}(\sigma)$ is greater than zero , we know that, after the consultation, the first examination of the blood sample starts roughly 5 h later. Table 1 summarizes the implication of the behavior between two stage classes based on the inter-stage performance metrics.

Table 1. Implication of behavior between two stage classes S_1 and S_2 in a trace σ based on the inter-stage performance metrics.

Metrics Relation	Implication of Stage Behavior in a Trace σ
$lf_{(S_1,S_2)}(\sigma) > 0$	S_2 starts after S_1 terminates permanently
$fl_{(S_1,S_2)}(\sigma) > 0$	S_2 terminates permanently before S_1 starts
$lf_{(S_1,S_2)}(\sigma) < 0 \wedge ff_{(S_1,S_2)}(\sigma) > 0$	S_2 *may* be executed in parallel with S_1

3.4 Visualize Stage Performance Metrics

We introduce two visualizations, *stage performance evolution* and *stage performance summary*. The first one demonstrates the evolution of performance over stages executed, allowing for further diagnosis. The latter one summarizes the statistics of the performance of all the stage classes defined. This section introduces the visualizations and demonstrates how the analysis can be performed with the visualizations using an event log L and four stage classes $\mathcal{S}=\{Apply, Claim, Travel, Declare\}$.

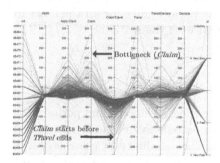

Fig. 5. Stage performance evolution.

	Apply	Claim	Travel	Declare
FT-Sum	28762 days 20:39:48	92302 days 12:04:24	54477 days 11:00:00	90329 days 09:02:08
FT-Minimum	0 days 00:00:01	0 days 00:00:50	0 days 00:00:00	0 days 00:00:03
FT-Maximum	263 days 21:33:09	314 days 07:06:04	1102 days 00:00:00	455 days 05:40:06
FT-Median	1 days 20:41:16	66 days 02:19:54	4 days 00:00:00	10 days 07:51:44
FT-Average	4 days 02:01:58	70 days 05:53:26	7 days 17:03:41	16 days 05:19:26
FT-S.D.	9 days 11:15:56	45 days 19:49:45	28 days 09:15:06	22 days 10:09:51
FT-Count	7042	1314	7065	5569
CT-Sum	25889 days 00:27:42	19979 days 06:21:56	54477 days 11:00:00	73348 days 12:34:50
CT-Minimum	0 days 00:00:01	0 days 00:00:02	0 days 00:00:00	0 days 00:00:01
CT-Maximum	263 days 21:33:09	170 days 21:42:44	1102 days 00:00:00	429 days 02:42:28
CT-Median	1 days 16:55:03	7 days 06:24:00	4 days 00:00:00	7 days 04:36:51
CT-Average	3 days 13:10:23	9 days 20:40:28	7 days 17:03:41	9 days 21:51:18
CT-S.D.	7 days 14:27:03	12 days 07:46:11	28 days 09:15:06	14 days 09:53:38
CT-Count	7295	2026	7065	7401

Fig. 6. Stage performance summary.

Stage Performance Evolution. It may occur that some cases execute some stages while others do not. We consider that the execution of stage classes reflects the business context. It is not reasonable to compare the performance of the cases without identifying different scenarios. Thus, we visualize the performance based on different *types* of cases according to the stage classes executed.

For each combination of stage classes executed, we visualize the stage performance metrics for the cases executing all the stage classes in the combination using parallel coordinates as shown in Fig. 5 [7]. The leftmost coordinate is the organization handling the cases and the rightmost one is the total case throughput time classified into *Very Slow*, *Slow*, *Fast*, and *Very Fast*. Between the two coordinates, the performance metrics of each trace $\sigma \in \{\sigma | \forall \sigma \in L \forall S \in \mathcal{S}, \gamma_\sigma(S) \neq \emptyset\}$ is plotted with a horizontal folded line in the order of $ct^{sum}_{Apply}(\sigma)$, $lf_{(Apply, Claim)}(\sigma)$, $ct^{sum}_{Claim}(\sigma)$, $lf_{(Claim, Travel)}(\sigma)$, $ct^{sum}_{Travel}(\sigma)$, $lf_{(Travel, Declare)}(\sigma)$, $ct^{sum}_{Declare}(\sigma)$ in the figure. The visualization can be applied interactively as below:

- The order of the coordinates can be flexibly arranged and the metrics of every stage class or between two stage classes can be changed to the flow time or other metrics, which allows for exploring the behavior of the stage performance from different angles.
- Depending on the use cases, the leftmost and rightmost coordinates may be replaced with any case attributes for analyzing the relationships between the attributes, e.g., the financial costs of handling a case, and the stage performance evolution.
- The scale of the coordinates for the performance metrics can be set the same (*absolute*) for identifying the bottlenecks, or the maximum value for each metrics (*relative*) such that the cause of the bottlenecks may be diagnosed.

Figure 5 shows the visualizations using relative performance with the analysis. Suppose one assumes that the stages are executed one after another, i.e., no parallelism of stages. In Fig. 5, the bottleneck and the most severe deviation are identified based on the stage performance distribution of the cases.

Stage Performance Summary. To have an overview of the performance of all the stage classes defined, we summarize the performance for all the cases in L. As shown in Fig. 6, the summary is presented with the statistics of $FT^{stat}_S(L)$ and $CT^{stat}_S(L)$ for every stage class $S \in \mathcal{S}$.

4   Evaluation

With the aim of supporting analysts to identify the bottlenecks and perform the diagnosis of a complex process, we conduct a comparative evaluation based on two criteria: the ease of use of a method and the reliability of the metrics. A method that requires much preparation, manipulation of an event log, or the domain knowledge hampers an analyst to perform an effective analysis. The metrics that contains only a few measurements may cause misleading conclusion of the bottlnecks. In the evaluation, we conduct experiments by applying various techniques to perform analysis at a coarser granular level of a process. The techniques are compared with the following questions:

- How much is the manipulation of an event log required to analyze the process performance at a coarser granular level?

– How much domain knowledge is required for the abstraction of an event log?
– How reliable is the resulting performance metrics?

To the best of our knowledge, the stage performance evolution proposed is the only visualization that supports our goal for such analysis. Therefore, we conduct the experiments by using various event abstraction techniques of which our visualization is applied on top. Table 2 lists the implementations of the techniques evaluated with their abbreviations for convenience. Except for the proposed approach[2], other techniques are available in ProM [24]. This section presents the evaluation from the aspects of the ease of use and the reliability by analyzing two event logs, *PermitLog* [23] and *BPIC15_1* [22].

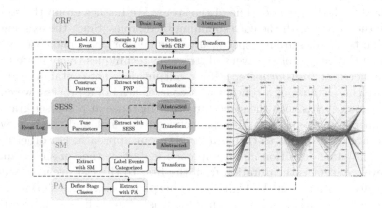

Fig. 7. A schematic overview of analyzing performance at a coarser granular level using different event abstraction techniques.

Table 2. Overview: Techniques used for Experiments.

Techniques	Abbreviation
- Proposed Approach	PA
- Abstract Event Labels using linear chaining [18, 19] Conditional Random Field (GRMM)	CRF
- Log Abstraction - Abstract Log based on Patterns [16]	PNP
- Session-based Log Abstraction [12]	SESS
- Stage Mining (SM) [15]	SM

---

[2] The implementation and the datasets used for experiments are in https://github.com/chiaoyunli/spm.

4.1   Evaluation on Ease of Use

The ease of use of a tool is evaluated from two aspects, the amount of the domain knowledge required and the necessity of the manipulation of an event log. The inputs and outputs of the event abstraction techniques vary. Therefore, for each technique, we manipulate the event logs for performance analysis at the abstracted level if necessary.

External Effort Required. Figure 7 presents the overview of the steps to analyze the performance at the coarser granularity level using the techniques. The dashed line indicates the data flow and the solid line refers to the control flow. Each box represents a step and the steps that require human intervention are emphasized with the green outline of the steps. We further group the steps and annotate the groups with the corresponding techniques. Since the existing event abstraction techniques are not specifically designed for performance analysis, we manipulate the output of the event abstraction techniques to compute the performance metrics (*Transform* step). If the output does not contain the attribute to indicate the instance of a concept at a coarser level of a process, we consider the continuous events with the same targeting instance at a coarser granularity level as an instance, i.e., the duration between the first and the last event of such instance corresponds to the *cycle time* in our approach. For other metrics, we apply the same definition in our approach, e.g., flow time of an instance of a higher level concept is the duration from the first to the last event of which the activities are contained in the high-level concept identified. As shown in Fig. 7, the proposed approach, i.e., PA, requires the least steps and does not require any transformation for the performance visualization. Note that other performance analysis techniques can be applied to analyze other aspects of the performance. In this case, our approach can, alternatively, generate an event log consisting of the events in the stage instances and the *Transform* step should be applied like the other techniques.

Domain Knowledge Required. The domain knowledge required for each technique varies. For example, to train a prediction model, CRF requires an event log with every event being labeled; for PNP, a coarse-granular instance is extracted with a pattern of the activities. To compare the domain knowledge quantitatively, we calculate the percentage of the activities required to extract a coarse-granular instance. Table 3 shows how the domain knowledge is required for every technique evaluated and the corresponding number of activities in the input of the techniques in the experiments. Our approach outperforms the CRF and PNP. However, it is inferior to SESS and SM since the two techniques are unsupervised. Nevertheless, SESS requires exhaustively tuning of the parameters and the results are non deterministic. SM, as presented in the next section, cannot guarantee the availability of the results.

Table 3. Overview: Techniques used for Experiments.

	Domain Knowledge Required	#Activities Required (%) (*PermitLog/BPIC15_1*)
CRF	All events labeled with the coarse-granular instance for training	1/1
PNP	Behavior of the activities of every concept at a coarser granular level	0.71/0.9
PA	Start and end activities of a stage class	0.37/0.78
SESS	Parameters tuning	0/0
SM	Minimum number of activities in a concept at a coarser granular level	0/0

4.2 Evaluation on Metrics Reliability

We perform analysis using the methods based on the steps illustrated in Fig. 7. To evaluate the results, generally speaking, the accuracy is an ideal indicator of the reliability of the results. However, due to the assumptions of different techniques, e.g., some are supervised while others are unsupervised approaches, it is unfair to compare the accuracy for the reliability. Therefore, the experiments are conducted on a best effort basis and we consider the number of measurements included as the indicator for the reliability of the metrics, i.e., the more measurements and cases used to compute a performance metrics, the more reliable the results are. Note that, except for SM of which the results are unavailable, all the techniques require a user to determine the number of concepts in a coarser granular level, i.e., the number of stage classes in terms of the proposed approach. Therefore, for the concepts at a coarser granular level, we define four concepts for a travel reimbursement management process for *PermitLog* [8] and nine phases which are implied in activity code in a dutch municipality for *BPIC15_1* [6]. The quality of the results are examined from two perspectives, whether the number of the concepts identified matches with the number of the concepts defined and the amount of the measurements.

Table 4 presents the performance statistics with the number of the measurements for the cycle time and the cases executing the concepts at a coarser granular level, i.e., $\#ft$. For both event logs, CRF and SESS cannot extract the exact number of concepts defined. CRF identifies too many concepts which include the events that the technique fails to predict (None) using *PermitLog* and too less concepts using *BPIC15_1*. SESS extracts less clusters despite the fact that the numbers of clusters desired are specified with the parameter. Therefore, only PNP generates the same number of concepts at a coarser granular level as specified. However, only the results using *PermitLog* are available while they are inferior to the proposed approach in terms of the number of measurements included. To conclude, the proposed approach provides the most reliable metrics compared with the other techniques in the experiments.

Table 4. Number of measurements per high-level concept identified using *PermitLog* and *BPIC15_1*. NaN indicates that the results are unavailable.

〚*PermitLog*〛

High-Level Concept Identified (#ct/#ft)

CRF	PNP	PA	SESS	SM
- Apply (7911/7062)	- Apply (7911/7062)	- Apply (7911/7062)	- Start trip+ (5406/3965)	NaN
- Claim (1715/1336)	- Claim (1605/1296)	- Claim (2026/1314)	- Permit FINAL_APPROVED	
- Travel (7843/7065)	- Travel (6331/633)	- Travel (7065/7065)	by SUPERVISOR+ (5715/4095)	
- Declare (5980/5718)	- Declare (5043/4963)	- Declare (7401/5569)	- Request Payment+ (10512/5856)	
- None (1276/1276)				

〚*BPIC15_1*〛

High-Level Concept Identified (#ct/#ft)

CRF	PNP	PA	SESS	SM
- Phase 1 (29/29)	NaN	- Phase 0 (1992/1199)	- register submission date	NaN
- Phase 2 (29/29)		- Phase 1 (3967/1119)	request+complete (901/670)	
- Phase 3 (193/178)		- Phase 2 (2727/969)	- enter senddate decision environmental	
- Phase 4 (200/178)		- Phase 3 (2573/1028)	permit+complete (1498/948)	
- Phase 5 (180/176)		- Phase 4 (3397/925)	- registration date publication+complete (105/102)	
- Phase 8 (1027/1027)		- Phase 5 (2054/899)	- enter senddate procedure	
		- Phase 6 (1/1)	confirmation+complete (100/97)	
		- Phase 7 (138/138)	- enter senddate acknowledgement+complete (106/102)	
		- Phase 8 (156/153)	- generate publication document decision	
			environmental permit+complete (154/147)	
			- create subcases completeness+complete (18/18)	

4.3   Experiments Summary

We perform a comparative evaluation by analyzing stage performance using various techniques. We compare the ease of the use of the techniques and the reliability of the resulting performance metrics. In terms of the ease of use, our approach requires the least effort from a user. However, we still require some domain knowledge in comparison with the unsupervised techniques. The reliability of the metrics is based on whether the number of the concepts at a coarser granular level is the same as specified and the number of measurements. The proposed approach outperforms all the other techniques evaluated. To conclude, the results show that our approach meets the balance between the ease of use and the reliability of the metrics.

4.4   Threats to Validity

The existing techniques are not designed for analyzing the performance at a high level of a process. Therefore, some information that is required to compute the duration of a coarse-granular instance, i.e., the start and complete time of the instance, is left for users to determine. Consider two interleaving instances of two concepts at a coarser granular level. Such behavior may result in multiple cycle time for each instance in the *Transform* step. However, in fact, only two measurements should be extracted. Thus, despite the best effort to apply the techniques, the results may not be accurate due to the manipulation.

For the proposed approach, the implementation allows an analyst to define only the stage classes with the distance and the mapping of events configured as default. However, there may be some scenarios where the parameters may not be defined easily and, thus, require further effort to configure the parameters to obtain reliable results. In addition, the performance of stage instances is aggregated at the case level. Which metrics makes sense for the analysis depends on the context. For example, in terms of stage instances of a stage class executed in parallel, the average cycle time may not be a reasonable choice for some processes. Nevertheless, consider the scenario in Fig. 3, the average can be used to compute the costs for hiring the staff in the laboratories. Such decision requires analysts to be aware of the context.

5   Conclusion

The diagnosis of inefficiencies requires performance metrics provided based on interpretable results. We elevate the analysis to the stage level and visualize the performance accordingly. Existing techniques are insufficient for stage performance analysis. The evaluation shows that combining existing techniques requires additional manipulation of an event log and domain knowledge from a user. Moreover, the results may be unreliable or unavailable. We propose an approach that supports performance analysis at the stage level by extracting events that are critical for the metrics. As such, our approach minimizes the effort from users while providing the most reliable results compared to the existing works. Meanwhile, the technique can be flexibly combined with other visualization to analyze other aspects of a process. To facilitate the analysis at the stage level, further research aims at automatic identification of stage classes.

References

1. van der Aalst, W., Unterberg, D.T.G., Denisov, V., Fahland, D.: Visualizing token flows using interactive performance spectra. In: International Conference on Applications and Theory of Petri Nets and Concurrency (2020)
2. Denisov, V., Belkina, E., Fahland, D., van der Aalst, W.: The performance spectrum miner: visual analytics for fine-grained performance analysis of processes. In: BPM (Dissertation/Demos/Industry) (2018)
3. Denisov, V., Fahland, D., van der Aalst, W.: Unbiased, fine-grained description of processes performance from event data. In: International Conference on Business Process Management (2018)
4. Dumas, M., La Rosa, M., Mendling, J., Reijers, H.A.: Fundamentals of Business Process Management (2018)
5. Günther, C.W., Rozinat, A.: Disco: Discover your processes. BPM (Demos) (2012)
6. van der Ham, U.: Benchmarking of five dutch municipalities with process mining techniques reveals opportunities for improvement (2015)
7. Haziza, D., Rapin, J., Synnaeve, G.: Hiplot, interactive high-dimensionality plots. https://github.com/facebookresearch/hiplot (2020)
8. Hobeck, R., et al.: Performance, variant, and conformance analysis of an academic travel reimbursement process (2020)
9. Hornix, P.T.: Performance analysis of business processes through process mining. Master's thesis, Eindhoven University of Technology (2007)

10. Kasim, T., Haracic, M., Haracic, M.: The improvement of business efficiency through business process management. Econ. Rev. J. Econ. Bus. **16**(1), 31–43 (2018)
11. Leemans, S.J., Fahland, D., van der Aalst, W.: Discovering block-structured process models from event logs containing infrequent behaviour. In: International Conference on Business Process Management (2013)
12. de Leoni, M., Dündar, S.: Event-log abstraction using batch session identification and clustering. In: Proceedings of the 35th Annual ACM Symposium on Applied Computing (2020)
13. Li, C.Y., van Zelst, S.J., van der Aalst, W.: A generic approach for process performance analysis using bipartite graph matching. In: International Conference on Business Process Management (2019)
14. Mannhardt, F., Tax, N.: Unsupervised event abstraction using pattern abstraction and local process models. arXiv preprint arXiv:1704.03520 (2017)
15. Nguyen, H., Dumas, M., ter Hofstede, A.H., La Rosa, M., Maggi, F.M.: Stage-based discovery of business process models from event logs. Inf. Syst. (2019)
16. Nguyen, H., Dumas, M., ter Hofstede, A., La Rosa, M., Maggi, F.: Business process performance mining with staged process flows. In: International Conference on Advanced Information Systems Engineering (2016)
17. Song, M., van der Aalst, W.: Supporting process mining by showing events at a glance. In: Proceedings of the 17th Annual Workshop on Information Technologies and Systems (2007)
18. Tax, N., Sidorova, N., Haakma, R., van der Aalst, W.: Mining process model descriptions of daily life through event abstraction. In: Proceedings of SAI Intelligent Systems Conference (2016)
19. Tax, N., Sidorova, N., Haakma, R., van der Aalst, W.: Event abstraction for process mining using supervised learning techniques. In: Proceedings of SAI Intelligent Systems Conference (2016)
20. van der Aalst, W.: Process Mining: Data Science in Action. Springer, Heideberg (2016). https://doi.org/10.1007/978-3-662-49851-4
21. van der Aalst, W., Adriansyah, A., van Dongen, B.: Replaying history on process models for conformance checking and performance analysis. Data Mining and Knowledge Discovery, Wiley Interdisciplinary Reviews (2012)
22. van Dongen, B.: BPI challenge 2015 municipality 1 (2015). https://doi.org/10.4121/uuid:a0addfda-2044-4541-a450-fdcc9fe16d17
23. van Dongen, B.: BPI challenge 2020: Travel permit data (2020). https://doi.org/10.4121/uuid:ea03d361-a7cd-4f5e-83d8-5fbdf0362550
24. van Dongen, B., de Medeiros, A.K.A., Verbeek, H., Weijters, A., van der Aalst, W.: The prom framework: a new era in process mining tool support. In: International conference on application and theory of petri nets (2005)

AudioLens: Audio-Aware Video Recommendation for Mitigating New Item Problem

Mohammad Hossein Rimaz[1](✉) (iD), Reza Hosseini[2], Mehdi Elahi[3] (iD), and Farshad Bakhshandegan Moghaddam[4]

[1] Technical University of Kaiserslautern, Erwin-Schrödinger-Str 52, 67663 Kaiserslautern, Germany
mrimaz@rhrk.uni-kl.de

[2] Vaillant Group Business Services, Berghauser Str. 63, 42859 Remscheid, Germany
seyed-reza.hosseini@vaillant-group.com

[3] University of Bergen, Fosswinckelsgt. 6, 5007 Bergen, Norway
mehdi.elahi@uib.no

[4] University of Bonn, Regina-Pacis-Weg 3, 53113 Bonn, Germany
farshad.bakhshandegan@uni-bonn.de

Abstract. From the early years, the research on recommender systems has been largely focused on developing advanced recommender algorithms. These sophisticated algorithms are capable of exploiting a wide range of data, associated with video items, and build quality recommendations for users. It is true that the excellency of recommender systems can be very much boosted with the performance of their recommender algorithms. However, the most advanced algorithms may still fail to recommend video items that the system has no form of representative data associated to them (e.g., tags and ratings). This is a situation called *New Item* problem and it is part of a major challenge called *Cold Start*. This problem happens when a new item is added to the catalog of the system and no data is available for that item. This can be a serious issue in video-sharing applications where hundreds of hours of videos are uploaded in every minute, and considerable number of these videos may have no or very limited amount of associated data.

In this paper, we address this problem by proposing recommendation based on novel features that do not require human-annotation, as they can be extracted completely automatic. This enables these features to be used in the cold start situation where any other source of data could be missing. Our proposed features describe audio aspects of video items (e.g., energy, tempo, and danceability, and speechiness) which can capture a different (still important) picture of user preferences. While recommendation based on such preferences could be important, very limited attention has been paid to this type of approaches.

We have collected a large dataset of unique audio features (from Spotify) extracted from more than 9000 movies. We have conducted a set of experiments using this dataset and evaluated our proposed recommendation technique in terms of different metrics, i.e., Precision@K, Recall@K, RMSE, and Coverage. The results have shown the superior performance

© Springer Nature Switzerland AG 2021
H. Hacid et al. (Eds.): ICSOC 2020 Workshops, LNCS 12632, pp. 365–378, 2021.
https://doi.org/10.1007/978-3-030-76352-7_35

of recommendations based on audio features, used individually or combined, in the cold start evaluation scenario.

Keywords: Recommender systems · Audio visual · Multimedia · Cold start

1   Introduction

YouTube, as an instance of popular video-sharing web and mobile applications, has about 1.5 billion active users who consume incredible number of 5 billion videos per day[1]. Hence, it is not uncommon to observe confused video consumers with problem in deciding what to watch from a missive volume and variety of videos [3]. Recommender Systems can cope with this problem by supporting the users when making decision on what to watch [27,30,34]. Recommender systems can build personalized video suggestions based on the *particular* interests of users for videos and find what can better match users' needs and constraints [33,35]. Over the many years, wide range of video recommendation algorithms have been proposed and evaluated presenting excellency in performance. These algorithms can receive a variety of data sources, e.g., content-associated data (tags), and generate personalized recommendations on top of this data [2,10,20,30,39].

While the performance of these recommender algorithms can impact the quality of the generated recommendations, however, any type of algorithms may fail to generate relevant recommendations of video items which have no or very limited amount of associated data [17,29,37,42]. This is a situation known as *New Item Cold Start* problem, which typically occurs when a new item is added to the catalog of the system and no input data is available for that item [14, 15,25]. This is a major problem in video-sharing applications, such as YouTube where hundreds of hours of videos are uploaded in every minute, by millions of active video makers[2].

Furthermore, collecting the traditional types of content-associated data, that are typically represented by semantic attributes (e.g., tags), requires either a group of experts or a network of users [6,7,13,32,43]. This indeed is an expensive process and needs human efforts. Then recommendations based on these costly semantic attributes still may not properly capture the true users' preferences, e.g., the user tastes associated with audio characteristics of videos.

In addressing this problem, this article investigates the potential behind different types of audio features representative of video content in building quality recommendations for users. We have exploited two different audio features that can be extracted completely *automatic* without any need for costly *manual* human annotation. Hence they can be exploited by any content-based recommender algorithm capable of incorporating them in the recommendation process.

---

[1] https://www.omnicoreagency.com/youtube-statistics.
[2] http://tubularinsights.com/hours-minute-uploaded-youtube/.

We have compared quality of recommendation based on the (automatic) audio features against other types of (automatic) features and (manual) tags. The comparisons have been conducted with respect to various evaluation metrics (i.e., Precision@N, Recall@N, RMSE, and Coverage) using a large dataset of more than ≈18M ratings obtained from a large network of ≈162K users who provided the ratings for ≈9K movies.

The overall results of the evaluation have shown the consistent superiority of the recommendations based on our novel audio features over the traditional tags.

- we propose a novel technique for video recommendation based on audio features (e.g., energy, tempo, and danceability, and speechiness) that can be extracted automatically, without any need for costly human-annotation;
- we will publish a large dataset[3], which is the most important contribution of this paper, that contains a wide range of audio features (collected from Spotify) for 9,104 movies, linked directly with the user ratings and tags (+ other descriptors such as visual features);
- we have evaluated the recommendations based on novel audio features in cold start scenarios, when features are used *individually* or when used in *combination* with other features; we tested the recommendation quality exploiting using millions of ratings given by hundreds of thousands of users;

2   Related Works

This work is related to two research fields, i.e., the *Cold Start* problem and *Audio-aware* Recommendation Systems.

One of the major problems of recommender systems in general is the cold start problem, i.e., when a new user or a new item is added to the catalog and the system does not have sufficient data associated with these users/items [4]. In such a case, the system cannot properly recommend existing items to a new user (new user problem) or recommend a new item to the existing users (new item problem) [1]. In video domain, one of the effective approaches that can tackle the cold start problem exploit different forms of video content for generating recommendation [2]. Such video content can be manually added, e.g. tags [18,28], or automatically extracted, e.g., visual descriptors [5,12,23,26,36].

Another form of content data that can be used for video recommendation is based on audio descriptors [38]. Very limited works have focused on investigating such type of descriptors and their potential in representing user preferences. As an example, in [31] the correlation between user music taste and his/her personality has been discussed. Several medium and weak correlations between music audio features and personality traits have been shown, and their results have provided useful insights into the relationship between the personality and the music preference. Moreover, authors in [21] have collected a dataset of movies and television shows matched with subtitles and soundtracks and analyzed the

---

[3] https://github.com/mhrimaz/audio-lens.

relationship between story, song, and the user taste. However, they have taken a non-personalized approach and used IMDb ratings. [44] has investigated the effect of the movie soundtrack search volume on the movie revenue in different time periods. It has shown that the online search volume of a movie soundtrack has an effect on the movie revenue. [19] has investigated the relationship between the musical and visual art preferences, and the role of personality traits in predicting preferences for different musical styles and visual art motives. Beside this, [11] recommender system has integrated the some forms of deep learning features as well as block-level and i-vector audio features of more than 4,000 movie trailers.

This work differs from the prior works in different aspects. In terms of dataset, prior works extracted the audio features from movie trailers or short clips (e.g., in [11]), while in our dataset, the audio features have been extracted from original score soundtracks for full-length movies. Even though in [21] the authors take a similar approach, however, their focus is not really personalization. Moreover, we cover almost double in number of items. Second, in our experiments, we use the recently released MovieLens25M dataset, with much larger number of ratings. Finally, unlike previous datasets, e.g., introduced by [11,21], our data went through extensive manual checks, and errors have been corrected with careful expert checks.

3 Proposed Method

3.1 Data Collection Process

We did the data collection process in two phases. In the first phase, we queried albums in Spotify with a specific pattern "{movie_name} (Original Motion Picture Soundtrack) {year}". This naming pattern is quite prevalent within the music industry, and many publisher's releases follow this naming convention. This phase was completely automated using the Spotify Search API[4] to find a Spotify identifier (Spotify ID) for each movie. Each Spotify ID could represent an album or a playlist. However, there are several shortcomings to this approach. First, many albums do not follow this naming convention (e.g., "Toy Story (Soundtrack)"). Second, some movies do not have any related published album, whereas their soundtrack is a playlist in Spotify. To alleviate this problem, and enhance the quality of our dataset, in the second phase, we carefully checked each individual entry, manually. A team of 7 trained person taught to check the matching manually. Several criteria and identifier factors have been used to check the correctness of matching. First and foremost, the album's poster and the movie's poster should usually look identical or share some common elements. Moreover, composer information and track names checked against various online resources, including IMDb's soundtrack section and Wikipedia. In some few cases, the decision is inconclusive, which in such cases, we simply removed the entry from the dataset. We manually checked the corresponding movie or

---

[4] https://developer.spotify.com/documentation/web-api/.

playlist Spotify identifier for missing popular movies with the highest number of ratings in the IMDb platform. We decided to use IMDb since many new releases may have very low number of ratings in MovieLens25M [22][5] dataset released on January 2019. With the advent of sophisticated signal processing techniques, automatically extracting musical and vocal features from a full-length movie would be possible for real-world recommender systems. Since this is out of the scope of this research, we used already existed Spotify API. By having a manual checking procedure, we are ensuring to have a high quality and error-prone dataset for further researches.

3.2 Dataset Description

Our dataset provides a link between every movie and its corresponding soundtrack in Spotify (using Spotify ID). We found the Spotify ID for 9,104 movies. These movies received 18,745,630 ratings from 16,254 users. For each Spotify ID, we could find a corresponding album or playlist, which contains the number of included music tracks. Using the unique ID, we could collect the representing audio features provided by Spotify Audio Feature API[6]. The following list, briefly explains our collected audio features:

- *f1*: **Acousticness** is a confidence measure from 0.0 to 1.0 (high confidence) of whether the track is acoustic.
- *f2*: **Danceability** describes how suitable a track is for dancing based on a combination of musical elements including tempo, rhythm stability, beat strength, and overall regularity. The value is in the range of [0,1].
- *f3*: **Energy** is a measure from 0.0 to 1.0 and represents a perceptual measure of intensity and activity. Features contributing to this attribute include dynamic range, perceived loudness, timbre, onset rate, and general entropy. Typically, energetic tracks feel fast, loud, and noisy. For example, death metal has high energy, while a Bach prelude scores low on the scale.
- *f4*: **Instrumentalness** predicts whether a track contains no vocals. Rap or spoken word tracks are clearly "vocal". The closer the instrumentalness value is to 1.0, the greater likelihood the track contains no vocal content. Values above 0.5 are intended to represent instrumental tracks, but confidence is higher as the value approaches 1.0.
- *f5*: **Liveness** shows the presence of an audience in the recording. Higher liveness values represent an increased probability that the track was performed live. A value above 0.8 provides strong likelihood that the track is live.
- *f6*: **Loudness** is the overall loudness of the entire track in decibels (dB) ranging typically between −60 and 0 db. Loudness is the quality of a sound that is the primary psychological correlate of physical strength (amplitude).

---

[5] https://grouplens.org/datasets/movielens/25m/.
[6] https://developer.spotify.com/web-api/get-audio-features.

- **f7**: **Popularity** of a track is a value between 0 and 100, and is based on the total number of plays the track has had and how recent those plays are.
- **f8**: **Speechiness** detects the presence of spoken words in a track. The more exclusively speech-like the recording (e.g. talk show, audio book, poetry), the closer to 1.0 the attribute value.
- **f9**: **Tempo** is the speed or pace of a given piece and is the overall estimated tempo of a track in beats per minute.
- **f10**: **Track Duration** is the duration of the track in milliseconds.
- **f11**: **Valence** is a measure from 0.0 to 1.0 describing the musical positiveness conveyed by a track. More positive tracks (e.g. happy, cheerful, euphoric) have higher valence sound, while tracks with low valence sound more negative (e.g. sad, depressed, angry).
- **f12**: **Key** is the estimated overall key of the track. The values ranging from 0 to 11 mapping to pitches using standard Pitch Class notation[7] (E.g. 0 = C, 1 = C-sharp/D-flat, 2 = D, and −1 if no key was detected).
- **f13**: **Mode** indicates the modality (major is 1 and minor is 0) of a track, the type of scale from which its melodic content is derived. Note that the major key (e.g. C major) could more likely be confused with the minor key at 3 semitones lower (e.g. A minor) as both keys carry the same pitches.
- **f14**: **Time Signature** specifies how many beats are in each bar (or measure). It ranges from 3 to 7 indicating time signatures of "3/4", to "7/4".

3.3   Recommendation Algorithm

We adopted a classical "K-Nearest Neighbor" content-based algorithm. Given a set of users $u \in U$ and a catalogue of items $i \in I$, a set of preference scores r_{ui} provided by user u to item i has been collected. Moreover, each item $i \in I$ is associated to its feature vector $\boldsymbol{f}_i$. For each couple of items i and j, the similarity score s_{ij} is computed using *cosine similarity*. For each item i the set of its nearest neighbors closer that a specified threshold NN_i is built, Then, for each user $u \in U$, the predicted preference score $\hat{r}_{ui}$ for an unseen item i is computed as follows

$$s_{ij} = \frac{\boldsymbol{f}_i^T \boldsymbol{f}_j}{\boldsymbol{f}_i \boldsymbol{f}_j} \qquad \text{and} \qquad \hat{r}_{ui} = \frac{\sum_{j \in NN_i, r_{uj} > 0} r_{uj} s_{ij}}{\sum_{j \in NN_i, r_{uj} > 0} s_{ij}} \qquad (1)$$

3.4   Baselines

We have compared our proposed recommendation technique (*AudioLens*) against recommendation based on a range of automatic and manual features. For automatic features, that can be used in cold start situation, we considered recommendation based on **Musical Keys** and **Visual features**. Musical keys can be also collected from Spotify and be a informative descriptor of the musics composed for movies. Visual features is a novel form of content descriptors that

---

[7] https://en.wikipedia.org/wiki/Pitch_class.

has been shown to be effective in cold start situation. In our experiment, we used a recent dataset $MA14KD^8$ that have shown promising results in recommender systems [16]. In addition, we combined both audio and visual features and formed **Hybrid features** in order to compare the recommendation based on these features used individually or in combination. All of these features can be extracted automatically and adopted for recommendation in cold start situation. For the sake of comparison, we consider recommendation based on manual **Tags** which certainly need human-annotation and may be missing in cold start situation. However, this form of recommendation can still be included as a traditional baseline in our experiment.

4   Experimental Result

4.1   Evaluation Methodology

For evaluation, we followed a methodology similar to the one proposed by [9]. We used a large rating dataset, i.e., MovieLense25M with 25M ratings, and filtered out users who have rated at least 10 relevant items (i.e., items with ratings equal or higher than 4). This ensured us that each user has a minimum number of favorite items. Then we randomly selected 4000 users for our experiment. For each selected user, we choose 2 items with rating equal or higher than 4 (forming a favorite set of items). Then we randomly add 500 items not rated by the user to this set. After that we predict the ratings for all the 502 movies using the recommender system and order them according to the predicted ratings. For each $1 \leq N \leq 502$, number of hits will be the number of favorite movies appear in top N movies (e.g. 0, 1 or 2). Assume T is the total number of favorite items in the test set for all selected users ($T = 8000$ in our case), then:

$$recall@N = \frac{\#hits}{T} \quad \text{and} \quad precision@N = \frac{\#hits}{N \cdot T} = \frac{recall@N}{N} \quad (2)$$

In addition to these metrics, we also computed *Root Mean Squared Error (RMSE)*, i.e., the rating prediction error, and *Coverage* [38], i.e., the proportion of items over which the system is capable of generating recommendations [24].

4.2   Experiment A: Exploratory Analysis

In experiment A, we performed a set of exploratory analysis. Due to the space limit, we focus on reporting some interesting results we observed by analyzing the time evolution of the audio features over the history of (sound) cinema. For that, we computed the yearly average of every audio features for the period of 1940 to 2020. Figures 1 and 2 illustrate the obtained results. Interestingly, there

---

[8] https://zenodo.org/record/3266236#.Xx7hLPgzako.

are two opposite trends in the evolution of the audio features over time, i.e., a positive trend (for audio features such as **Energy**, **Danceability**, and **Tempo** shown in Fig. 1), and a negative trend (for audio features such as **Liveness**, **Acousticness**, and **Instrumentalness** shown in Fig. 2). These trends indicate that while, over the history of cinema, the musics of the movies have become more energetic with higher tempo (and perhaps more danceable), at the same time, the musics are also losing their liveness, acousticness, and instrumentalness.

Another interesting observation is that, according to our collected audio features, the musics of the newer movies (produced after 2000) have different characteristics compared to the older movies (produced before 2000). In earlier years of cinema, the musics of the movies illustrate more diversity in terms of our audio features. This could mean that composers have been making a more similar type of music for newer movies. In addition, the observed trend for newer movies goes slightly into the opposite direction compared to the older movies (e.g., see the u-turn in Fig. 2-middle, around 2000s). This might be due to the fact that the music production has encountered a big shift in 2000s with the introduction of *digital* composition techniques[9]. We could not present all figures for the other audio features, due to space limit. However similar trends have been observed for them.

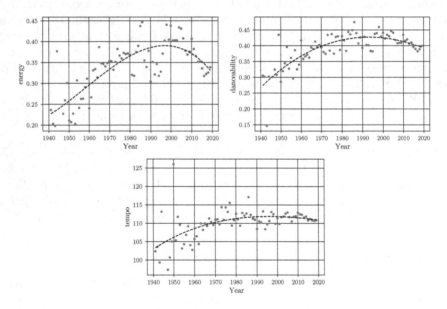

Fig. 1. Time evolution of **Energy** (left), **Danceability** (right), and **Tempo** (bottom) audio features over history of cinema.

---

[9] https://www.filmindependent.org/blog/know-score-brief-history-film-music/.

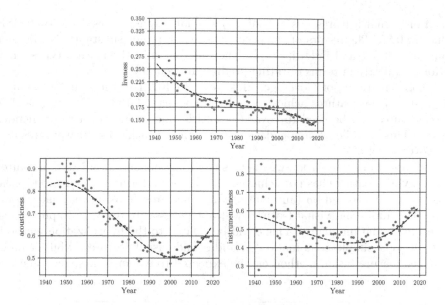

Fig. 2. Time evolution of **Liveness** (top), **Acousticness** (left), and **Instrumental-ness** (right) features over history of cinema. Time evolution of **Liveness** (top), **Acousticness** (left), and **Instrumentalness** (right) features over history of cinema.

4.3   Experiment B: Recommendation Quality

In experiment B, we evaluated our audio-aware recommendation technique (*AudioLens*) and compared it against different baselines, e.g., recommendation based on other automatic features (i.e., musical key, visual features, and hybrid features) as well as recommendation based on manual tags (see Sect. 3.1 for more details). Figure 3 and Fig. 4 illustrate the results.

In terms of the precision@N, as shown in Fig. 3 (left), the best results have been consistently achieved by AudioLens, i.e., our proposed recommendation approach based (automatic) audio features. The precision value of AudioLens is 0.0023, 0.0023, 0.0022, 0.0022 for recommendation sizes (N) of 5, 10, 15, and 20, respectively. The second best approach is recommendations based on visual features which achieves precision of 0.0017, 0.0018, 0.0019, and 0.0019 for growing recommendation sizes of 5, 10, 15, and 20. The worst results have been achieved for recommendation based on (manual) tags with values of 0.0008, 0.0010, 0.0011, and 0.0012 for different recommendation sizes.

In terms of Recall@N, similar results have been observed, as depicted in Fig. 3 (right). Again, recommendation based on (automatic) audio features (AudioLens) obtains the best results, visual features are the second best, and again, the worst results achieved by recommendation based on tags.

In terms of RMSE, presented in Fig. 4 (left), our proposed recommendation technique based on the audio features (AudioLens) achieves superior results compared to the other features, with RMSE values of 0.83. Recommendation based on visual features has also obtained relatively good results with RMSE values of 0.86. The results of the other features were not substantially different from each other, and indeed, despite the differences in the feature types, they perform similarly in terms of rating prediction accuracy.

Finally, in terms of Coverage, illustrated in Fig. 4 (right), all (automatic) audio and visual features achieves the best coverage of 100%. This means that these features can be used to cover the entire item catalog of a recommender system. This is while recommendation based on tags achieves the worst results, i.e., coverage of 93%.

An important observation we made is that, recommendation based on (automatic) hybrid features has not achieved a superior performance compared to the recommendation based on (automatic) audio features. This means that a combining the audio and visual features will not necessarily result in improvement on recommendation quality. This could be related to the hybridization method, as we used a simple combination of audio and visual features, while a more advanced feature fusion method can be expected to enhance these outcomes.

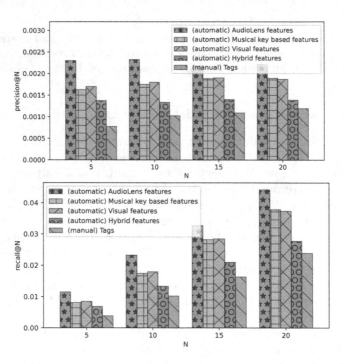

Fig. 3. Quality of movie recommendation, based on different content features, w.r.t, **Precision (top)** and **Recall (bottom)**

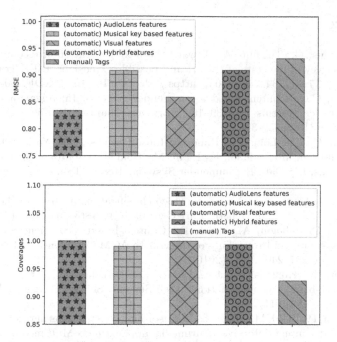

Fig. 4. Quality of movie recommendation, based on different content features, w.r.t, **RMSE (top)** and **Coverage (bottom)**

5   Conclusion

This paper addresses the cold start problem by proposing a recommendation technique based on *audio* features that can be automatically extracted with no need for human involvement. These novel features can represent video items when neither any rating nor any tag is available for a new video item. We have conducted a preliminary experiments to better investigate the potential power of these audio features in generating video recommendation and compared the results against user tags labeled manually. The experiment has been conducted using our new dataset with novel audio features extracted from more than 9000 movies. The results of the experiment have shown consistent superiority of these audio features in generating relevant recommendation and hence effectively dealing with the cold start problem.

Our plans for future work includes building a mobile recommender system with a specific design that adopts novel interface elements [8] for explaining the audio features to the user. We also plan to elicit user-generated video content from other video sharing social networks (e.g., Instagram). We also plan to obtain the implicit preferences of music listeners through their facial appearance using recent findings [40] that have shown correlation between peoples musical preferences and their facial expressions [41].

References

1. Adomavicius, G., Tuzhilin, A.: Toward the next generation of recommender systems: a survey of the state-of-the-art and possible extensions. IEEE Trans. Knowl. Data Eng. **17**(6), 734–749 (2005). https://doi.org/10.1109/TKDE.2005.99
2. Aggarwal, C.C.: Content-based recommender systems. In: Aggarwal, C.C. (ed.) Recommender Systems, pp. 139–166. Springer, Cham (2016). https://doi.org/10.1007/978-3-319-29659-3_4
3. Anderson, C.: The Long Tail. Random House Business, New York (2006)
4. Bakhshandegan Moghaddam, F., Elahi, M.: Cold start solutions for recommendation systems. Big Data Recommender Systems, Recent Trends and Advances IET (2019)
5. Brezeale, D., Cook, D.J.: Automatic video classification: a survey of the literature. IEEE Trans. Syst. Man Cybern. Part C Appl. Rev. **38**(3), 416–430 (2008)
6. Cantador, I., Bellogín, A., Vallet, D.: Content-based recommendation in social tagging systems. In: Proceedings of the Fourth ACM Conference on Recommender Systems, pp. 237–240. ACM (2010)
7. Cantador, I., Konstas, I., Jose, J.M.: Categorising social tags to improve folksonomy-based recommendations. Web Semant. Sci. Serv. Agents World Wide Web **9**(1), 1–15 (2011)
8. Cremonesi, P., Elahi, M., Garzotto, F.: User interface patterns in recommendation-empowered content intensive multimedia applications. Multimedia Tools Appl. **76**(4), 5275–5309 (2016). https://doi.org/10.1007/s11042-016-3946-5
9. Cremonesi, P., Koren, Y., Turrin, R.: Performance of recommender algorithms on top-n recommendation tasks. In: Proceedings of the Fourth ACM Conference on Recommender Systems, pp. 39–46 (2010)
10. De Gemmis, M., Lops, P., Semeraro, G., Basile, P.: Integrating tags in a semantic content-based recommender. In: Proceedings of the 2008 ACM Conference on Recommender Systems, pp. 163–170. ACM (2008)
11. Deldjoo, Y., Constantin, M.G., Eghbal-Zadeh, H., Ionescu, B., Schedl, M., Cremonesi, P.: Audio-visual encoding of multimedia content for enhancing movie recommendations. In: Proceedings of the 12th ACM Conference on Recommender Systems, RecSys 2018, New York, NY, USA, pp. 455–459. Association for Computing Machinery (2018). https://doi.org/10.1145/3240323.3240407
12. Deldjoo, Y., Elahi, M., Cremonesi, P., Garzotto, F., Piazzolla, P., Quadrana, M.: Content-based video recommendation system based on stylistic visual features. J. Data Semant., 1–15 (2016)
13. Di Noia, T., Mirizzi, R., Ostuni, V.C., Romito, D., Zanker, M.: Linked open data to support content-based recommender systems. In: Proceedings of the 8th International Conference on Semantic Systems, pp. 1–8. ACM (2012)
14. Elahi, M.: Empirical evaluation of active learning strategies in collaborative filtering. Ph.D. thesis, Ph.D. Dissertation. Free University of Bozen-Bolzano (2014)
15. Elahi, M., Braunhofer, M., Gurbanov, T., Ricci, F.: User preference elicitation, rating sparsity and cold start (2018)
16. Elahi, M., Hosseini, R., Rimaz, M.H., Moghaddam, F.B., Trattner, C.: Visually-aware video recommendation in the cold start. In: Proceedings of the 31st ACM Conference on Hypertext and Social Media, pp. 225–229 (2020)
17. Elahi, M., Ricci, F., Rubens, N.: A survey of active learning in collaborative filtering recommender systems. Comput. Sci. Rev. **20**, 29–50 (2016)

18. Enrich, M., Braunhofer, M., Ricci, F.: Cold-start management with cross-domain collaborative filtering and tags. In: Huemer, C., Lops, P. (eds.) EC-Web 2013. LNBIP, vol. 152, pp. 101–112. Springer, Heidelberg (2013). https://doi.org/10.1007/978-3-642-39878-0_10

19. Ercegovac, I.R., Dobrota, S., Kuščević, D.: Relationship between music and visual art preferences and some personality traits. Empirical Stud. Arts **33**(2), 207–227 (2015). https://doi.org/10.1177/0276237415597390

20. Gedikli, F., Jannach, D.: Improving recommendation accuracy based on item-specific tag preferences. ACM Trans. Intell. Sys. Technol. (TIST) **4**(1), 11 (2013)

21. Gillick, J., Bamman, D.: Telling stories with soundtracks: an empirical analysis of music in film. In: Proceedings of the First Workshop on Storytelling, New Orleans, Louisiana, pp. 33–42. Association for Computational Linguistics, June 2018. https://doi.org/10.18653/v1/W18-1504. https://www.aclweb.org/anthology/W18-1504

22. Harper, F.M., Konstan, J.A.: The movielens datasets: history and context. ACM Trans. Interact. Intell. Syst. **5**(4) (2015). https://doi.org/10.1145/2827872

23. Hazrati, N., Elahi, M.: Addressing the new item problem in video recommender systems by incorporation of visual features with restricted Boltzmann machines. Expert Syst. **38**, e12645 (2020)

24. Herlocker, J.L., Konstan, J.A., Terveen, L.G., Riedl, J.T.: Evaluating collaborative filtering recommender systems. ACM Trans. Inf. Syst. **22**(1), 5 53 (2004). https://doi.org/10.1145/963770.963772

25. Hornick, M.F., Tamayo, P.: Extending recommender systems for disjoint user/item sets: the conference recommendation problem. IEEE Trans. Knowl. Data Eng. **8**, 1478–1490 (2012)

26. Hu, W., Xie, N., Li, Zeng, X., Maybank, S.: A survey on visual content-based video indexing and retrieval. Trans. Sys. Man Cyber Part C **41**(6), 797–819 (2011). https://doi.org/10.1109/TSMCC.2011.2109710

27. Jannach, D., Zanker, M., Felfernig, A., Friedrich, G.: Recommender Systems: An Introduction. Cambridge University Press, Cambridge (2010)

28. Liang, H., Xu, Y., Li, Y., Nayak, R.: Tag based collaborative filtering for recommender systems. In: Wen, P., Li, Y., Polkowski, L., Yao, Y., Tsumoto, S., Wang, G. (eds.) RSKT 2009. LNCS (LNAI), vol. 5589, pp. 666–673. Springer, Heidelberg (2009). https://doi.org/10.1007/978-3-642-02962-2_84

29. Lika, B., Kolomvatsos, K., Hadjiefthymiades, S.: Facing the cold start problem in recommender systems. Expert Syst. Appl. **41**(4), 2065–2073 (2014)

30. Lops, P., de Gemmis, M., Semeraro, G.: Content-based recommender systems: state of the art and trends. In: Ricci, F., Rokach, L., Shapira, B., Kantor, P.B. (eds.) Recommender Systems Handbook, pp. 73–105. Springer, Boston (2011). https://doi.org/10.1007/978-0-387-85820-3_3

31. Melchiorre, A.B., Schedl, M.: Personality correlates of music audio preferences for modelling music listeners. In: Proceedings of the 28th ACM Conference on User Modeling, Adaptation and Personalization, UMAP 2020, New York, NY, USA, pp. 313–317. Association for Computing Machinery (2020). https://doi.org/10.1145/3340631.3394874

32. Milicevic, A.K., Nanopoulos, A., Ivanovic, M.: Social tagging in recommender systems: a survey of the state-of-the-art and possible extensions. Artif. Intell. Rev. **33**(3), 187–209 (2010)

33. Resnick, P., Varian, H.R.: Recommender systems. Commun. ACM **40**(3), 56–58 (1997). https://doi.org/10.1145/245108.245121

34. Ricci, F., Rokach, L., Shapira, B.: Recommender systems: introduction and challenges. In: Ricci, F., Rokach, L., Shapira, B. (eds.) Recommender Systems Handbook, pp. 1–34. Springer, Boston, MA (2015). https://doi.org/10.1007/978-1-4899-7637-6_1

35. Ricci, F., Rokach, L., Shapira, B., Kantor, P.B. (eds.): Recommender Systems Handbook. Springer, Boston (2011). https://doi.org/10.1007/978-0-387-85820-3

36. Rimaz, M.H., Elahi, M., Bakhshandegan Moghadam, F., Trattner, C., Hosseini, R., Tkalčič, M.: Exploring the power of visual features for the recommendation of movies. In: Proceedings of the 27th ACM Conference on User Modeling, Adaptation and Personalization, pp. 303–308 (2019)

37. Rubens, N., Elahi, M., Sugiyama, M., Kaplan, D.: Active learning in recommender systems. In: Ricci, F., Rokach, L., Shapira, B. (eds.) Recommender Systems Handbook, pp. 809–846. Springer, Boston (2015). https://doi.org/10.1007/978-1-4899-7637-6_24

38. Schedl, M., Zamani, H., Chen, C.-W., Deldjoo, Y., Elahi, M.: Current challenges and visions in music recommender systems research. Int. J. Multimed. Inf. Retr. **7**(2), 95–116 (2018). https://doi.org/10.1007/s13735-018-0154-2

39. Shepitsen, A., Gemmell, J., Mobasher, B., Burke, R.: Personalized recommendation in social tagging systems using hierarchical clustering. In: Proceedings of the 2008 ACM Conference on Recommender Systems, pp. 259–266. ACM (2008)

40. Tkalčič, M., Maleki, N., Pesek, M., Elahi, M., Ricci, F., Marolt, M.: A research tool for user preferences elicitation with facial expressions. In: Proceedings of the Eleventh ACM Conference on Recommender Systems, pp. 353–354. ACM (2017)

41. Tkalčič, M., Maleki, N., Pesek, M., Elahi, M., Ricci, F., Marolt, M.: Prediction of music pairwise preferences from facial expressions. In: Proceedings of the 24th International Conference on Intelligent User Interfaces, IUI 2019, New York, NY, USA, pp. 150–159. Association for Computing Machinery (2019). https://doi.org/10.1145/3301275.3302266

42. Vlachos, M., Duenner, C., Heckel, R., Vassiliadis, V.G., Parnell, T., Atasu, K.: Addressing interpretability and cold-start in matrix factorization for recommender systems. IEEE Trans. Knowl. Data Eng. **31**, 1253–1266 (2018)

43. Wang, L., Zeng, X., Koehl, L., Chen, Y.: Intelligent fashion recommender system: fuzzy logic in personalized garment design. IEEE Trans. Hum.-Mach. Syst. **45**(1), 95–109 (2015)

44. Xu, H., Goonawardene, N.: Does movie soundtrack matter? The role of soundtrack in predicting movie revenue. In: Siau, K., Li, Q., Guo, X. (eds.) 18th Pacific Asia Conference on Information Systems, PACIS 2014, Chengdu, China, 24–28 June 2014, p. 350 (2014). http://aisel.aisnet.org/pacis2014/350

Scalable Online Conformance Checking Using Incremental Prefix-Alignment Computation

Daniel Schuster[1]([⊠]) [iD] and Gero Joss Kolhof[2]

[1] Fraunhofer Institute for Applied Information Technology,
Sankt Augustin, Germany
daniel.schuster@fit.fraunhofer.de
[2] RWTH Aachen University, Aachen, Germany
gero.kolhof@rwth-aachen.de

Abstract. Conformance checking techniques aim to collate observed process behavior with normative/modeled process models. The majority of existing approaches focuses on completed process executions, i.e., offline conformance checking. Recently, novel approaches have been designed to monitor ongoing processes, i.e., online conformance checking. Such techniques detect deviations of an ongoing process execution from a normative process model at the moment they occur. Thereby, countermeasures can be taken immediately to prevent a process deviation from causing further, undesired consequences. Most online approaches only allow to detect approximations of deviations. This causes the problem of falsely detected deviations, i.e., detected deviations that are actually no deviations. We have, therefore, recently introduced a novel approach to compute exact conformance checking results in an online environment. In this paper, we focus on the practical application and present a scalable, distributed implementation of the proposed online conformance checking approach. Moreover, we present two extensions to said approach to reduce its computational effort and its practical applicability. We evaluate our implementation using data sets capturing the execution of real processes.

Keywords: Process mining · Conformance checking · Process monitoring · Event streams · Streaming platform

1 Introduction

To achieve operational excellence, accurate knowledge of the different processes executed within one's company is of utmost importance. Today's information systems accurately track and store the executions of said processes, i.e., *event data*. The field of *process mining* [3] deals with the analysis of such event data to increase the overall knowledge and insights about the execution of a process.

© Springer Nature Switzerland AG 2021
H. Hacid et al. (Eds.): ICSOC 2020 Workshops, LNCS 12632, pp. 379–394, 2021.
https://doi.org/10.1007/978-3-030-76352-7_36

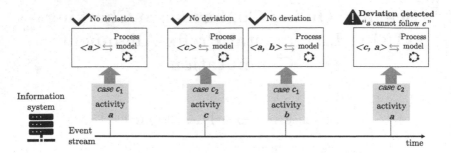

Fig. 1. Idea of online conformance checking. Events are observed over time, e.g., activity "a" was executed for process instance "c_1". Upon receiving an event, we check if the newly observed event causes a deviation w.r.t. a reference model [21]

In conformance checking [9], the goal is to assess to what degree a given process model describes the captured event data, i.e., allowing us to assess compliance and adherence to predefined policies. Most approaches in conformance checking are offline techniques. Thus, historical event data are extracted and analyzed. Whereas this type of analysis is helpful to gain a better understanding of a process' execution and to audit the past execution, it does not allow one to actively intervene in running process executions. Consider Fig. 1, where we visualize the idea of process monitoring. By applying online conformance checking, we are actively able to detect and pinpoint faulty process executions and communicate such deviations to the process owner.

Work covering the notion of online conformance checking is roughly subdivided into two categories. First, techniques that indicate (non-)conformance on the basis of abstractions of the process models and/or the event data [6–8]. Secondly, techniques that do not work on abstractions of the input [21,22]. Clearly, abstractions help as a first indicator of (non-)compliance, however, they remain inaccurate, e.g., "wrongly detected" deviations. Whereas the techniques computing exact results enable us to detect and understand non-conformance issues more accurately. However, for practical applications one needs to resort to approximation schemes because of the computational time involved.

In previous work [21], we presented an approach for incrementally computing prefix-alignments on event streams, i.e., an exact technique to detect deviations. Moreover, we showed that the approach outperforms an existing state-of-the-art approximation technique [22]. The main focus of the previous work, however, is on the theoretical foundations. We showed that we can efficiently compute prefix-alignments on an event stream by continuing a shortest path problem based on an extended search space upon receiving a new event. In this paper, we focus on the practical application of the presented approach. Therefore, we present a scalable and distributed implementation of the approach. In addition, we present extensions that improve the calculation time of the approach to enhance its practical applicability. Finally, we demonstrate the practical applicability of our approach by applying it to event streams from real-life processes.

The remainder is structured as follows. In Sect. 2, we present related work. In Sect. 3, we present preliminaries. In Sect. 4, we present an implementation of the incremental prefix-alignment computation and extensions improving the practical applicability by reduced computational effort. In Sect. 5, we evaluate our implementation. Section 6 concludes this paper.

2   Related Work

Process mining comprises a variety of different techniques such as: process discovery, conformance checking, process enhancement and enrichment techniques. For an overview, we refer to [3]. Next, we focus on conformance checking.

Token-based replay [20] and footprint comparison [3] are one of the first techniques in the area of conformance checking. Both techniques have drawbacks described in [3]. Therefore, alignments have been introduced [2,4] that map traces onto an acceptable path of a given process model. Moreover, alignments indicate mismatches between observed and modeled behavior. The problem of finding an alignment was shown to be reducible to a shortest path problem [4,5].

The previously mentioned techniques are designed for offline usage. Thus, deviations can only be detected post-mortem, i.e., after the process instance has already finished. In [22] an approach was presented to monitor ongoing process executions based on event streams. Essentially, a framework was introduced that computes prefix-alignments for ongoing processes each time a new activity has been performed. Moreover, the framework includes multiple options to decrease computational effort; in return, false negatives in terms of deviation detection may arise. In [21] it was shown that prefix-alignments can be computed in an online setting by continuing a shortest path search on an extended search space upon receiving a new event. Moreover, this approach guarantees optimal prefix-alignments. Thus, false deviations w.r.t. deviation detection cannot occur.

In [8] an approach was presented that computes the conformance on event streams, too. In contrast to the prefix-alignment approaches, conformance of a process execution is computed based on behavioral patterns that describe control flow relations between activities. Furthermore, this approach is suited for partial and for already running process executions where past information on such process executions is not available. In return, it uses abstractions, i.e., behavioral patterns, of the process model and the event log that leads to a loss of expressiveness in the deviation explanation. Another approach calculates an extended transition system for a given process model in advance. Such extended transition system allows for replaying the ongoing process [6]. Costs are attached to the edges in an extended transition system, and replaying a divergent, non-compliant process instance leads to costs greater than zero.

3   Preliminaries

A multiset B over a set X contains an arbitrary number of each element in X. The set of all possible multisets over a set X is denoted by $\mathcal{B}(X)$.

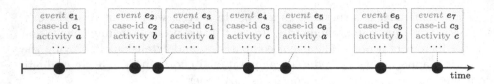

Fig. 2. Visualization of an event stream

For instance, $[x^5, y] \in \mathcal{B}(\{x, y, z\})$ contains 5 times x, once y and no z. A sequence σ of length n over a set X assigns an element to each index in $\{1, \ldots, n\}$, i.e., $\sigma : \{1, \ldots, n\} \rightarrow X$. We let $|\sigma|$ denote the length of σ. The set of all possible sequences over a set X is written as X^*, e.g., $\langle a, a, b \rangle \in \{a, b, c, d\}^*$.

3.1 Event Data and Event Streams

Today's information systems deployed in organizations capture the execution of (business) processes in great detail. These systems record the executed activity, the corresponding process instance within the activity was executed and potentially many other attributes. We refer to such data as *event data*.

In this paper, we assume an (infinite) event stream. Each event contains information as described above, i.e., it describes the execution of an activity within a process instance/case. In this paper, however, we are only interested in the label of the executed activity, the case-id of the corresponding process instance and the order of events. Consider Fig. 2 for an example event stream. Next, we formally define an event stream.

Definition 1 (Event; Event Stream). *Let $\mathcal{C}$ denote the universe of case identifiers and $\mathcal{A}$ the universe of activities. An event e describes the execution of an activity $a \in \mathcal{A}$ in the context of a process instance identified by $c \in \mathcal{C}$. An event stream S is a sequence of events, i.e., $S \in (\mathcal{C} \times \mathcal{A})^*$.*

3.2 Process Models

Process models allow us to describe process behavior. In this paper, we focus on *sound Workflow nets* [1]. Workflow nets (WF-nets) are a subclass of *Petri nets* [18] and sound WF-nets are a subclass of WF-nets with preferred *behavioral properties* guaranteeing the absence of deadlocks, livelocks and other anomalies.

A Petri net $N = (P, T, F, \lambda)$ consists of a set of *places* P, *transitions* T and *arcs* $F = (P \times T) \cup (T \times P)$ connecting places and transitions. Given the universe of activities $\mathcal{A}$, the *labeling function* $\lambda : T \rightarrow \mathcal{A} \cup \{\tau\}$ assigns an (possibly invisible, i.e., τ) activity label to each transition. For instance, $\lambda(t_1) = a$ and $\lambda(t_2) = \tau$ (Fig. 3).

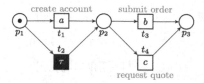

Fig. 3. Example WF-net N_1 with visualized initial marking $M_i=[p_1]$ and final marking $M_f=[p_3]$ describing a simplified ordering process

Fig. 4. Three possible alignments for N_1 (Fig. 3) and the trace $\langle a, b, c \rangle$

A state of a Petri net is defined by its *marking* M that is defined as a multiset of places, i.e. $M \in \mathcal{B}(P)$. Given a Petri net N and a marking M, a *marked net* is written as (N, M). We write M_i/M_f to represent the initial/final marking.

For $x \in P \cup T$, we define the set of all elements having an incoming arc from x, i.e., $x\bullet = \{y \in P \cup T \mid (x, y) \in F\}$. Symmetrically, we define $\bullet x = \{y \in P \cup T \mid (y, x) \in F\}$. Transitions allow for changing the state of a Petri net. Given a marking $M \in \mathcal{B}(P)$, we call a transition t *enabled* if all incoming places contain at least one token, i.e., $\forall p \in \bullet t\, (M(p) > 0)$. We write $(N, M)[t\rangle$ if t is enabled in marking M. An enabled transition can be *fired*. Such firing leads to a state change, i.e., a new marking $M' \in \mathcal{B}(P)$, where $M'(p) = M(p) + 1$ if $p \in t\bullet \setminus \bullet t$, $M'(p) = M(p) - 1$ if $\bullet t \setminus t\bullet$ and $M'(p) = M(p)$ otherwise. Given a sequence of transitions $\sigma \in T^*$, we write $(N, M) \xrightarrow{\sigma} (N, M')$ to denote that firing the transitions in σ leads to M'.

A *WF-net* is a Petri net with a unique source place p_i, i.e. $\bullet p_i = \emptyset$ and sink place p_o, i.e. $p_o \bullet = \emptyset$ that form the initial/final marking, i.e., $M_i = [p_i]/M_f = [p_o]$. Moreover, all transitions and places are on a path from source to sink.

3.3 Alignments

Alignments. [4] allow to compare observed behavior with modeled behavior. Consider Fig. 4 in which we present three possible alignments for the trace $\langle a, b, c \rangle$ and the WF-net N_1. The first row of an alignment (ignoring the skip symbol $\gg$) corresponds to the trace, and the second row corresponds to a sequence of transitions (ignoring $\gg$) leading from the initial to the final marking. Each column represents an alignment-*move*. We distinguish three move types. A *synchronous move* (gray) matches an observed activity to the execution of a transition, e.g., the first move of the first alignment (Fig. 4) with $\lambda(t_1) = a$. *Log moves* (black) indicate that an observed activity is not re-playable in the current state of the process model. *Model moves* (white) indicate that the execution of a transition cannot be mapped onto an observed activity. We further differentiate *invisible* and *visible model moves*. An invisible model move consists of an invisible transition, e.g., for the first model move in the third alignment (Fig. 4) $\lambda(t_2) = \tau$.

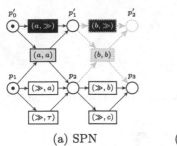
(a) SPN

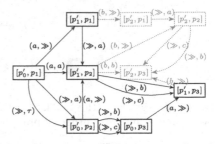
(b) SPN's state space/search-space for shortest path search

Fig. 5. SPN of $\langle a, b \rangle$ and WF-net N_1 (Fig. 3)

Invisible model moves do not represent a deviation. In contrast, visible model moves represent a deviation, e.g., the second model move of the third alignment.

Next to alignments, there is the concept of prefix-alignments, which are a relaxed version of conventional alignments. The first row of a prefix-alignment (ignoring $\gg$) also corresponds to the trace, but the second row corresponds to a sequence of transitions (ignoring $\gg$) leading from the initial marking to a marking from which the final marking can be still reached. Hence, prefix-alignments are suited to compare ongoing processes with a reference model.

In general, we aim to minimize log and visible model moves. Since multiple (prefix-)alignments exist for a given model and trace, costs are assigned to alignment moves. The *standard cost function* assigns cost 0 to synchronous and invisible model moves, and cost 1 to log and visible model moves. A (prefix-)alignment with minimal costs is called *optimal*. The computation of an optimal (prefix-)alignment is reducible to a shortest path problem [4]. Therefore, a *synchronous product net* (SPN) is created for a given trace and WF-net, e.g., as in Fig. 5. For a formal definition of a SPN, we refer to [4]. Note that each transition corresponds to a (prefix-)alignment move. Figure 5b shows the corresponding SPN's state space, on which the shortest path search is executed. Note that each edge represents an alignment move and, hence, has assigned costs. In the given example, the initial state/marking for the shortest path search is $[p'_0, p_1]$ and the goal states are all states/markings containing p'_2.

4   Incremental Prefix-Alignment Computation

In this section, we initially present the main idea of incrementally computing prefix-alignments on an event stream. Next, we introduce the proposed implementation of our approach. Subsequently, we present two extensions to improve the practical applicability of the approach.

4.1   Background

The core idea of incrementally computing prefix-alignments is to continue a shortest path search on an extended search space upon receiving a new event. For each process instance, we cache its current SPN and extend it upon receiving a new event. Moreover, we store intermediate search-results and reuse them when we continue the search. Next, we briefly list the main steps of the algorithm. Assume an event stream and a reference process model as input.

1. We receive a new event. For instance, consider (c_1, b) describing that the activity b was executed for the process instance identified by case-id c_1.
2. We extend the SPN for process instance c_1 by the new activity b. For example, assume that we previously received the event (c_1, a). Consider Fig. 5a showing the extension of the SPN highlighted by dashed gray elements. Note that when extending a SPN by a new activity, new transitions are added representing a log move on the new activity and potential synchronous moves.
3. We continue the search for a shortest path on the state space of the extended SPN from previously cached intermediate search-results, i.e., states already explored/investigated.
4. We return the prefix-alignment and cache the search-results.

For a detailed overview, we refer to [21]. Next, we introduce the proposed implementation of the incremental prefix-alignment computation approach.

4.2   Implementation

In this section, we present a horizontally scalable and fault-tolerant online conformance checking application based on *Apache Kafka* [15]. In the following, we explain the system's architecture using Fig. 6.

Apache Kafka is a distributed log processing/streaming system that is designed for handling high throughput of event data while maintaining low latency. Kafka implements the publish/subscribe pattern [14]. The key property of a publisher is that the messages are not directed to a specific subscriber. Instead, a middleware-component stores the produced messages and categorizes them according to given criteria. Independent of the publishers, subscribers can process published messages. In our implementation, Kafka acts as the middleware component for event storage and offers APIs for publishing event data and subscribing to it. In Kafka's terminology, *producers* correspond to publishers and *consumers* to subscribers. A producer, in our case a single event stream, sends a message to one of potentially many *Kafka Brokers*. Brokers, which are usually distributed over several physical nodes, form a *Kafka cluster*. In Fig. 6, we depict a Kafka cluster with three brokers.

Kafka stores messages, i.e., the events capturing the execution of process instances, in topics. To avoid a broker holding all messages, which is impossible in some cases, topics can be split into any number of *partitions*. In our implementation, the events from the event stream are published to the *event-topic*.

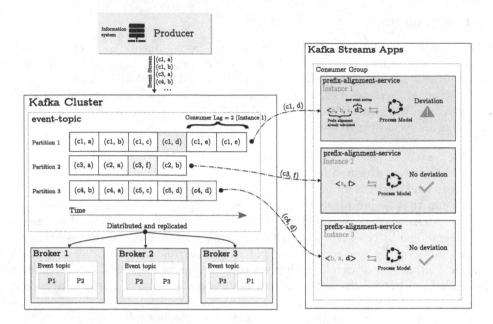

Fig. 6. Architecture overview of the proposed implementation

The structure of this topic is depicted in Fig. 6 within the dashed box in the Kafka Cluster. Further, we assume that the topic is divided into three partitions. Partitions are visualized as arrays where each cell contains a message, i.e., an event containing a case-id and an activity label. The first tuple entry refers to a case identifier of a process instance which is also the key of the message. The second entry is the payload consisting of the activity label. Kafka's default partitioning strategy ensures that messages with the same key are written to the same partition. Hence, events belonging to the same case are routed to the same partition. Moreover, it is shown that the topic called *event-topic* is replicated once. Thus, there are a total of six partitions, which are evenly distributed among the brokers (indicated by boxes labeled with P1–P3).

Kafka achieves horizontal scalability through consumers, who read messages according to their order in the corresponding partition. When a consumer group subscribes to a topic, Kafka assigns each partition to a member in the group s.t. each partition is processed by exactly one consumer. Thus, the various consumers can process a topic in parallel. Hence, the maximum degree of parallelism of an application is determined by the number of partitions of a topic. Consequently, the way to scale a system that uses Kafka is to simply add more consumers to a group and increase the number of partitions.

Kafka Streams is a client library used to build services that stream data from Kafka topics. Consider the three instances of the *prefix-alignment-service* in Fig. 6. Each instance calculates prefix-alignments for the process instances whose events are written to the assigned partitions. Since the *event-topic* is divided into three partitions, each partition is assigned to one instance.

4.3   Direct Synchronizing

In this section, we introduce an extension of the proposed implementation to improve the calculation time. In particular, we explain how we can skip shortest path searches in cases where we can immediately return the shortest path on an extended search space upon receiving a new event.

The idea of direct synchronizing is to skip shortest path searches in cases where we can simply extend the previous calculated prefix-alignment by a synchronous move on the newly observed activity. Hence, we execute the first and second step as described in Sect. 4.1. Next, we check if we can add a synchronous move without executing a shortest path search. We depict this check in Algorithm 1.

Algorithm 1: Direct synchronizing

input: $N=(P,T,F,\lambda)$ // reference process model, i.e., a sound WF-net

$\quad\quad\sigma=\sigma'\cdot\langle a\rangle\in\mathcal{A}^*$ with $a\in\mathcal{A}$ // extended trace

$\quad\quad\overline{\gamma}\in\big((\mathcal{A}\cup\{\gg\})\times(T\cup\{\gg\})\big)^*$ // previous prefix-alignment for σ' and N

begin

1 $\quad$ $S=(P^S,T^S,F^S,\lambda^S)$ with $M_i \leftarrow$ create/extend SPN for N and σ;

2 $\quad$ $\sigma_{\overline{\gamma}} \leftarrow$ extract sequence of transitions from T^S corresponding to $\overline{\gamma}$;

3 $\quad$ let M' s.t. $(S,M_i) \xrightarrow{\sigma_{\overline{\gamma}}} (S,M')$;

4 $\quad$ **for** $(t',t)\in T^S$ **do** // iterate over transitions from SPN S

5 $\quad\quad$ **if** $(S,M')[(t',t)\rangle \land \lambda^S((t',t))=(a,a)$ **then**

$\quad\quad\quad$ // transition (t',t) is enabled and represents a synchronous move on the

$\quad\quad\quad$ new activity a

6 $\quad\quad\quad$ **return** $\overline{\gamma}\cdot\big\langle(a,(t',t))\big\rangle$;$\quad\quad\quad$ // append sync. move to prefix-alignment

7 $\quad$ apply standard approach;$\quad\quad\quad\quad\quad$ // direct synchronizing is not possible

---

As input we assume a reference process model N, the extended trace σ where σ' represents the previous trace for the corresponding case, and the previously calculated prefix-alignment $\overline{\gamma}$ of the trace σ' and N. First, we create/extend the SPN for the extended trace and the reference process model (line 1). Next, we translate the previous prefix-alignment $\overline{\gamma}$ to a sequence of transitions in the SPN S (line 2). Note that this is always possible because every (prefix-)alignment corresponds to a sequence of transitions in the corresponding SPN. Moreover, since we always extend the SPN upon receiving a new event, such sequence in the extended SPN exists that corresponds to $\overline{\gamma}$. Given the sequence of transitions $\sigma_{\overline{\gamma}}$, we determine the state M' where the previous search stopped (line 3). Next, we check if we can directly execute a transition representing a synchronous move for the new activity a in this state (line 5). If true, we simply extend the previous prefix-alignment by a synchronous move and return (line 6). Otherwise, we apply the standard approach, i.e., the third and fourth step described in Sect. 4.1.

The main reason why it is beneficial to do the presented pre-check before actually continuing the shortest path search is to avoid heuristic recalculations. Such heuristic function is part of the used heuristic search algorithm to increase search efficiency and estimates for each state the costs to reach a goal state. Since the search space gets extended, and also the goal states are different in each incremental search, such heuristic recalculations are needed for each incremental search. As shown in [21], heuristic recalculations involve a high calculation effort. Thus, avoiding trivial shortest path searches can potentially speed up the prefix-alignment calculation.

4.4 Prefix Caching

In this section, we introduce prefix-caching for the incremental prefix-alignment approach. In an online environment, where multiple process instances of the same process are running, it is likely that the sequences of activities performed is similar to some degree. Thus, one wants to avoid recalculating prefix-alignments for event sequences that were already observed in the past. By applying prefix-caching, we avoid solving identical shortest path problem multiple times for process instances that share a certain prefix.

Table 1. Conceptual idea of prefix-caching

Processed events	Cached prefixes
(c_1, a)	$\langle a \rangle$
$(c_1, a),(c_2, a)$	$\langle a \rangle$
$(c_1, a),(c_2, a),(c_2, b)$	$\langle a \rangle, \langle a, b \rangle$
$(c_1, a),(c_2, a),(c_2, b),(c_1, b)$	$\langle a \rangle, \langle a, b \rangle$

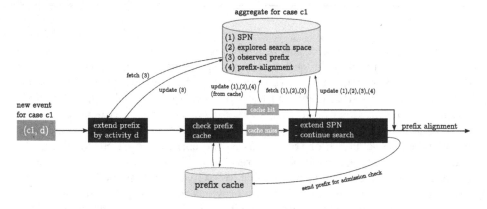

Fig. 7. Overview of prefix-caching within our proposed implementation

For instance, assume the event stream $\langle (c_1, a), (c_2, a), (c_2, b), (c_1, b), \dots \rangle$. Consider Table 1, where we show the cached prefixes while processing the sample event stream. Per cached prefix we save intermediate search results representing the current state of the search, i.e., already explored states from the search space. Moreover, we save the SPN and the prefix-alignment per cached prefix. Hence, we are either able to immediately return the prefix-alignment if we have calculated it already for a given prefix or to continue the search. In the given example (Table 1), we can skip two shortest path problems and return immediately a prefix-alignment, i.e., upon receiving the event (c_2, a) and (c_1, b).

In Fig. 7, we depict the prefix caching approach integrated within the proposed implementation. Assume a new event (c_1, d) arrives. First, we fetch the prefix observed so far for case c_1 and extend it by the new activity d. Note that the aggregate contains all information stored for a given case. Next, we check the prefix-cache if we have processed the given prefix before. If this is the case, we have a *cache hit*. Hence, we update the SPN, the already explored search space and the prefix-alignment in the aggregate of the given case c_1. Thus, we immediately return a prefix-alignment and do not solve a shortest path problem. If we have a *cache miss*, we apply the standard approach, i.e., we fetch the SPN, extend it by the new activity and continue the shortest path search from the already explored search space (Sect. 4.1). At this stage, we can also optionally use the direct synchronizing approach presented in Sect. 4.3 by first checking if we can directly append a synchronous move to the previous prefix-alignment. Finally, we update the aggregate of the case and return a prefix-alignment. Moreover, we send the new observed prefix to the prefix cache.

Since available memory is finite, the size of the cache has to be limited. Consequently, a cached prefix has to be replaced when a new prefix is written to the cache that has already reached its maximum capacity. For instance, popular cache replacement algorithms are: *least recently used (LRU)*, *most recently used (MRU)* and *least frequently used (LFU)* [19]. Many extensions to these general strategies have been proposed that address problems such as large memory consumption for metadata. One example is the strategy *TinyLFU* [13]. In summary, the strategy evaluates for a new cache candidate whether it should be added to the cache at the expense of deleting an already cached candidate. For the prefix caching we use an *in-process* TinyLFU cache for each node in the cluster. Thus, for each node in the Kafka cluster, an independent cache is maintained.

Alternatively, a *distributed cache* can be used. However, this requires that the prefix-alignments, including the whole search status, have to be serialized and transported over the network. This introduces a significant overhead due to network latency and high computational cost for serialization.

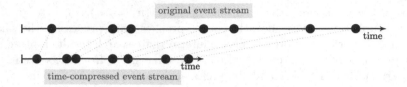

Fig. 8. Visualization of a time-compressed event stream. Each dot represents an event. The original event stream is compressed by 50% s.t. that relative time distance between the events is remained w.r.t. the original event stream

5 Experimental Evaluation

In this section, we present an experimental evaluation of the proposed implementation and the two presented extensions. First, we describe the experimental setup. Subsequently, we discuss the results.

Table 2. Specifications of the hardware used in the experimental setup

Node	Memory	CPU
1–2	128 GB RAM	2x Xeon 5115 Gold @ 2.40 GHz base
3–5	512 GB RAM	2x Xeon 5115 Gold @ 2.40 GHz base

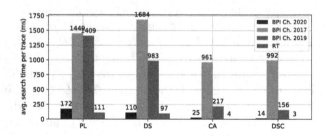

Fig. 9. Average computation time (ms) per trace

5.1 Experimental Setup

In the conducted experiments, we use publicly available, real-life event logs [10–12,17]. From the event logs, we generate an event stream by emitting the events according to their timestamps. Since the used event logs cover a large time span, we apply time-compression as visualized in Fig. 8. Moreover, we discovered a reference model for each event log with the Inductive Miner [16]. We conducted experiments for the two presented extensions presented in Sect. 4:

1. **PL:** plain version (Sect. 4.1 and Sect. 4.2)
2. **DS:** extension direct synchronizing (Sect. 4.3)
3. **CA:** extension prefix caching (Sect. 4.4)
4. **DSC:** both extensions, i.e., direct synchronizing and prefix caching

We use a five node Kafka cluster with each broker running on a separate physical machine. Consider Table 2 for detailed specifications. In cases where we use prefix-caching, we set the cache size to 100 prefixes per instance.

5.2   Results

Consider Fig. 9 showing the average computation time per trace for the four different versions. For all tested event logs, we observe that applying the two proposed extensions, i.e., DSC, leads to a significant speed-up of the calculation compared to the original PL version. Moreover, we observe synergetic effects of both proposed extensions, i.e., DS and CA compared to DSC, for all logs except for BPI Ch. 2017. Interestingly, we observe that for the BPI Ch. 2017 log, the extension DS performs worse than PL. This can be explained by the fact that for this event log, direct synchronization could be applied only in a few cases. Hence, the additional check, i.e., line 1–6 in Algorithm 1, is causing the higher computation time. However, for the other tested event logs, both extensions significantly reduce the calculation time.

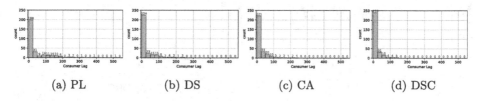

| (a) PL | (b) DS | (c) CA | (d) DSC |

Fig. 10. Distribution of consumer lag count for the four algorithm-variants using the BPI Ch. 2020 domestic log [12] with a replay time of 10 min

In Fig. 10, we show the distribution of the consumer lag for the four variants. We observe that the extensions, especially DSC including both extensions, lead to a significant improvement of the consumer lag, i.e., number of queued states for consumption (reconsider Fig. 10). Note that we replay the log in only 10 min whereas the original log spans 890 days, i.e., the time difference between the earliest and latest event in the event log. Regarding the consumer lag distribution, we observe similar results for the other tested event logs.

In Figs. 11, 12, 13 and 14 we depict the flow of *messages in* and *messages consumed* for the different event logs. Per event log, we compare the PL version, which serves as a baseline, and the DSC version, which includes both proposed

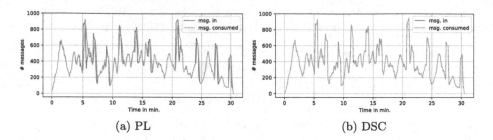

Fig. 11. *Messages in* vs. *messages consumed* for RT event log

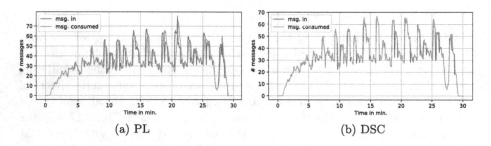

Fig. 12. *Messages in* vs. *messages consumed* for BPI Ch. 2019 event log

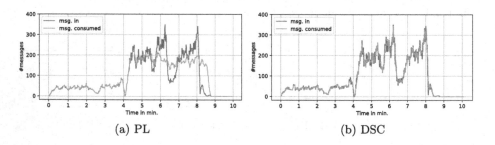

Fig. 13. *Messages in* vs. *messages consumed* for BPI Ch. 2020 event log

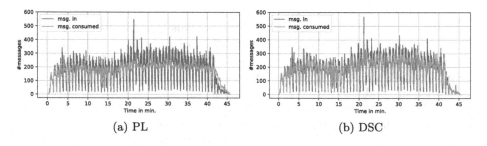

Fig. 14. *Messages in* vs. *messages consumed* for BPI Ch. 2017 event log

extensions. In general, we observe that the flow of *messages in*, i.e., incoming events from the stream, and the flow of *messages consumed*, i.e., processed events, overlaps much more for the DSC version compared to the PL version. Hence, the consumer lag of the DSC version is significantly lower compared to the PL version. This shows that the two proposed extensions significantly reduce the computational effort.

6 Conclusion

In this paper, we presented an implementation for scalable online conformance checking based on incremental prefix-alignment calculation. The proposed implementation is based on Apache Kafka, a streaming platform that is widely used within industry. Therefore, the paper offers an important basis for the industrial application of online conformance checking techniques. Moreover, we presented two extensions that significantly reduce the computational effort. Our conducted experiments show that the proposed implementation combined with the presented extensions is capable to efficiently process real-life event streams.

Open questions that need to be addressed in future work include the problem of deciding when a process instance/case is considered to be complete. This decision is necessary because we will eventually be required to delete aggregates, i.e. the current search state stored for each process instances (Fig. 7), due to limited memory resources of the cluster. However, this is a general problem within the area of online process mining and, hence, outside the scope of this paper.

References

1. van der Aalst, W.M.P.: The application of petri nets to workflow management. J. Circuits Syst. Comput. **8**(1), 21–66 (1998)
2. van der Aalst, W.M.P., Adriansyah, A., van Dongen, B.F.: Replaying history on process models for conformance checking and performance analysis. Wiley Interdisc. Rew. Data Mining Knowl. Discov. **2**(2), 182–192 (2012)
3. van der Aalst, W.M.P.: Process Mining - Data Science in Action. Springer, Heidelberg (2016). https://doi.org/10.1007/978-3-662-49851-410.1007/978-3-662-49851-4
4. Adriansyah, A.: Aligning observed and modeled behavior. Ph.D. thesis, Eindhoven University of Technology (2014)
5. Adriansyah, A., van Dongen, B.F., van der Aalst, W.M.: Memory-efficient alignment of observed and modeled behavior. BPM Center Report, p. 03 (2013)
6. Burattin, A., Carmona, J.: A framework for online conformance checking. In: Teniente, E., Weidlich, M. (eds.) BPM 2017. LNBIP, vol. 308, pp. 165–177. Springer, Cham (2018). https://doi.org/10.1007/978-3-319-74030-0_12
7. Burattin, A., Sperduti, A., van der Aalst, W.M.P.: Control-flow discovery from event streams. In: Proceedings of the IEEE Congress on Evolutionary Computation, CEC 2014, Beijing, China, 6–11 July 2014, pp. 2420–2427. IEEE (2014)
8. Burattin, A., van Zelst, S.J., Armas-Cervantes, A., van Dongen, B.F., Carmona, J.: Online conformance checking using behavioural patterns. In: BPM (2018)

9. Carmona, J., van Dongen, B.F., Solti, A., Weidlich, M.: Conformance Checking - Relating Processes and Models. Springer, Cham (2018). https://doi.org/10.1007/978-3-319-99414-7

10. van Dongen, B.: Dataset BPI challenge 2019 (2019). https://doi.org/10.4121/uuid:d06aff4b-79f0-45e6-8ec8-e19730c248f1

11. van Dongen, B.: BPI challenge 2017 (2017). https://doi.org/10.4121/uuid:5f3067df-f10b-45da-b98b-86ae4c7a310b

12. van Dongen, B.: BPI challenge 2020: Domestic declarations (2020). https://doi.org/10.4121/uuid:3f422315-ed9d-4882-891f-e180b5b4feb5

13. Einziger, G., Friedman, R., Manes, B.: Tinylfu: a highly efficient cache admission policy. ACM Trans. Storage 13(4), 1–31 (2017)

14. Eugster, P.T., Felber, P.A., Guerraoui, R., Kermarrec, A.M.: The many faces of publish/subscribe. ACM Comput. Surv. (CSUR) 35(2), 114–131 (2003)

15. Kreps, J., Narkhede, N., Rao, J., et al.: Kafka: a distributed messaging system for log processing. Proc. NetDB 11, 1–7 (2011)

16. Leemans, S.J.J., Fahland, D., van der Aalst, W.M.P.: Discovering block-structured process models from event logs containing infrequent behaviour. In: Lohmann, N., Song, M., Wohed, P. (eds.) BPM 2013. LNBIP, vol. 171, pp. 66–78. Springer, Cham (2014). https://doi.org/10.1007/978-3-319-06257-0_6

17. de Leoni, M.M., Mannhardt, F.: Road traffic fine management process (2015). https://doi.org/10.4121/uuid:270fd440-1057-4fb9-89a9-b699b47990f5

18. Murata, T.: Petri nets: properties, analysis and applications. Proc. IEEE 77(4), 541–580 (1989)

19. Podlipnig, S., Böszörmenyi, L.: A survey of web cache replacement strategies. ACM Comput. Surv. (CSUR) 35(4), 374–398 (2003)

20. Rozinat, A., van der Aalst, W.M.P.: Conformance checking of processes based on monitoring real behavior. Inf. Syst. 33(1), 64–95 (2008)

21. Schuster, D., van Zelst, S.J.: Online process monitoring using incremental state-space expansion: an exact algorithm. In: Fahland, D., Ghidini, C., Becker, J., Dumas, M. (eds.) BPM 2020. LNCS, vol. 12168, pp. 147–164. Springer, Cham (2020). https://doi.org/10.1007/978-3-030-58666-9_9

22. van Zelst, S.J., Bolt, A., Hassani, M., van Dongen, B.F., van der Aalst, W.M.P.: Online conformance checking: relating event streams to process models using prefix-alignments. Int. J. Data Sci. Anal. 8(3), 269–284 (2017). https://doi.org/10.1007/s41060-017-0078-6

Bringing Cognitive Augmentation to Web Browsing Accessibility

Alessandro Pina[1], Marcos Baez[2(✉)], and Florian Daniel[1]

[1] Politecnico di Milano, Milan, Italy
alessandro.pina@mail.polimi.it, florian.daniel@polimi.it
[2] LIRIS – University of Claude Bernard Lyon 1, Villeurbanne, France
marcos.baez@liris.cnrs.fr

Abstract. In this paper we explore the opportunities brought by cognitive augmentation to provide a more natural and accessible web browsing experience. We explore these opportunities through *conversational web browsing*, an emerging interaction paradigm for the Web that enables blind and visually impaired users (BVIP), as well as regular users, to access the contents and features of websites through conversational agents. Informed by the literature, our previous work and prototyping exercises, we derive a conceptual framework for supporting BVIP conversational web browsing needs, to then focus on the challenges of automatically providing this support, describing our early work and prototype that leverage heuristics that consider structural and content features only.

Keywords: Chatbots · Conversational web browsing · Heuristics · Web accessibility

1 Introduction

Accessing the Web has long relied on users to correctly process and interpret visual cues in order to have a proper user experience. Web browsers as well as information and services on the Web are optimised to make full use of user's visual perceptive capabilities for organising, delivering and fulfilling their goals. This introduces problems for blind and visually impaired people (BVIP) who due to genetic, health or age-related conditions are not able to effectively rely on their visual perception [6].

Assistive technology such as screen readers have traditionally supported BVIP users in interacting with visual interfaces. These tools exploit the *accessibility tags* used by Web developers and content creators in order to facilitate access to information and services online, typically by reading out the elements of the website sequentially from top to bottom (see Fig. 1). They are usually controlled with a keyboard, offering shortcuts to navigate and access content. The challenges faced by BVIP in browsing the Web with this type of support is well documented in the literature, ranging from websites not designed for accessibility [10,18] to limitations of screen reading technology [3,21,29].

© Springer Nature Switzerland AG 2021
H. Hacid et al. (Eds.): ICSOC 2020 Workshops, LNCS 12632, pp. 395–407, 2021.
https://doi.org/10.1007/978-3-030-76352-7_37

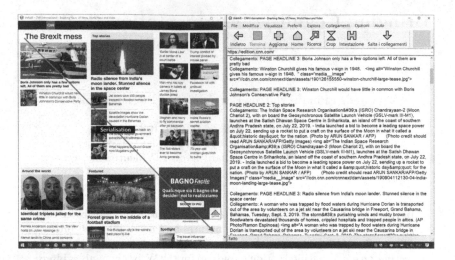

Fig. 1. Example serialisation of a website with by a screen reader. HTML elements are read typically from top to bottom, as informed by the website HTML structure

Cognitive augmentation has been regarded as a promising direction to empower populations challenged by the traditional interaction paradigm for accessing information and services [7]. Conversational browsing is an emerging interaction paradigm for the Web that builds on this promise to enable BVIP, as well as regular users, to access the contents and services provided by websites through dialog-based interactions with a conversational agent [4]. Instead of relying on the sequential navigation and keyboard shortcuts provided by screen readers, this approach would enable BVIP to express their goals by directly "talking to websites". The first step towards this vision was to identify the conceptual vocabulary for augmenting websites with conversational capabilities and explore techniques for generating chatbots out of websites equipped with bot-specific annotations [13].

In this paper we take a deeper dive into the opportunities of cognitive augmentation for BVIP by building a conceptual framework that takes the lessons learned from the literature, our prior work and prototyping exercises to highlight areas for conversational support. We then focus on the specific tasks that are currently served rather poorly by screen readers, and describe our early work towards a *heuristic-based* approach that would leverage visual and structural properties of websites to translate the experience of graphical user interface into a conversational medium.

2   Conceptual Framework

2.1   Motivating Scenario

Before introducing the main concepts, we illustrate our vision by describing an example interaction of a BVIP looking for information on COVID-19 on a local

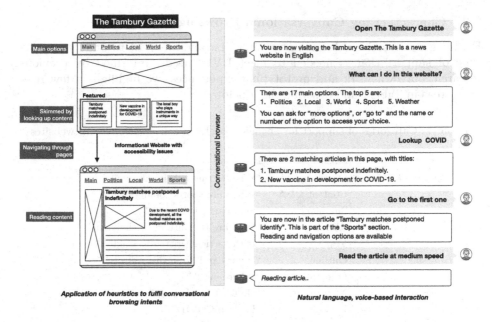

Fig. 2. Example conversational browsing on an information-intensive website.

newspaper. Peter, 72, is a visually impaired man affected by Parkinson disease, who keeps hearing about the new virus COVID-19 on TV and wants to be updated constantly about the recent news from his favorite local newspaper, The Tambury Gazette. However, his experience with screen readers has been poor and frustrating, often requiring assistance from others to get informed.

The vision is to enable users like Peter to browse the Web by directly "talking" to websites. As seen in Fig. 2, the user interacts with the website in dialog-based voice-based interactions with a conversational agent (e.g., Google Assistant). The user can start the session by searching for the website, or opening it up directly if already bookmarked. Once open, the user can inquire about the relevant actions that are available in its current context (e.g., *"What can I do in this website"*), which are automatically derived by the conversational agent based on heuristics. Instead of sequentially going through the website, the user can lookup for specific information within the website matching his interests (e.g., *"Lookup COVID"*). The user can then follow up on the list of resulting articles and chose one to be read out. As part of these interactions, the user can use the voice commands to navigate and get oriented in the website.

The above illustrates the experience of browsing a website by leveraging natural language commands that improve over the keyboard-based sequential navigation of websites. As we will see, more advanced support can be provided by leveraging the contents and application-specific domain knowledge, but in this work we focus on improving on the features provided by screen readers, making no assumptions about compliance with accessibility and bot-specific annotations.

2.2 Characterising Conversational Browsing Support

Enabling conversational browsing requires first and foremost to understand the type of support that is needed to meet BVIP needs. Informed by previous research, our own work and prototyping experiences, we highlight a few relevant areas in Table 1 and describe them below.

Table 1. Categories of support for engaging BVIP in conversations with websites

Category	Skills	Examples
Metadata & content	Overview	"What is this website about?"
	Content Q&A	"When are sports coming back?"
	Summary	"Summarise the article?"
	About	"Who are the authors of this article?"
	Yes/No	"Is the article written in English?"
Browsing	Outline	"What can I do in this website?"
	Orientation	"Where am I?"
	Navigation	"Go to the main page"; "Next article"
	Lookup	"Lookup COVID"
	Reading	"Read article"; "Stop reading"
Workflows	Element-specific	"Fill out the form"
	App-specific	"Post a new comment on the news article"
Operations	Open	"Open The Tambury Gazette"
	Search	"Search for The Tambury Gazette"
	Bookmark	"Bookmark page The Tambury Gazette"
	Speech	"Increase speech rate"
	Verbosity	"Turn on short interactions"

Conversational Access to Content and Metadata. BVIP should be able to satisfy their information needs without having to sequentially go through all the website content and structure – a process that can be time consuming and frustrating for screen reader users [18]. This support is rooted in the ongoing efforts in conversational Q&A [12] and document-centered digital assistant [17]. The idea is to support BVIP users to perform natural language queries (NLQ) on the contents of websites, and to inquire about the properties defined in the website's metadata. For example, a BVIP user might request an *overview* of the website (e.g., "What is this website about?"), engage in *question & answering*, with questions that can be answered by referring directly to the contents of the website (e.g., "When are the sports coming back?"), and ask for *summaries* of the contents of the website, its parts or responses from the agent (e.g., "Summarise the article"). Users might also ask *about the properties* and metadata of the artefacts, such as last modification, language, authors (e.g., "Who are the authors of this article?"), or simply engage in *yes/no* questions on metadata and content (e.g., "Is the document written in English?").

Conversational Browsing. BVIP should be allowed to explore and navigate the artefacts using natural language, so as to support more traditional information seeking tasks. The idea is to improve on the navigation provided by traditional screen readers, which often require learning complex shortcuts and lower level knowledge about the structure of artefact (e.g., to move between different sections), by allowing users to utter simpler high level commands in natural language. This category of support is inspired by the work in Web accessibility, in using spoken commands to interact with non visual web browsers [3,28] and conversational search [26,27]. For example, BVIP should be able to inquire about the website organization and get an *outline* (e.g., "What can I do in this website?"), and *navigate* through the structure of the website and even across linked webpages (e.g., "Go to the main page"), and being able to get *oriented* during this exploratory process (e.g., "Where am I?"). The user should also be able to *lookup* for relevant content to avoid sequentially navigating the page structure (e.g., "Lookup COVID").

Conversational User Workflows. BVIP should also be able to enact user workflows by leveraging the features provided by the website. This is typically done by the users, enacting their plan by following links, filling out forms and pressing buttons. This low level interactions have been explored by speech-enabled screen readers such as Capti-Speak [3], enabling user to utter commands (e.g., "press the cart button", "move to the search box"). We call these *element-specific* intents. In our previous work we highlighted the need for supporting *application-specific* intents, i.e., intents that are specific to the offerings of a website (e.g., "Post a new comment on the news article") and that would trigger a series of low-level actions as a result. In our approach such experience required bot-specific annotations [13]. The automation of such workflows has also been explored in the context of Web accessibility. For example, Bigham et al. [9] introduced the trailblazer system, which focused on facilitating the process of creating web automation macros, by providing step by step suggestions based on CoScript [19]. It is also the focus of the research in robotic process automation [20].

Conversational Control Operations. BVIP should be able to easily access and personalise the operational environment. This goes from simple operations to support the main browsing experience, such as searching and opening websites and managing the bookmarks, to personalising properties of the voice-based interactions. Recent works in this context have highlighted the importance of providing BVIP with higher control over the experience. Abdolrahmani et al. [1] investigated the experience by BVIP with voice-activated personal assistance and reported that users often feel responses being too verbose, frustrated at interacting at a lower pace than desired, or not able adapt interactions to the requirements of social situations. It has been argued [11] that guidelines by major commercial voice-based assistants fail to capture preferences and experience of BVIP, used to faster and more efficient interactions with screen readers. This calls for further research into conversation design tailored to BVIP.

2.3   Approaches and Challenges

There are many challenges in delivering the type of support required for conversational browsing. As discussed in our prior work [13], this requires deriving two important types of knowledge:

- **domain knowledge**: it refers to the knowledge about the type of functionality and content provided by the website, and that will inform the agent of what should be exposed to the users (e.g., intents, utterances and slots);
- **interaction knowledge**: it refers to the knowledge about how to operate and automate the browsing interactions on behalf of the user.

Websites are not equipped with the required conversational knowledge to enable voice-based interaction, which have motivated three general approaches.

The **annotation-based approach** provides conversational access to websites by enabling developers and content producers to provide appropriate annotations [4]. Early approaches can be traced back to enabling access to web pages through telephone call services via VoiceXML [22]. Another general approach to voice-based accessible information is to rely on accessibility technical specifications, such as Accessible Rich Internet Applications (WAI-ARIA) [15], but these specifications are meant for screen reading. Baez et al. [4] instead propose to equip websites with bot-specific annotations. The challenge in this regard is the adoption by annotations by developers and content producers. A recent report[1] analysing 1 million websites reported that a staggering 98.1% of the websites analysed had detectable accessibility errors, illustrating the extent of the adoption of accessibility tags and proper design choices on the Web.

The **crowd-based approach** utilizes collaborative metadata augmentation approaches [8,25], relying instead on the crowd to "fix" accessibility problems or provide annotations for voice-based access. The Social Accessibility project [24] is one of these initiatives whose database supports various non visual browsers. Still, collaborative approaches require a significant community to be viable, and even so the numbers of services and the rate at which they are created make it virtually impossible to cover all websites.

Automatic approaches have been used to support non-visual browsing and are based on heuristics and algorithms. The approaches in this space have focused on automatically fixing accessibility issues (e.g., page segmentation [14]), deriving browsing context [21] or predicting next user actions based on current context [23]. These approaches, however, have not focused on enabling conversational access to websites.

All of the above tell us of the diverse approaches that can support the cognitive augmentation of websites to enable voice-based conversational browsing. In this work we explore automatic approaches, which have not been studied in the context of conversational access to websites.

---

[1] https://webaim.org/projects/million/.

3   A Heuristic-Based Approach

In this work we focus on heuristics that enable voice-based *navigation* and *access to content* of websites that are *information intensive*. We focus on these set of features (Table 1, "Browsing category") as they i) match the level of support expected but poorly served by screen readers, and ii) are highly impacted by accessibility errors in websites.

3.1   Requirements

From the conceptual framework, it becomes clear that enabling BVIP to browse websites conversationally requires us to:

- Determine the main (and contextual) offerings of the website
- Identify the current navigation context
- Enable navigation through meaningful segments of the website
- Allow for scanning and search for information in the website

Determining the offerings of the website can be done by leveraging the components used in graphical user interfaces to guide users through their offerings: menus. Menus have specific semantic tags in HTML (<nav>) and roles as part of the technical specifications for web accessibility (WAI-ARIA) that allow screen readers to identify the main navigation links in a website. They also rely on distinctive visual and structural properties (e.g., styles and position) to make them easily identifiable by sighted users. For example, they tend to be more prominent, towards the top and repeat across all pages in the website. Instead, more localised options are typically embedded in the content (e.g., links) or located within the same section of the page. Advanced models have relied in such visual properties to derive the role of rendered components in websites [2].

Identifying and keeping track of navigation context is supported in different ways by visual web browsing. In a website, this can be provided by design e.g., by implementing navigation breadcrumbs that explicitly render the navigation path that took users to their current context. It is also supported by Web browsers as part of the navigation history, allowing users to go back an forth in their navigation path but without illustrating it explicitly. In the context of a dialog, we can leverage this browsing history (available as a Web API) along with the conversation history to resolve the current browsing context based on navigation path (e.g., current page) and previous choices (e.g., name of links selected).

Enabling navigation requires supporting browsing activities across different pages in the website, and therefore identifying relevant links and their target components. In visual browsing, the identification of the target components are typically done *visually* by sighted users by relying on the layout of websites and their own goals. That is, when opening a news article, sighted users can focus their attention on the content of the article, ignoring other components such as headers, menus and ads. Given the proper accessibility tags, screen readers can allow users to (manually) identify their targets by skipping regions of the website. To provide a proper experience, it is fundamental to have meaningful segmentation of the website according to visual properties, and identifying target segments during navigation based on navigation context (e.g., as in done in [21]). Segmentation techniques have been widely studied in accessibility research and could be leveraged for this purpose.

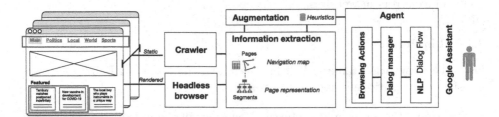

Fig. 3. Pipeline for augmenting information-intensive websites with conversational capabilities by leveraging heuristics.

Searching within a website is not a particularly challenging feature. The challenge lies again in segmenting the resulting elements and contextualising the guidance based on the type of visual element (e.g., paragraph → reading; links → navigation). However, in accessibility this is associated to the *non visual scanning* task, i.e., efficiently finding relevant information among many relevant ones, which has motivated several techniques, including the use of multiple concurrent audio channels [16], that should be considered as potential techniques.

3.2   Prototype Implementation

We implemented a prototype as an exercise into understanding and informing i) the type of support required in voice-based interactions, as well as ii) their technical requirements. The main focus was to establish a pipeline that can take voice commands, and fulfill them based on an (evolving) set of heuristics. In doing so, we faced the following architectural constraints:

- The agent needs to serve multiple websites, as with a regular web browser
- The agent needs to support conversational browsing intents, and the experience need to be optimised for "reading" content
- Processing times should be minimised for a meaningful user experience
- The agent needs to support dynamic web pages

The resulting pipeline is illustrated in Fig. 3. In a nutshell, the pipeline takes a website and its static and dynamic content to create internal representations that can be leveraged by the heuristics to serve the predefined browsing intents. In the following we detail on this pipeline.

Crawling and Data Scraping. The first component in the pipeline is in charge of obtaining the static and dynamic contents and structure of the website for further processing and analysis. The process starts with the input URL to fetch the static HTML of each page in the website. It performs a *breadth-first search* of the website's tree structure, identifying in each page all the hyperlinks to visit. This process is performed the first time the website is accessed, and it is cached (with an expiration date) for later use. The crawling runs in the background, stopping at a configurable depth d in the tree or when a number of web pages p have been processed. This process is implemented with *Scrapy*[2], a Python framework used for large scale web crawling.

While accessing the static version of a website ensures higher performance by reducing rendering times and allowing faster website-level analyses, it does not necessarily represent the actual content presented to the user, since part of it can be dynamically generated. For this reason, we complement the "quick glance" provided by the crawling process by accessing the rendered version of the website on demand – meaning the actual pages the user navigates to. The implementation relies on *Selenium*, a powerful tool for automated testing in web browsers, running *Mozilla Firefox* in headless mode to access the rendered version of the websites, with extensions such as AdBlocker and i-dont-care-about-cookies.eu to speed up rendering and loading times.

Information Extraction and Augmentation. This component takes in input the *website-level* information and the more detailed and accurate *page-level* information from rendered pages to build internal representations of the website. The website-level information is leveraged to build the navigation graph of the website, and calculate basic metrics on the structure (e.g., popularity: the number of times a link is referenced) that can later inform the heuristics. Basic metadata is extracted but the static HTML is not further processed at this stage. Then, when the user navigates to specific page (a node in the navigation graph) the rendered version of the website is requested, and the actual content and structural properties of the website are analysed. The page is represented as a tree-structure, much like the DOM but where each node is a meaningful segment of the website, as derived by the segmentation heuristics (described later). The contents of the nodes are extracted and cleaned to make them reading friendly (e.g., inline links replaced by placeholders and offered separately). The implementation of this component relied on the *BeautifulSoup*[3] Python package to analyse the HTML code and scrape data.

Computing Heuristics. The current prototype implements simple heuristics that serve as placeholders that will allow more comprehensive tests of the entire pipeline. An example of such heuristics, for the identification of the main offerings of the website, is based on the observation that links in the main menu tend to be at the top of the page and present across the entire website. We therefore leveraged on the navigation map and the calculated popularity metric for each

---

[2] https://scrapy.org/.

[3] https://www.crummy.com/software/BeautifulSoup/bs4/doc/.

node (i.e., how many times the link is referenced), weighted by the position attribute of the link element in the rendered website (e.g., thus discerning links in footer and headers) to rank the links. We do not currently perform segmentation, and the segmentation placeholder just leverages existing region landmarks.

Other features of Table 1 currently rely on existing cognitive services. For example, for providing *Summary*, we rely on Aylien Text Analysis API[4], which along with Fortiguard Web Filter (see Footnote 4) augments the information about the website with extra metadata (e.g., language and topic of the website). The search feature are provided by Google Search.

Conversational Agent. The browsing experience is ultimately delivered through *Google Assistant*, which was chosen as the voice-based service. This service provides a conversational medium and performs the speech-to-text and text-to-speech transformations to and from the natural language processing unit. We relied on *Dialogflow*[5] as natural language platform, where the intents for serving the conversational browsing needs were defined. These include the Browsing, and a few of Operations and Metadata & Content from Table 1. The webhooks to handle the fulfillments ultimately pointed to our Python server. The source code of our prototype is available at https://github.com/Shakk17/WebsiteReader.

3.3   Preliminary Evaluation

A preliminary evaluation of the system was performed so as to assess the technical performance of the tool, and gain insights on the structure of websites and the challenges they present to our heuristic-based approach.

A total of 30 websites were selected from Alexa's top ranking[6], taking 5 websites from each of the six categories typically associated with information-intensive websites: Newspapers, Sports, Reference, Health, Society and Science. We tested the accessibility compliance of these top websites with the WAVE accessibility tool[7]. This revealed that only 4 out of the 30 websites were free of accessibility errors, which further illustrates the challenges to our approach and to assistive technology in general.

In this exploratory run, we evaluated the performance of the simple heuristic for inferring the offerings of the website. To do this, we first manually analysed each website to identify the links from the menus (the offerings). These actual links were then compared against the output of the heuristic, which was we set to return a maximum of 30 links (threshold), to compute precision and recall.

The results showed that the heuristic is effective in identifying relevant links (recall = 0.79) but less precise in determining the number of links to recommend (precision = 0.42). However, this is mainly due to the static threshold and the highly wide range of menu size and complexity in websites (from 4 to 40 links).

---

[4] https://fortiguard.com/webfilter.
[5] https://dialogflow.com/.
[6] https://www.alexa.com/topsites/category/Top.
[7] https://wave.webaim.org.

Indeed, the precision was much higher when the number of recommended links approached the number of actual links in the menu.

Our observations running these tests tell us that the solution goes beyond more intelligent heuristics and cut-off values for links. The analysis revealed the complexity of menus in websites – some of then with dozens of hierarchical links – which motivates an exploration into new approaches to presenting and discovering available offerings conversationally. The exploration of conversational patterns for menu-access as well as heuristics for identifying global and local intents (links) emerge are interesting areas for future research.

4   Discussion and Future Work

In this paper we have explored the opportunities of cognitive augmentation and automation to support BVIP in browsing information-intensive websites. The approach is based on the notion of enabling dialog-based interaction with websites, mediated by a voice-based conversational agent.

These opportunities were materialised in a conceptual framework that summarised based on literature review, our prior work and prototyping exercises, the categories of support to be addressed to enable conversational browsing by BVIP. These include the ability to interact with the contents of the website, support more traditional browsing tasks, automating user workflows and managing the entire operating environment of the browsing experience. The infrastructure of the Web today is not equipped serve these needs, but we have shown that cognitive computing can enable and augment the existing foundation – much as with cognitive process augmentation [7] – to help address these needs. Existing research and techniques in accessibility can greatly kick-start these efforts.

It is however clear that automation alone cannot fulfill this vision. Delivering a proper conversational experience, under the limitations and constraints posed by the problem, would require addressing technical issues as well as implementing dialog patterns that can reduce their impact and provide guidance. Equipping website with conversational knowledge, reflecting the intended conversational experience, appears to be key in this regard. Understanding the correct trade-off between what should be explicitly annotated and can be automatically derived are among the challenges to be addressed.

As part of our ongoing work we are planning to integrate a pool of existing algorithms and heuristics developed in the accessibility community and setup benchmarks to understand their suitability and performance. We are also planning user studies to understand the impact of different dialog patterns, associated with different levels of explicit and implicit conversational knowledge. The long-term vision is to integrate conversational capabilities into systems of any kind, a problem that we have seen emerging and gaining traction in recent years [5].

References

1. Abdolrahmani, A., Kuber, R., Branham, S.M.: "Siri talks at you" an empirical investigation of voice-activated personal assistant (VAPA) usage by individuals who are blind. In: Proceedings of the 20th International ACM SIGACCESS Conference on Computers and Accessibility, pp. 249–258 (2018)
2. Akpinar, M.E., Yeşilada, Y.: Discovering visual elements of web pages and their roles: users' perception. Interact. Comput. **29**(6), 845–867 (2017)
3. Ashok, V., Borodin, Y., Puzis, Y., Ramakrishnan, I.: Capti-speak: a speech-enabled web screen reader. In: W4A, p. 22. ACM (2015)
4. Baez, M., Daniel, F., Casati, F.: Conversational web interaction: proposal of a dialog-based natural language interaction paradigm for the web. In: Følstad, A., Araujo, T., Papadopoulos, S., Law, E.L.-C., Granmo, O.-C., Luger, E., Brandtzaeg, P.B. (eds.) CONVERSATIONS 2019. LNCS, vol. 11970, pp. 94–110. Springer, Cham (2020). https://doi.org/10.1007/978-3-030-39540-7_7
5. Baez, M., Daniel, F., Casati, F., Benatallah, B.: Chatbot integration in few patterns. IEEE Internet Comput. (2020)
6. Barreto, A., Hollier, S.: Visual disabilities. In: Yesilada, Y., Harper, S. (eds.) Web Accessibility. HIS, pp. 3–17. Springer, London (2019). https://doi.org/10.1007/978-1-4471-7440-0_1
7. Barukh, M.C., et al.: Cognitive augmentation in processes. In: Aiello, M., Bouguettaya, A., Tamburri, D.A., van den Heuvel, W.-J. (eds.) Next-Generation Digital Services - Essays Dedicated to Mike Papazoglou. LNCS, vol. 12521, pp. 123–137. Springer, Cham (2021). https://doi.org/10.1007/978-3-030-73203-5_10
8. Bigham, J.P.: Accessmonkey: enabling and sharing end user accessibility improvements. ACM SIGACCESS Access. Comput. **89**, 3–6 (2007)
9. Bigham, J.P., Lau, T., Nichols, J.: Trailblazer: enabling blind users to blaze trails through the web. In: IUI, pp. 177–186. ACM (2009)
10. Bigham, J.P., Lin, I., Savage, S.: The effects of not knowing what you don't know on web accessibility for blind web users. In: Proceedings of the 19th International ACM SIGACCESS Conference on Computers and Accessibility, pp. 101–109. ACM (2017)
11. Branham, S.M., Mukkath Roy, A.R.: Reading between the guidelines: how commercial voice assistant guidelines hinder accessibility for blind users. In: The 21st International ACM SIGACCESS Conference on Computers and Accessibility, pp. 446–458 (2019)
12. Braun, D., Mendez, A.H., Matthes, F., Langen, M.: Evaluating natural language understanding services for conversational question answering systems. In: Proceedings of the 18th Annual SIGdial Meeting on Discourse and Dialogue, pp. 174–185 (2017)
13. Chittò, P., Baez, M., Daniel, F., Benatallah, B.: Automatic generation of chatbots for conversational web browsing. arXiv preprint arXiv:2008.12097 (2020)
14. Cormier, M., Moffatt, K., Cohen, R., Mann, R.: Purely vision-based segmentation of web pages for assistive technology. Comput. Visi. Image Underst. **148**, 46–66 (2016)
15. Diggs, J., Craig, J., McCarron, S., Cooper, M.: Accessible rich internet applications (WAI-ARIA) 1.1 (2016). Accessed 24 Apr 2016
16. Guerreiro, J., Gonçalves, D.: Scanning for digital content: how blind and sighted people perceive concurrent speech. ACM Trans. Access. Comput. (TACCESS) **8**(1), 1–28 (2016)

17. ter Hoeve, M., Sim, R., Nouri, E., Fourney, A., de Rijke, M., White, R.W.: Conversations with documents: an exploration of document-centered assistance. In: Proceedings of the 2020 Conference on Human Information Interaction and Retrieval, pp. 43–52 (2020)
18. Lazar, J., Allen, A., Kleinman, J., Malarkey, C.: What frustrates screen reader users on the web: a study of 100 blind users. Int. J. Hum.-Comput. Interact. **22**(3), 247–269 (2007)
19. Leshed, G., Haber, E.M., Matthews, T., Lau, T.: Coscripter: automating & sharing how-to knowledge in the enterprise. In: CHI, pp. 1719–1728. ACM (2008)
20. Mahala, G., Sindhgatta, R., Dam, H.K., Ghose, A.: Designing optimal robotic process automation architectures. In: Kafeza, E., Benatallah, B., Martinelli, F., Hacid, H., Bouguettaya, A., Motahari, H. (eds.) ICSOC 2020. LNCS, vol. 12571, pp. 448–456. Springer, Cham (2020). https://doi.org/10.1007/978-3-030-65310-1_32
21. Mahmud, J.U., Borodin, Y., Ramakrishnan, I.: Csurf: a context-driven non-visual web-browser. In: Proceedings of the 16th International Conference on World Wide Web, pp. 31–40. ACM (2007)
22. Oshry, M., Auburn, R., Baggia, P., Bodell, M., Burke, D., Burnett, D., et al.: Voice extensible markup language (VoiceXML) 2.1. W3C recommendation (2007)
23. Puzis, Y., Borodin, Y., Puzis, R., Ramakrishnan, I.: Predictive web automation assistant for people with vision impairments. In: Proceedings of the 22nd International Conference on World Wide Web, pp. 1031–1040 (2013)
24. Sato, D., Takagi, H., Kobayashi, M., Kawanaka, S., Asakawa, C.: Exploratory analysis of collaborative web accessibility improvement. ACM TACCESS **3**(2), 5 (2010)
25. Takagi, H., Kawanaka, S., Kobayashi, M., Itoh, T., Asakawa, C.: Social accessibility: achieving accessibility through collaborative metadata authoring. In: ACM SIGACCESS Conference on Computers and Accessibility, pp. 193–200. ACM (2008)
26. Trippas, J.R.: Spoken conversational search: information retrieval over a speech-only communication channel. In: Proceedings of the 38th International ACM SIGIR Conference on Research and Development in Information Retrieval, p. 1067 (2015)
27. Trippas, J.R., Spina, D., Cavedon, L., Sanderson, M.: How do people interact in conversational speech-only search tasks: a preliminary analysis. In: Proceedings of the 2017 Conference on Conference Human Information Interaction and Retrieval, pp. 325–328 (2017)
28. Vesnicer, B., Zibert, J., Dobrisek, S., Pavesic, N., Mihelic, F.: A voice-driven web browser for blind people. In: Eighth European Conference on Speech Communication and Technology (2003)
29. Zhu, S., Sato, D., Takagi, H., Asakawa, C.: Sasayaki: an augmented voice-based web browsing experience. In: Proceedings of SIGACCESS 2010. ACM (2010)

Towards Knowledge-Driven Automatic Service Composition for Wildfire Prediction

Hela Taktak[1,2(✉)], Khouloud Boukadi[1], Chirine Ghedira Guégan[2],
Michael Mrissa[3], and Faïez Gargouri[1]

[1] MIRACL Laboratory, FSEG Sfax, Sfax University, Sfax, Tunisia
`khouloud.boukadi@fsegs.usf.tn, faiez.gargouri@isims.usf.tn`
[2] Lyon University, Lyon 3 University, LIRIS UMR5205, Lyon, France
`{hela.taktak1,chirine.ghedira-guegan}@univ-lyon3.fr`
[3] InnoRenew CoE, University of Primorska, Livade 6, 6310 Izola, 6000 Koper,
Slovenia
`michael.mrissa@innorenew.eu`

Abstract. Wildfire prediction from Earth Observation (EO) data has
gained much attention in the past years, through the development of connected sensors and weather satellites. Nowadays, it is possible to extract
knowledge from collected EO data and to learn from this knowledge
without human intervention to trigger wildfire alerts. However, exploiting knowledge extracted from multiple EO data sources at run-time and
predicting wildfire raise multiple challenges. One major challenge is to
provide dynamic construction of service composition plans, according to
the data obtained from sensors. In this paper, we present a knowledge-
driven Machine Learning approach that relies on historical data related
to wildfire observations to guide the collection of EO data and to automatically and dynamically compose services for triggering wildfire alerts.

Keywords: Machine Learning · Fire prediction · Service composition

1 Introduction

Wildfire, or wildland fire, is a regularly spotted critical phenomenon that can
massively damage human lives, infrastructures, agriculture, and forest ecosystems. It has negative implications on air and water quality, and soil integrity.
Recent estimations based on Earth Observation (EO) satellites of the global

This work was financially supported by the "PHC Utique" program of the French Ministry of Foreign Affairs and Ministry of higher education and research and the Tunisian
Ministry of higher education and scientific research in the CMCU project number
17G1122. The authors acknowledge the European Commission for funding the InnoRenew CoE project (Grant Agreement #739574) under the Horizon2020 Widespread-
Teaming program and the Republic of Slovenia (Investment funding of the Republic
of Slovenia and the European Union of the European regional Development Fund).

© Springer Nature Switzerland AG 2021
H. Hacid et al. (Eds.): ICSOC 2020 Workshops, LNCS 12632, pp. 408–420, 2021.
https://doi.org/10.1007/978-3-030-76352-7_38

burned area is around 420 Mha [9]. Several advances have been made in fire observations based on physics-based simulators and remote-sensing technologies, such as satellites (e.g. NASA TERRA), and fire detection sensors (e.g. Visible Infrared Imaging Radiometer Suite (VIIRS)), that continuously monitor vegetation distribution and changes. Our PREDICAT (PREDIct natural CATastrophes) project[1], collects EO data from several data sources, such as the National Oceanic and Atmospheric Administration (NOAA) which focuses on climate and oceans, and the Observatory of Sahara and Sahel (OSS), which relies on remote sensing technologies and Internet of Things (IoT) devices for climate parameters sensing. Exploiting EO data rises multiple challenges, ranging from data collection through service composition[2] to triggering wildfire alerts. The issue is that triggering wildfire alerts, is generally realized, once the collection of all the EO data is available. Existing wildfire detection and prediction systems mainly focus on accurately building feature observations and manually defining domain-rules. Indeed, several Web service composition approaches have been developed to implement data collection and trigger alerts, based on optimization, decision-making methods [10,12], and semantics [2,8]. While producing candidate solutions, most service composition techniques neglect knowledge from previously called services or from the application domain. Existing approaches start from a user request and try to find out a set of optimal services, based on functional and non-functional parameters. However, the service composition process for wildfire prediction should be data-driven, which means that the service calls alternatives are chosen at run-time depending on the data received from the previously called services.

Therefore, in this paper, we propose a bottom-up approach to guide the EO data collection from IoT devices based on a knowledge-driven process, and dynamically compose services accessing IoT data for wildfire prediction. We apply Machine Learning (ML) techniques [1], to guide the service composition process for EO collection and fire prediction by learning from the data itself and from historical data related to fire observations. Our solution exploits a service-based combination of ML and knowledge engineering methodologies, to automatically and dynamically compose services for wildfire prediction. Our contribution can be summarized as follows: we describe our knowledge-driven service composition approach through a system architecture for wildfire prediction, including (1) a dynamic and knowledge-driven construction of services composition scheme to organize the flow of services, based on a Prediction Module. (2) a classification of the EO historical data collections, by a Learning Module. (3) EO data collection, according to the most important feature related to fires to alert the scientists, by an Awareness Module. (4) the generation of alert risk patterns, by the Prediction Module. The remainder of this paper is structured as follows: Sect. 2 overviews related works dealing with Web service composition. Section 3 describes the global system architecture for prediction. Section 4 defines the pre-

---

[1] https://sites.google.com/view/predicat/predicat.

[2] Service composition is the combination of a set of the smallest services forming a more complex service to meet users' complex requirements.

diction flow. Section 5 demonstrates the applicability of our approach. Finally, we conclude and present our future work in Sect. 6.

2   Related Work

Web service composition for IoT environments has been often employed to access multiple IoT devices [18,19]. Authors in [21] proposed a distributed social network approach for IoT management and service composition. An agent-based middleware was proposed in [22], to handle service composition of logistics services in IoT. Asghari et al. [20] proposed a systematic literature review on service composition approaches in IoT. They aimed to analyze and categorize IoT service composition methods into two main categories focusing on functional and non-functional properties. Other efforts were oriented towards semantic Web service composition [2,3,8], and QoS-aware web service composition [10–12]. Many approaches employed Evolutionary Computing (EC) to automate the generation of composition solutions and the optimization of the QoS web service composition by handling the large search spaces of services. Genetic Algorithm (GA) is used in [10], to propose a composition solution. Genetic Programming is employed in [13] and [11] to find near-optimal solutions using a fitness function. Although these solutions are a promising direction, efforts are still needed for enforcing composition constraints and also, for optimizing the quality of solutions. Another group of approaches for Qos-aware Web service composition employ the Particle Swarm Optimisation (PSO), which searches for near-optimal solutions by avoiding producing invalid solutions [12]. All these QoS-aware service composition approaches do not consider the semantic matchmaking quality of service compositions. Therefore, another group of researches focused on the semantic Web service composition using ontology-based semantics, such as OWL-S, WSML, and SAWSDL [14], to semantically represent the knowledge conveyed in these Web services [5]. To sum-up, although a large number of approaches for semantic web service composition and QoS-aware service composition exist, they all propose solutions based on the expression of a user query and not on domain knowledge. Furthermore, these approaches do not dynamically adapt the composition schema when new knowledge is learned, to guide the data collection process in a knowledge-capable manner.

3   System Architecture for Prediction

Figure 1 depicts our layered prediction system architecture [4] that includes five layers: Semantic layer, Knowledge layer, Application layer, Service Composition layer, and Service layer. The Semantic layer contains the domain and source ontologies. It aims to semantically describe the domain services and their related data sources. Hence, the domain ontology represents the environmental domain concepts related to services for EO data collection. Furthermore, the source ontology represents the quality of the data sources accessed by their related services[3]. The Knowledge layer consists of collecting the EO data from

---

[3] Qualities of the data sources and services are out of the scope of this paper.

diverse environmental data sources, provided by the OSS. The environmental data sources include data from several devices (i.e.: sensors, connected objects, satellites, etc.). In our case, the knowledge-base contains the main features of interest of the environmental domain related to the fire occurrences.

The Application layer which constitutes the main focus of this paper, sums-up our vision of how to predict wildfire occurrences based on knowledge and learning methodologies. The Application layer uses the EO data from the knowledge-base. This layer encompasses the Prediction Module, which is composed of two sub-modules: the Learning Module and the Awareness Module. The Prediction Module produces alert risk patterns. The Learning Module builds the prediction model by applying a classification ML algorithm. It takes as inputs the set of features of interest that shapes the fire (i.e.: temperature, wind speed, humidity, wind direction, etc.) and produces as output the class of fire danger (i.e.: Moderate, Low, High, Very High, Extreme, Catastrophic, etc.). The supervised ML algorithm in the Learning Module generates the decision tree (DT), which is the fire prediction model. This model determines the danger classes of fire, in the leaf-nodes of the DT. The Awareness Module handles the search for the important features and the traversal branches of the generated DT. It consists of traversing all the necessary branches of the DT, from the important feature node till the leaf-node determining the danger class fire. The first step in the Awareness Module is to determine the most important feature, upon a classification task of all the features of interest. The most important feature is determined by computing the highest score among all the features of interest based on RF feature selection method. More details about this method are given in Sect. 5.1. In the Service layer, the most important feature is mapped to an abstract service, which is a class interface representing its functionality and designed independently from particular implementations of services. The abstract service of the most important feature is then mapped to its service instance. This service is invoked, executed, and returns a value. The returned service value is then, compared to the node threshold of the most important feature. Upon this comparison, the set of the traversed branches, from the most important feature node to the leaf-node, is determined, as a second step in the Awareness Module. The set of the traversed branches determines whether to traverse the left sub-tree or the right sub-tree of the most important feature node. Thus, the traversed path is guided by the most important feature. If the most important feature is present in multiple nodes in the prediction model, then, we have multiple paths traversing each of these nodes. Hence, we have multiple services composition schemas in the Service Composition layer, each of which is built according to the following processes. Each node in the constructed path traversing the most important feature node are mapped to abstract services. These abstract services constitute one of the abstract composition scheme, in the Service Composition layer. In order to instantiate these services, within the Service layer, the Features Matching Module, maps each abstract service representing a feature node, with its suitable service description from the service registry. Thus, all the abstract composition schema in the Service Composition layer are mapped to schema composition with concrete services. In fact, the service registry encompasses services descriptions expressed semantically as detailed in our previous work [6]. Thus, all the invoked services from the service registry, in the Service layer, are enhanced with EO linked-data. The orchestration of these services execution follows the ordered set of decision nodes traversed from the root until the leaf-node, and passing by the most important feature node. Otherwise, each value of a feature node is determined by the execution of the suitable environmental Web service accessing to its related environmental data source within the PREDICAT project.

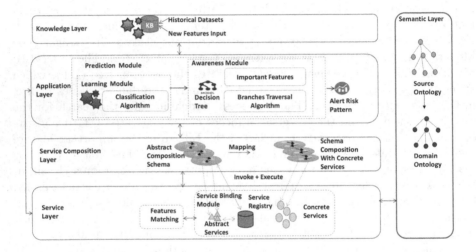

Fig. 1. The prediction system architecture.

4   The Wildfire Prediction Flow

Figure 2 presents the proposed prediction flow related to wildfire alerts along with its different phases. In fact, the novelty of our proposed flow is that it allows to trigger alerts to scientists, transparently and without any human intervention nor a user request. The first phase in this flow is to collect the EO data stored in the knowledge-base. 1-EO data-acquisition phase is realized by the IoT devices that belong among others to the OSS system. The second phase consists in 2-Data Preparation, which takes as input the data from the knowledge-base and splits it into two sets: the training data-set and the testing data-set. The Model Input Data consists of the set of features of interest, to which is applied the ML algorithm. The third phase is 3-Prediction Model Building one, which consists of performing the ML algorithm and produces the prediction model, which is the decision tree (DT). This phase relates to the Learning Module, detailed in the previous section. A DT is a tree structure that consists of multiple internal-nodes and leaf-nodes [15]. Each internal node represents a single category; each branch of a node represents one possible value or a set of possible values of the category, and each leaf-node represents a class label. DT uses a tree structure to represent the rules between independent and dependent variables. Each node has a threshold compared with (i.e.: $<=$, $>$).

The Random Forest (RF) ML classifier is an ensemble learning method developed by constructing multiple DTs [16]. In the training process, an RF applies a bagging technique to bootstrap instances and selects a random subset of features. A set of DTs is then constructed based on each set of bootstrap instances with a subset of features. Once the set of trees is constructed, a prediction regarding unseen samples can be generated by selecting the majority class of individual trees. Once the prediction model is generated, all the needed features of interest defined in the decision nodes are ready to be extracted. In the fourth phase 4-Prediction Model Deployment, each extracted feature of interest represents an abstract service that will be part of the abstract composition schema. The design of the abstract service composition schema is handled by the Awareness Module which is detailed in Sect. 5.1. Furthermore, it is worthy to note that we can have multiple designed abstract composition schema as far as there are multiple nodes representing the most important feature of interest in the DT, determined by the Awareness Module, and guiding the EO data collection in the model prediction generated upon the Prediction Model Building Phase. See Sect. 5.1 for additional details given in the presented algorithms. A Feature matching process is applied to each extracted feature, and which is mapped to an abstract service. The list of services along with their related descriptions are stored in a service registry. In the fifth 5-Service Composition phase, each abstract service is instantiated and executed. This execution is handled by the Composition Execution Engine. The orchestration of the execution of these services follows the order of traversing the decision nodes defined by the traversal algorithm, in the Awareness Module. Otherwise, each executed decision node, based on its value, determines which next decision node and its service instance to be executed. Furthermore, the execution of multiple instantiated composition schema is realized in parallel. Upon the execution of the schema composition, an alert is then, transmitted to the scientist.

5   Implementation and Evaluation

In this section, we present the feasibility of our fire model prediction in the guidance of the building of the service composition scheme. In particular, we demonstrate the effectiveness of our model. First, we provide an implementation example of our model with the Random Forest (RF) algorithm to perform the fire prediction. Then, we provide some algorithms showing details of the DT traversal from the most important feature of interest in the model prediction, which will guide the search for the rest of the features, in the tree. Thus, the generated paths across the DT are mapped to service composition schemas. Second, we focus on the evaluation of our built-model, by comparing it with other classifier algorithms, through the examination of the different performance indicators'.

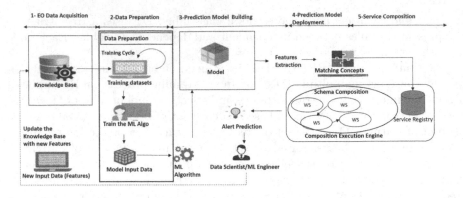

Fig. 2. The prediction flow.

5.1   Implementation

As aforementioned, we focus the implementation details on the two supervised classifiers: The Random Forest (RF) and the Decision Tree (DT). We used the built-in implementation of the RF and the DT algorithms, from the free software ML library "sklearn". Furthermore, we used Python programming language for the implementation and the "export_graphviz" module for the visualization of the generated DT in both classifiers. Each generated DT represents the fire model prediction. We, then, provide an evaluation based on performance indicators to choose the relevant classifier. In the following, we first introduce the used datasets, then, present the implementation results.

Data-Sets. We used data[4] related to weather recorded by the OSS as a set of inputs and stored in the knowledge-base. In particular, we used hourly data recorded between 1st November 2017 and 31st March 2019. This period of time includes a wide range of temperature ($°C$), relative humidity (%), wind speed (km/h), wind direction (°) and, drought factor values relevant to fire weather considerations. Furthermore, we used the McArthur Forest Fire Danger Index (FFDI) and its rating namely, the fire danger rating scale for forest (FFDR) as output used by our ML algorithm. The fire danger index is determined by the calculation of the FFDI according to the equation defined in [17]. FFDR is defined by the following classes: catastrophic (>100), extreme (75–99), severe (50–75), very high (25–49), high (12–24), and low-moderate (0–11). All these classes are considered as output to the ML algorithm and stored in the knowledge-base. We considered about 1500 tuples in our knowledge-base when performing the ML algorithm. We split the data-set into 70% for training and 30% for testing. Furthermore, in order to avoid overfitting and determine the optimal model performances, we used the "GridSearchCV" function from the "sklearn" library, to

---

[4] https://docs.google.com/spreadsheets/d/1v-46-KMHtErt3IGigFsusk7Fnp61DKvct Ms9KMH_a-E/edit?usp=sharing.

tune the model hyperparameters for both RF and DT classifiers. It consists in using a subset of the training collection as a validation dataset. We considered the following hyperparameters. For cv=5 in the DT classifier: max_depth=10, criterion='entropy' and, min_samples_split=2. For cv=3 in the RF classifier: criterion='gini', max_depth=10 and, n_estimators=90. Furthermore, we used the Amicus fire knowledge-base[5], which is a free suite of tools to simulate the calculation of the FFDI index.

Learning Module. As aforementioned, the main objective of this module is to learn from the EO data itself and the historical EO data collected by the IoT devices. This module generates the fire model prediction. For the construction of the latter, we chose two decision tree algorithms: the Random Forest (RF) [16] and the Decision Tree (DT) [15]. This choice is explained by the fact that the decision tree algorithms are effective in that they provide human-readable rules of classification. We performed tests on both classifiers to choose the relevant one. Section 5.2 provides evaluation details. After performing an ML classifier algorithm, an extraction of the produced fire model prediction decision tree is depicted in Fig. 3. Each classifier produces a fire model prediction, each of which represents a decision tree (DT).

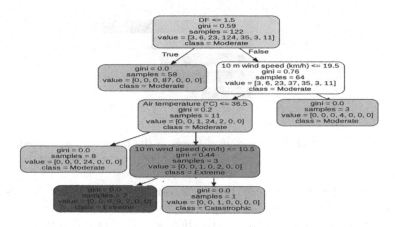

Fig. 3. Extraction from the decision tree.

Awareness Module. The Awareness Module encompasses two algorithms: the first one determines the most important feature which indicates its impact on the model compared to the rest of the features, and the second one determines the path to traverse in the DT, from the most important feature node till the leaf-node containing the fire danger class. Figure 4 depicts the relative important features, taking into consideration the set of the used features. We observed

---

[5] https://research.csiro.au/amicus/.

that the drought factor feature (DF) has the highest importance value. This value is determined by at first, applying the RF feature selection method [7], which produces a list of scored features within the prediction model. This list is denoted "L" int the Algorithm 1. Second, by applying the maximum equation on these values to determine the most important feature in our model prediction. Algorithm 1[6] is a pseudo-code presenting details about determining the most important feature, which is the feature DF in our model prediction. Furthermore, once the most important feature is determined, the idea is to map the feature tag to its suitable service. This service is executed in order to have the DF value. According to the returned value by the DF service, the DF node will guide the tree traversal to search for the other features in the DT.

Algorithm 1. CMIF+ES.

Begin
Let L ← Search for features importance //List of the important features
DF ← $\max_{i \in L}\{importance(i)\}$ //The most important feature in our model prediction
Val ← Execute the Service Having DF as a tag//Exec of the important Feature service
End.

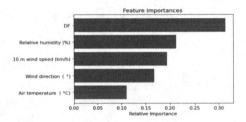

Fig. 4. The set of the important features in the fire model prediction.

Then comes the generation of the tree traversing the DF node which is composed on the one hand, of the path departing from the DF node to the root, and on the other hand, of the sub-tree of the DF node. To do so, Algorithm 2 reuses from Algorithm 1 the most important feature (e.g.: DF) and its related value 'Val'. As a first step, in order to search for the DF node in the DT, the algorithm determines the nearest node to the root whose feature is DF. Furthermore, it generates the path from the DF node to the root. As a second step, according to the returned value 'Val', this latter is compared to the 'DF_Threshold' indicated in the chosen DF node in the DT. Thus, the algorithm decides which sub-tree to extract (i.e.: the left sub-tree or the right sub-tree). Afterwards, the path and the sub-tree are merged to generate the tree that traverses the DF node. In fact, the DF node, according to its value, guides the ordered connections to the root

---

[6] Computing the most important feature and executing its service.

Algorithm 2. Construction of the tree: path departing from the important feature to the root and, the sub-tree of the important feature.

Input: Tree //The generated decision tree of the model prediction
Output: Tree //Concatenation of the path departed from the important feature to the root and its extracted sub-binary tree
Begin
Node ← Nearest Node to the root whose feature is DF
Path ← Path from DF to the root
if (DF_Threshold ≤ Val) then   SubTree ← Left_SubTree of DF
    else SubTree ← Right_SubTree of DF
end if
Tree ← Path + SubTree //Fusion of the path and the sub-tree
return Tree
End.

---

and to the leaf-nodes. Thus, the DF node impacts on the dynamic generation of the service composition scheme. In case if we have multiple DF nodes in the DT, then we have multiple paths traversing each of these nodes. Therefore, we present Algorithm 3, which defines a pseudo-code managing the generation of multiple paths traversing the multiple DF nodes in the DT. This algorithm reuses the generated binary decision tree of the model prediction and the extracted sub-tree determined according to the DF value in Algorithm 2. The idea, when having multiple DF nodes in the DT, is to only prune the extracted sub-tree from the binary tree and return the new tree. This way, the other non extracted sub-tree will be used, or another path traversing another DF node will be used. The sub-tree pruning is realized 'i' times until reaching a 'Stop_Condition'. The value of the 'Stop_Condition' (e.g.: 10) is to be fixed at the beginning of the experimentation by the experimental user of the PREDICAT platform, to define the maximum number of service composition schemas supported to be run in parallel depending on the capacity of the PREDICAT platform. The generated paths constitute the possible constructed abstract service composition schemas. The execution of these composition schema is realized in parallel by instantiating the services at run-time. An alert is, then, triggered. In the next section, we provide the evaluation of the fire prediction model based on comparative performance measures computed from both previously detailed classifiers.

Algorithm 3. Generating multiple paths traversing multiple DF nodes.

Input: Binary Tree //The generated DT of the model prediction from Algorithm 2
Stop_Condition←10//10 supported generated paths to be executed in parallel in the platform
Output: Tree //The generated tree
Begin
Repeat
Tree ← Binary Tree \{$SubTree(i)$} //Generate a new tree by eliminating the subTree of the important feature extracted in Algorithm 2
Until (i > Stop_Condition) //Stop_Condition is to be fixed limiting the number of the generated //paths traversing the important feature DF.
return Tree
End.

5.2   Evaluation Metrics

Several metrics for the evaluation of the performance of the classifier from the literature can be used. In our experiments, we considered four commonly used metrics, which are accuracy, precision, recall, and f1-score. These latter are indicators to measure the performance of our prediction models.

Table 1. System performance measures

ML classifier	Accuracy %	Precision	Recall
Random Forest	86	0.862	0.862
Decision Tree	84.06	0.840	0.840

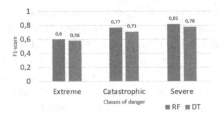

Fig. 5. Classes of fire danger chart.

To assess the evaluation of our fire prediction model, we used the most two popular ML classifiers which allow evaluating the RF classifier against the Decision Tree classifier. Moreover, we used performance measures computed for both ML classifier models. Furthermore, we considered the classes: Catastrophic, Extreme, and Severe as the most important classes of fire danger triggering alerts, and we measured the f1-score related to these classes, for each of the classifiers. According to results in Fig. 5, we noticed that f1-score values in the RF classifier, all the danger classes advance those in the Decision Tree classifier. These results related to the most important danger classes show relevant values for prediction. Moreover, according to our experiments on our fire prediction model generated by the RF classifier, in Table 1, showed a high accuracy value, which is in advance to the one in the Decision Tree classifier.

6   Conclusion

In this paper, we proposed an approach that combines ML and knowledge-driven engineering to dynamically compose services from sensor data for wildfire predictions. The predicted alerts help scientists to anticipate and to manage fire in threatened areas. We evaluated our wildfire model prediction through several experiments, which showed relevant values with respect to the most important classes of fire danger. Future work includes exploring optimal services composition along with optimal services selection based on the quality of data and quality of services.

References

1. Mitchell, T.: Machine Learning. Publisher McGraw-Hill, New York (1997). ISBN 0070428077ISBN 0070428077
2. Boustil, A., Maamri, R., Sahnoun, Z.: A semantic selection approach for composite web services using OWL-DL and rules. Serv. Oriented Comput. Appl. **8**, 221–238 (2014)
3. Rodriguez-M, P., Pedrinaci, C., Lama, M., Mucientes, M.: An integrated semantic web service discovery and composition framework. IEEE Trans. Serv. Comput. **9**, 537–550 (2016)
4. Masmoudi, M., et al.: PREDICAT: a semantic service-oriented platform for data interoperability and linking in earth observation and disaster prediction. In: Conference on Service-Oriented Computing Applications (SOCA), Paris, pp. 194–201 (2018)
5. Bartalos, P., Bielikova, M.: Automatic dynamic web service composition: a survey and problem formalization. Comput. Inf. **30**(4), 793–827 (2011)
6. Taktak, H., Boukadi, K., Mrissa, M., Ghedira, C., Gargouri, F.: A model-driven approach for semantic data-as-a-service generation. In: IEEE International Conference on Enabling Technologies: Infrastructure for Collaborative Enterprises - WETICE, France (2020)
7. Robin Genuer, R., Poggi, J.M., Tuleau-Malot, C.: Variable selection using random forests. J. Pattern Recognit. Lett. **31**, 2225–2236 (2010)
8. Bansal, S., Bansal, A., Gupta, G., Blake, M.B.: Generalized semantic Web service composition. J. Serv. Oriented Comput. Appl. **10**, 111–133 (2016)
9. Giglio, L., Boschetti, L., Roy, D.P., Humber, M.L., Justice, C.O.: The collection 6MODIS burned area mapping algorithm and product. Remote Sens. Environ. **217**, 72–85 (2018)
10. Gupta, I.K., Kumar, J., Rai, P.: Optimization to quality-of-service-driven web service composition using modified genetic algorithm. In: International Conference on Computer, Communication and Control, pp. 1–6 (2015)
11. Ma, H., Wang, A., Zhang, M.: A hybrid approach using genetic programming and greedy search for QoS-aware web service composition. In: Hameurlain, A., Küng, J., Wagner, R., Decker, H., Lhotska, L., Link, S. (eds.) Transactions on Large-Scale Data- and Knowledge-Centered Systems XVIII. LNCS, vol. 8980, pp. 180–205. Springer, Heidelberg (2015). https://doi.org/10.1007/978-3-662-46485-4_7
12. Sawczuk da Silva, A., Mei, Y., Ma, H., Zhang, M.: Particle swarm optimisation with sequence-like indirect representation for web service composition. In: Chicano, F., Hu, B., García-Sánchez, P. (eds.) EvoCOP 2016. LNCS, vol. 9595, pp. 202–218. Springer, Cham (2016). https://doi.org/10.1007/978-3-319-30698-8_14
13. Yu, Y., Ma, H., Zhang, M.: An adaptive genetic programming approach to QoSaware web services composition. In: 2013 IEEE Congress on Evolutionary Computation, pp. 1740–1747 (2013)
14. Petrie, C.J.: Web Service Composition. Springer, Cham (2016). https://doi.org/10.1007/978-3-319-32833-1

15. Murthy, S.K.: Automatic construction of decision trees from data: a multi-disciplinary survey. J. Data Min. Knowl. Disc. **2**, 345–389 (1998)
16. Breiman, L.: Random forests. J. Mach. Learn. **45**, 5–32 (2001)
17. Sharples, J.J., McRae, R.H.D., Weber, R.O., Gill, A.M.: A simple index for assessing fire danger rating. Environ. Model. Softw. **24**, 764–774 (2009)
18. Urbieta, A., González-B, A., Mokhtar, S., Hossain, M., Capra, L.: Adaptive and context-aware service composition for IoT-based smart cities. Future Gener. Comput. Syst. **76**, 262–274 (2017)
19. Deng, S., Xiang, Z., Yin, J., Taheri, J., Zomaya, A.Y.: Composition-driven IoT service provisioning in distributed edges. IEEE Access **6**, 54258–54269 (2018)
20. Asghari, P., Rahmani, A., Javadi, H.H.S.: Service composition approaches in IoT: a systematic review. J. Netw. Comput. Appl. **120**, 61–77 (2018)
21. Chen, G., Huang, J., Cheng, B., Chen, J.: A social network based approach for IoT device management and service composition. In: IEEE World Congress on Services, pp. 1–8 (2015)
22. Yang, R., Li, B., Cheng, C.: Adaptable service composition for intelligent logistics: a middleware approach. In: Conference on Cloud Computing and Big Data, pp. 75–82 (2015)

Eyewitness Prediction During Crisis via Linguistic Features

Suliman Aladhadh[(✉)]

Department of Information Technology, College of Computer,
Qassim University, Buraydah, Saudi Arabia
s.aladhadh@qu.edu.sa

Abstract. Social media is one of the first places people share information about serious topics, such as a crisis event. Stakeholders, including the agencies of crisis response, seek to understand this valuable information in order to reach affected people. This paper addresses the problem of locating eyewitnesses during times of crisis. We included published tweets of 26 crises of various types, including earthquakes, floods, train crashes, and others. This paper investigated the impact of linguistic features extracted from tweets on different learning algorithms and included two languages, English and Italian. Better results than the state of the art were achieved; in the cross-event scenario, we achieved F1-scores of 0.88 for English and 0.86 for Italian; in the split-across scenario, we achieved F1-scores of 0.69 for English and 0.89 for Italian.

Keywords: Machine learning · Text mining · Social media · Eyewitness

1 Introduction

During such an event people often share information about that event on social media, making these platforms the first alarm for significant events. This is particularly true for crisis, such as earthquake, flood, pandemic, etc. On social media users are classified based on their location (i.e. close or far from the event location), users posting information on a particular event who live in or close to the event area are known as "eyewitnesses". Eyewitnesses are important as they are able to provide first-hand information about that event. Geo-location information in social media is the most direct way to locate eyewitness. However, this information is rare on social media platforms (e.g. <1% in Twitter) so alternative methods are needed to find them.

Little research has been conducted into defining eyewitness in social media, even though such research is important in critical times such as crisis events, including pandemic like coronavirus disease [6]. Agencies of disaster response have began to include social media as important sources of information to reach affected people, and these research aim to locate authors from the same place of the event. However, eyewitnesses are very rare in social media, for example among tweets of 26 crisis events only 8% were labeled as being from an eyewitness [10].

© Springer Nature Switzerland AG 2021
H. Hacid et al. (Eds.): ICSOC 2020 Workshops, LNCS 12632, pp. 421–429, 2021.
https://doi.org/10.1007/978-3-030-76352-7_39

Different methodologies have been used for defining eyewitnesses, including semantic, source-based features, user-based features, networking features, and other metadata such as number of Retweet, URL, etc. In this study we aimed to investigate the impact of using a language features approach to improve the performance of machine learning model, by employing Linguistic Inquiry and Word Count (LIWC). Linguistic features have previously been found to improve performance, as detailed in the next section.

The research question of this study is:

– How do linguistic features improve detection of eyewitnesses during crisis?

2   Related Work

Defining the eyewitness during crisis is challenging as less than 1% of tweets are geo-tagged. Many researchers have tried to find alternative ways to reach users from the ground. Morstatter et al. [8] aimed to differentiate between tweets from the affected and different locations. They built a model to predict tweets of an affected location (eyewitnesses) and collected geo-tagged tweets from the United States (US) only related to two crisis event in the US (2013 Boston Marathon Bombing and 2012 Hurricane Sandy). The authors classified the tweets into two types (inside and outside region) and employed Naive Bayes to train a machine learning model to predict tweet type. The model was built using a number of linguistics features, including unigrams & bigrams, Part of Speech (POS) and shallow parsing to identify the tweet semantics (e.g. verbs categories and name entities). The accuracy of predicting tweets as inside or outside affected region was 0.831 for the Boston bombing and 0.882 for Hurricane Sandy. This model relied only on linguistic features and was able to achieve good results. However, the study had a number of limitations. For example, they included only geo-tagged tweets but the reality is that most eyewitness in social media are not geo-tagged, in fact this is very rare. In addition, only two events were included, both from the same country and the events belonged to different crisis types, while in the real scenario a model performance differ when apply for different crisis types. Moreover, Language differences have a large influence on user location prediction in social media [4], yet this study only included English language. These limitations make it hard to generalise findings as the following research has shown [13].

Tanev et al. [13] trained a model to predict eyewitness during crisis, these crisis related to different types such as floods, wildfires, earthquakes, etc. where occurred in different countries. Available data from 26 crisis tweets were annotated as eyewitness or not. English and Italian languages were included to measure the influence of language on model accuracy, and the authors used a set of language features such as lexical, stylistic and word capitalisation and tweet features such as hashtag, mention, etc. They compared performance of three different classifiers: Naive Bayes, SVM and Random Forest. In the scenario of training and testing (classical) of the same event types, Naive Bayes achieved the best result for Italian language, with an accuracy of 0.69, and Random Forest was the best for English at 0.79. However, in the realistic (harder) scenario the results were poor with 0.19 for English and 0.38 for Italian.

Pekar et al. [12] used the same data used by Tanev et al. [13] and trained four different classifiers using mixed linguistic (e.g. lexical, grammatical, semantic, etc.) and other metadata features (e.g. hashtags, mentions, retweet, etc.). In both scenarios (classical and realistic), results were poor and performance was very low. In the first scenario, the best result was 0.40 with Random Forest and in the second scenario, was <0.10 with SVM.

Pekar et al. [11] predicted the information types related to crisis events, including eyewitnesses. They used six different features to train their models, including unigrams, bigrams, POS, hashtags, RT and URL counts. The authors classified these features into two different views: lexical features and grammatical features with metadata, and used SVM and Maximum Entropy algorithms for the classification process. The best results achieved was around 0.25 with SVM classifier via the second view (grammatical features with metadata). The results indicated a very low accuracy.

In the aforementioned research, we observed that when a small number of features related to language was added to the models, performance increased. So, we suppose that inclusion of more linguistic features will improve the existing models and result in much higher performance.

3 Methodology

We used the linguistic features of semantic or stylistic as the indicator for eyewitness detection. This is for many reasons, including:

- The linguistic features of tweets have been found to be an important predictor for eyewitness detection in social media [8].
- The language of the tweets is influenced by user location [3,4].
- The text is available in all tweets, whereas other features may be absent.
- The influence of linguistic features have prominent affect on precision and recall [12].

Linguistic Inquiry and Word Count (LIWC): In order to study the influence of only linguistic features on eyewitness prediction in social media, we need to generate a large number of these features. We used LIWC, which is a tool for semantic analysis of text that counts words of different psychological categories [14]. Its dictionary of categories include almost 6,400 words and word stems for English and other languages, including Italian. Use of LIWC is common in social media data analyses in different areas such as demographics [9], during crisis [7,15].

The categories we used are listed in Table 1 and have been used in previous research. All categories, except the last one in the table, have subcategories. For example, affect feature is a main category, which includes two subcategories: positive and negative emotions; negative emotion includes three subsub categories:

Table 1. The LIWC general categories used to perform analyses on tweet content.

Feature	Example
Function words	it, to, no, very
Affect words	happy, cried
Social words	mate, talk, they
Cognitive processes(cogproc)	cause, know, ought
Perceptual processes (percept)	look, heard, feeling
Biological processes (bio)	eat, blood, pain
Core drives and Needs (Drives)	ally, win, superior
Relativity (relativ)	area, bend, exit
Informal language (informal)	damn, btw, umm
Authentic, Pronoun, Word count(WC), Qmark, Exclam	

anxiety, anger, and sadness. In total there are 93 LIWC features. The full list of both general and sub-categories are available online.[1]

Bigram: In addition to LIWC the 93 features, we included bigram as an extra feature, as it can greatly increase the performance of previous models [8,12].

3.1 Dataset

In this study, we used labeled data for 26 different types of crisis events, including earthquakes, train crashes and floods. The data is available for research and called CrisisLexT26 [10]. The crisis events occurred between 2012 and 2013, and included the 2012 Italian earthquake, 2013 Queensland floods and 2013 Australia bushfire. English and Italian tweets were included and identified by self identification tool language filtering [5].

In total, there were 24,589 labelled tweets, with 2,193 authored by eyewitnesses.

Since the data was unbalanced between eyewitness and non-eyewitness tweets, we created a balanced version between the two by under-sampling non-eyewitness tweets (negative) (50/0) randomly before of training and testing, as per the methods of [2,13]. We located 2811 English tweets and 1575 Italian tweets. The size of our data was much bigger than the data used in [13], as their model needs an extra meta data of the tweets from Twitter, and many of these tweets do not exist anymore. They therefore needed to exclude a large number of eyewitness tweets.

---

[1] http://liwc.wpengine.com.

4   Experiment

We use four different algorithms to compare performance, including: Random Forest (RF), k-Nearest Neighbors (KNN), Naive Bayes (NB) and Support Vector Machine (SVM). We used these four classifiers as they achieved the best results in previous studies, and to enable comparison of our results with previous research (Tables 2 and 3). We used $\approx$ in [11,12] because they did not give the absolute number in their papers.

There are two common scenarios to evaluate the classification accuracy of our model. In the first scenario, all datasets were randomly split into training and testing sets in a 10 cross folds validation. This scenario ensures that the features distribution is same in training and testing. Table 2 presents the results of the first scenario from previous research. This scenario is called cross-event.

The second scenario was designed to reflect the reality, i.e. that the training and testing datasets are different. The tweets were split in a way that meant the crisis types included in the training dataset were different to those in the testing dataset. This scenario, called split-across, is well know as the harder one in which to achieve good results. Table 3 reports the results of previous studies that included split-across scenarios. In the testing phase, we applied this scenario on three different crisis types: flood, train crash and earthquake. This allowed us to observe the impact of crisis type on performance.

Table 2. Results of the previous studies used CrisisLexT26 in the first scenario

Used by	Evaluation metrics	RF	KNN	NB	SVM	MaxEnt
Pekar et al. [12]	Precision	$\approx$0.60	$\approx$0.60	$\approx$0.80	$\approx$0.60	$\approx$0.60
	Recall	$\approx$0.20	<0.10	<0.10	$\approx$0.20	$\approx$0.10
	F1	0.40	<0.10	<0.10	$\approx$0.30	$\approx$0.20
Pekar et al. [11]	Precision	–	–	–	$\approx$0.40	$\approx$0.30
	Recall	–	–	–	<0.20	$\approx$0.10
	F1	–	–	–	0.20	$\approx$0.20
Tanev et al. [13] English	Precision	0.81	–	0.70	0.80	–
	Recall	0.78	–	0.84	0.75	–
	F1	0.79	–	0.77	0.77	–
Tanev et al. [13] Italian	Precision	0.57	–	0.64	0.58	–
	Recall	0.83	–	0.75	0.65	–
	F1	0.68	–	0.70	0.61	–

5   Results and Discussion

Table 4 shows precision, recall and F1-measure for the four classifiers RF, KNN, NB and SVM, by English and Italian tweets. For each language we reported the results of the two scenarios. In the second scenario, the crisis types in the test were not included in the training, we performed the testing on three different types of crisis (earthquake, flood and train crash). We did this to make sure that crisis type had no effect on testing, as some crisis types are close to each other in terms of distribution of features and which the previous studies did not consider. For Italian tweets we did not include train crash due to lack of data.

Table 3. Results of the previous studies used CrisisLexT26 in the second scenario

Used by	Evaluation metrics	RF	KNN	NB	SVM	MaxEnt
Pekar et al. [12]	Precision	≈0.30	<0.20	<0.20	≈0.60	0.60
	Recall	≈0.00	≈0.00	≈0.00	<0.10	<0.10
	F1	<0.10	≈0.00	≈0.00	<0.10	<0.10
Pekar et al. [13] English	Precision	0.12	–	0.17	0.12	–
	Recall	0.30	–	0.68	0.34	–
	F1	0.17	-	0.20	0.17	–
Tanev et al. [13] Italian	Precision	0.19	–	0.24	0.24	–
	Recall	1.00	–	0.98	1.00	–
	F1	0.32	–	0.36	0.39	–

We now compare our overall best results to the previous studies that used the same data for the same purpose. In the cross-event scenario for English, we obtained 0.88 F1-score by SVM, compared to 0.40 and 0.79 by [12,13], respectively. In the split-across scenario, for English the best results were 0.69 F1-score in train crash and earthquake crisis by NB.

For Italian language, the F1-score for cross-event scenario was 0.87 by NB, compared to 0.70 by [13], while for the split-event scenario the best result was 0.89 F1-score for earthquake and 0.39 for flood. As can be seen, there is a big difference between results of the split-event (flood and earthquake), which highlights how the performance can differ when applied to other crisis types. However, in the case of English, the F1-scores for the three different crisis events in split-event were close to each other, unlike those for Italian. We believe this is due to a limited dataset of Italian tweets in flood crisis; most of the Italian tweets were related to the earthquake crisis that happened in Italy in 2012.

There is a well known issue in tweet classification in general when applied to different domains called prediction overestimation [1]. We observed a drop in precision for English between across and split scenarios for all classifiers and two classifiers for Italian. However, the drop observed in our results is much smaller than that of previous results [13]. For example, in previous research the average drop was 0.70 points in English, compared to 0.35 in our study. However, the

Table 4. Our models results for the two scenarios Cross-Event and Split-Across, for two languages English and Italian.

		RF			KNN			NB			SVM		
		Precision	Recall	F1	Precision	Recall	F1	Precision	Recall	F1	Precision	Recall	F1
English	Cross-Event	0.80	0.84	0.82	0.85	0.79	0.82	0.87	0.84	0.86	0.90	0.86	**0.88**
	Split-Across (Flood)	0.50	0.97	0.66	0.50	1.00	**0.67**	0.52	0.91	0.66	0.50	0.98	0.66
	Split-Across (Train Crash)	0.50	1.00	0.67	0.50	1.00	0.67	0.55	0.92	**0.69**	0.50	0.95	0.67
	Split-Across (Earthquake)	0.49	0.97	0.65	0.50	1.00	0.67	0.62	0.78	**0.69**	0.49	0.93	0.64
Italian	Cross-Event	0.90	0.83	0.86	0.85	0.87	0.86	0.90	0.84	**0.87**	0.90	0.83	0.86
	Split-Across (Flood)	0.31	0.17	0.22	0.47	0.28	**0.35**	0.44	0.14	0.21	0.33	0.03	0.06
	Split-Across (Earthquake)	0.75	0.94	0.83	0.66	0.94	0.78	0.94	0.85	**0.89**	0.92	0.87	0.88

drop in precision in Italian is smaller than that for English in both ours and previous study results, although the drop in our results for Italian (0.20 points) was much smaller than that in previous findings (see Tables 2 and 3).

6   Conclusion

In this study, we examined the following research question:

– How do linguistic features improve detection of eyewitness in social media during crisis events?

We investigated the effectiveness of using linguistic features only for identifying eyewitness in crisis event. By employing the text analysis tool LIWC, 93 features were generated in addition to bigram. We found that using linguistic features greatly improved performance in prediction of eyewitness in social media. The results of this study outperformed previous studies' results in both scenarios, especially in split-across (the hardest one). Moreover, the results confirm findings from previous studies on the importance of linguistic features for eyewitness detection in social media [8]. The results of this paper go beyond previous research by showing that use of a large number of linguistic features increase performance significantly. The results of the previous studies which used the same dataset were included in the paper, to make the results comparable with other research and make the impact clear.

In future research, other languages will be studied and further analysis on features importance, as that can help to understand which features have the greatest impact on the performance of the model. Also, combining other linguistic features with other types of features including network and other metadata will help to understand the impact of different types of features on locating eyewitness in social media.

References

1. Aladhadh, S., Zhang, X., Sanderson, M.: Tweet author location impacts on tweet credibility. In: Proceedings of the 2014 Australasian Document Computing Symposium, p. 73. ACM (2014)
2. Armstrong, C.L., McAdams, M.J.: Blogs of information: how gender cues and individual motivations influence perceptions of credibility. J. Comput. Mediated Commun. **14**(3), 435–456 (2009)
3. Boididou, C., Papadopoulos, S., Kompatsiaris, Y., Schifferes, S., Newman, N.: Challenges of computational verification in social multimedia. In: Proceedings of the 23rd International Conference on World Wide Web, pp. 743–748. ACM (2014)
4. Castillo, C., Mendoza, M., Poblete, B.: Information credibility on twitter. In: Proceedings of the 20th International Conference on World Wide Web, pp. 675–684. ACM (2011)
5. Castillo, C., Mendoza, M., Poblete, B.: Predicting information credibility in time-sensitive social media. Internet Res. **23**(5), 560–588 (2013)
6. Counts, S., Fisher, K.: Taking it all in? visual attention in microblog consumption. ICWSM **11**, 97–104 (2011)
7. Dedoussis, E.: A cross-cultural comparison of organizational culture: evidence from universities in the arab world and Japan. Cross Cultural Manage. Int. J. **11**(1), 15–34 (2004)
8. Flanagin, A.J., Metzger, M.J.: The role of site features, user attributes, and information verification behaviors on the perceived credibility of web-based information. New Media Soc. **9**(2), 319–342 (2007)
9. Fogg, B., et al.: What makes web sites credible?: a report on a large quantitative study. In: Proceedings of the SIGCHI Conference on Human Factors in Computing Systems, pp. 61–68. ACM (2001)
10. Freeman, K.S., Spyridakis, J.H.: An examination of factors that affect the credibility of online health information. Techn. Commun. **51**(2), 239–263 (2004)
11. Ghosh, S., Sharma, N., Benevenuto, F., Ganguly, N., Gummadi, K.: Cognos: crowdsourcing search for topic experts in microblogs. In: Proceedings of the 35th International ACM SIGIR Conference on Research and Development in Information Retrieval, pp. 575–590. ACM (2012)
12. Google Social Search: Official Blog (2011). http://bit.ly/2tm4LXJ
13. Gottfried, B.Y.J., Shearer, E.: News use across social media platforms 2016. Pew Research Center **2016** (2016)
14. Gupta, A., Kumaraguru, P.: Credibility ranking of tweets during high impact events. In: Proceedings of the 1st Workshop on Privacy and Security in Online Social Media, p. 2. ACM (2012)
15. Gupta, A., Kumaraguru, P., Castillo, C., Meier, P.: Tweetcred: a real-time web-based system for assessing credibility of content on twitter. In: Proceedings of the 6th International Conference on Social Informatics (SocInfo). Barcelona, Spain (2014)
16. Han, B., Cook, P., Baldwin, T.: Text-based twitter user geolocation prediction. J. Artif. Intell. Res. **49**, 451–500 (2014)
17. Hofstede, G.: Cultures and organizations: software of the mind (1991)
18. Hofstede, G.: Dimensionalizing cultures: he hofstede model in context. Online Read. Pychol. Culture **2**(1), 8 (2011)
19. Hong, L., Convertino, G., Chi, E.H.: Language matters in twitter: a large scale study. In: ICWSM (2011)

20. Imran, M., Castillo, C.: Towards a data-driven approach to identify crisis-related topics in social media streams. In: Proceedings of the 24th International Conference on World Wide Web, pp. 1205–1210. ACM (2015)
21. Kang, B., Höllerer, T., O'Donovan, J.: Believe it or not? analyzing information credibility in microblogs. In: Proceedings of the 2015 IEEE/ACM International Conference on Advances in Social Networks Analysis and Mining 2015, pp. 611–616. ACM (2015)
22. Kwak, H., Lee, C., Park, H., Moon, S.: What is twitter, a social network or a news media? In: Proceedings of the 19th International Conference on World Wide Web, pp. 591–600. ACM (2010)
23. Morris, M., Counts, S., Roseway, A.: Tweeting is believing?: understanding microblog credibility perceptions. In: CSCW, pp. 441–450 (2012)
24. Mourad, A., Scholer, F., Sanderson, M.: Language influences on tweeter geolocation. In: Jose, J.M., et al. (eds.) ECIR 2017. LNCS, vol. 10193, pp. 331–342. Springer, Cham (2017). https://doi.org/10.1007/978-3-319-56608-5_26
25. Obeidat, B., Shannak, R., Masa'deh, R., Al-Jarrah, I.: Toward better understanding for Arabian culture: implications based on Hofstede's cultural model. Eur. J. Soc. Sci. **28**(4), 512–522 (2012)
26. Olteanu, A., Vieweg, S., Castillo, C.: What to expect when the unexpected happens: Social media communications across crises. In: Proceedings of the 18th ACM Conference on Computer Supported Cooperative Work & Social Computing, pp. 994–1009. ACM (2015)
27. Pal, A., Counts, S.: What's in a@ name? how name value biases judgment of microblog authors. In: ICWSM (2011)
28. Poblete, B., Garcia, R., Mendoza, M., Jaimes, A.: Do all birds tweet the same?: characterizing twitter around the world. In: Proceedings of the 20th ACM CIKM International Conference on Information and Knowledge Management, pp. 1025–1030. ACM (2011)
29. Rosa, K.D., Shah, R., Lin, B., Gershman, A., Frederking, R.: Topical clustering of tweets. In: Proceedings of the ACM SIGIR: SWSM (2011)
30. Sakaki, T., Okazaki, M., Matsuo, Y.: Earthquake shakes twitter users: real-time event detection by social sensors. In: Proceedings of the 19th International Conference on World Wide Web, pp. 851–860. ACM (2010)
31. Schmierbach, M., Oeldorf-Hirsch, A.: A little bird told me, so i didn't believe it: twitter, credibility, and issue perceptions. Commun. Q. **60**(3), 317–337 (2012)
32. Wagner, C., Liao, V., Pirolli, P., Nelson, L., Strohmaier, M.: It's not in their tweets: modeling topical expertise of twitter users. In: Privacy, Security, Risk and Trust (PASSAT), 2012 International Conference on and 2012 International Confernece on Social Computing (SocialCom), pp. 91–100. IEEE (2012)
33. Weerkamp, W., Carter, S., Tsagkias, M.: How people use twitter in different languages. (1), 1 (2011)
34. Wilson, M.E.: Arabic speakers: language and culture, here and abroad. Topics Lang. Disord. **16**(4), 65–80 (1996)
35. Yang, J., Counts, S., Morris, M.R., Hoff, A.: Microblog credibility perceptions: comparing the USA and China. In: Proceedings of the 2013 Conference on Computer Supported Cooperative Work, pp. 575–586. ACM (2013)
36. Yang, J., Morris, M.R., Teevan, J., Adamic, L.A., Ackerman, M.S.: Culture matters: a survey study of social q&a behavior. In: Fifth International AAAI Conference on Weblogs and Social Media (2011)

Smart Data Integration and Processing on Service Based Environments (STRAPS 2020)

STRAPS 2020: 2nd International Workshop on Smart daTa integRation And Processing on Service-based environments

Preface

More than ever, reducing the cost of data integration by efficiently evaluating queries is a significant challenge, given that today the economic cost in computing cycles (see your cloud invoice), the energy consumption, and the performance required for some critical tasks have become important. Besides, new applications require solving even more complex queries, including millions of sources and data with high volume and variety levels. These new challenges call for intelligent processes that can learn from previous experiences, that can adapt to changing requirements and dynamic execution contexts.

The second edition of the workshop (STRAPS 2020) aimed at promoting scientific discussion on the way data produced under different conditions can be efficiently integrated to answer simple, relational, analytical queries. These queries must cope with quality preferences associated with providers, algorithms, and data trust. New scales in volume, , and value related to integrated data collections require adapted solutions providing computing, storage, and processing services deployed on different highly distributed infrastructures and target architectures. With services, data, and algorithms stemming from different and potentially vast numbers of providers, properties like provenance, quality, and trust arise as crucial properties to be quantified, evaluated, and exposed to data consumers. How can data integration in such conditions be smart? This was the central question discussed by workshop participants.

The second edition of the workshop accepted five full research papers (acceptance rate of 38%) focusing on important and timely research problems, and hosted two keynotes:

- **Building edge and fog applications on the FogStore platform**, David Bermbach, TU-Berlin, Germany.
- **Enabling Interactivity between Human and Artificial Intelligence**, Behrooz Omidvar-Tehrani, Naver Labs, France.

Papers were evaluated under a blind evaluation process through three evaluation rounds by three domain experts who were members of the workshop Program Committee. We are thankful to the Program Committee members for performing a lengthy evaluation process that ensured the accepted papers' quality. Papers presented experience reports in real-life application settings addressing large scale data integration issues guided by SLA, quality, trust, and privacy and performed through services/microservices-based systems on cloud and multi-cloud architectures.

Genoveva Vargas-Solar
Chirine Ghedira Guégan
Nadia Bennani

On the Definition of Data Regulation Risk

Guillaume Delorme[1,2](✉) ⓘ, Guilaine Talens[2], Eric Disson[2], Guillaume Collard[1], and Elise Gaget[1]

[1] Solvay, 190 Avenue Thiers, 69006 Lyon, France
{guillaume.delorme,guillaume.collard,elise.gaget}@solvay.com,
guillaume.delorme@univ-lyon3.fr
[2] Jean Moulin University, iaelyon School of Management, Magellan, 6 Cours Albert Thomas, 69008 Lyon, France
{guilaine.talens,eric.disson}@univ-lyon3.fr

Abstract. The rapid development of Information and Communication Technologies (ICT) has led to firms embracing data processing. Scholars and professionals have developed a range of assessments and management methodologies to better answer the needs for trust and privacy in ICT. With the ambition of establishing trust by reinforcing the protection of individuals' rights and privacy, economic interests and national security, policy makers attempt to regulate data processing through enactment of laws and regulations. Non-compliance with these norms may harm companies which in turn need to incorporate it in their risk assessment. We propose to define this new class of risk: "Data Regulation Risk" (DRR) as "a risk originating from the possibility of a penalty from a regulatory agency following evidence of non-compliance with regulated data processing and/or ICT governances and processes and/or information technologies and services". Our definition clarifies the meaning of the defined terms in a given context and adds a specific scope to facilitate and optimize decision-making.

Keywords: Data regulation risk · Trust · Information system risk management · Privacy · Information security

1 Introduction

The rapid development of Information and Communication Technologies (ICT) has led to a worldwide increase in data generation and valuation by individuals, public and private entities. These trends have engendered cyber attacks threatening the confidentiality, integrity and availability of information systems and data. As a direct consequence, the need has emerged for reinforced, long lasting trust among the different market actors.

In attempting to limit risks related to data processing and information systems, policy makers enact laws and regulations that aim to promote trust through appropriate security, privacy and reliability. As defined per Kosseff [1], "Cybersecurity law promotes the confidentiality, integrity, and availability of public and private information, systems, and networks, through the use of forward-looking regulations and incentives, with the goal of protecting individual rights and privacy, economic interests, and national security".

H. Hacid et al. (Eds.): ICSOC 2020 Workshops, LNCS 12632, pp. 433–443, 2021.
https://doi.org/10.1007/978-3-030-76352-7_40

This contribution will not be limited to cybersecurity laws but will incorporate laws containing sections related to regulating data processing even if their primary purpose is not cybersecurity.

The implementation of all the requirements that may be found in the laws can be challenging for organizations subject to multiple laws. Since failing to comply is sanctioned, understanding the risk the laws may represent is vital. As new laws are continuously enacted, the complexity of complying with the different requirements increases and so does the risk of non-compliance.

We introduce a new risk definition: Data Regulation Risk (DRR) which originates from the possibility of a penalty from a regulatory agency following evidence of non-compliance with a norm governing data processing and/or ICT governances and processes and/or information technologies and services. We then discuss the need for adequate methodologies, frameworks and tools for risk managers to effectively assess DRR and ensure compliance at an acceptable cost.

This contribution is structured as follows. Section 2 will present trust as the foundation of data regulations. Section 3 will first introduce how these laws create a new risk for organizations that we call DRR. Based on a literature review, Sect. 4 will discuss the limits of existing framework and risk management methodology in regards to Data Regulation Risk management.

2   Data Regulation as the Baseline for Trust

Organizations operate in uncertain highly competitive markets requiring them to identify, assess, mitigate and, if needed, take risks. They elaborate their risk strategies considering their relationships and available information on other markets actors. Crafting such strategies involve managing future uncertain events. Market actors must find ways to establish and ensure lasting trust among them by affecting the level of uncertainty [2].

2.1   Defining Trust

Scholars from diverse disciplines have discussed and presented insightful views regarding the causes, nature and effect of trust in different contexts and areas of research. Various contrasting concepts of trust have been developed across disciplines and domains resulting in distinctive definitions [3]. Among them, the operational and internal definitions of trust dominate. The former relates to game theory [4] implying rational decision making process and risk aversion based on foreseen gains and losses. The latter describes trust as a state of belief referring to one's acceptance to vulnerability based on its belief or positive expectations in regards to the motivations and behaviors of others [5]. Uncertainty decrease is then correlated with the increase in communication and exchange of information [6].

Trust as the Intersection of Privacy, Security and Reliability. As raised by Gefen and all. [7], trust evolves and can grow over time. Trust is then considered a requirement for a stable relationship by affecting one's risk appetite [8]. Camp presented the three dimensional concept of trust which defines it as the intersection of privacy, security and reliability [9]. By focusing on the existence of a risk and not its quantification, this operational definition is based on risks rather than risk perception.

On the basis of the three dimensional concept of trust, security is not privacy but a means to provide the ability to generate privacy by enabling the control of digital information. Security by itself therefore does not necessarily imply privacy nor trust [5]. Similarly, security is not reliability but a means to provide resilience contributing to the belief in the integrity or authority of the trusted party. Finally, security is not a separable element of trust. In other words, not only trust includes technological challenges but it also requires a deep understanding of the interactions and motivations of the involved parties to building it as well as the human concept of trust and privacy [9].

Trust in IT Context. In the context of cloud services, data processing and more broadly speaking IT-related topics, trust might be referred to as increasing positive predictability. It is then reached by ensuring sufficient security, accuracy, transparency and account-ability regarding data processing. Trust draws the ambition of reaching a perceived risk level sufficiently low for an organization to use third parties' services or a consumer to entrust organizations with the handling of its personal data. As perfect competition is not realistic, reaching an absolute uncertainty free state is not possible [10]. Organizations must therefore reduce uncertainty by increasing predictability through commitment, transparency and security. They may develop and implement controls to prevent unde-sired and harmful events which reinforce overall confidence if combined with adequate communication and commitment [5].

2.2 Data Regulation Addressing the Three Aspects of Trust

In order to achieve the protection of individual's rights and privacy, economic interests and national security, policy makers have enacted countless laws and regulations [1]. The focal point of these regulations is promoting trust through the confidentiality, integrity, and availability of public and private information, systems and networks. Policy makers can ensure the expected state of trust through different ways: regulating data processing or the use of technology and service.

Privacy and Security. Policy makers and governments attempt to protect individuals' rights and privacy by regulating individuals' data processing as well as empowering individuals to take ownership of their personal data and rights. Their goals are to guar-antee the confidentiality of personal data by setting security requirements and ensure full transparency over their processing, access or disclosure. Moreover, they ensure personal data integrity by providing individuals the right to access their personal data and ask for correction if needed. Trust is then obtained by increasing transparency, empowerment of individuals and information sharing.

In Europe, the General Data Protection Regulation (GDPR) is a legal framework that sets the rules and guidelines for the collection and processing of personal information from individuals [11]. Article 1 of GDPR states that the purpose of the regulation is to "lay down rules relating to the protection of natural persons with regard to the processing of personal data and rules relating to the free movement of personal data, protecting fundamental rights and freedoms of natural persons and in particular their right to the protection of personal data while ensuring the free movement of personal data within the European Union" [11]. Similar privacy regulations have been enacted over the world such as the California Consumer Privacy Act (CCPA) or the Personal Data Protection Act in Singapore (PDPA) [12, 13].

Reliability and Security. Data processing encompasses among others the collection, consultation, use, disclosure, storage, erasure of data [11]. Policy makers may regulate these actions in multiple ways. The data access and disclosure may be supervised, controlled and limited. Access to regulated data may be limited to certain nationality, to a need-to-know, to the geographical localization of the requestor accessing the data or to the data storage location. Limiting and controlling data processing by controlling the access, or storage in a fragmented IT with plenty of parties inevitably reinforces the trust of the different market's actors.

For instance, the EAR states that export any item subject to the EAR to another country or reexport any item of U.S.-origin may require a license and therefore a prior authorization from the Bureau of Industry and Security (BIS). In addition, the EAR also forbids access to EAR items and related data to specific countries and end users [14].

Use of certain types of technology can be fully allowed, restricted or forbidden by regulations. Since only trusted technology is allowed in the processing of such regulated data, it reinforces the trust in the different actors by increasing transparency and security. One key component in ensuring the security of information systems and thus data, is encryption technology. Use of encryption technology can be prohibited, subject to conditions such as providing the keys to the government, or subject to prior government approval [15]. For instance, U.S. regulations such as Defense Federal Acquisition Regulation Supplement (DFARS) [16], International Traffic in Arms Regulations (ITAR) [18] and EAR only allows encryption modules validated by the Cryptographic Module Validation Program. Laws may also supervise export or import of certain technologies such as "dual-use" technologies [14] which may restrict firms' access to specific technology.

Some regulations may limit the choice of a supplier or a service provider as they require compliance with specific certifications or place restrictions on certain nationalities of the provider in addition to technological requirements [11, 14]. A service provider's information systems may also be conditioned to specific security requirements, restrictive internal processes, certification or government approvals [14].

3   A Risk Originating from Data Regulation Enforcement

Policy makers establish through legal texts, different sets of rules and obligations that both public and private organizations must follow. Following the basis of reduction theory [17] and information theory [18], policy makers ensure trust through greater exchange of information, decrease of uncertainty and increase of perceived predictability. For the purpose of this contribution, this section will focus on laws regulating data processing, information technology and services. Finally we present a definition of a new multi-disciplinary risks' class inherent to Data Regulation.

3.1   Enforcing Data Regulation

It is a necessity for companies to understand the different legal requirements to ensure their compliance and reach the Data Regulations' objectives. Provided with official technical documentations, guidelines and reference points, companies are able to meet these requirements by reducing their margin of errors and misinterpretation [11, 14]. The possibility of foreseen penalties in case of non-compliance force organizations to consider and properly assess Data Regulation Risk.

Assessing the Non-compliance. To ensure the respect of the laws, policy makers may force the company to be audited from either external or regulatory agencies. The audit may arise prior to an event at the discretion of the authorities or after evidence of a security breach to determine whether the company was compliant at the time the breach occurred or not. Only regulatory agencies or appointed institutions can perform such audits and are not always made to notify a given company ahead of time. In case of litigation with a third party or an individual, the authorized authority may request a company audit at its sole discretion. Policy makers may also make companies accountable for assessing the compliance of their service providers by foreseeing audit clauses in their contractual agreements [11]. In addition to the audit right, laws may embed a voluntary disclosure clause forcing the enterprise to disclose security breach within a limited time after their discovery [11]. Finally, laws may also foster company self-denouncement for or individual denunciation of its company non-compliance [14, 19, 20].

The Penalties Resulting from Non-compliance. Once non-compliance is proved and known to a regulatory agency, the company may be sanctioned. Based on the interpretation of the law, jurisprudence and foreseen penalty in the text, a company may be sanctioned economically in a monetary fine or incapacity to perform activities in markets for a period of time [16]. In addition to this, data breach disclosure may be required by authorities, forcing firms to notify impacted individuals [11]. A disclosure may lead to indirect consequences such as an impact on the company's reputation, loss of trust or impact on stocks price [21].

Another impact for individuals is the personal risk they take when making decisions that are subject to compliance issues. Individuals are made to engage their personal liability and are personally accountable for the decisions taken while performing their job. Purposely failing to comply with or violating regulatory compliance may not only expose the company to administrative penalties but also the individual at the origin of the non-compliance to criminal penalties [14, 19, 20].

The Documentation to Meet the Requirements. Laws and regulations do not always define how to implement the necessary controls but may only refer to appropriate technical and organizational measures that are to be defined by each enterprise. To avoid misinterpretation and provide organizations with the freedom to adapt the technical requirements to cope with evolving technologies, regulations may refer to additional documents such as framework, certifications or guidelines [13, 14, 16]. These documents may change over time and may originate from appointed public entities such as the Singaporean Personal Data Protection Commission [22], the French Data Protection Commission (Commission Nationale de l'Informatique et des Liberté) [23], the American National Institute of Standards and Technology (NIST) [24], governments, or other private organizations like the International Organization for Standardization (ISO) [25]. The documents address various points such as technological solution [13], monitoring of activities [24], governance, roles and responsibilities [11, 20] or even audit methodology SOX AS5 [26]. Organization may refer to the available documents to guide them in implementing adequate controls and ensure compliance.

3.2 Data Regulation Risk

The above sections allow us to identify that first, the different data regulations promote trust while ensuring the protection of individuals' rights and privacy, economic interest or national safety. Second, the Data Regulation frame organizations' data processing, internal governance and processes along with the use of technology and services. Finally, these norms ensure their application through diverse means and force organizations' compliance or risk sanction from a regulatory agency.

A Unique Risk. The norms foresee two different types of sanction in case of non-compliance: business sanctions and criminal charges. The penalties may only be pronounced by a norm authority, is context dependent and depends on an external appreciation of the norm.
Indeed, Data Regulation Risk does not originate from classic IT related risks such as data breaches but from the possibility of being sanctioned following a proven noncompliance. Only the awareness of the failure to comply with the different restrictions on data processing, internal governance and processes, information technologies and services may lead to a penalty. In other words, a security breach may not necessarily lead to penalties if compliance to the law is proven. In addition, based on jurisprudence and interpretations, non-compliance may not necessarily result in penalties. For instance, the absence of prejudice caused by a data breach may not trigger a GDPR penalty. Non-compliance that does not lead to a sanction may therefore not constitute a DRR. DRR then derives from the possibility of a penalty and not the non-compliance, nor a security breach or legal uncertainty. Only penalties originating from a regulatory agency may be the source of DRR, despite that risk is based on external and internal factors. Internal factors may be the decision not to comply with the norm, voluntary disclosure or involuntary failure to comply. External factors may be the disclosure of a data breach, the result of an external audit, etc.

The Definition. DRR arises from specific laws that seek to address IT risk to protect individuals' rights and privacy, economic interests and national security. The risk is therefore intrinsically linked to data processing, information technologies and services. Legal norms that are not regulating data or the technology and services around them are therefore excluded from the scope of DRR. Penalties can only be pronounced based on documented non-compliance of a company. They are based on antecedent which are events occurring under the scope of a law and that are proved and known. They only occur if a company fails to demonstrate its compliance and if the regulatory agency is aware of the established non-compliance. Penalties or any kind of negative impact that are not resulting from the evidence of a firm non-compliance are therefore not in the scope of DRR. In summary, a DRR is a risk originating from the possibility of a penalty from a regulatory agency following evidence of non-compliance with a norm governing data processing and/or ICT governances and processes and/or information technologies and services.

4   Addressing Data Regulation Risk

This section contains a brief description of the existing work focusing on IT frameworks, information security and risk management. This literature review aims at highlighting the impact and contribution of our definition in respect to existing works.

4.1   Frameworks to Address Data Regulation Risk

Following the guidelines of Reconstructing the Giant: On the Importance of Rigor in Documenting the Literature Search Process [26], we used a wide scope of sources to cover relevant publications in cybersecurity, compliance, information risk management, information security and legal disciplines. We focused on relevant publications from business, public administration and academia querying for the keywords: Information Security, IT Framework, IT compliance, Information Security Risk Management (ISRM) and regulatory risk in Google Scholar.

Frameworks Specificities. This literature review aims at highlighting the impact and contribution of our definition in respect to existing works. We therefore conducted an exhaustive review including English literature only with the following selective criterion: articles discussing one or more industry leading IT frameworks (limits, scope, deployment and implementation, selection), articles comparing IT frameworks (complementarity, overlaps and controls mapping) and articles discussing Information Security Risk Management.

As a result of our literature review, we constructed Table 1 to evaluate the adequacy of industry leading frameworks with the mandatory elements present in our definition. Numerous works can be found on how to efficiently map different frameworks to regroup similar control, ensure a broader coverage and optimize the integration costs. [27–30].

Table 1. Topics addressed by Frameworks

Framework	Data processing			ICT governance	ICT processes	Information technologies and services		
	Access management	Usage	Storage			Technologies	Services	Security
COBIT	-	-	-	++	++	-	-	-
ITIL	-	-	-	+	++	+	++	+
ISO	+	-	+	++	++	++	+	++
NIST	++	+	+	+	-	++	-	++
CSA	++	+	++	-	+	++	++	++

Caption:
++ Fully Addressed
+ Partially Addressed
- Not Addressed

Frameworks may be sorted according to the requirements for IT controls infrastructure into four major components: Governance addressed by the Control Objectives for

Information and Related Technologies (COBIT), IT Services such as the IT Infrastructure Library (ITIL), Information Security and IT Operations addressed respectively by the ISO and NIST. [28]. Finally, the Cloud Security Alliance (CSA) provides practical recommendations on reducing the associated risks when adopting cloud computing [30, 31].

Frameworks Selection. Organizations find themselves in need to adopt internationally accepted frameworks or best practices to adroitly address their compliance and ISRM challenges. Each Framework brings specific value to the organization as they tend to focus on one specific component of an organization. Sometimes frameworks however encroach on topics addressed by more suited ones. The analysis presented in A comparative Review of Cloud Security Proposals with ISO/IEC 27002 highlights that an integral security model is needed in the cloud computing area [30]. The analysis concludes that no existing frameworks can provide sufficient assurances to its users due to existing gaps in regards to the risk and threats introduced by cloud computing. Al-Ahmad and Mohammad deduced from their analysis that "each framework has its strengths and weaknesses that promote or limit its adoption" [32]. The existence of many frameworks and standards enable organizations to effectively address ISRM challenges.

Plethora of studies discuss the selection and adoption of one or more frameworks by organization [28, 29, 32]. Both academia and public institutions show focal points to be considered prior to implementing such frameworks. Among them, the type or risks and threats addressed by considering the business size and sector, the cost of implementation, the required skills, industry adoption and recognition, available support and implementation guidelines or the degree of customization [24, 25, 27–30, 33].

4.2   Related Works on Risk Management and Frameworks

Risk Analysis on the Upstream Preparation of Framework Implementation.
Organizations face risks originating from various factors diminishing the relevance of a universal methodology. Classic risk management methodologies tend to lack adaptability and turned out to be less effective than initially foreseen [34]. As a result, risks analysis methodologies evolved to become domain and context specific [35]. With a need for recurrent analysis [36], they address different and distinct needs [37]. The limited scope of the specific risk analysis along with the high implementation costs and the difficulty to compare them may however negatively affect how organizations value them [38–40].

In information security, risk management originates from frameworks and directives created by governments or military institutions to ensure their own infrastructures security [38]. ISRM tends to focus on technology [39, 40] and the management of the known and identified threats towards organizations' assets. The interdependency of organizations' information systems, their environment and the need for long lasting trust makes efficient risk management mandatory [41, 42].

Properly identifying risks is essential as it lays down the foundation for their management. The resulting knowledge base is used all along the following phases of the risks management endeavor. Organizations must therefore consider the granularity, exhaustiveness and accuracy of the risk factors' identification [43]. Indeed, the efficiency of the implemented controls and the adequacy of the processes lay on the knowledge obtained from the risk analysis [37, 38].

To effectively address Data Regulation Risk, organizations must then clearly identify and extract the different requirements present in the laws and regulations governing their activities. Their identification is therefore a prerequisite to manage DRR despite the chosen risk management methodology.

The Challenges of DRR Identifications. Data Regulation Risk is context specific and depends on one organization's markets, geographical presence and jurisdictions. To be efficiently addressed and managed, DRR requires an in depth analysis involving a broad set of skills. They are usually fragmented across the organization's departments such as legal, business, IS security, IT operational, Human Resources, Finance, etc. Knowledge sharing among employees is crucial for ensuring information security and risk management [44]. Experts have to aggregate the requirements the organization has to comply with. They first have to determine the different regulations they are subject to. Then, they have to identify the required changes into the different policies, standards, guidelines or security controls within the organization. This task may be challenging as "A significant issue is that current risk identification perspectives are focused upon technical infrastructures" [45]. There is a need for studies of data collection methods and techniques for making Information Security Risk Analysis more efficient [46]. As pointed by Aven, "risk assessments are well established in situations with considerable data and clearly defined boundaries" [47]. There is a need to further consider multidisciplinary risk classes such as the presented Data Regulation Risk. There seems to be a potential for improvement in the way these risks are identified and addressed.

5 Conclusion and Perspectives

Research undertaken in this article shows the need for organizations to consider the risk inherent to Data Regulations. Establishing and preserving trust is a priority for most organizations. Governments are enacting laws and regulations to ensure a security baseline to protect individuals, economic interests and national security while safeguarding trust among the market's actors. The adoption of different internationally recognized frameworks, standards and guidelines help organizations to reach Data Regulations' requirements, to implement effective security controls and to reinforce trust. Despite important contributions supporting organizations in assessing and managing their risks, there is a need for methodologies and models to identify multi-disciplinary risks like Data Regulation Risk. We are currently in the process of co-building such methodology with Solvay, a worldwide company specialized in chemistry and subject to many norms governing its data processing, ICT governances and processes, information technologies and services.

References

1. Kossef, J.: Defining cybersecurity law. Iowa Law Rev. **103**(1), 985–1031 (2017)

2. Knight, F.: Risk, Uncertainty and Profit. Houghton Mifflin, Boston (1921)
3. Gambetta, D.: Can we trust trust. In: Trust: Making and Breaking Cooperative Relations, vol. 13, pp. 213–237 (2000)
4. Myerson, R.B.: Game Theory. Harvard University Press (2013)
5. Flowerday, S., von Solms, R.: Trust: an element of information security. In: Fischer-Hübner, S., Rannenberg, K., Yngström, L., Lindskog, S. (eds.) SEC 2006. IIFIP, vol. 201, pp. 87–98. Springer, Boston, MA (2006). https://doi.org/10.1007/0-387-33406-8_8
6. Pearce, W.B.: Trust in interpersonal communication. Speech Monogr. **41**(3), 236–244 (1974)
7. Gefen, D., Rao, V.S., Tractinsky, N.: The conceptualization of trust, risk and their relationship in electronic commerce: the need for clarification. IEEE Computer Society (2002)
8. Mayer, R.C., Davis, J.H., Schoorman, F.D.: An integrative model of organizational trust. Acad. Manag. Rev. **20**(3), 709–734 (1995)
9. Camp, L.J.: Designing for trust. In: Falcone, R., Barber, S., Korba, L., Singh, M. (eds.) TRUST 2002. LNCS, vol. 2631, pp. 15–29. Springer, Heidelberg (2003). https://doi.org/10.1007/3-540-36609-1_3
10. Humphrey, J., Schmitz, H.: Trust and Inter firm relations in developing and transition economies. J. Dev. Stud. **34**(4), 33–61 (1998)
11. Regulation (EU) 2016/679 of the European Parliament and of the Council of 27 April 2016 on the protection of natural persons with regard to the processing of personal data and on the free movement of such data, and repealing Directive 95/46/EC (General Data Protection Regulation) Official Journal L (2016)
12. California Consumer Privacy Act of 2018 (CCPA): California Civil Code, section 1798.100 (2018)
13. Personal Data Protection Act of 2012 (PDPA): Parliament of Singapore, No. 26 (2012)
14. Export Administration Regulation (EAR), 15 C.F.R. § 730 et seq. https://www.bis.doc.gov/index.php/regulations/export-administration-regulations-ear. Accessed 31 Jan 2020
15. Saper, N.: International cryptography regulation and the global information economy. Northwest. J. Technol. Intellect. Prop. **11**(7), 673–688 (2013)
16. Defense Federal Acquisition Regulation Supplement (DFARS). https://www.federalregister.gov/defense-federal-acquisition-regulation-supplement-dfars. Accessed 02 Jan 2020
17. Berger, C.: Uncertainty reduction theory. Dalam Griffin, EM A First Look at Communication Theory. Edisi, 6 (2006)
18. Shannon, C.E.: A mathematical theory of communication. Bell Syst. Tech. J. **27**(3), 379–423 (1948)
19. COUNCIL REGULATION (EC) No 428/2009 of 5 May 2009 setting up a Community regime for the control of exports, transfer, brokering and transit of dual-use items (EU dual-use), last consolidated version 2018/15/12
20. International Traffic in Arms Regulations (ITAR). https://www.pmddtc.state.gov/ddtc_public?id=ddtc_kb_article_page&sys_id=%2024d528fddbfc930044f9ff621f961987. Accessed 06 Jan 2020
21. Gordon, L., Loeb, M., Zhou, L.: The impact of information security breaches: has there been a downward shift in costs? J. Comput. Secur. **19**(1), 33–56 (2011)
22. Personal Data Protection Commission. https://www.pdpc.gov.sg/. Accessed 06 Jan 2020
23. Commission Nationale de l'Informatique et des Liberté. https://www.cnil.fr/fr. Accessed 06 Jan 2020
24. National Institute of Standards and Technology (NIST). https://www.nist.gov/. Accessed 06 Jan 2020
25. International Organization for Standardization (ISO). https://www.iso.org/home.html. Accessed 01 June 2020
26. Vom Brocke, J., et al.: Reconstructing the giant: on the importance of rigour in documenting the literature search process. In: ECIS Proceedings, p. 161 (2009)

27. Sheikhpour, R., Modiri, N.: An approach to map COBIT processes to ISO/IEC 27001 information security management controls. Int. J. Secur. Appl. **6**(2), 13–28 (2012)
28. Al-Ahmad, W., Mohammad, B.: Can a single security framework address information security risks adequately. Int. J. Digi. Inf. Wireless Commun. **2**(3), 222–230 (2012)
29. Schlarman, S.: Selecting an IT control framework. Inf. Syst. Secur. **16**(3), 147–151 (2007)
30. Rebollo, O., Mellado, D., Fernandez-Medina, E.: A comparative review of cloud security proposals with ISO/IEC 27002. In: WOSIS, pp. 3–12 (2011)
31. Cloud Security Alliance (CSA). https://cloudsecurityalliance.org/. Accessed 09 Sept 2020
32. Al-Ahmad, W., Mohammad, B.: Addressing information security risks by adopting standards. Int. J. Inf. Secur. Sci. **2**(2), 28–43 (2013)
33. Eloff, J., Eloff, M.: Information security architecture. Comput. Fraud Secur. **11**(1), 10–16 (2005)
34. Komljenovic, D., Gaha, M., Abdul-Nour, G., Langheit, C., Bourgeois, M.: Risks of extreme and rare events in Asset Management. Saf. Sci. **88**(1), 129–145 (2016)
35. Tixier, J., Dusserre, G., Salvi, O., Gaston, D.: Review of sixty two risk analysis methodologies of industrial plants. J. Loss Prev. Process Ind. **15**(4), 291–303 (2002)
36. Koivisto, R., et al.: Integrating FTA and risk assessment methodologies. In: FTA, pp. 37–38 (2008)
37. Saleh, M.S., Alfantookh, A.: A new comprehensive framework for enterprise information security risk management. Appl. Comput. Inform. **9**(2), 107–118 (2011)
38. Spears, J.L.: A holistic risk analysis method for identifying information security risks. In: Dowland, P., Furnell, S., Thuraisingham, B., Wang, X.S. (eds.) Security Management, Integrity, and Internal Control in Information Systems. IIFIP, vol. 193, pp. 185–202. Springer, Boston, MA (2005). https://doi.org/10.1007/0-387-31167-X_12
39. Gerber, M., Von Solms, R.: Management of risk in the information age. Comput. Secur. **24**(1), 16–30 (2005)
40. Dor, D., Elovici, Y.: A model of the information security investment decision-making process. Comput. Secur. **63**(1), 1–13 (2016)
41. Halliday, S., Badenhorst, K., Von Solms, R.: A business approach to effective information technology risk analysis and management. Inf. Manag. Comput. Secur. **4**(1), 19–31 (1996)
42. Ross, R.: Managing enterprise risk in today's world of sophisticated threats: a framework for developing broad-based, cost-effective information security programs. In: EDPACS, vol. 35, no. 2, pp. 1–10 (2007)
43. Jallow, A., Majeed, B., Vergidis, K., Tiwari, A., Roy, R.: Operational risk analysis in business processes. BT Technol. J. **25**(1), 168–177 (2007)
44. Sohrabi Safa, N., Von Solms, R., Furnell, S.: Inf. Comput. Secur. **56**(1), 70–82 (2016)
45. Shedden P., Smith, W., Ahmas, A.: Information security risk assessment: towards a business practice perspective. In: AISMC, pp. 119–130 (2010)
46. Wangen, G.: Information security risk assessment: a method comparison. Computer **50**(4), 52–61 (2017)
47. Terje, A.: Risk assessment and risk management: review of recent advances on their foundation. Eur. J. Oper. Res. **253**(1), 1–13 (2016)

Classifying Micro-text Document Datasets: Application to Query Expansion of Crisis-Related Tweets

Mehrdad Farokhnejad[1(⊠)], Raj Ratn Pranesh[2],
and Javier A. Espinosa-Oviedo[3]

[1] Univ. Grenoble Alpes, Grenoble INP, CNRS, LIG, Grenoble, France
`Mehrdad.Farokhnejad@univ-grenoble-alpes.fr`
[2] Birla Institute of Technology, Mesra, Ranchi, India
[3] University of Lyon, LIG-LAFMIA, Lyon, France
`javier.espinosa-oviedo@univ-lyon2.fr`

Abstract. Twitter is an active communication channel for spreading information during crises (e.g., earthquake). To exploit this information, civilians require to explore the tweets produced along a crisis period. For instance, for getting information about crisis' related events (e.g. landslide, building collapse), and their associated relief actions (e.g., gathering of food supply, search for victims). However, such Twitter usage demand significant effort and answers must be accurate to support the coordination of actions in response to crisis events (e.g., avoiding a massive concentration of efforts in only one place). This requirement calls for efficient information classification so that people can perform agile and useful relief actions. This paper introduces an approach based on classification and query expansion techniques in the context of micro-texts (i.e., tweets) search. In our approach, a user's query is rewritten using a classified vocabulary derived from top-k results, to reflect her search intent better. For classification purpose, we study and compare different models to find the one that can best provide answers to a user query. Our experimental results show that the use of Multi-Task Deep Neural Network (MT-DNN) models further improves micro-text classification. Also, the experimental results demonstrate that our query expansion method is effective and reduces noise in the expanded query terms when looking for crisis tweets on Twitter datasets.

Keywords: Crisis computing · Tweets classification · Query expansion · Microblog retrieval

1 Introduction

With the rapid development of social networks and the Internet, social media are being used more and more for communicating, tracking, and extracting

This work was partially funded by the Iranian Ministry of Science, Research and Technology through the fellowship of Mehrdad Farokhnejad.

H. Hacid et al. (Eds.): ICSOC 2020 Workshops, LNCS 12632, pp. 444–456, 2021.
https://doi.org/10.1007/978-3-030-76352-7_41

information about currently occurring or recently passed crises. However, important information is hidden within a large volume of irrelevant and noisy content [16]. Twitter is one of the popular microblog service providers which people at the scene of a disaster post information about the disaster on it. This citizen-generated data provides information about the need and availability of resources at the affected locations which humanitarian organizations can use this information to provide relief. For instance, for getting information about crisis' related events (e.g. landslide, building collapse), and their associated relief actions (e.g., gathering of food supply, search for victims). However, such Twitter usage demand significant effort and answers must be accurate to support the coordination of actions in response to crisis events (e.g., avoiding a massive concentration of efforts in only one place). This requirement calls for efficient information classification so that people can perform agile and useful relief actions.

Despite the potential benefits, it is increasingly difficult to accurately and thoroughly obtain useful information from massive microblog datasets using traditional information retrieval models. The size of the microblog texts that contain few semantic information increases the difficulty of analysing their content. Also, since microblog retrieval only uses the search keywords provided by the users, there is a considerable risk that query terms fail to match any word observed in relevant tweets. The existing research shows that searchers supply two or three query terms on average [18], which is a short number and these terms can only express a small part of the user's information needs. To overcome mismatch problem (i.e., how to retrieve concise documents, which might be conceptually relevant, but do not explicitly contain some or all of the query terms) query expansion techniques [21] provide alternatives.

Query expansion (QE) techniques refer to the process of reformulating queries with additional terms that better define the information needs of the user [1]. Query expansion approaches rewrite the original query by adding other relevant keywords or suggesting additional appropriate keywords. Classical QE techniques have achieved good results in traditional text retrieval tasks. However, directly applying these methods to microblog information retrieval cannot achieve the desired performance, given the characteristics of the posts [12,21]. To overcome the limitations of existing methods, we propose a classification based method to extract relevant keywords for query expansion. In this method, the aim is to find expanded query terms from top documents. To do this, we firstly preprocess an initial query and classify it; then we extract frequent terms from top results. The idea is that we only consider top results that are in the same class as a query to extract the expanded terms. For classification purpose, we study and compare different models namely, Support Vector Machines (SVM), Naive Bayes (NB) and Random Forest (RF); the convolutional neural model (CNN) and the Multi-Task Deep Neural Network (MT-DNN) to find the best model for classifying crisis micro-texts. We conducted experiments to assess the models on a crisis tweet dataset [7,14].

The main contributions of this paper include: (1) the leverage of a novel approach to extract relevant terms to expand the original query, which can better reflect users' search intent, and (2) the implementation and comparison of several classification techniques. The objective was to study and compare different models for addressing the classification of crisis-related tweets. We show that we obtain competitive results with other works addressing crisis tweets analysis [7,14], and we achieve to obtain better results applying MT-DNN for tweets classification in a novel and original manner. Besides, the experimental results demonstrate that our query expansion method is effective and reduces noise in the expanded query terms, which improves the accuracy of microblog retrieval.

The remainder of the paper is organized as follows. Section 2 discusses related work and compares our work with approaches addressing crisis micro-text classification and query expansion. Section 3 describes the general micro-texts classification study expressed as a classification problem addressed using supervised and deep learning models. Section 4 describes the experimental settings, datasets and discusses the obtained results. Section 5 concludes the paper and discusses future work.

2   Related Work

Millions of people use social media platforms, and the amount of data they produce is enormous. Researchers are continuously working on developing systems which can efficiently process the human-generated data during events like disasters to use them for building solutions that can save millions of lives.

Studies have shown how crisis data can be beneficial and crucial in analysing and collecting insight during and after a disaster. In the paper [19], the authors proposed a match discovering system for mapping the disaster aid messages and victims problem reports. Authors in the [2], analysed the social media data generated during the occurrence of a disaster. Paper [17] presented a classification system for identifying the type of disaster tweet. Research in the field of query processing can be classified, based on the source of expansion terms, into three groups: query expansion based on relevance feedback, query expansion based on local analysis, and query expansion based on global analysis [22]. Query expansion based on relevance feedback utilises feedback from the initial retrieval to enrich the original query. Query expansion based on local analysis is also known as pseudo-relevance feedback method. Specifically, the retrieval system assumes that the first k documents returned are relevant documents and query expansion words extract from the top k retrieved documents. Query expansion based on global analysis aims to mine the relevance difference among words, and treats the most relevant words as complements to the query.

The traditional text retrieval field applies the query mentioned above expansion methods. However, it is not easy to achieve the desired performance by directly using these methods in microblog retrieval [10,22]. The reason is that there is a large number of network vocabularies in microblogs and the junk text, without any useful information. Because of these factors, if top-ranked microblogs, returned by the initial search, are not relevant, microblog query expansion through pseudo-relevance feedback will be of little use.

Our work integrates a tweet classification process to retrieve documents that are in the query's class to extract relevant keywords for query expansion. For classifying crisis-related data, various machine learning algorithms and their performance have been proposed [5,7,8]. In [6], authors have shown DNN outperforms the traditional models in most of the tasks. The results of applying CNN for analysing crisis data [13,14] have surpassed the traditional machine learning models by a significant margin. The authors proposed the semantically-enhanced duel-CNN with two layers in [4].

Our work applied and compared the techniques previously used to classify micro-texts, and particularly crisis tweets related to disasters. We reproduced existing experiments like [7,14]. Seeking for better performance with the datasets we used, we applied MT-DNN. The application of MT-DNN in this context is novel and original and has led to promising results. Moreover, based on our preliminary experiments, we observed that query expansion based on the classification method obtains better candidate expansion words, which are semantically close to the user query.

3   Query Expansion Based on Classification Results

Our approach is calibrated to explore disaster management datasets (e.g. earthquake, flooding, fire) produced by social media. We use prepared tweet disaster datasets ready to be explored. We focused on expanding queries looking for micro-texts (i.e., tweets) related of two classes: events which represent situations produced during the disaster life cycle (e.g., someone looks for shelter, a building has been damaged); and actions performed in response to events (a hotel is providing shelter for victims, people is approaching a damaged building to search victims).

Figure 1 illustrates the proposed framework to find the relevant terms to expand a query. It consists of two phases. The first phase pre-processes an initial query to rewrite it by extending it with relevant terms. The second is devoted to classifying the query to determine its type (i.e., event, action) inspired in [15,20].

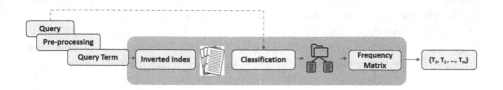

Fig. 1. Classification based query expansion

3.1   Phase A

Phase (**A**) consists of the following steps:

(i) The original query is preprocessed and cleaned removing stop words and symbols. (ii) Then the query is classified into two classes (event, action). The classified query is used to obtain a set of relevant tweets from a large unlabelled tweet corpus using an inverted indexed matrix consisting of terms extracted from the tweet corpus. (iii) Once we have a set of relevant tweets, we classify them using our classification language models and select the tweets that belong to the user query class. (iv) We obtain the **m** top frequent keywords out of the classified relevant tweets (in step (iii)). We have elaborated our detailed approach for phase A in Algorithm 1. The following paragraphs give details of the most relevant steps of the algorithm.

Algorithm 1: Finding Relevant Terms

Data: (Crisis Related Tweets, User Query)
Result: Expanded Query Terms
begin

$Cleaned\_Data \longleftarrow Data\_Cleaning(CrisisRelatedTweets)$
$Indexed\_Data \longleftarrow Inverted\_Index(Cleaned\_Data)$
$Query\_Keywords \longleftarrow PreProcessing(UserQuery)$
for $Keyword \in Query\_Keywords$ **do**
  $Initial\_Result \longleftarrow Finding\_tweets\_contain\_keyword(Keyword)$

$Query\_Label \longleftarrow MTDNN\_Classifier(UserQuery)$
for $Tweet \in Initial\_Result$ **do**
  if $MTDNN\_Classifier(Tweet) == Query\_Label$ **then**
    $Final\_Result \longleftarrow Tweet$

$ExpandedQueryTerms \longleftarrow Term\_Frequency(Final\_Result)$

Classifying Micro-texts. The data science workflow implementing the classification phase applies different machine learning and deep learning models [7, 14]. The workflow splits into three groups of activities: (1) data preparation; (2) classification and (3) assessment. The activities of group 2 and 3 are specialised into the following activities: (2.1) Creation of a baseline applying supervised learning models (i.e., Support Vector Machines (SVM), Random Forest(RF) and Naive Bayes (NB) as classic classifiers) and (3.1) their assessment. (2.2) Classification with no prior knowledge and (3.2) assessment. (2.3) Classification based on MT-DNN that looks for a better classification score and (3.3) assessment. Assessment activities enable the comparison of the performance of the models according to their accuracy, to choose the one that provides the best results for rewriting the queries.

Classification Baseline. As said before, for the classification step, our objective was to identify 2 classes: events (situations coming up in disaster) and actions (reactions performed in response to events).

We first implemented a supervised learning classification that maps an input (tweet) to an output (label) based on labelled crisis data available on crisis NLP website [7]. For example, the tweet "#BREAKING New Injury Numbers 172 injured, 7 fractures, 1 critical #napaquake" is related to the concepts of death and accident, so it is mapped to the class *event*. In contrast, the tweet "Full statement by Napa Valley Vintnerson new #earthquake relief find, with a link for making donations." concerns Non-Governmental Organisations (NGO) efforts and donations, so it is mapped to the class *action*. We used three supervised learning algorithms, namely, Support Vector Machines (SVM), Random Forest (RF) and Naive Bayes (NB) as classic classifiers.

Convolutional Neural Networks. The activities of the data science workflow specialised on Deep neural networks (DNNs) [9] were designed as follows. We used a Convolutional Neural Network (CNN) which a deep learning network consisting of an input layer, multiple convolution layers and an output layer. For applying CNN in NLP tasks, like tweets classification, we used previously computed token sequences as input to the CNN. Then, CNN filters preform as n-grams over continuous representations. These n-grams filters are combined by subsequent network layers, namely the dense layers. CNN can learn the features and distinguish them automatically, and therefore, it does not require hand-engineered features. This saves human effort and time and eliminates the need for prior knowledge. A distributed word representation and generalisation feature effectively utilise the already used labelled data from the other event. This increases the efficiency of the classification process on new data. It removes the need to use manually craft features as it learns automatically latent features as distributed dense vectors, which generalise well and have shown to benefit various NLP tasks [14].

Looking for Better Classification Results. We used Multi-Task Deep Neural Networks (MT-DNN) [11] to classify the tweets looking for better classification results. MT-DNN is based on knowledge distillation which is a process of transferring the knowledge from a set of the larger, complicated model(s) to a lighter compact, easier to deploy single model, without significant loss in performance. In MT-DNN Lexicon Encoder l_1 and Transformer Encoder l_2 are the shared layer. The input sentence: $X = \{x_1, x_2, \cdots, x_m\}$ is a sequence of tokens of length m. Then the lexicon encoder maps X into a sequence of input embedding vectors, one for each token, constructed by summing the corresponding word, segment, and positional embeddings. In layer l_2 a multilayer bidirectional Transformer encoder is used to map the input representation vectors l_1 into a sequence of contextual embedding vectors. MT-DNN learns the representation using multi-task objectives, in addition to pre-training. The MT-DNN model is shown in Fig. 2.

We fine-tuned the MT-DNN codebase to perform the specific task of single sentence classification. The input X (a sentence) is first represented as a sequence of embedding vectors, one for each word, in l_1. Then the Transformer encoder captures the contextual information for each word and produces the shared contextual embedding vectors in l_2 (l_2 is a layer above l_1). Finally, the additional

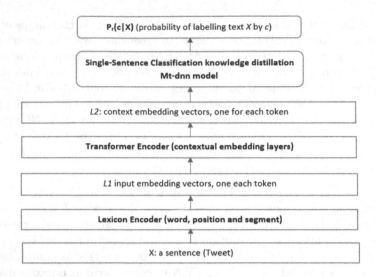

Fig. 2. MT-DNN model for representation learning.

task-specific layers generate task-specific representations (single sentence classification task in our case), followed by the process of knowledge distillation. The logistic regression with softmax predicts the probability that tweet(X) is labelled as class c.

Finding Relevant Terms Set. In the following paragraph, we describe the design of our proposed system framework, as shown in Fig. 1. Given a set of microblog corpus. We first perform data cleaning and indexing into the database. A query term will then be matched with the tweet index to retrieve initial result set. The query and initial results are then classified and only those results, which are in the query class, used to expand the initial query with more relevant and frequent terms. The logical flow of the process is detailed in Algorithm 1.

3.2   Phase B

Phase (**B**) consists of the following steps: (v) Obtain the a sentence-level vector for the user query by utilising the crisisNLP pretrained word embedding via word2vec method. (vi) Compute the similarity between the query vector obtained in step (v) compute m keyword vectors obtained in step (iv). Select the expansion words with the highest similarity as the query expansion words. (vi) Use the top similar keywords to expand the user query using query expansion technique.

4   Experiments

We conducted experiments for finding the best classification model to classify crisis related tweets as micro-text documents and then we used the best model to set up experiments for our query expansion method[1].

We used crisis NLP labelled data sets contain approximately 50k labelled tweets and consist of various event types such as earthquakes, floods, typhoons, etc.

The tweets are initially labelled into various informative classes (e.g., urgent needs, donation offers, infrastructure damage, dead or injured people) and one not-related or irrelevant class. The objective of the experiment was to find out the model that can further and best classify the tweets into event and action classes to have a vocabulary depicting respectively emerging situations (events like water and shelter shortage) and performed actions during a disaster (relief like water delivered to a given area, several rooms available for families).

4.1   Data Pre-processing

Pre-processing was required before using Tweets to address issues that characterize them and thereby produce a clean dataset.

Data Cleaning. In our experiment, we considered that tweet texts are brief, irregular expressions, noisy, unstructured, and often containing misspellings and grammatical mistakes with words out of the dictionary. We removed blank rows, changed all the text to lowercase, removed URLs, re-tweets and user-mentions. Then we moved towards tokenization that broke each tweet in the corpus into a bag of words. Followed by removal of English stopwords, non-numeric and special characters and perform word-stemming/lemmatization. WordNetLemmatizer required pos tags to understand if the word is a noun or verb or adjective (by default, it was set to "noun").

Indexing Data Collections. As a result of indexing the cleaned tweets collection, we created an inverted index matrix that represents the content of the collection. An inverted index is a dictionary where each word is associated with a list of document identifiers in which that word appears. It enables agile access to the position within a document in which a term appears. Indeed, this structure allows avoiding making quadratic the running time of token comparisons. So, instead of comparing, record by record, each token to every other token to see if they match, the inverted indices are used to look up records that match on a particular token.

---

[1] https://github.com/MehrdadFarokhnejad/Classifying_Tweeter_Crisis_Related_Data.

Word Embedding Initialisation. In our CNN experiment setting, we have used crisis embedding to initialise the embedding at the beginning of the experiment. Crisis embedding is a 300-dimensions domain-specific embedding created by [13] trained on 20 million crisis-related tweets using the Skip-gram model of the *word2vec* tool from a large corpus of disaster-related tweets. The corpus contains 57, 908 tweets and 9.4 million tokens.

4.2   Applying Classification Techniques

We performed our experiments using three models with the following hyper-parameters initialisation.

a. Machine Learning Model: Chosen parameter values for SVM - regularization were (C) = 100, kernel type = 'rbf', gamma value = 0.1. For Random Forest- max depth = 5, n estimators = 10, max features = 1.

The label encodes the target variable - This is done to transform Categorical data of string type in the data set into numerical values. Next step is word vectorization by using TF-IDF Vectorizer - This is done to find how important a word in a document is in comparison to the corpus. After fitting the data, we run the machine learning algorithm to check accuracy.

b. CNN Model. Values of parameters: Filters no. = 250, Pool size = 2, Hidden size = 128, Kernel size = 3.

We have re-implemented the CNN and Crisis embedding model from [14] to compare it with the other models. We used a multilayer perceptron with a CNN.

c. MT-DNN Model. Values of parameters: Learning rate = 5e–5, global gradient clipping = 1.0, Learning gamma = 0.1, epoch = 30, Variable batch sizes = (16,32,64).

We applied the latest Microsoft MT-DNN [11] model on our data set looking for better classification performance.

4.3   Data Sets and Labels

We have used the CrisisNLP data set for our classification task and measuring the accuracy of all the three models. We modified the labels of tweets such that each crisis event data set has two labels - Event (a situation produced during a disaster) and Action (represents reactions to events). Table 1 shows the number of tweets for each set. The class "Event" includes tweets which subject is related to any occurrence or incidence happening during or after the crisis. For example, "damage happened to a building" or "people are trapped in buildings in downtown". For "Action" we consider those tweets that focus on operations taking place during or after the crisis. Such as government or NGOs providing help to the affected people.

We performed a set of experiments on California and Nepal earthquake, Typhoon Hagupit and Pakistan Flood data sets (see Table 2). The distribution of data is shown in Table 2: train (70), validation (10) and test sets (20). Column Labels show the total number of annotations for each class.

Table 1. Description of the classes in the data sets.

Class	Total label	Description
Event	1869	Tweets reporting occurrence and happening of events during the crisis. Reports deaths, injuries, missing, found, or displaced people, infrastructure and utilities damage
Action	2684	Tweets reporting responses and measures taken by people during crisis. Messages containing donations or volunteering offers also sympathy-emotional support

Table 2. Class distribution of events under consideration and all other crises.

Class	Nepal	California	Typhoon	Pakistan
Event	688	574	271	356
Action	1535	255	462	432
Total	2203	829	733	788

4.4 Classification Results

Table 3 reports the performance of the five models applied to the California, Nepal, Hagupit and Pakistan crisis datasets. Note that for a given model, dataset quality across different disaster events is not similar and also tweets were noisy and after cleaning there can be some data loss which could affect the contextual understanding of models. Hence the models learn and generalise better events with higher data quality and uniform class label distribution. Note that among all the machine learning models (i.e., SVM, RF and NB), the accuracy score of SVM is comparatively higher than RF and NB. For example, the accuracy score of the California dataset using SVM is **88.17**, whereas the accuracy score for RF and NB are **87.36** and **86.56**, respectively. For the CNN model with crisis embedding, the accuracy scores for the California dataset is **90.13** and for Nepal **88.62**. This is also true for the Typhoon Hagupit and Pakistan Flood datasets, with SVM core accuracy of **87.61** and **92.82**, respectively.

CNN model outperforms the machine learning models in terms of accuracy score. For the MT-DNN model the accuracy scores were: for California crisis = **91.72** and for Nepal crisis = **90.31**. For Typhoon Hagupit = **91.22** and for Pakistan Flood = **95.72**. We can see that the MT-DNN model surpasses the machine learning and the CNN model and have the best accuracy score.

Several works have done crisis tweets classification research [7,14]. Existing work has addressed different labels using several classes applying RF, linear

Table 3. Accuracy score of SVMs, RF, NB, CNN and MT-DNN with respect to crisis tweet data.

Data sets	SVM	NB	RF	CNN	MT-DNN
California	88.17	86.56	87.36	90.13	**91.72**
Nepal	87.21	83.83	86.95	88.62	**90.31**
Hagupit	87.61	80.75	85.43	89.31	**91.22**
Pakistan	92.82	91.25	90.64	93.34	**95.72**

regression (LR), SVM and CNN. The best accuracy results were obtained with CNN on the Nepal and California Earthquakes, datasets the typhon Hagupit and the cyclone PAM (resp. 86,89, 81.21, 87,83 and 94,17). Our binary classification leads to acceptable accuracy results ranging from 87.36, 86.95, 85.43 and 90.64.

Our results in Table 3 show that MT-DNN model performs better than CNN and ML models used in previous experiments, however from computing complexity point of view, since CNN performance was very close to MT-DNN we can also use CNN instead of MT-DNN which will save computation power and cost.

4.5   Models Compared: Classification vs Non-classification

In this section, we have presented an ablation study in which we have compared the performance of our proposed classification based query expanding method against the traditional query expanding method. We use the available crisisNLP pre-trained word embedding via word2vec method [7] to obtained query and expansion terms vectors. In the vector space model, all queries and terms are represented as vectors in dimensional space 300. Documents similarity is determined by computing the similarity of their content vector. To obtain a query vector, we represent keywords in user queries as vectors, and then sum all the keyword vectors followed by averaging them. For our analysis, we calculated the average similarity between the query vector and 'm' keyword vectors obtained for a given query by using the formula 1.

$$Similarity(Candidate\_terms, Query) = \frac{\sum_{i=1}^{m}(Cosine(Query\_vector, Term\_vector[i]))}{m} \quad (1)$$

where 'm' is a hyper-parameter in query expansion-based retrieval, which shows the number of expansion terms (ET). Using as reference the studies [3,23], we set the number of expansion terms to 10, 20 and 30 (ET@10, ET@20, ET@30). We repeat this task for 100 queries and report the mean of average of each **ET@** set in Table 4. The experimental results show that the expanded query terms obtained from the classified query expansion model are more similar and relevant than the non-classification model. The ET@10, ET@20 and ET@30 scores of our proposed classification model surpassed the transition non-classification based model. Also, we observe that when we set the number of expansion terms to 10, we achieve the best performance.

Table 4. The mean average of Cosine Similarity (MACS) between query and expanded query terms with and without classification model.

Query expansion model	ET@10	ET@20	ET@30
Classification	**0.420**	**0.377**	**0.371**
Non-classification	0.401	0.366	0.369

5   Conclusions and Future Work

This paper introduced a classification based query expansion method. For classification purpose, various machine learning algorithms studied and were validated through experiments using crisis tweet datasets that compared the performance of the applied models. Also, we showed that query expansion base on classification method obtains better candidate expansion words, which are semantically close to the user query. We are currently exploring more robust and advanced NLP models for processing and analyzing crisis data to improve the achieved results. Our future work includes the use of classified vocabularies for exploring data collections using different techniques like queries as answers, query morphing and query by example.

References

1. Abberley, D., Kirby, D., Renals, S., Robinson, T.: The THISL broadcast news retrieval system (1999)
2. Acar, A., Muraki, Y.: Twitter for crisis communication: lessons learned from Japan's Tsunami disaster. Int. J. Web Based Communities **7**(3), 392–402 (2011)
3. Azad, H.K., Deepak, A.: Query expansion techniques for information retrieval: a survey. Inf. Process. Manage. **56**(5), 1698–1735 (2019)
4. Burel, G., Alani, H.: Crisis event extraction service (CREES)-automatic detection and classification of crisis-related content on social media (2018)
5. Cameron, M.A., Power, R., Robinson, B., Yin, J.: Emergency situation awareness from twitter for crisis management. In: Proceedings of the 21st International Conference on World Wide Web, pp. 695–698. ACM (2012)
6. Collobert, R., Weston, J., Bottou, L., Karlen, M., Kavukcuoglu, K., Kuksa, P.: Natural language processing (almost) from scratch. J. Mach. Learn. Res. **12**, 2493–2537 (2011)
7. Imran, M., Mitra, P., Castillo, C.: Twitter as a lifeline: human-annotated Twitter corpora for NLP of crisis-related messages. In: Proceedings of the Tenth International Conference on Language Resources and Evaluation (LREC 2016). European Language Resources Association (ELRA), Paris, France, May 2016
8. Imran, M., Mitra, P., Srivastava, J.: Cross-language domain adaptation for classifying crisis-related short messages. arXiv preprint arXiv:1602.05388 (2016)
9. Kim, Y.: Convolutional neural networks for sentence classification. arXiv preprint arXiv:1408.5882 (2014)
10. Li, L., Xu, G., Yang, Z., Dolog, P., Zhang, Y., Kitsuregawa, M.: An efficient approach to suggesting topically related web queries using hidden topic model. World Wide Web **16**(3), 273–297 (2013)

11. Liu, X., He, P., Chen, W., Gao, J.: Multi-task deep neural networks for natural language understanding. In: Proceedings of the 57th Annual Meeting of the Association for Computational Linguistics, pp. 4487–4496. Association for Computational Linguistics, Florence, Italy, July 2019. https://doi.org/10.18653/v1/P19-1441. https://www.aclweb.org/anthology/P19-1441

12. Miyanishi, T., Seki, K., Uehara, K.: Improving pseudo-relevance feedback via tweet selection. In: Proceedings of the 22nd ACM International Conference on Information & Knowledge Management, pp. 439–448 (2013)

13. Nguyen, D.T., Joty, S., Imran, M., Sajjad, H., Mitra, P.: Applications of online deep learning for crisis response using social media information. arXiv preprint arXiv:1610.01030 (2016)

14. Nguyen, D.T., Mannai, K.A.A., Joty, S., Sajjad, H., Imran, M., Mitra, P.: Rapid classification of crisis-related data on social networks using convolutional neural networks. arXiv preprint arXiv:1608.03902 (2016)

15. Palen, L., Vieweg, S.: The emergence of online widescale interaction in unexpected events: assistance, alliance & retreat. In: Proceedings of the 2008 ACM Conference on Computer Supported Cooperative Work, pp. 117–126. ACM (2008)

16. Priya, S., Bhanu, M., Dandapat, S.K., Ghosh, K., Chandra, J.: TAQE: tweet retrieval-based infrastructure damage assessment during disasters. IEEE Trans. Comput. Soc. Syst. **7**(2), 389–403 (2020)

17. Sakaki, T., Okazaki, M., Matsuo, Y.: Earthquake shakes Twitter users: real-time event detection by social sensors. In: Proceedings of the 19th International Conference on World Wide Web, pp. 851–860. ACM (2010)

18. Spink, A., Wolfram, D., Jansen, M.B., Saracevic, T.: Searching the web: the public and their queries. J. Am. Soc. Inform. Sci. Technol. **52**(3), 226–234 (2001)

19. Varga, I., et al.: Aid is out there: looking for help from tweets during a large scale disaster. In: Proceedings of the 51st Annual Meeting of the Association for Computational Linguistics (Volume 1: Long Papers), pp. 1619–1629 (2013)

20. Vieweg, S., Hughes, A.L., Starbird, K., Palen, L.: Microblogging during two natural hazards events: what Twitter may contribute to situational awareness. In: Proceedings of the SIGCHI Conference on Human Factors in Computing Systems, pp. 1079–1088. ACM (2010)

21. Wang, Y., Huang, H., Feng, C.: Query expansion based on a feedback concept model for microblog retrieval. In: Proceedings of the 26th International Conference on World Wide Web, pp. 559–568 (2017)

22. Xu, B., Lin, H., Lin, Y., Xu, K., Wang, L., Gao, J.: Incorporating semantic word representations into query expansion for microblog information retrieval. Inform. Technol. Control **48**(4), 626–636 (2019)

23. Zhai, C., Lafferty, J.: Model-based feedback in the language modeling approach to information retrieval. In: Proceedings of the Tenth International Conference on Information and Knowledge Management, pp. 403–410 (2001)

Data Centered and Usage-Based Security Service

Jingya Yuan⬛, Frédérique Biennier(✉) ⬛, and Nabila Benharkat⬛

University of Lyon, CNRS, INSA-Lyon, LIRIS, UMR 5205, Lyon, France
{jingya.yuan,frederique.biennier,nabila.benharkat}@liris.cnrs.fr

Abstract. Protecting Information Systems (IS) relies traditionally on security risk analysis methods. Designed for well-perimetrised environments, these methods rely on a systematic identification of threats and vulnerabilities to identify efficient control-centered protection countermeasures. Unfortunately, this does not fit security challenges carried out by the opened and agile organizations provided by the Social, Mobile, big data Analytics, Cloud and Internet of Things (SMACIT) environment. Due to their inherently collaborative and distributed organization, such multi-tenancy systems require the integration of contextual vulnerabilities, depending on the a priori unknown way of using, storing and exchanging data in opened cloud environment. Moreover, as data can be associated to multiple copies, different protection requirements can be set for each of these copies, which may lead the initial data owner lose control on the data protection. This involves (1) turning the traditional control-centered security vision to a dynamic data-centered protection and even (2) considering that the way a data is used can be a potential threat that may corrupt data protection efficiency. To fit these challenges, we propose a Data-centric Usage-based Protection service (DUP). This service is based on an information system meta-model, used to identify formally data assets and store the processes using copies of these assets. To define a usage-entered protection, we extend the Usage Based Access Control model, which is mostly focused on managing CRUD operations, to more complex operation fitting the SMACIT context. These usage rules are used to generate smart contracts, storing usage consents and managing usage control for cloud services.

Keywords: Privacy · Data-driven organization · Blockchain · GDPR · Usage governance

1 Introduction

The explosion of Social networks, Mobile environment, big data Analytics, Cloud computing or Internet of Things, known as SMACIT services, has created unprecedented opportunities and fundamental security and privacy challenges. First, asset identification is more complex than in classical Information Systems (IS for short) as data are not just "big" but also unstructured and multi-modeled [1]: this makes harder identifying precisely potential vulnerabilities and threats. Second, the complexity of the software stack involves integrating specific security requirements related to the cloud technology [2] and to mobile access to this cloud architecture [3]. Third, the intrinsic openness of

© Springer Nature Switzerland AG 2021
H. Hacid et al. (Eds.): ICSOC 2020 Workshops, LNCS 12632, pp. 457–471, 2021.
https://doi.org/10.1007/978-3-030-76352-7_42

SMACIT systems involves integrating different stakeholders, sharing data and processes without clear and common security policies. This context leads to inconsistent protection as each party deploys and manages its own security strategy. Fourth, analytic processes are also threating agents, carrying privacy vulnerabilities: whereas privacy-preserving processes, as anonymization algorithms, are designed for well-identified datasets [4], analytic processes mixing different "protected" data sets may lead to identify users from initially anonymized data.

To overcome these limits/mitigating these risks, technical solutions such as the integration of privacy preserving techniques in big data [5] have been introduced. Nevertheless, privacy intrusion may also occur in the domains of social web, consumer and business analytics [6] and governmental surveillance. To fit the extended, collaborative and sharing organization model carried out by SMACIT, legal regulation has evolved. The European Union General Data Protection Regulation (GDPR) [7] equilibrates relationships between end-users and data consumers, allowing analytic purpose "by default" and empowering users to manage usage rights on their own data, constraining consumers to report any data breach and to enhance transparency [8]. In brief, GDPR integrates rights and obligations for both data provider and data consumer, increasing the call for "proving" fair consents and fair usages in highly interconnected and distributed environments.

This leads to different challenges to adapt these Cloud-based services security and privacy. First, data providers must manage the way they dispatch copies of their data among different service providers, keeping information on the usage they granted. Second, data provider must integrate a precise usage description while defining their terms of service. Third, service providers also need to track any action they have on a data to prove the "fair usage" on it. To address these challenges, illustrated by a motivating example (Sect. 2), we first explore related works (Sect. 3) before defining (Sect. 4) our Distributed Data centered Usage Protection service, extending Usage-based access purpose to business knowledge. This architecture is evaluated Sect. 5.

2   Motivating Example

Our motivating example integrates collaborative business and end-users' interactions. It relies on an online-shopping platform (called later Online Shopping) which proposes "manufactured on demand" products from different suppliers (such as Company A and Company B) to clients. These different partners share and exchange product information or client personal information depending on the business process requirements. To reduce the carbon footprint, Company A uses the 3D printers hosted by company B to "manufacture" the product as close as possible to the client. Online Shopping also shares data with MyAnalytics company which uses the customer data to establish recommendations and provide marketing analysis to Online Shopping company.

Alice browses the Online Shopping platform. Online shopping platform collects Alice's view history, connection information as well as other information related to the product she intends to buy. Consequently, Alice will have to "share" different personal information such as traces of her on-line activity, financial information, her address and what she has bought with Online Shopping platform, MyAnalytics (which can also mix these information with other sources) and Company A and B (the product suppliers).

Online Shopping platform is responsible of Alice's personal information protection. According to the GDPR, Online Shopping may also have to prove that it uses and protects this information according to the exact Terms of Service (ToS). While exchanging data with its partners (MyAnalytics, Company A or B), Online shopping must check the business purpose of the external service requesting information to verify if this is allowed according to Alice's consent. Online Shopping has also to transfer this ToS to the service provider.

Alice also interacts with other online platforms, Personal Information Management Systems (PIMSs), social networks, etc. using her own computer or smart phone. Protecting consistently her personal information is difficult for Alice as she interacts with different systems, providing their own ToS. Moreover, she cannot get any information to check whereas her personal information is really protected and if it is used according to the ToS she accepted.

Based on this motivating example, we identify different requirements. First, Alice needs to describe globally her information system, including protection requirements for each logical asset and manage assets' replications. Second, Online Shopping needs to get Alice's consent, defining precise "fair usages" and to log the different operations and exchanges associated to the information it gets. These requirements call for a data-centric usage control security service, able to manage data protection and usage control according to the approved Terms of Service.

3 Related Works

Collaborative Business as well as SMACIT services provide collaboration contracts or Terms of Service (ToS), defining rather precisely the way information will be used and shared. For example, Terms of Service Didn't Read (see http://tosdr.org) provides evaluation of services privacy level based on different criteria such as ToS readability, business transfer, licensing changes notification, anonymity and tracking features, right of end-users to leave the service. Focusing on data consumers, TM Forum has proposed an evaluation template[1] to establish these ToS and rate the privacy and protection score of new SaaS services. These ToS and collaboration contracts state rather precisely business purpose and information sharing strategies. Unfortunately, access control functions, such as Access control lists, Role based access control and even Usage based access control...) do not use these "business" motivations.

Moreover, SMACIT changes data usage: data analytics and data mining processes extract knowledge and generate new data to serve business goals, creating privacy breaches as new personal information can be generated, tracking users according to data they leave in daily transactions [9], reducing human beings to a collection of data [10]. To face this risk, GDPR empowers users with their personal data protection, requiring service providers to state the usages they will achieve for a particular data, manage user consents accordingly and report any security failure to the data owner. This involves to report (and prove) both any actions on data to show that the real usage complies with the usages that the data provider has accepted. Security events (namely security service

---

[1] https://www.tmforum.org/resources/technical-report-exploratory-report/tr232-privacy-score-for-a-service-v1-0-2/.

deployment and security breach identification) must also be reported. To fit the GDPR requirements, several works have been developed: [11] integrates GDPR obligations in traditional Enterprise Architecture model by identifying both information and processing categories whereas [12] proposes to manage data collection and track data flows between stakeholders. Moreover, several works take advantage of the Blockchain immutability property to allow proving fair usages: [13] manages access control function thanks to smart contracts, [14] manages data encryption key to protect data access, [15] manage user consents in Blockchain whereas [16] uses the Blockchain to track data accountability and provenance, or even track usage operation thanks to smart contracts generated according to the data usage policy [17].

Although these works provide a rich background to protect data, none of them allows integrating business information nor managing a consistent protection on the multiple copies of a same data "consumed" by different cloud services, exchanged and shared by different cloud stakeholders.

4   A Distributed Usage Control Architecture

Based on this rich background, on GDPR constraints and on requirements picked from our motivating example, we identify three main challenges to provide a consistent protection to data in opened SMACIT services environment. First, protection requirements must be managed consistently, although the same asset is replicated in different information systems. Second, fair usage involves integrating business knowledge to identify precisely the way assets are used. Third, usage governance and tracking are requested by both data provider and data consumer to "prove" that data are used and protected according to what has been approved by the data provider and the data consumer.

We take advantage of the service-oriented security architecture that outsources security requirements and protection means in policies to build our Data-centered Usage-based Protection Service (DUP service for short), fitting these requirements. Plugged on Information Systems thanks to an Information System interface component, our DUP service relies on three interconnected core components (see Fig. 1):

– First, the Information System Meta-model is in charge of storing the Information System description including the logical data assets, Business processes and their associated protection policies as well as business agreements between parties. It also describes the way this Information System is implemented.
– Second, the Usage Rule Manager is in charge of describing usage consents. Consents are defined by setting proper asset protection mechanisms and defining the operations that can be used on a given asset.
– Third, the Usage Governance manager is in charge of publishing these consents in a blockchain to support consent immutability and to track the operation execution so that asset protection and usage can be governed.

4.1   Information System Meta-model

To manage the current asset protection and allowed usages, we design an Information System (IS for short) meta-model to store the IS collaborative ecosystem description, identifying the different service providers and their relationships, the different data

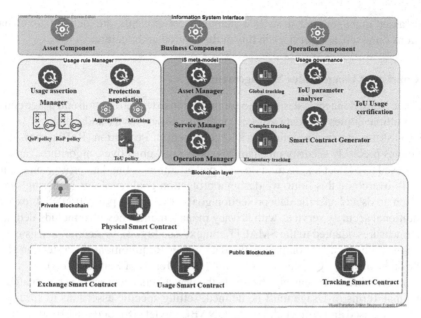

Fig. 1. DUP service global organization

assets description and the description of their replicated content. More precisely, this information system meta-model is organized in three main layers:

– *The organization layer* gathers information on the different actors (human being/enterprise/organizational unit) that own, store or process logical assets with meta-data description. Contracts, including security agreements are associated to the relationships between actors. Actors are also associated to different business areas/organizational units according to their competencies or to the Business Processes and business activities they have to manage.
– *The "logical" layer* gathers description of both logical data asset and the way they may be processed. Logical data assets are described thanks to meta-data and are associated to security requirements depending on their sensitivity level. These data assets can be combined to set more complex asset, aggregating atomic ones. Processes are defined thanks to abstract services, described thanks activity patterns related to their business domain. Each of these abstract services is also characterized by its interface, described thanks to data objects associated to logical assets.
– *The implementation layer* stores the identification of the real physical instances associated to the data object. These physical copies of logical assets are called "containers". These "containers" are consumed by "concrete services" associated to IT or manual application services. Each of these "concrete" services is provided by an actor and has its own security policy which may be inherited from the data consumer's generic security policy, enriched with the target usage of the data object. Paying attention to the cloud-based multi-tenants' implementation, each concrete service deployment may use other support services (cloud or network services), using a "support" relationship.

An asset manager and a service manager components are defined to manage operations on assets and services in this multi-layered architecture.

4.2 Consistent Usage Rules Management

The Usage Rule Manager is designed to support a consistent protection and usage control on the multiple copies of a data asset according to its sensitivity and value.

Each asset protection is defined in a Requirement of Protection (RoP) policy. While [19] defines precisely security mechanisms, protocols, objectives, algorithms, and credentials used to protect a system, it doesn't pay attention to the data protection requirements. To overcome this limit, we design a protection goal ontology, gathering system protection goals and specific data protection goals. These data protection goals expands the traditional security services with privacy management. They also include dedicated countermeasures adapted to the SMACIT context (see Fig. 2). Protection consistency is continuously checked by comparing the data owner Requirement of Protection (RoP) and the data consumer Quality of Protection (QoP)/Terms of Service (ToS).

As security breaches may also be due to the way assets are processed and shared, we assume that Terms of Usage must be defined, setting specific assertions to specify the way assets are used and protected. The UCONABC model [18] relies on a usage ontology to define rights that are associated to basic usage operations (i.e. Create, Read, Update, Delete), subjects who will get a usage right, objects which define the asset on which the usage right is granted, obligations associated to protection means and restrictions associated to time or environment constraints. We expend this ontology (see Fig. 3) to

- Integrate more complex data related operations, to define more precisely the way data are processed, exchanged, replicated and stored
- Integrate organizational knowledge to define the subject (a user, a group or an organizational entity)
- Integrate business knowledge to define more precisely the usage context, specifying a process motivation, process purpose and business areas.

Fig. 2. Protection ontology

Our consent assertion model is based on our ontology. Formally, a consent is associated to a set of Usage Control Assertions (UCA) doubled approved by the asset owner (who will provide the asset) and by the asset consumer (who will use the asset). As

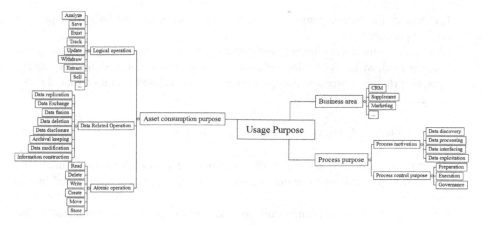

Fig. 3. Usage ontology

usage can be defined globally, an assertion development process is set to allow the data consumer generate more precise usage assertions from an original consent. Each of these assertions is defined as a tuple (Eq. 1):

$$UCA = (AS, AO, S, O, U, PO, OSP, CSP) \tag{1}$$

- AS defines the assertion status, i.e. if the assertion is originated from the original consent or has been inferred from another assertion
- AO defines the asset owner related to the rule. This owner is specified as a set of two attributes:

  - Assignee attribute defines the organizational entity or simple user owning the asset
  - Assignee status defines if it is the original owner or a delegatee representing the owner

- S is the subject, i.e. the party that will get the right on the asset. This subject can be an organizational entity, a simple user or an IT business service in charge of a part of a business process. Similarly to the asset's owner description, it includes 2 attributes:

  - Assignee attribute defines the organizational entity or simple user that may use the asset
  - Assignee status defines if it is the original owner or a delegatee representing the initial asset consumer

- O is the object, i.e. the exchanged asset which usage is regulated by the assertion. This exchanged asset is associated to a unique identifier shared by both parties and related to the corresponding logical assets stored in each party information system
- U is the usage purpose regulated by the assertion. It is specified as a set of attributes:

  - BuP denotes the business purpose. It can refer either to a business area or to a more precise business activity

- PrP denotes the process purpose (including the process motivation and the process control purpose)
- ACP denotes the asset consumption purpose, it can refer to a physical operation (such as read, write…) or to logical asset consumption operation defined as logical operation (transfer, process, store, show) and or data related operation (data fusion, data removal, data replication…)

- PC is the Protection Context. It is defined by a set of attributes:

  - PG denotes the protection goal, i.e. the security service (confidentiality, availability, integrity, non repudiation) or the quality of service that must be provided by the subject
  - CCtX denotes the countermeasures context, i.e. the set of countermeasures that must be deployed to provide a consistent protection.

- OSP defines the Asset Owner Signing Party. It is specified thanks to 2 attributes

  - OSC is the signing owner authentication certification. When an assertion is inferred from a global one, this attribute refers to the "parent assertion"
  - OSK is the owner signature parameters.

- CSP defines the Asset Consumer Signing Party. It is specified thanks to 2 attributes

  - CSC is the signing consumer authentication certification. When an assertion is inferred from a global one, this attribute refers to the "parent assertion"
  - CSK is the consumer signature parameters.

Thanks to the business usage specification stored in the assertion and to process description stored in the meta-model, Terms of Usage assertions are inferred by the data consumer to define more precise assertions associated to the different services supporting the business activities.

From our motivating example, when Alice consents to share contact information with OnLineShopping (OLS for short) while ordering a product as a gift to be delivered to her brother Bob, two usage assertions are originally defined:

- One is defined for Alice's contact information that will be used by the ordering and billing processes (Eq. 2)
- The other is defined for Bob's contact information that will be used by the delivery process. This assertion accepts that this last asset can be shared with other parties involved in the delivery process (Eq. 3).

$$
\begin{aligned}
UCA1 = (&AS = \text{Original, } AO = \text{Alice, } S = \text{OLS, } O = \text{Alice's contact information Id,} \\
&U\text{-}BuP = \text{Order, } U\text{-}ACP = \text{Process, } PC\text{-}PG = \text{Confidentiality} + \text{integrity,} \\
&OSP\text{-}OSC = \text{Direct, } OSP\text{-}OSK = \text{Alice key, } CSP\text{-}CSC = \text{direct, } CSP\text{-}CSK = \text{OLS key}) \quad (2)
\end{aligned}
$$

UCA2 = (AS = Original, AO = Alice, S = OLS, O = Bob's contact information Id,

U - BuP = Deliver, U - ACP = delegate, U - ACP = Process, PC - PG = Confidentiality + integrity,

OSP - OSC = Inferred, OSP - OSK = Bob's consent assertion key,

CSP - CSC = direct, CSP - CSK = OLS key)   (3)

OnLine Shopping infers these assertions to generate sub-consent assertions associated to more precise activities and sub-contractors. For example, OnLineShopping will extract the Deliver Business Usage Purpose from UCA2 to select the corresponding Business Process and its related sub-processes from its information system meta-model. This extracted Delivery Business Process consists into two sub-processes: the order delivery preparation process managed by OnLine Shopping and the Client Final Delivery process managed by Company B. As UCA 2 includes a "delegate" right, a new usage assertion formalizing the inferred consent between OnLine shopping and Company B can be created. This assertion will allow Company B processing Bob's contact information, paying attention to Confidentiality and Integrity requirements. This last consent will be signed according to UCA2 and by company B (Eq. 4).

UCA3 = (AS = Inferred, AO = OLS, S = CompanyB, O = Bob's contact information Id,

U - BuP = Delivery, U - ACP = Process, PC - PG = Confidentiality + integrity,

OSP - OSC = Inferred, OSP - OSK = UCA2 - key,

CSP - CSC = direct, CSP - CSK = CompagnyB key)   (4)

By setting a process that generates fine-grained usage assertions from a global consent, users can have a more comprehensive view on the rights they grant while protecting efficiently their assets. These Terms of Usage assertions are stored locally by each party (data owner and data consumer) in their own Information System meta-model. This makes more efficient ToU information retrieval to evaluate more efficiently the current asset protection. To ensure the consent immutability and provide usage tracking features, these usage assertions are also processed and stored in a Blockchain thanks to smart contracts. This is achieved by the usage governance manager.

4.3 Usage Governance Manager

The Usage Governance manager takes advantage of Blockchain Smart Contracts to store authorized usage, provide proofs of consents associated to data usage and proofs of usage, tracking when a usage right is used. To this end, we define three different types of Smart Contracts:

- The Exchange Smart Contract is used to store the consent assertions. It is stored in a public blockchain. It uses the asset owner and asset consumer key information to sign the authorization specified the assertion. It stores the set of data that can be exchanged and used for a given usage purpose. It generates usage tokens that can be decrypted by the data consumer to get the consent proof.
- The Usage Smart Contract is used to generate a precise usage authorization. It provides a function that can be invoked by the cloud service to get the data it requires. It checks if

the requested data is a part of the exchanged data for a given usage and stores the exact requestor identity to generate a token that can be decrypted by the Asset Owner to certify that the requested data and the requestor fit the consent that has been provided. This smart contract generates events that are consumed by the physical Smart Contract to allow and log the exact access operation on the physical asset stored in the container.

- The Physical Smart Contract provides access operations (i.e. service functions with basic operations) to physical copy of the data asset (i.e. the container) and generates the logfiles with physical event when these operations are invoked. By this way usage tracking can be implemented.

The Smart Contract Factory is in charge of generating these different smart contracts thanks to patterns, extracting precise parameters from the IS meta-model according to what has been defined in the associated assertion. This smart contract factory consists in three smart contract generators associated to the three types of Smart Contract.

The exchange smart contract generator is used to generate the exchange smart contract, storing the approved consent assertion and "representing" this consent when sub-assertions are inferred. It generates a pair of keys (public key and private key) associated to the smart contract, used to hash the assertion and generate an encrypted token, merging both the assertion hashed value and the assertion signatures from data owner and data provider thanks to its private key. It sends safely its public key (that can be used to decrypt this token) to the data owner and data consumer, encrypting it with the public encryption key they provide in the assertion. Note that the Smart Contract token will be used to authenticate the requesting party invoking the usage smart contract "on behalf" on the authorization granted by the assertion this smart contract stores. It also provides an assertion extraction function that can be invoked by the data owner or the data provider to get the complete assertion value.

The usage smart contract generator is used to store more precisely which cloud service will be authorized to access to a subset of assets defined in the exchange smart contract. This generator is invoked, using an exchange smart contract token. It takes advantage of the meta-model to select the convenient services supporting the corresponding business activities and operations and data they need. The Usage Smart Contract generator extracts the assertion from the exchange smart contract. If the requested data fits the exchanged asset definition, a data object is created. A similar process, taking advantage of the business description included in our ToU ontology to check if the applying service fits the business context. Then for each pre-approved service, it generates the corresponding Usage Smart Contracts, storing the more precise usage assertion and the associated consent assertion. Similarly to the exchanged smart contract, a pair of keys (public key and private key) are generated to identify the usage smart contract. The private key is used to hash the usage assertion and generate an encrypted token merging both the assertion hashed value and the assertion signatures from data owner and data provider. It publishes safely its public key (that can be used to decrypt this token) by encrypting it with the key associated to target service.

The physical smart contract generator is used to store the different physical operation authorization. It provides different authorization functions, each of them associated to a particular operation achieved on a given container associated to the data object. This physical smart contract also stores a log of the operation execution. It takes advantage of the meta-model to select the container associated to the data object requested by the concrete service. It also provides a copy of the consent approval on behalf of which the access is granted to the concrete service. Similar to the exchanged smart contract, it generates a pair of keys (public key and private key), uses the private key to hash the assertion and generate the encrypted merging both the assertion hashed value and the assertion signatures from data owner and data provider thanks to its private key. It publishes safely its public key (that can be used to decrypt this token) by encrypting it with the key associated to target concrete service.

5   Evaluation

To evaluate our DUP service, we developed a prototype integrating two parts: the ToU negotiation component and the Blockchain based control system. The ToU negotiation component integrates (1) the IS meta-model from which requirements of protection and quality of protection/required usages are extracted and (2) the Usage Assertion generation component. This ToU negotiation component integrates different Java components have been developed to support policy evaluation, aggregation and comparison. To this end, we use JENA API and Jena-arq API to model policy files. SPARQL queries are generated and launched on these policy files to get these description in the form of Java Objects. These policies are then combined with the Business Process description to generate the different usage assertions. Our tests have shown that the policy aggregation process execution time is a maximum of 200 ms for data involved in 100 services, each of them exhibiting a 4 to 5 assertions policy.

The Blockchain part has been deployed using Ganache which provides an Ethereum sandbox to support Blockchain smart contract development and test. This involves that performance indicator does not reflect real deployment Quality of Service. The smart contract factory is defined as a smart contract, providing functions to create smart contracts according to the different patterns. Each pattern is associated to a set of paramaters. These parameters are extracted from the IS meta-model according to the usage assertion.

These developments have shown that the necessary business knowledge can be extracted from our meta-model to generate the Terms of Usage consent and generate the different usage assertions before setting the different smart contracts. Figure 4 presents the smart contract generated for the UCA3 assertion.

To evaluate our solution, we compare its abilities to other works presented in the state of the art according to both security management and GDPR requirements (Table 1). We identify 4 comparison criteria to evaluate our Terms of Usage (ToU) ontology:

- Countermeasure scope: defines which part of the security solution is concerned: data storage, infrastructure security, communication security, access control.
- Control object: defines the attributes which are used to describe 'Rights', 'Obligation' and 'Condition'.These attributes may deal with operation conditions (including time, location...), basic CRUD operations or business usage purpose.
- Subject Attributes: defines the attributes of the party requiring the access (i.e. the subject) used in the access control process. It may be a role, a group identification (such as friend, stranger or acquaintance usually used in social media) or a well identified entity

```
contract ExchangeSmartContract{
// The subject identity prove
    string constant identification_prove='CompanyB';
// the  Terms of Usage ToU_Token between OLS and CompanyB
    string constant Terms_of_usage_Token='ToU_Token(ToUID)';
// Object definition
    String exchangedAsset="Bob's contact informationId";
//define the Online-shopping's address in the public blockchain
    address private owner;
//the business purpose which is the relationship among logical asset and specified service function
    string Business_domain="Delivery";
//The asset status for the container
    string TrustCertification='Inferred';
//Broadcast to blockchain network the transaction among two data objects (accounts).
    event DueUsage(address indexed_from, address indexe_to,struct Process,
    string Business_domain, string TrustCertification);
//Verify Business purpose confirms Terms of Usage
    function BusinessPurposeVerification(string para)public returns(string Terms_of_usage_Token){
//verify CompanyB identity
    require (StakeholderAuthentication(para) == true);
//send to notfiy OLS's tracking manager for gloabl tracking and usage manager for usage smart contract
    owner.call("Business_Service_Selection(Logical_asset_ID)",
    "Global_tracking(Logical_asset_ID,ToUID)");
//specify each business purpose with the grant exact service functions
    emit DueUsage(owner,msg.sender, s, Business_domain, TrustCertification, exchangedAsset);
    return Terms_of_usage_Token;
    }
    ...
}
```

Fig. 4. Example of exchange smart contract associated to the UCA3 assertion

Table 1. Comparison of different ontologies

Ref.	Countermeasure scope	Control object	Subject Attribute
[20]	Access control	Trust	Role
[21]	Access control	Operation Condition	Role
[22]	Service selection	Quality of Service	Service attributes
[23]	Access control	Trust + Service	Role and reputation
[24]	Access control	Purpose + Service	Role
[25]	Access control	Knowledge	Roles and social relationship
[26]	Access control	Object	NA
[27]	Access control	Operation condition	Role
[28]	Service selection	Authorization predicates	Temporal of logic actions state
[19]	Service selection	Security Information	NA
Our ToU	Infrastructure + access control + business purpose	Logical data and physical copies	Individual/organization/entity

We also compare our DUP service with others according to GDPR requirements according to GDPR requirements (Table 2) thanks to 4 criteria:

- Usage scope: identifies if business purpose is taken into account or not
- Consent management: defines if the consent is stored or propagated
- Tracking: defines if operations and/or data provenance can be tracked
- Data life-long protection means that the data usage limitations and reporting can be achieved even after the data has been transmitted to another party

Table 2. Comparison of our system with other systems fitting GDPR requirements

Ref.	Usage scope	Consent management	Tracking	Life-long protection
[13]	No	Yes	Partly for right transfer	Partly: shared policy
[14]	No	Managed by the subject	Key exchange	No
[15]	No	Picked from the blockchain	No	No
[16]	No	Yes	Data forwarding operations	No
[29]	No	Managed by the subject	Data operation	No
Our DUP	Yes	Yes	Data exchange and some operations	Yes

6 Conclusion

In this paper, we have presented a data-centered security service allowing the definition of a data-centered protection. It integrates both traditional security policy and usage control. This service relies on an extended Terms of Usage ontology integrating these different parameters and on a multi-layer's architecture used to describe the distributed information system. Our security service takes advantage of this multilayer architecture and on Blockchain smart contracts to manage consistent authorization granting, data exchange and real usage tracking. By this way a consistent life-long protection can be governed on the different instances of a logical asset.

Our prototype has shown our architecture consistency. Further works will focus on usage and security governance components to support adaptive and smart protection process for smart data depending on the context.

References

1. Gupta, B.B., Yamaguchi, S., Agrawal, D.P.: Advances in security and privacy of multimedia big data in mobile and cloud computing. Multimedia Tools Appl. **77**(7), 9203–9208 (2018)

2. Feng, D.G., Zhang, M., Zhang, Y., Xu, Z.: Study on cloud computing security. J. Softw. **22**(1), 71–83 (2011)
3. Suo, H., Liu, Z., Wan, J.F., Zhou, K.: Security and privacy in mobile cloud computing. In: 9th International Wireless Communications and Mobile Computing Conference (IWCMC), pp. 655–659 (2013)
4. Patil, D., Mohapatra, R.K., Babu, K.S.: Evaluation of generalization based K-anonymization algorithms. In: 2017 Third International Conference on Sensing, Signal Processing and Security (ICSSS), pp. 171–175 (2017)
5. Jain, P., Gyanchandani, M., Khare, N.: Big data privacy: a technological perspective and review. J. Big Data **3**(1), 25 (2016)
6. Smith, M., Szongott, C., Henne, B., Voigt, G.V.: Big data privacy issues in public social media. In: 6th IEEE International Conference on Digital Ecosystems and Technologies (DEST), pp. 1–6 (2012)
7. GDPR, Art.4. http://www.privacy-regulation.eu/en/article-4-definitions-GDPR.html
8. Acquisto, G.D., Ferrer, J.D., Kikiras, P., Torra, V., Montjoye, Y.A., Bourka, A.: Privacy by design in big data: an overview of privacy enhancing technologies in the era of big data analytics, pp. 1–80 (2015)
9. Sullivan, K.M.: Under a watchful eye: incursions on personal privacy. In: The War on Our Freedoms: Civil Liberties in an Age of Terrorism, vol. 128, pp. 131(2003)
10. Kuner, C., Cate, F.H., Millard, C., Svantesson, D.J.B.: The challenge of 'big data' for data protection (2012)
11. Burmeister, F., Drews, P., Schirmer, I.: A privacy-driven enterprise architecture meta-model for supporting compliance with the general data protection regulation. In: Proceedings of the 52nd Hawaii International Conference on System Sciences (2019)
12. Cha, S.C., Yeh, K.H.: A data-driven security risk assessment scheme for personal data protection. IEEE Access **6**, 50510–50517 (2018)
13. Di Francesco Maesa, D., Mori, P., Ricci, L.: Distributed access control through Blockchain technology (2017)
14. Wirth, C., Kolain, M.: Privacy by Blockchain design: a Blockchain-enabled GDPR-compliant approach for handling personal data. In: Proceedings of 1st ERCIM Blockchain Workshop 2018. European Society for Socially Embedded Technologies (EUSSET) (2018)
15. Truong, N.B., Sun, K., Lee, G.M., Guo, Y.: GDPR-compliant personal data management: a blockchain-based solution. IEEE Trans. Inf. Forens. Secur. **15**, 1746–1761 (2019)
16. Neisse, R., Steri, G., Nai-Fovino, I.: A blockchain-based approach for data accountability and provenance tracking. In: Proceedings of the 12th International Conference on Availability, Reliability and Security, pp. 1–10 (2017)
17. Kelbert, F., Pretschner, A.: Data usage control enforcement in distributed systems. In Proceedings of the Third ACM Conference on Data and Application Security and Privacy, pp. 71–82 (2013)
18. Park, J.H., Sandhu, R.: The UCONABC usage control model. ACM Trans. Inf. Syst. Secur. **7**(1), 128–174 (2004)
19. Kim, A., Luo, J., Kang, M.: Security ontology for annotating resources. In: Meersman, R., Tari, Z. (eds.) OTM 2005. LNCS, vol. 3761, pp. 1483–1499. Springer, Heidelberg (2005). https://doi.org/10.1007/11575801_34
20. Hu, Y.J., Guo, H.Y., Lin, G.D.: Semantic enforcement of privacy protection policies via the combination of ontologies and rules. In: 2008 IEEE International Conference on Sensor Networks, Ubiquitous, and Trustworthy Computing (SUTC 2008), pp. 400–407 (2008)
21. Nejdl, W., Olmedilla, D., Winslett, M., Zhang, Charles C.: Ontology-based policy specification and management. In: Gómez-Pérez, A., Euzenat, J. (eds.) ESWC 2005. LNCS, vol. 3532, pp. 290–302. Springer, Heidelberg (2005). https://doi.org/10.1007/11431053_20

22. Sodki, C., Badr, Y., Biennier, F.: Enhancing web service selection by QoS-based ontology and WS-policy. In: Proceedings of the 2008 ACM Symposium on Applied Computing, pp. 2426–2431 (2008)
23. Choi, C., Choi, J.H., Kim, P.K.: Ontology-based access control model for security policy reasoning in cloud computing. J. Supercomput. **67**(3), 711–722 (2014)
24. Garcia, D., et al.: Towards a base ontology for privacy protection in service-oriented architecture. In: IEEE International Conference on Service-Oriented Computing and Applications (SOCA), pp. 1–8 (2009)
25. Masoumzadeh, A., Joshi, J.: OSNAC: an ontology-based access control model for social networking systems. In: 2010 IEEE Second International Conference on Social Computing, pp. 751–759. IEEE (2010)
26. Liu, C.L.: Cloud service access control system based on ontologies. Adv. Eng. Softw. **69**, 26–36 (2014)
27. Tsai, W.T., Shao, Q.: Role-based access-control using reference ontology in clouds. In: 2011 Tenth International Symposium on Autonomous Decentralized Systems, pp. 121–128. IEEE (2011)
28. Zhang, X., Park, J., Parisi-Presicce, F., Sandhu, R.: A logical specification for usage control. In: Proceedings of the Ninth ACM Symposium on Access Control Models and Technologies, pp. 1–10 (2004)
29. Kaaniche, N., Laurent, M.: A Blockchain-based data usage auditing architecture with enhanced privacy and availability. In: 2017 IEEE 16th International Symposium on Network Computing and Applications (NCA), pp. 1–5. IEEE (2017)

XYZ Monitor: IoT Monitoring
of Infrastructures Using Microservices

Marc Vila[1,2(✉)], Maria-Ribera Sancho[1,3], and Ernest Teniente[1]

[1] Universitat Politècnica de Catalunya, Barcelona, Spain
{marc.vila.gomez,maria.ribera.sancho,ernest.teniente}@upc.edu
[2] Worldsensing, Barcelona, Spain
mvila@worldsensing.com
[3] Barcelona Supercomputing Center, Barcelona, Spain
maria.ribera@bsc.es

Abstract. One of the main features of the Internet of Things (IoT) is the ability to collect data from everywhere, convert this data into knowledge, and then use this knowledge to monitor about an undesirable situation. Monitoring needs to be done automatically to be practical and should be related to the ontological structure of the information being processed to be useful. However, current solutions do not allow to properly handle this information from a wide range of IoT devices and also to be able to react if a certain value threshold is exceeded. This is the main purpose of XYZ Monitor, the system we propose here: to monitor IoT devices so that it can automatically react and notify when a given alarm is detected. We deal with alarms defined by means of business rules and allow setting ontological requirements over the information handled.

Keywords: IoT · Monitoring · API · Microservices · Framework

1 Introduction

Several authors recognize the IoT to be one of the most important developments of the 21st century [14]. According to them, the IoT represents the most exciting technological revolution since the Internet because it brings endless opportunities and impact in every corner of our planet. IoT devices are used as human consumables such as wearables or health trackers; but they are also key to the success of industrial applications such as Smart Cities, Industry 4.0, Smart Energy, Connected Cars or Healthcare. I.e., almost all industrial environments are currently highly dependant from the IoT.

IoT devices and systems are intended to collect and process data from the least expected places, and its expansion is allowing to operate sensors in a wide range of applications; energy management, mobility, manufacturing, Smart Cities [18] or healthcare, where there is the need of services able to monitor the medical condition of a patient [7]; or even operated in private use at home, for example to monitor the home safety.

© Springer Nature Switzerland AG 2021
H. Hacid et al. (Eds.): ICSOC 2020 Workshops, LNCS 12632, pp. 472–484, 2021.
https://doi.org/10.1007/978-3-030-76352-7_43

Thus, one of the inherent capabilities of such IoT systems is the ability to automatically monitor the information associated to the raw data they are processing. This has to be achieved by transforming raw data into relevant knowledge of the system domain, and then specifying conditions over this knowledge, referred to as alarms, that allow identifying and handling undesirable situations.

Microservices Architecture is in the core of providing a solution for such monitoring systems. It is an architectural style which promotes developing an application "as a suite of small services, each running in its own process and communicating with lightweight mechanisms with the others" [6]. There is a need for managing IoT platforms with these architectures to facilitate IoT development in itself, improving scalability, interoperability and extensibility [17].

This is particularly important when automatic IoT monitoring is concerned, since microservices and IoT share a lot in common in terms of architectural goals [3]. However, to our knowledge, previous work used to assume monolithic architectures [7,8,11], without taking into account all the benefits of a microservices orientation. An exception can be found in [4], which implements a service-oriented architecture (SOA) for monitoring in agriculture; and in [10] that defines microservices architecture about security monitoring in public buildings.

Summarizing, we can see that previous proposals dealing with microservices, monitoring and IoT consider only very specific domains, and are intended to monitor a certain data, from a particular sensor to achieve a single given response. I.e., they are tailor-made and there is no general purpose domain independent proposal.

Our work is related to the industrial research and innovation at Worldsensing (www.worldsensing.com), which focuses on the monitoring of industrial environments through IoT systems. In this sense, one of the company's main goals is to develop a generic environment (assuming different device types and data from different providers) able to monitor systems and alarms, as defined through customizable business rules, and based on a microservices architecture. This is, in fact, the main contribution of this paper. The system we have developed, called XYZ Monitor, is Open Source, easy to use and applicable to different domains.

2   Related Work

We distinguish between IoT in monolithic and IoT in microservice architectures, and monitoring in IoT in both architectures.

2.1   IoT in Monolithic Architectures

In these architectures, systems are built, tested, and deployed as one large body of code, as a unique solution [12]. This is the classical way of building applications in software deployments. Everything is unified and, thus, there is no modularity.

Among the relevant work in this area we may find [5], which focuses on a Service Oriented Middleware; [2] which designs a Service Oriented Architecture (SOA) for wireless sensor networks; and [15] that expands the SOA concept with cloud-based Publish/Subscribe Middleware and also implements the Web of Things (WoT) concept.

The main drawback of these solutions is that of successfully handling IoT environments which are increasingly complex, with many kinds of devices that are heterogeneous as far as their use and operation. Monolithic systems have their main limitations here. Everything is linked, if there is a change, improvement or correction to make, even if it is minimal, the whole system has to be deployed, tested and restarted. If one part of the system stops working, it is very likely that the whole system will stop working. Moreover, their reusability is very limited.

2.2   IoT in Microservices Architecture

The microservices architecture emerged as a solution to overcome previous drawbacks [12]. In this architecture, systems are developed as a set of self-contained components, or loosely coupled services, also called microservices. Each microservice encapsulates its logic to implement a single business function, and communication is done through web interfaces (APIs). This approach has contributed to improved fault isolation, simplicity in understanding the system, technology flexibility, faster technical deployments, scalability, and reusability [13].

Several works are intended to provide IoT solutions through microservices architecture. [9] explores how the service-oriented architecture paradigm may be revisited to address challenges posed by the IoT for the development of distributed applications. [3] investigates patterns and best practices used in microservices and analizes how they can be used in the IoT. [17] proposed an architecture of a microservices based middleware, to ensure cohesion between different types of devices, services and communication protocols. [1] proposes a modular and scalable architecture based on lightweight virtualization, with Docker. [16] proposes an open microservices system framework for IoT applications. [19] provides an environment to transform automatically functionalities from IoT devices to a Service Oriented Architecture based IoT services.

2.3   Monitoring in IoT

Some proposals have also been devoted to monitoring on IoT. [7] discusses the integration of IoT devices for health monitoring. [8] extends the previous concept to include safety protocols for data transmission also in healthcare. [11] proposes an IoT-based solution to monitor Smart Cities environments. However, all these proposals are based on monolithic architectures.

Moving to non-monolithic architectures, [4] implements a SOA for monitoring in agriculture; while [10] develops a microservices based architecture for a monitoring system to improve the safety of public buildings. It is worth noting that both applications are very domain-specific and that they handle only very concrete devices.

Summarizing, the use cases are very specific. Input elements to the systems are very limited, only certain devices are available. And once the data is in the system, it follows a closed monitoring flow. They include a business rules with notification system. But, this is also closed to modifications and cannot be changed externally.

3 XYZ Monitor System Overview

Our goal is aimed at overcoming the limitations of previous proposals. With this purpose, we have built the *XYZ Monitor* system, which is able to monitor data from different IoT devices and reporting if certain conditions over this data are met. Input data is generic and extensible to changes to support different elements. IoT devices communicate with the system via HTTP calls, through APIs. Once the data is in the system, it is analyzed and monitored. If the user has specified an alarm by means of a business rule, it will be monitored by sending notifications to an email address indicated in the system. XYZ Monitor relies on a microservices architecture, with the advantages that this entails.

XYZ Monitor gets the input of data from the devices themselves since we naturally assume that the real-world objects, sensors and devices can autonomously infer their state and submit this information to the service. This is a feasible assumption in the context of the IoT, where environmental data can be collected by the objects, which can then infer their own state.

Once in the system, this data will serve us to monitor the behaviour of the devices and also to activate alarms if certain conditions over the data are met.

3.1 Conceptual Overview

The workflow of our proposal is summarized in Fig. 1. Initially, the *Data Collection* component receives information inputs from the different devices. These devices are onboarded in the *Data Management* system by the user. These data can be visualised on the platform itself as passive monitoring, *Monitoring* component. At the same time that the data is being received, the system, performs *Data Analysis*. This is done through business rules, predefined by the platform user, it is also called alarm monitoring. When an alarm occurs, i.e. a business rule detects that something is not right, like a value out of range, the system warns through the *Notification* component, automatically.

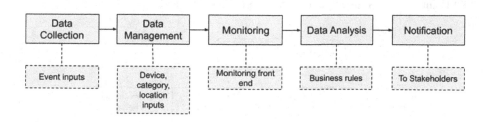

Fig. 1. Workflow of *our proposal*

One of the main features of our system is its ability to handle alarms that are defined by means of business rules, specified over the conceptual ontology of the information handled. This conceptual ontology is defined with UML in Fig. 2 and allows abstracting the concepts of interest from its technological implementation, thus providing independence between data gathering and processing.

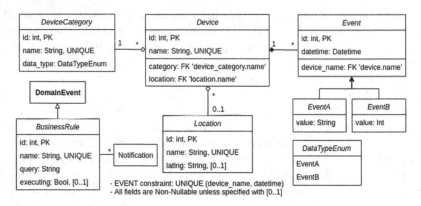

Fig. 2. Ontology of the information handled by XYZ

An *event* is an input of data sent by an IoT device to our system. We assume two different types of *events*, although our proposal is easily extensible by assuming additional subclasses of *events* with different types. A *device* is the smart object intercepting the *events* (such as a sensor or a thermometer). Each *device* has a location in coordinates format or it is named with a label. A *device* belongs to a *device category* which manages it and serves to indicate what type of *events* that *device* is sending. Finally, a *business rule* (also called alarm) allows stating a complex condition to be monitored over the data stored by means of a query and also notify to whoever has been determined in the system.

3.2   Architectural Overview

Our system has been designed to operate through a microservices architecture. Each functional module is isolated and communicates with each other over HTTP interfaces. This architecture is specified in Fig. 3.

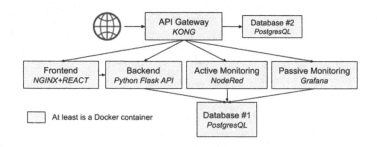

Fig. 3. Microservice architecture of *our proposal*

When a user or an IoT device wants to communicate with our system, our API Gateway will be in the front-line. An API Gateway takes all the HTTP requests from a client determines which services are needed and then routes them to the appropriate microservice. It translates between web protocols and web-unfriendly protocols, used internally. Among the existing, Netflix Zuul, Kong, Apache Apisix, etc., we have chosen Kong (https://konghq.com/kong/), an Open Source Software, simple to configure, it works well and meets our needs.

Among the existing OSS data visualization systems for monitoring, *e.g.* Kibana, Prometheus, Grafana or Chronograf, we decided to use Grafana (https://grafana.com) because it is easy to use and fits well with most types of IoT data we are dealing with. We are able to have a monitoring and visualization system with the data reported by the devices.

Our system incorporates an element that allows it to react to certain circumstances. Node-Red (https://nodered.org), its OSS, and permits creating flows for actions through custom JavaScript functions. Within this system we have created a flow to check these circumstances or business rules.

Also a frontend has been developed in React (https://reactjs.org), which is an OSS JavaScript framework. The backend is done with Flask (https://flask.palletsprojects.com), an OSS micro framework in Python. As database, PostgreSQL (https://postgresql.org) is used, an OSS object-relational database.

The system architecture has been deployed under microservices by the means of Docker (https://docker.com) and Docker-compose (https://docs.docker.com/compose/). Docker is an open source project providing a systematic way to automate the faster deployment of Linux applications inside portable containers.

4 Proposal in a Nutshell

4.1 Data Collection

Our system is easily extensible. Therefore, to exemplify its input we have chosen to distinguish between two types of inputs available for the system. If the user has the need to add another type of input, it can be done easily. This is why all input elements have a superior element, which generalizes their values. More precisely, XYZ Monitor is able to handle two types of possible elements: *EventA* and *EventB*, respectively. Both kind of inputs are specializations of input type *Event*, which is a communication trigger that is the base type for a device reading.

Events. An *Event* includes the information of the device to which the data is being input (*device_name*), and also the date of sample collection (*datetime*); but it does not have the ability to know what value that sample has. The value of the sample is obtained from the classes inherited from *Event*. Which can be *EventA* or *EventB*. For this, an API (`/api/events`) has been developed to gather *Events* sent to the system. It is defined following the JSON Schema[1] specification:

---

[1] JSON Schema - A Media Type for Describing JSON Documents: https://tools.ietf.org/html/draft-handrews-json-schema-01.

Listing 1.1. Event JSON Schema simplification

```
1   "event": {
2     "type": "object",
3     "properties": {
4       "device_name": {
5         "type": "string",
6         "$ref": "#/definitions/device/properties/name"
7       },
8       "datetime": {
9         "type": "string",
10        "format": "date-time",
11        "example": "2020-04-01T10:54:03+00:00"
12      }
13    },
14    "required": [ "device_name", "datetime" ] }
```

EventA is an input value in the form of a string, i.e. it can be any alphanumeric element. For example, "heat", "1.0", "1", etc.

Listing 1.2. EventA JSON Schema simplification

```
1   "event_a": {
2     "type": "object",
3     "allOf": [  { "$ref": "#/definitions/event" },
4                 { "properties": {
5                   "value": {
6                     "type": "string",
7                     "example": "1"
8                   }
9                 },
10                "required": [ "value" ] } ] }
```

EventB events are defined in a similar way, but the input value that is in the form of a number, so it can have numerical values, even with decimal values.

XYZ Monitor shows the events that have been correctly received by the system (see Fig. 4). In particular, XYZ monitor provides for each event: its identifier, device to which it reports, value and date of reception.

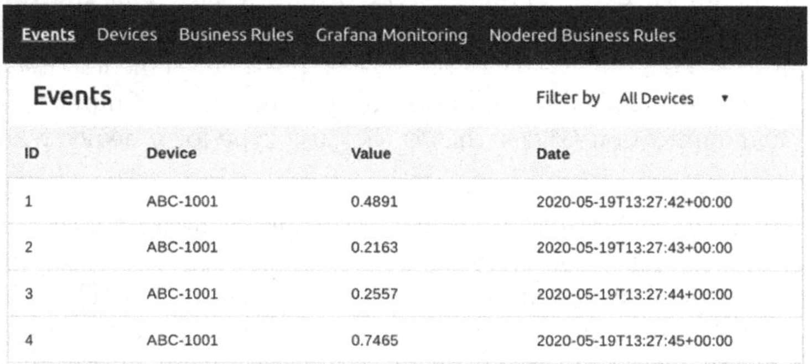

| Events | Devices | Business Rules | Grafana Monitoring | Nodered Business Rules |

Events   Filter by   All Devices ▾

ID	Device	Value	Date
1	ABC-1001	0.4891	2020-05-19T13:27:42+00:00
2	ABC-1001	0.2163	2020-05-19T13:27:43+00:00
3	ABC-1001	0.2557	2020-05-19T13:27:44+00:00
4	ABC-1001	0.7465	2020-05-19T13:27:45+00:00

Fig. 4. Frontend showing the Events yet in our system

4.2   Data Management

We have seen how to send an Event to the system. Next we will see how to indicate that a certain event is from a specific device. And how the other attributes of a device are handled.

Devices. The system is prepared to handle an infinite amount of devices. A device can be an element of the world as well as virtual. Infinite elements can coexist. Since the same user is in charge of registering them in the system.

A *Device* can have only one point of measuring, which is why it is associated with a type of reading, the *category* field. But also a *Device* can be located into an *Area*, this field is optional and its named *Location*. For this purpose, an API (`/api/devices`) has been developed, it that acts as an on-boarder to the system. This is achieved by means of a JSON schema similar to the previous ones.

XYZ Monitor shows the existing devices as illustrated in Fig. 5. For each device, our system provides its name, its category, the type of event it reports and its location. Moreover, on the right side of the figure, we show how Grafana is reporting in real time the events that are being received in that device.

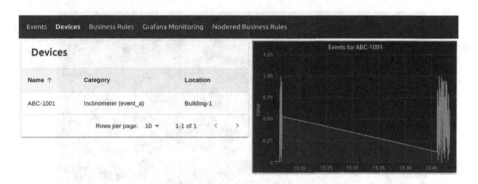

Fig. 5. Frontend with a Device and its Events, showing Grafana to monitor it

Device Categories. At this point we already know how to insert a *Device* and its *Events* into our system. However, we have not yet defined how to differentiate *Events* from one *Device* to another. The *category* field inside the *device* component gives us a small clue. A *Category* is a set of possible types of elements defined by its *name* and the *data_type* field, which is a string enumeration of the possible types of *Events* available in the system (*EventA* or *EventB*). This is achieved by means of a JSON schema similar to the previous ones and it is accessible through the endpoint (`/api/device-categories`).

Locations. As mentioned, a *Device* has the possibility of being located in one area. The location of a device is given by a *name*, in addition, there is the possibility of including coordinates, *latlng*, in order to have a precise location, in *latlng* format. Can be accessed in the API endpoint (`/api/locations`). This is achieved by means of a JSON schema similar to the previous ones.

4.3   Monitoring

At this point, our system is able of receiving external, sensor-based, information and classify it accordingly, by type of data, category and also by its location. To observe the data, the system has been equipped with a data visualization environment for the sake of monitoring.

Figure 6 shows the visualization system, built with Grafana `/grafana`. It is displaying all the *Events* that have been received from the ABC-1001 *device*, which are of type *EventA* on the date from 14:49 to 15:49 of 19/05/2020.

It can be observed that the data has arrived well between 14:49 - 14:50, 15:16 - 15:18 and 15:48 - 15:49, while no data has been received for the other ranges. The system therefore draws a continuous line with no variation between them to note that there is no data there.

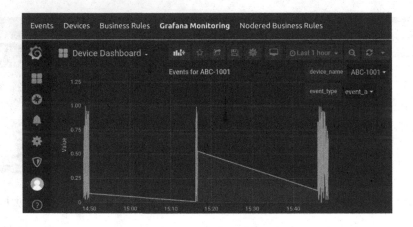

Fig. 6. Grafana panel showing the Device' Events in our system

The graph in Fig. 6 is generated from an *SQL query*, specified in Listing 1.3. This SQL query is a traditional query against a PostgreSQL database. The advantage of running SQL queries in Grafana is that the text starting with $ are the variables, and these can be sent by the browser to the system at the time of accessing the URL, which allows to dynamically generate these kind of graphs.

`/grafana/device-dashboard?var-device_name=ABC-1001&var-event_type=event_a`

Listing 1.3. SQL query to show Device's data in Grafana

```
SELECT  event.datetime as "time",
        $event_type.value,
        event.device_name
FROM  event
INNER JOIN  device ON event.device_name=device.name
INNER JOIN  $event_type ON event.id=$event_type.id
WHERE  event.device_name = '$device_name'
```

4.4 Business Rules

Figure 7 shows the flow of the *Business Rules* to achieve notifications. It is implemented with the Node-Red tool, which works by execution flows. The whole processing of *Business Rules* is based in a single flow, /nodered endpoint.

The execution is as follows: every 5 s (timer box) a query is prepared (query box) and the database that contains all the data is queried (Postgres #1 box), to check if there exists any *Business Rules* in *executing* state. The result is checked to see if there are any (prep array-loop and array-loop boxes). In the case that there is no *business rule* or its is not in *executing* state, the flow ends here. In the case of having *active Business Rules*, the system checks one by one (prepare BR query box) and executes the query against the database (Postgres #2 box) to check for alarms. If there is any, an email is prepared (generate email message box) and sent through the *Notification* system (notification box).

A *Business Rule* is identified by a *name* and it contains a *query* field where the designer specifies the SQL query identifying the alarm. For instance: SELECT * FROM Event where device_name = 'ABC-1001'. And the *executing* field is where the *Business Rule* can be started or stopped temporarily, and act as an *status* of it. The query system is an SQL call, because in this way, the options for the user are not limited. This is achieved by means of a JSON schema similar to the previous ones. Business rules can be specified on: *Device Categories*, *Device Locations*, *Devices*, *Events* and any kind of combinations between them.

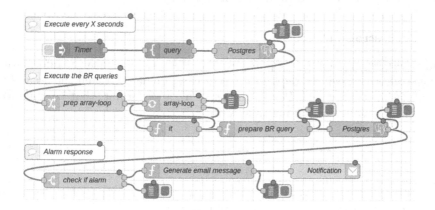

Fig. 7. Our flow to check Business Rules in Node-Red

4.5   Notifications

The notification system has been implemented in the Node-Red microservice, as an independent code block. It supports sending of notifications via email to a predefined recipient added in the system. These emails are sent via SMTP (Simple Mail Transfer Protocol), and its content warns about an alarm; informing regarding the device, the date and the values that provoked it.

Listing 1.4. SMTP Email sample sent from our Node-red

```
FROM:  sender@example.com
TO:  receiver@example.com
DATE: Wed., 29 Apr. 13:02
SUBJECT: New alarm from Node-Red
BODY:
    There is an alarm, the following elements are involved:
    Device: ABC-1001
    Value: 0.10
    Datetime: Tue Apr 29 2020 13:01:12 GMT+0200 (CEST)
```

5   Experimentation

The XYZ Monitor system has been implemented under a microservices architecture. Specifically on Docker and Docker-compose. An overview of the *Docker* recipe that makes the system work is available in Sect. 5.1.

The project code is Open Sourced and it is available in GitHub, at www.github.com/worldsensing/xyz-iot-monitoring. The project has been tested under Ubuntu 18, Linux. Although it should be compatible with any Linux distribution. Making small changes, it could be used in MacOS or even Windows, but this is out of the scope of the project since Worldsensing' main systems run on Linux. Inside the repository, the steps to follow to setup, initialize and run the whole system work are clearly stated. Everything is prepared so that the modifications required to make it work are minimal since the scripts we have prepared take care of most of the work to setup the environment.

5.1   Orchestration

A microservices architecture requires having, at least, a file with the information of the services that will be deployed. In our case, we have written the orchestration file using *Docker* and *Docker-compose* in *.yaml* format, that contains the instructions to compose the multiple services[2].

---

[2] This listing is simplified a lot to avoid making it extremely long here. The actual one can be found in the root folder of the repository.

Listing 1.5. Docker-compose orchestration

```
1  services:
2    # API Gateway - Base Image - Kong 2
3    # Frontend - Custom - Nginx 1.17.9 + Alpine + React 16.13
4    # Monitoring - Custom - Grafana 6.6.2
5    # Business Rules - Base - Node-red
6    # Backend API - Custom - Python 3.8.2 + Alpine 3.11 + Flask
7    # Main Database - Base Image - PostgresQL 11.7
```

6 Conclusions and Further Work

We have presented the *XYZ Monitor* system as an extensible solution to successfully handle general purpose alarms defined over different kinds of devices in an IoT environment. In our system, alarms are defined by means of business rules specified over the ontological structure of the information handled by these devices. The solution we have developed is based on microservices architecture, to facilitate the assignment of responsibilities among the components involved in alarm monitoring. Our solution is fully Open Source and it is publicly available.

As further work, we plan to enrich further the ontological structure of the information and to develop techniques to incrementally compute whether an alarm has been activated. In addition to observing how the system behaves when there are many devices sending information at the same time. Overall, the final goal of this work is to put this system into practice at Worldsensing.

Acknowledgements. This work is partially funded by Industrial Doctorates from Generalitat de Catalunya (DI-2019, 2017-SGR-1749). Also with the support of inLab FIB at Universitat Politècnica de Catalunya and Worldsensing S.L. The REMEDiAL project (Ministerio de Economia, Industria y Competitividad, TIN2017-87610-R) has also contributed. We thank the anonymous reviewers for their valuable comments.

References

1. Alam, M., Rufino, J., et al.: Orchestration of microservices for IoT using Docker and edge computing. IEEE Commun. Mag. **56**(9), 118–123 (2018)
2. Avilés-López, E., García-Macías, J.: TinySOA: a service-oriented architecture for wireless sensor networks. Serv. Oriented Comput. Appl. **3**, 99–108 (2009)
3. Butzin, B., Golatowski, F., et al.: Microservices approach for the Internet of Things. In: 21st International Conference on Emerging Technologies and Factory Automation (ETFA), pp. 1–6 (2016)
4. Cambra, C., Sendra, S., et al.: An IoT service-oriented system for agriculture monitoring. In: International Conference on Communications (ICC), pp. 1–6 (2017)
5. Caporuscio, M., Raverdy, P., et al.: ubiSOAP: a service-oriented middleware for ubiquitous networking. IEEE Trans. Serv. Comput. **5**(1), 86–98 (2012)
6. Fowler, M., Lewis, J.: Microservices, a definition (2014). http://martinfowler.com/articles/microservices.html. Accessed 12 Aug 2020
7. Hassanalieragh, M., et al.: Health monitoring and management using Internet-of-Things (IoT) sensing with cloud-based processing: opportunities and challenges. In: ICSOC 2015, pp. 285–292 (2015)

8. Hossain, M.S., Muhammad, G.: Cloud-assisted industrial Internet of Things (IIoT) - enabled framework for health monitoring. Comput. Netw. **101**, 192–202 (2016)

9. Issarny, V., Bouloukakis, G., Georgantas, N., Billet, B.: Revisiting service-oriented architecture for the IoT: a middleware perspective. In: Sheng, Q.Z., Stroulia, E., Tata, S., Bhiri, S. (eds.) ICSOC 2016. LNCS, vol. 9936, pp. 3–17. Springer, Cham (2016). https://doi.org/10.1007/978-3-319-46295-0_1

10. Mongiello, M., Nocera, F., et al.: A microservices-based IoT monitoring system to improve the safety in public building. In: SpliTech, pp. 1–6 (2018)

11. Montori, F., Bedogni, L., et al.: A collaborative Internet of Things architecture for smart cities and environmental monitoring. IEEE Internet Things J. **5**(2), 592–605 (2018)

12. Namiot, D., Sneps-Sneppe, M.: On micro-services architecture. Int. J. Open Inf. Technol. **2**(9), 24–27 (2014)

13. Newman, S.: Building Microservices, 1st edn. O'Reilly Media Inc. (2015)

14. SmartDataCollective: IoT is the most important development of the 21st century (2018). https://www.smartdatacollective.com/iot-most-important-development-of-21st-century. Accessed 06 Sept 2020

15. Soldatos, J., et al.: OpenIoT: open source Internet-of-Things in the cloud. In: Podnar Žarko, I., Pripužić, K., Serrano, M. (eds.) Interoperability and Open-Source Solutions for the Internet of Things. LNCS, vol. 9001, pp. 13–25. Springer, Cham (2015). https://doi.org/10.1007/978-3-319-16546-2_3

16. Sun, L., Li, Y., et al.: An open IoT framework based on microservices architecture. China Commun. **14**(2), 154–162 (2017)

17. Vresk, T., Čavrak, I.: Architecture of an interoperable IoT platform based on microservices. In: MIPRO 2016, pp. 1196–1201 (2016)

18. Zanella, A., Bui, N., et al.: Internet of Things for smart cities. IEEE Internet Things J. **1**(1), 22–32 (2014)

19. Zhao, Yu., Zou, Y., Ng, J., da Costa, D.A.: An automatic approach for transforming IoT applications to RESTful services on the cloud. In: Maximilien, M., Vallecillo, A., Wang, J., Oriol, M. (eds.) ICSOC 2017. LNCS, vol. 10601, pp. 673–689. Springer, Cham (2017). https://doi.org/10.1007/978-3-319-69035-3_49

Multi-cloud Solution Design for Migrating a Portfolio of Applications to the Cloud

Shubhi Asthana[1]([✉]), Aly Megahed[1], and Ilyas Iyoob[2]

[1] IBM Research - Almaden, San Jose, CA, USA
{sasthan,aly.megahed}@us.ibm.com
[2] IBM Services, Austin, TX, USA
iiyoob@us.ibm.com

Abstract. Migrating applications to the cloud is rapidly increasing in many organizations as it enables them to take advantages of the cloud, such as the lower costs and accessibility of data. Moreover, such organizations typically try to avoid sticking to a single cloud provider and rather prefer to be able to spread out their applications across different providers. However, there are many challenges in achieving this. First, many of the applications that are required to be moved to the cloud might be legacy applications that do not have good documentation, and so it is not trivial to even assess whether it is feasible to move them to the cloud or not. Moreover, such legacy applications might need a significant architecture overhaul to achieve the task of moving them to the cloud. Large client may have significant percentage of applications in this category. So, one has to evaluate cloud feasibility and understand whether there is a need to re-architect application based on what services providers are able to offer. Second, clients usually define multiple features, encryption/security level, and other service level requirements they expect in the providers they will migrate each of their applications to. Thus, choosing the right providers for different application is another challenging task here. In this work-in-progress paper, we present a novel methodology for preparing such a cloud migration solution, where we perform text mining on application data to evaluate cloud-migration feasibility and then recommend the optimal solution using a mathematical optimization model. We illustrate our approach with an example use case.

Keywords: Cloud computing · Multi-cloud · Cloud feasibility · Text mining · Cloud migration · Optimization

1 Introduction and Related Work

The use of cloud computing is increasing rapidly in many organizations [1]. Moving an application to the cloud enables organization to make use of the advantages of the cloud like elasticity [2], lower costs, and accessibility of data. To further utilize these advantages, achieve maximum flexibility, and avoid "concentration risk" (putting too many application eggs in one cloud basket), organizations recently tend to spread their applications across different cloud provider. According to a recent study [3], 85% of

© Springer Nature Switzerland AG 2021
H. Hacid et al. (Eds.): ICSOC 2020 Workshops, LNCS 12632, pp. 485–494, 2021.
https://doi.org/10.1007/978-3-030-76352-7_44

enterprises operate in a multi-cloud environment. That is why different cloud vendors nowadays are enabling mix and match of cloud services across different clouds.

However, there are many challenges in achieving this. First, there are usually a lot of legacy applications that lack proper documentation. This makes it hard to even assess whether it is feasible to migrate such applications to the cloud or not. Also, sometimes these applications require an overall architecture overhaul, which again, suffer from the lack of documentation of the current architecture. Despite to momentum to shift to multi cloud, the cost-benefit analysis models illustrating the business impact of cloud adoption are still a significant risk factor [4]. It is sometimes challenging to redesign the current IT infrastructure to meet the requirement before moving to the cloud. Cloud providers charge customers on a variable cost pay-as-you-go basis determined by the number of users and their volume of transactions [5]. Organizations are not readily willing to pay extra for the additional cost [6].

Second, clients typically define requirements related to service levels they expect for each of their migrated applications. Determining the cloud migration solution that agrees to all these requirements while still be applicable, is not trivial and cannot be achieved by non-analytical/manual ways.

To overcome these drawbacks and challenges, we propose a methodology that recommends the optimal set of cloud providers and creates a multi-cloud solution for the client. We evaluate the optimal set of cloud providers based on best the fit between recommendations from historical data as well as a decision optimization solution framework. We also provide recommendation for multiple applications at once instead of just one application as the current literature. Lastly, we provide recommendation based on data collected from the client and processing it automatically instead of going back and forth by humans.

The rest of this paper is organized as follows: In Sect. 2 we discuss the prior art and in Sect. 3, we discuss our methodology. In Sect. 4, we provide a detail proof-of-concept implementation for it, then we end it with providing our conclusions, ongoing, and future work in Sect. 5.

2 Prior Art

Prior research and analytical work of finding cloud providers using structured application data as well as optimizing cloud solution design together with migration have been done in this area. Pamami et al. [7] shows a framework to create a generic reference for process of cloud migration while Iqbal et al. [8] discusses different cloud migration strategies and models, right from evaluating performance to choosing a cloud provider. Iyoob et al. [9] proposes a cloud comparison engine that maps application specifications to cloud services pricing for specific cloud offerings. In [10], Iyoob et al. detail a system for auto-prioritization of workload migration to cloud while in [11], they present data-driven cloud workload screening. Yang [12] shows a hybrid cloud solution design for genomics Next Generation Sequencing (NGS) service, which is streamlined for this particular service. Megahed et al. proposes an optimal approach for cloud solution design that satisfies client requirements and cloud offering constraints for an application in [13, 14], though they do not account for the different constraints of choosing different cloud providers in the solution.

Moreover, there is a numerous amount of prior art in applying analytical and optimization techniques to different cloud computing problems. For example, migrating virtual machines, applications to cloud environment have been discussed in [15–18]. Cloud elasticity optimization in software as a service cloud computing have been explored in [19–22]. Teyeb et al. [23] proposed another optimization model for dynamic placement of virtual machines in cloud data centers. Other works that provide other analytical approaches for different problems, though still in the cloud computing arena, are the ones in [24–26]. Amato et al. [27] and Iyoob et al. [28] provide analytical works that have solved problems in the cloud, including multi-cloud. However, they do not include any works that involve multi-cloud solutions.

In the aforementioned existing state of the art, we observe a few drawbacks. First, the time taken to gather all the features of the applications is quite inefficient, as it takes a lot of time and resources. Second, there is no analytical automated way of efficiently recommending a multi-cloud solution, where the current solutions rely on manual, inefficient evaluation of possible available cloud providers and thus requires a lot of back and forth with the client. That is, the prior art discussed above, as well as other works not reviewed above, focus on different aspects of cloud computing optimization and analytics, rather than multi-cloud solutions.

3   Methodology

We first provide an overview of our two-step approach in Sect. 3.1, and then detail each of these steps in Sect. 3.2.

3.1   Overview of Our Approach

Our methodology is a four-step approach that aims at preparing a multi-cloud solution for a client portfolio of applications. We start with collecting meta data for each application that the client wants to migrate to cloud. These lists of applications together with their details like users, geography, security, software platforms etc. are the inputs to our model. We then perform text mining on this semi-structured text data for each application. Next, we evaluate the cloud feasibility of each application using a dependency graphs approach. The output of this step is to come up with a rank of applications showing how feasible it is to move each application to the cloud. We evaluate the features of the applications and build a matrix of applications versus service cloud providers based on different factors such as cost, security features, service level requirements, etc.

Lastly, we build an optimization model that finds the optimal set of solution providers for the different applications that are deemed cloud-migration-feasible. The model puts in consideration the different requirements of the client. Our output is a competitive solution comprising of different applications and their corresponding service providers, fulfilling the requirements of the client. Figure 1 outlines the four steps of our method 1. In the next section, we detail each step in our method.

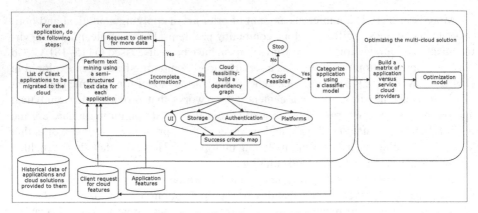

Fig. 1. Outline of our methodology

3.2 Detailed Steps of Our Approach

In this section, we provide a detailed description of the four steps of our approach as follows.

Perform Text Mining on Data for Each Application. In this step, we first collect meta-data for each application. For each client application that the client wants to migrate to cloud, we screen applications and collect application metadata features like number of users, geography of users, investment on application, security and compliance agreement, geography of database where data is stored, software platforms, authentication server, server, hardware, network setup, reliability, scalability, etc. Also, we collect client request for features that should be in the cloud for this application, e.g., security compliance, cost budget for cloud usage.

Next, we perform text mining on the loosely semi-structured document which has data collected about the application. Text mining can be performed by identifying tags/keywords for each feature and labelling them. We extract texts and map them to relevant tags, to identify completeness of data using these tags. For example:

Text Items:

1. Number of users for XYZ application is 455 in Europe.
2. Application requires software Java SDK v6.0 or below.

Sample Rules to extract data from text and ensure completeness of information:

A. Number of Users → 455: Text Item 1 → Extracted tag: users.
B. Investment on Application → ?: Information incomplete.
C. Software Platform → Java: Extracted tag: software.

We have a pre-set list of tags that are required for any specific model that we check for completeness while we apply our methodology. Particularly, after the text mining step of our methodology, if we are unable to extract the data for a particular tag, we deem it to be missing and we get back to the user with such missing data if any. Thus, we evaluate completeness of data using the necessary tags that are required to evaluate the application.

Evaluate Cloud Feasibility and Rank the Applications. In this step, we evaluate feasibility of cloud migration of applications and rank them based on effort required to migrate each one of them. Here we train a graph machine learning model and build a dependency graph. The nodes of the graph correspond to the tags/keywords that we extracted in the earlier text mining step. We then stack the nodes' feature vectors into a design graph and then train a classifier like random forest on a subset of data to build this graph. Our model classifies each tag under parent nodes like web, authentication, user interface, platforms, data storage, etc. The model is visualized in the form of a graph, and we are able to identify the upward and downstream dependencies showing different levels of interconnections.

The interconnections can be between components like databased sharing software and network services, LDAP services, platforms which are software version specific etc.

Here we consider the feasibility of cloud based on number of services or components that have minimum upward and downward dependencies. Examples of application dependencies include similar services that are pulling data from the application, common database for lightweight directory access protocol and backend data, etc.

We can rank the application based on rules like:

- Applications that have immediate business need to scale and are running out of capacity
- Applications that have architectural flexibility
- Applications that require global scale e.g., marketing and advertising applications
- Lower priority to applications that require specialized hardware to function

We leverage the dependencies between the different components of the application identify the inter-dependencies. This inter-dependency as well as ranking of applications and application data enough to get a clear picture of the effort required to migrate the application to cloud, and its feasibility.

Build a Matrix of Application versus Service Cloud Provider. In this step, using application features, we use a classifier model to categorize the applications into different categories–

- Based on historical recommendations and current framework
- Based on security type of data
- Based on similar dependencies in applications
- Based on applications using similar platforms
- Based on applications with similar architectures including microservices, cloud native, mobile etc.
- Based on ease of cloud migration (depending on effort to migrate applications)

For this, we use a weighting function based on importance of features (e.g. cost, QoS, coverage, security etc.). After creating the weighting function, we can easily prepare a solution comprising of services from different cloud providers with minimum cost and maximum coverage of services. This is done by preparing a matrix of applications versus different service cloud providers on metrics like cost, QoS, coverage, security etc. as illustrated in Fig. 2. We build the matrix in a simplistic manner, where we take the available cloud providers as fields in the columns, and the different applications in the rows

Application Characteristics	Service Provider 1	Service Provider 2	Service Provider 3
Cost			
QoS			
Coverage			
Security			

Fig. 2. Matrix of applications versus service cloud provider.

Build an Optimization Model. In this step, we build an optimization model for the current set of applications. Here we try to optimize the cloud service providers based on constraints like:

- Minimum set of service providers
- Maximum set of coverage
- Service providers which fulfill security requirements of data
- Service provider providing services based on geographical region
- Constraint where applications cannot be under the same provider or same geography.

Defining the set of applications as I, that of providers as J, and that of geographies as K, the decision variables of the optimization model are X_{ijk} and Y_j. Both sets of variables are binary, where X_{ijk} is 1 if application $i \in I$ is assigned to provider $j \in J$ at geography $k \in K$ and zero otherwise. Y_j is 1 if provider $j \in J$ is selected and zero, otherwise. The objective function minimizes the capacity of cloud provider selections and cost of application portfolio assignments. Besides the aforementioned constraints, other logical constraints are those assuring that each provider can be assigned to any application only if it has been selected.

4   An Example Use Case

In this section, we show an example use case of our method in Sect. 3. We consider a large IT company with global employees across 20 countries, having up to ~100 applications that it wants to migrate to cloud. Here, we want to clarify that we do have the data from different cloud providers, based on services provided and have its metrics like coverage, QoS, security etc. We did not include all the details of the data because of confidentiality reasons as well as space limitations. However, we did implement our approach and applied it to an application using real data.

For our implementation, we first understood which applications should be moved to the cloud. We also documented the applications that should be grouped together under same cloud provider, and which applications should be kept separate from each other.

There can be other constraints from the client like applications which can be under same cloud provider, but in different regions. We eliminated applications which cannot be moved to cloud because of organization policy or security. After screening all the applications, we put it in a semi structured document, with tags. Use of tags is important to understand if data about each application is complete and ready for assessment.

Next, we evaluated if application is feasible. For each application, we built the dependency graph around different parts of the application. We showed the sample dependency graph in Fig. 3. Our real-world example application XYZ has the following logical constructs that were used to evaluate the dependencies, risks, security and compliance requirements.

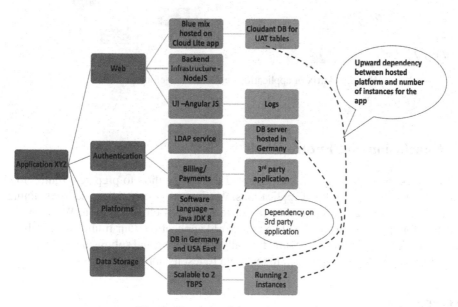

Fig. 3. Example of dependency graph.

Here the dotted lines indicated the identified upward and downstream dependencies showing different levels of interconnections of components. Next we evaluated applications using questions like:

- Does the application have immediate business need to scale and are running out of capacity? – *limited scalable storage up to 2 TBPS*
- Does it have architectural flexibility? – *Not much, it requires more storage capacity and more in-stances to make it scalable and reliable.*
- Number of upward and downward dependencies – *two internal, 1 external*

- Any dependency on 3rd party infrastructure/hardware or applications? – *Yes*

This evaluation was used to rank how cloud feasible the application is. After evaluating the cloud feasibility of the application, we built the matrix of application versus service cloud providers as shown in Fig. 4. Then, the optimization algorithm helped to optimize the recommended set of cloud providers which prepared the multi-cloud solution for the client.

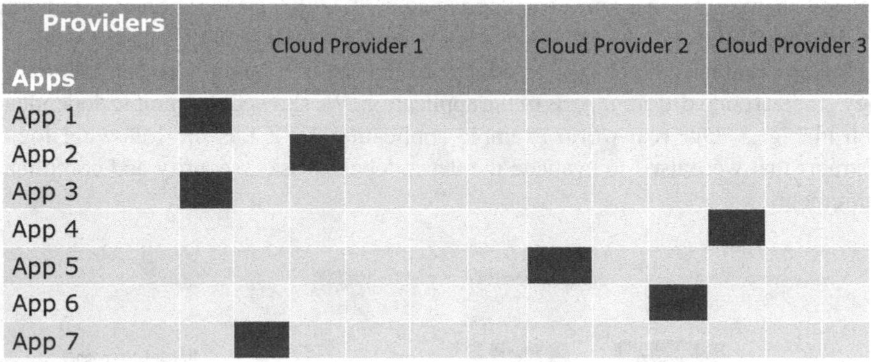

Providers Apps	Cloud Provider 1			Cloud Provider 2	Cloud Provider 3
App 1	■				
App 2		■			
App 3	■				
App 4					■
App 5				■	
App 6				■	
App 7		■			

Fig. 4. Matrix of application versus service cloud provider.

5 Conclusion and Future Work

In this work-in-progress paper, we presented a novel method to prepare a multi-cloud solution for a client portfolio of applications. We showed a method that used text mining, dependency graph, and optimization to create that multi-cloud solution. We showed a use case for applying our method to a real-world application. Our future work includes enhancing the text-mining procedure, formally formulating and solving the optimization model, and applying our approach to real-data to test its applicability.

References

1. Mohamed, M., Megahed, A.: Optimal assignment of autonomic managers to cloud resources. In: 2015 IEEE International Conference on Service Operations and Logistics, and Informatics (SOLI), pp. 88–93. IEEE (2015)
2. Megahed, A., Mohamed, M., Tata, S.: A stochastic optimization approach for cloud elasticity. In: 2017 IEEE 10th International Conference on Cloud Computing (CLOUD), pp. 456–463 (2017)
3. IBM blog-'New IBM services help companies manage the new multicloud world'
4. Litchfield, A., Althouse, J.: A systematic review of cloud computing, big data and databases on the cloud. In: Americas' Conference on Information Systems (AMCIS) (2014)
5. Newlin Rajkumar, V.: Security measures in cloud computing an extensive assessment. Int. J. Adv. Inf. Commun. Technol. **4**, 405–410 (2014)

6. Cranford, N.: Five challenges of cloud migration. RCR Wirel News (2017). https://www.rcr wireless.com/20171003/five-challenges-of-cloud-migration-tag27-tag99

7. Pamami, P., Jain, A., Sharma, N.: Cloud migration metamodel: a framework for legacy to cloud migration. In: 9th International Conference on Cloud Computing, Data Science & Engineering (Confluence), pp. 43–50 (2019)

8. Iqbal, A., Colomo-Palacios, R.: Key opportunities and challenges of data migration in cloud: results from a multivocal literature review. Procedia Comput. Sci. **164**, 48–55 (2019)

9. Ilyas, I., Modh, M.M.: Implementing comparison of cloud service provider package offerings. U.S. Patent 9,818,127, issued 14 November 2017 (2017)

10. Iyoob, I., Yan, A.M.: Assessment of best fit cloud deployment infrastructures. U.S. Patent 9,813,318, issued 7 November 2017 (2017)

11. Iyoob, I., Modh, M., Farooq, M.S.: Assessment of best fit cloud deployment infrastructures. U.S. Patent Application 14/140,443, filed 18 September 2014 (2014)

12. Yang, J.: Hybrid cloud computing solution for streamlined genome data analysis. In: 9th International Conference on Management of Digital EcoSystems, pp. 173–180 (2017)

13. Megahed, A., et al.: An optimization-based approach for cloud solution design. In: Panetto, H., et al. (eds.) OTM 2017. LNCS, vol. 10573, pp. 751–764. Springer, Cham (2017). https://doi.org/10.1007/978-3-319-69462-7_47

14. Megahed, A., Nazeem, A., Yin, P., Tata, S., Nezhad, H.R.M., Nakamura, T.: Optimizing cloud solutioning design. Future Gener. Comput. Syst. **91**, 407–424 (2019)

15. Singh, G., Malhotra, M., Sharma, A.: An agent based virtual machine migration process for cloud cnvironment. In: 2019 4th International Conference on Internet of Things: Smart Innovation and Usages (IoT-SIU), pp. 1–4 (2019)

16. Guillén, J., Miranda, J., Murillo, J.M., Canal, C.: Developing migratable multicloud applications based on MDE and adaptation techniques. In: Second Nordic Symposium on Cloud Computing & Internet Technologies, pp. 30–37 (2013)

17. Jamshidi, P., Pahl, C., Mendonça, N.C.: Pattern-based multi-cloud architecture migration. Softw.: Pract. Exp. **47**(9), 1159–1184 (2017)

18. Wang, K.: Migration strategy of cloud collaborative computing for delay-sensitive industrial IoT applications in the context of intelligent manufacturing. Comput. Commun. **150**, 413–420 (2020)

19. Stauffer, J.M., Megahed, A., Sriskandarajah, C.: Elasticity management for capacity planning in software as a service cloud computing. IISE Trans. **53**(4), 1–69 (2020)

20. Megahed, A., Mohamed, M., Tata, S.: Cognitive elasticity of cloud applications. U.S. Patent Application 15/814,608. International Business Machines Corp, (2019)

21. Coutinho, E.F., Neto, M.M., Moreira, L.O., de Souza, J.N.: Analysis of elasticity impact in hybrid computational clouds. In: Euro American Conference on Telematics and Information Systems, pp. 1–8 (2018)

22. Tyagi, N., Rana, A., Kansal, V.: Creating elasticity with enhanced weighted optimization load balancing algorithm in cloud computing. In: 2019 Amity International Conference on Artificial Intelligence (AICAI), pp. 600–604 (2019)

23. Teyeb, H., Hadj-Alouane, N.B., Tata, S., Balma, A.: Optimal dynamic placement of virtual machines in geographically distributed cloud data centers. Int. J. Coop. Inf. Syst. **26**(3), 1750001 (2017)

24. Routray, R., Megahed, A., Tata, S.: Cognitive classification of workload behaviors in multi-tenant cloud computing environments. U.S. Patent Application 16/051,192. International Business Machines Corp (2020)

25. Megahed, A., Mohamed, M., Tata, S.: Cognitive allocation of monitoring resources for cloud applications. U.S. Patent Application 16/147,136. International Business Machines Corp (2020)

26. Megahed, A., Routray, R., Tata, S.: Cognitive handling of workload requests. U.S. Patent Application 16/129,042. International Business Machines Corp (2020)
27. Amato, A., Venticinque, S.: Multiobjective optimization for brokering of multicloud service composition. ACM Trans. Internet Technol. (TOIT) **16**(2), 1–20 (2016)
28. Iyoob, I., Zarifoglu, E., Dieker, A.B.: Cloud computing operations research. Serv. Sci. **5**(2), 88–101 (2013)

Higher Order Statistical Analysis in Multiresolution Domain - Application to Breast Cancer Histopathology

Durgamahanthi Vaishali[(✉)], P. Vishnu Priya, Nithyasri Govind, and K. Venkat Ratna Prabha

Department of Electronics and Communication Engineering,
SRM Institute of Science and Technology, Vadapalani, Chennai 600 026, TN, India
vaishalb@srmist.edu.in

Abstract. Objective is to analyze textures in breast histopathology images for cancer diagnosis.

Background: It is observed that breast cancer has second highest mortality rate in women. Detection of cancer in early stages can give more treatment options and thus reduce the mortality rate. In cancer diagnosis using histopathology images, histologists examine biopsy samples based on cell morphology, tissue distribution, randomness in their growth or placements. These methods are time taking and sometime leads to incorrect diagnosis. These methods are highly subjective/arbitrary. The new techniques use computers, archived data and standard algorithms to provide fast and accurate results.

Material & Methods: In this work we have proposed a multiresolution statistical model in wavelet domain. The primary idea is to study complex random field of histopathology images which contain long–range and nonlinear spatial interactions in wavelet domain. This model emphasizes the contribution of Gray level Run Length Matrix (GLRLM) and related higher order statistical features in wavelet subbands. The image samples are taken from 'BreaKhis' database. The standard database generated in collaboration with the P&D Laboratory—Pathological Anatomy and Cytopathology, Parana, Brazil. This study has been designed for breast cancer histopathology images of ductal carcinoma. GLRLM feature dataset further classified by SVM classifier with linear kernel. The classification accuracies of signal resolution and multiresolution have been compared.

Results: The results show that the GLRLM based features provides exceptional distinguishing features for multiresolution analysis of histopathology images. Apart from recent deep learning method this study proposes use of higher order statistics to gain stronger image features. These features carry inherent discriminative properties. This higher order statistical model will be suitable for cancer detection.

Conclusion: This work proposes automated diagnosis. Tumor spatial heterogeneity is the main concern in analyzing, diagnosing and grading cancer. This model focuses on Long range spatial dependencies in heterogeneous spatial process and offers solutions for accurate classification in two class problems. The work describes an innovative way of using GLRLM based textural features to extract underlying information in breast cancer images.

© Springer Nature Switzerland AG 2021
H. Hacid et al. (Eds.): ICSOC 2020 Workshops, LNCS 12632, pp. 495–508, 2021.
https://doi.org/10.1007/978-3-030-76352-7_45

Keywords: Computer assisted diagnostics · CAD · Grey Level Run Length Matrix · GLRLM · Support vector machine · SVM · Texture analysis · Multiresolution analysis · Wavelet transforms

1 Introduction

In the recent times it has been observed that a significant percentage of women are getting diagnosed with breast cancer. Most women ignore the initial symptoms of breast cancer and find out only during the later stages. The lives are on the risk due to the delayed diagnosis and treatment of cancer. Researchers have conducted many tests and developed new techniques to identify and classify the various levels of cancer so the patient can be given early and accurate treatment [1, 2]. The breast cancer begins with uncontrolled mitosis and causes the accumulation of extra cells in the form of a tumor. To identify whether the tumor is malignant or not, a biopsy is performed. In olden times the pathologist first stains the tissue sample and then analyzes it with the microscope and gives their opinion. This process used to take a longer time and by the time the results are produced the cancer would have started damaging an adjacent area and making the further treatment complicated [3]. By obtaining the digitalized version of the stained images, 'histopathological images' and performing image analysis in multiresolution domain of wavelet subbands is the main concept behind this work. At each wavelet decomposition, the subbands are considered to extract the Grey Level Run Length Matrix (GLRLM) features. The dataset formed by GLRLM features later given for classification. This study considers SVM classifier [4, 5].

The rest of this article is organized as follows. Section 2 gives an overview of histopathology image analysis. Section 2.2 describes the dataset 'BreaKhis', Sect. 3 describes methodology. This section gives brief idea about the multiresolution analysis and GLRLM's higher ordered statistical feature in wavelet domain. It also describes the classification. Section 4 specifies experimentation and results. In Sect. 5 we draw the conclusion [6].

2 Histopathology

The term 'Histopathology' is formed of three Greek words; they are 'Histos' (Tissue), 'Pathos' (disease) and 'Logia' (the study of). So, it is the study of suffering tissues. A. D. Belsare et al. have reviewed several image processing techniques for histopathology images mainly for segmentation and disease classification [1, 4]. Histopathology involves the study of whole tissue, while another technique, called 'Cytopathology' which is the study of cell's nucleus seen in tissue.

2.1 Tissue Preparation: Virtual Slide

Tissue preparation is an important step in histopathology image analysis. As shown in Fig. 1, it consists of five steps, viz tissue collection, fixation, tissue embedding, sectioning, and staining. Once the tissue is removed through the needle biopsy it is fixed with paraffin wax to avoid further decay. Tissue embedding process dehydrates the section to perform slicing in thin section (3.5 μm approx.) These sections are placed on a glass slide. The visibility of the structures on the slide is improved by dying the sections in standard stains. The standard staining protocols are Hematoxylin & Eosin (H & E) and Immunohistochemical (IHC).

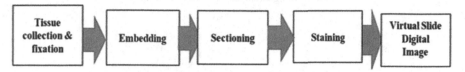

Fig. 1. Virtual slide preparation

H & E stained slides enhance the spatial nuclei features like size, shape, texture, spatial arrangements, tubule's and stromal etc. [1]. In IHC stained the information is given by color and intensity changes. Figure 2 shows H & E stained and IHC stained tissues [3, 4].

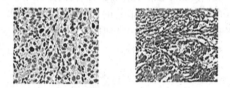

Fig. 2. H & E stained and IHC stained tissues

2.2 Image Dataset 'BreakHis' - Breast Cancer Histopathological Database

The Breast histopathology image data has been obtained from the standard Image database called 'BreakHis'. These are public resources for the researchers and pathologists. This database has been built in collaboration with the P&D Laboratory and Pathological Anatomy and Cytopathology, Parana, Brazil. BreaKhis is composed of 9109 microscopic images of breast tumor tissue collected from 82 patients. Tissue images are available at different magnification scales (×2, ×4, ×10, ×20 and ×40 etc.). Figure 3 shows Images at different scales. Images at ×20 magnification scale (0.049 mm/pixel) have been taken for analysis. BreakHis dataset is divided into two main groups: benign tumors and malignant tumors. Histologically benign is a term referring to an accumulation of cells to form tumor which is innocent and not invading into other parts of the organ while malignant (cancerous) tumor is accumulation of cells but they are growing faster and invading into nearby organs and destroying them as well. Figure 4 shows the sample Images of benign tissue and malignant tissue [4].

The dataset currently contains four histological distinct types of benign breast tumors: adenosis (A), fibroadenoma (F), phyllodes tumor (PT), and tubular adenoma (TA);

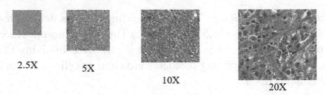

2.5X   5X

10X

20X

Fig. 3. Image database at different scales

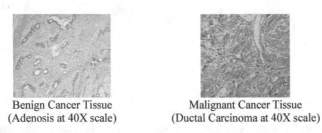

Benign Cancer Tissue   Malignant Cancer Tissue
(Adenosis at 40X scale)   (Ductal Carcinoma at 40X scale)

Fig. 4. Sample images

and four malignant tumors (breast cancer): Ductal carcinoma (DC), lobular carcinoma (LC), mucinous carcinoma (MC) and papillary carcinoma (PC). In this work we have considered ductal carcinoma and benign tumors of adenosis for test [7].

3 Methodology

3.1 Medical Image Analysis

Medical image analysis is a useful non-invasive tool for detecting and diagnosing various diseases accurately and efficiently. As a main application of digital image processing, medical image processing also involves the same techniques and operations such as image acquisition, storage, presentation, restoration, and communication. The medical imaging techniques are based on many scientific phenomena, image acquisition methods, and invasiveness to capture an image from a certain part of the body in a specific disease. Some popular imaging modalities are Microscopy (Digital microscopy, Histopathology) and endoscopy are listed under invasive methods while Radiography, X-Ray imaging, Computed Tomography (CT), Magnetic Resonance Imaging (MRI), Ultrasound imaging, Positron Emission Tomography(PET), and Mammography are non-invasive techniques. Figure 5 shows different imaging modalities used in cancer diagnostics.

3.2 Histopathology Image Processing

As shown in Fig. 6 Image analysis using computer system consists of image acquisition, pre-processing, features extraction and classification (Diagnosis). The preprocessing step is to improve the spatial information in an image, removes noise and in general it improves image quality. Image segmentation is *an* integral part of image pre-processing to find region of interest (ROI). In feature extraction process, the entire image is represented by

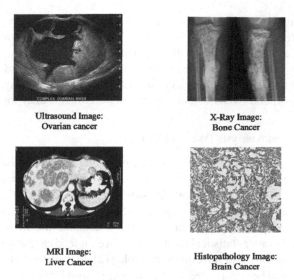

Ultrasound Image:
Ovarian cancer

X-Ray Image:
Bone Cancer

MRI Image:
Liver Cancer

Histopathology Image:
Brain Cancer

Fig. 5. Digital image matrix

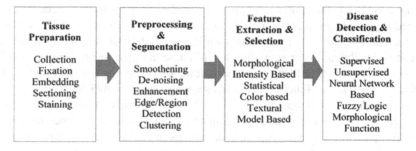

Fig. 6. Histopathology image processing

extracted features and the features are mainly morphological features, textural features, model-based statistical features.

These features are further used to identify and quantify deformities. The aim of the classification step is to extricate irregularities in the given set features and classify them into different classes [9, 10].

Figure 6 describes different techniques involved in pre-processing, feature extraction and classification [4]. The pre-processing and segmentation can be carried out smoothening, de-noising filters. Image enhancement can be done with many techniques like histogram equalization, thresholding, contrast stretching, bit-plane slicing etc. Many clustering algorithms have been developed for accurate image segmentation or Region of Interest (ROI) detection. The main aim of image analysis is to extract hidden information in the image. Many techniques have been evolved, such as shape related features extracted through morphological processing, statistical features of images have been processed with first, second, and higher order statistics and in model based analysis, the model features have been used for image analysis as well as synthesis etc. After performing feature extraction, selection of right features for the study purpose is carried out with different optimization techniques. The focus of optimization techniques is precise classification with size or dimensionality reduction. Many supervised and un-supervised classification algorithms have been designed in which dataset designed with extracted features get classified accurately. Some of them are Bayesian classifier, K-Nearest Neighborhood (KNN), Neural Networks classifier, Fuzzy logic etc. [11].

3.3 Multiresolution Analysis (Wavelet Filters)

In the past few decades, multiresolution image modeling techniques have gained popularity in cancer research because of the availability of better computational resources, faster algorithms and simple framework. At single resolution, it is difficult to interpret interactions between the pixels of heterogeneous, nonlinear and long-range textural patterns in histopathology images. In a fine-resolution domain, this study can be carried out only with higher-order statistical models. However, it leads to parameter estimation computationally intensive and sometimes unstable [12]. In wavelet domain a long-range correlation model can be constructed with a number of simpler models carrying shorter correlation lengths and with lesser computational effort. The disparity in tissue samples offers textural variations and quantifying these changes is a challenging task during cancer diagnosis [13]. In the multiresolution analysis, the image details are decomposed and analyzed at multiple levels of resolutions (scales). The large structures and high contrast images are analyzed at higher resolutions while a small/delicate structure with low contrast can be studied at lower resolution level. These multiple resolutions are achieved by decomposing an image using small basis function of varying frequency and limited duration (window). This function is called as a wavelet function. This wavelet function was first introduced by Mallat [13]. The wavelet domain analysis offers an exceptional mathematical tool to extract scale-dependent and location-dependent structures in breast cancer histopathology image samples. It also helps in identifying the global as well as local image properties. A 2-dimensional wavelet transform can be implemented using digital filtering followed by sub-sampling. Figure 7 shows Image Filtering and sub-sampling, Fig. 8 Image subbands details and Fig. 9 shows two level decomposition with DWT of benign tissue sample.

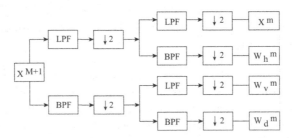

Fig. 7. Image filtering and sub-sampling

Equation 1 represents an approximation of image f (x, y) at decomposition level j0, and **Eq.** 2, represents horizontal, vertical, and diagonal detailing of an image

LL	HL	HL2	HJ1
LH	HH		
LH2		HH2	
LH1		HH1	

Fig. 8. Image subbands

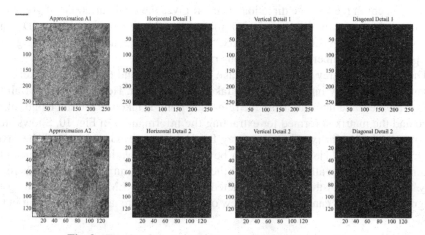

Fig. 9. Two level decomposition with DWT of benign tissue

at decomposition level j ≥ j0. **Approximated Image of subband at 'j0' level of decomposition**.

$$W_0(j_0, i, j) = 1/\text{sqrt}(MN) \sum_{x=1}^{M-1} \sum_{y=0}^{N-1} \{\{f(x, y)\varphi j0, i, j, (x, y)\}\} \quad (1)$$

Detailed Image at 'j0' level of decomposition

$$W_{k \cdot \%}(j, i, j) = 1/\text{sqrt}(MN) \sum_{x=1}^{M-1} \sum_{y=0}^{N-1} \{\{f(x, y)\psi k, j0, i, j, (x, y)\}\}_{K = (H, V, D)} \quad (2)$$

Wavelet Selection

The choice of the wavelet function is based on the computational complexity, ability to analyze the signal, and ability to compress or reconstruct the image with acceptable error. We have chosen Daubechies 4 (db4) wavelet for this work. The highest number of vanishing moments for a given support width and compactly supported wavelet with extremal phase are the two major properties of Daubechies 4 (db4). Associated scaling filters are minimum-phase filters in db4 [14].

3.4 Sub-band Modeling

At each level of wavelet decomposition level, one approximate image and three sub-band images have been created. Sub-bands show detailed properties. Each sub band is considered for GLRLM modeling and thus feature extraction.

3.4.1 Modelling with Grey Level Run Length Matrix (GLRLM)

GLRLM is a histogram of a 2-dimensional image where it appears in the form of a matrix that has the record of the various combinations of different values of grey levels that are present in the specific direction. Thus, it gives the total variations of grey value that runs in the image. In the case of a 3D image it has 13 directions and for a 2D image it has 4 directions. In GLRLM (i, j) matrix "i" represents the intensity in an image and "j" represents the number of homogeneous runs in an image.

The intensity of grey levels in an image is considered and the adjacent pixels with the same level of grey intensity are also taken into account. Here the length of all the pixels with the same intensity from the reference point is taken as the run length of the image and the matrix is formed for extracting the information. In Fig. 10. Shows steps in forming GLRL Matrix, where (a) Gray Level Image, (b) Gray levels of the pixels and (c) GLRLM [8, 11]. The matrix can be formed in the various directions from the reference points along all the angles and then the features can be extracted. There are many textural features that can be extracted from the GLRLM for distinguishing the malignant and benign cancer images, a few of them which are taken into account in this work.

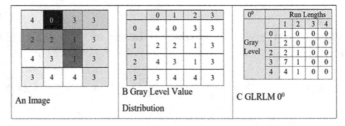

Fig. 10. GLRL Matrix (a) Gray Level Image (b) Gray levels of the pixels (c) GLRLM

We have considered seven statistical features derived by Galloway [8] and Chu et al. [11]. The first two features are based on run emphasis such as short run and long run (SRE and LRE). The next two are based on non-linearity existing in gray levels and run lengths (GLN and RLN). The last one is run percentage (RP). Chu et al. described some more features based on GLRLM, but this work refers low gray level runs emphasis (LGLRE) and high gray level runs emphasis (HGLRE). These features capture the discriminative property of a gray level run length distribution. The parameters (i, j) are associated with the gray level and measure of run length, respectively. 'M' represents a number of gray levels and 'N' represents the number of runs in 'n r' shows the over-all number of runs and 'n p' shows over-all number of pixels.

Short Runs Emphasis (SRE), it is the distribution of the number of short runs of grey level that is present in a homogenous image that is given for processing. During the formation of a matrix the SRE is calculated along with the other features.

Long Runs Emphasis (LRE), it is the number of long runs of grey that can be obtained from the image that is taken into consideration. The long runs are calculated from the reference point. **Gray-Level Non-Uniformity (GLNU)** it is used to identify the non-uniformity in the grey level from the image and that feature is extracted for the classification. **Run Length Non-Uniformity (RLNU)** used for extracting the information regarding the non-uniformity that can be obtained from the image while observing the run length from the reference point in the digital image. Run percentage (RP) is used to measure the percentage of homogeneity of the homogeneous runs that is observed from the image. **Low gray-level emphasis(LGRE)** there are two types in this one is the emphasis of the low gray level that is distributed along the short run and the other is along the long run from the reference point in the homogeneous image. **High grey level emphasis (HGRE)** is the high grey level that can be extracted along the short and long run of the homogeneous image. The distribution of the high grey level along the image in a specific direction is taken into account. Each of these parameters is used for the classification of the images in later stages. Table 1 shows GLRLM Features described by Galloway and Chu.

Table 1. GLRLM features

GLRLM Features	Standard Formulae
Short Run Emphasis	$SRE = \frac{1}{n_r}\sum_{i=1}^{M}\sum_{j=1}^{N}\{p(i,j)\}\frac{1}{j2} = \frac{1}{n_r}\sum_{j=1}^{N}\{p(j|\theta)\}\frac{1}{j2}$
Long Run Emphasis	$LRE = \frac{1}{n_r}\sum_{i=1}^{M}\sum_{j=1}^{N}\{p(i,j|\theta)\}*j^2 = \frac{1}{n_r}\sum_{j=1}^{N}\{p(j|\theta)\}*j^2$
Gray level non-uniformity	$GLN = \frac{1}{n_r}\sum_{i=1}^{M}(\sum_{j=1}^{N}\{p(i,j|\theta)\})*^2 = \frac{1}{n_r}\sum_{i=1}^{M}\{p_g(j|\theta)\}*^2$
Run Length non-uniformity	$RLN = \frac{1}{n_r}\sum_{i=1}^{N}1(\sum_{j=1}^{M}\{p(i,j|\theta)\})*^2 = \frac{1}{n_r}\sum_{j=1}^{N}\{p_r(i|\theta)\}*^2$
Run Percentage	$RP = n_r/n_p$
low gray level runs emphasis	$LGRE = \frac{1}{n_r}\sum_{i=1}^{M}\sum_{j=1}^{N}\{p(i,j|\theta)\}*\frac{i2}{j2} = \frac{1}{n_r}\sum_{j=1}^{N}\{p_g(i|\theta)\}*\frac{i2}{j2}$
High gray level runs emphasis	$HGRE = \frac{1}{n_r}\sum_{i=1}^{M}\sum_{j=1}^{N}\{p(i,j|\theta)\}i^2 = \frac{1}{n_r}\sum_{j=1}^{N}\{p_g(j|\theta)\}*j^2$

3.5 Classification

To distinguish the malignant structures in histopathology images, the datasets of GLRLM parameters have been taken for two-class classification problem. In supervised classification, classifiers have been trained with a set of images with class labels called as 'Training data'. After training the classifier, the classifier has been tested with 'Test Data'. In unsupervised classification, outcomes have been based on detailed analysis of image data without prior knowledge of data and classes [15–25].

Many evaluation strategies have been designed to select training and testing sets for classification. There are four major approaches for evaluation of classification system. (a) Using entire dataset for training as well as for testing, (b) Using individual datasets to train the classifier and to test the classifier later. (c) K-fold cross-validation, (d) Leave one out. It has been proved that more training data leads to a better system design while more test data leads to a reliable evaluation. The first evaluation approach is simple but computationally intensive and slow. In the second approach, the main issue is 'How to select the data for training purpose and for testing purpose?'. This work uses K-fold cross-validation technique on randomly selected training and testing dataset. In this technique, the dataset is partitioned into 'k' similar size small groups. Each time (k–1) groups are used to train the classifier and the rest are used for testing the performance of the classifier. This process is repeated for 'k' number of times. Finally, accuracy is calculated by taking the average of all accuracies computed in each repetition. Thus, in this evaluation system, each group has been used for testing the system [20, 21].

This work considered Support Vector Machine (SVM) classifiers for classification. In SVM classification, the classifier forms one or more than one hyper planes to provide maximum distance to divide feature data into the different classes. The optimal hyperplane is a linear decision boundary with the maximum margin between the vectors of different classes [15, 16]. This margin is optimally computed by critical components of the training data called as support vectors. SVM can classify miscellaneous data just by changing its kernel functions. These kernels are Linear, Polynomial, Radial Basis Function (RBF), Sigmoid functions. This work considered Linear kernel SVM classifier [17].

3.6 Experimentation

This work has been implemented on MATLAB platform, proven for its performance in technical computing. This research work has been carried out in three parts. The first part is for wavelet decomposition, second is for image modelling (parameter estimation) and third is for classification. We test our model on histopathology images of breast cancer (ductal carcinoma) diagnosis. Histopathological image database is taken from BrekHis.

In the first phase, we cropped a portion of size 512×512 pixels from all the sample images before pre-processing. Pre-processing removes the noise and improves the quality of sample image, thereby making it suitable for key feature extraction. The filters in pre-processing smoothen or sharpen the image so that the disease-specific features are enhanced and correctly extracted. After pre-processing, we performed wavelet decomposition using Dobechies-4 (db4) for multiresolution analysis. In the next step we have calculated GLRLM Matrix and seven mentioned parameters have been calculated for all subbands. We considered 125 malignant and 100 benign images for GLRLM parameter calculation.

The feature extracted in all four subband further given to SVM classifier. The linear kernel SVM classifier with the k-fold validation technique is implemented in this work. The 'k' value considered in this work is 10 [18, 22, 23].

Table 2 represents sample parameters of benign and malignant images and Fig. 11 is showing graph of clear discrimination of malignant and benign parameters. Figure 12 shows sample dataset constructed with second level decomposed wavelet subband image.

Table 2. GLRLM estimated parameters for benign and malignant image.

GLRLM features	Benign images	Malignant images
SRE	0.7396	0.8073
LRE	466.0709	408.6144
GLN	321.3386	302.2676
RP	0.1669	0.1901
RLN	1389.4629	1903.2376
LGRE	0.0344	0.0386
HGRE	139.6581	90.6119

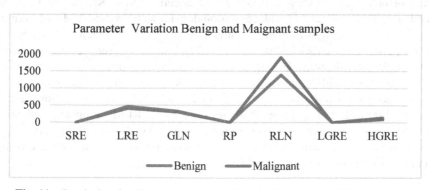

Fig. 11. Graph showing the parameters variation for Benign and Malignant images

4 Result and Discussions

In the proposed system for cancer detection, the GLRLM parameters have been considered in each subband to form the data sets for classification. Refer Fig. 12.

In recent times, CAD systems have been considered for a second opinion by most of the clinical community. It also reduces the time for diagnosis and enables early

0.5334	84.1919	596.4705	0.2925	1414.639	0.0359	136.224	1
0.5232	88.7903	350.1100	0.2820	1324.925	0.0410	120.789	1
0.5321	87.3366	373.7066	0.2852	1381.438	0.0391	126.817	1
0.5517	87.3554	405.7000	0.2882	1472.962	0.0395	138.770	1
0.5490	129.415	312.4399	0.2269	1124.487	0.0460	135.342	1
0.5336	96.6696	476.9113	0.2669	1270.232	0.0380	144.639	1
0.5335	83.2859	408.8382	0.2950	1432.288	0.0422	97.7530	1
0.5592	77.3121	415.8672	0.3159	1675.902	0.0395	109.365	1
0.5391	92.4269	461.0700	0.2739	1319.931	0.0392	129.587	1
0.5395	86.0932	354.4251	0.2881	1403.235	0.0406	119.980	1
0.5839	70.1069	587.7915	0.3438	1987.459	0.0398	84.9185	2
0.6049	72.2716	432.0817	0.3394	2063.168	0.0480	85.4543	2
0.5817	69.8633	628.0489	0.3447	1992.631	0.0410	79.0926	2
0.5838	69.9801	558.1562	0.3442	1980.222	0.0404	83.9750	2
0.5982	67.1957	607.5633	0.3571	2158.408	0.0326	118.860	2
0.5955	68.0716	591.3423	0.3531	2115.136	0.0357	99.0633	2
0.5587	76.1024	558.4618	0.3193	1684.827	0.0376	100.634	2
0.5931	68.6670	442.0366	0.3505	2085.930	0.0537	84.8939	2

Fig. 12. GLRLM sample dataset.

treatments on malignant patients. It is observed that conventional or classical methods of classification are taken over by Deep learning algorithms such as convolutional neural networks, Deep Neural Networks etc. due to high data handling capacity. The 83.34% classification accuracy is observed using CNN which includes a convolutional layer, small SE-ResNet module, and SVM by A. Chan et al. for two. The same method for grading of breast cancer they achieved 77.8% accuracy. MA. Kahya and team also worked on BreakHis dataset in multiresolution domain with Sparse SVM and Adaptive Sparse SVM considering different magnification scale class problem. They achieved accuracy in between 84.43% and 94.54% [26–27]. This study proposes use of stronger feature sets for classification using higher order statistics. Experimental results reveal that accuracy calculated with \single resolution GLRLM parameter is 87.32% using SVM classifier and classification accuracy calculated considering multiresolution subband dataset is 92.00%. The result shows that with the same GLRLM parameters with single resolution system are not able to discriminate benign and malignant tissue very well. In multiresolution domain with wavelet subbands, same GLRLM parameters are able to detect malignant tissue correctly (Table 3).

Table 3. Comparison of results with single resolution and multiresolution

Work domain	Classification accuracy with SVM
Single resolution GLRLM model	87.32%
Multiresolution wavelet domain GLRLM model	92.00%

At single resolutions identification of key features of malignant tissue becomes difficult because of Intra-tumour heterogeneity and long-range dependency. The wavelet sub-band images the correlation length get reduced by factor two at every decomposition and makes it easier to identify key features for discrimination. This gives better accuracies in multiresolution domain.

5 Conclusion

This work focuses on multi-resolution analysis and higher order statistical features with GLRLM of histopathology images to deal with intratumor heterogeneity and long-range dependency. Working in the wavelet domain for cancer diagnosis offers better accuracies It also offers stable and reliable statistical features. The GLRLM parameters in wavelet sub-bands are perfect with distinctive features in higher order statistical analysis. Results show that a wavelet based GLRLM model is better than single resolution GLRLM model. As a future work, we can consider polynomial kernel and Radial Basis Fiction (RBF) kernel SVM classifier for better accuracies. We can also consider multiple classifier with various classifier fusion schemes for better understanding of discriminative feature in GLRLM based higher order statistical Parameters.

Acknowledgment. The authors would like to thank SRM Institute of Science and Technology, Vadapalani, and Chennai for their continued support and encouragement during this research work.

Conflicts of Interest. The authors declare that there are no conflicts of interest regarding the Publication of this article.

References

1. Foran, D.J., Chen, W., Yang, L.: Automated image interpretation and computer-assisted diagnostics. Anal. Cell Pathol. (Amst) **34**(6), 279–300 (2011). https://doi.org/10.3233/acp-2011-0046
2. Belsare, A.D., Mushrif, M.M.: Histopathological image analysis using image processing techniques: an overview. Sig. Image Process.: Int. J. (SIPIJ) **3**(4), 22–31 (2012)
3. Chang, T., Kuo, C.: Texture analysis and classification with tree structured wavelet transform. IEEE Trans. Image Process. **2**, 429–441 (1993)
4. Vaishali, D., Ramesh, R., Christaline, J.A.: Performance evaluation of cancer diagnostics using autoregressive features with SVM classifier: applications to brain cancer histopathology. Int. J. Multimedia Ubiquit. Eng. **11**(6) 241–254 (2016)
5. Vaishali, D., Ramesh, R., Christaline, J.A.: Histopathology image analysis and classification for cancer detection using 2D autoregressive model. Int. Rev. Comput. Softw. **10**(2), 182–188 (2015)
6. Jain, A.K., Duin, R.P.W., Mao, J.: Statistical pattern recognition: a review. IEEE Trans. Pattern Anal. Mach. Intell. **22**(1), 4–37 (2000)
7. https://web.inf.ufpr.br/vri/databases/breast-cancer-histopathological-database-breakhis/
8. Galloway, M.M.: Texture analysis using gray level run lengths. Comput. Graph. Image Process. **4**, 172–179 (1975)
9. Mallat, S.: A theory for multiresolution signal decomposition: the wavelet representation. IEEE Trans. Pattern Anal. Mach. Intell. **11**, 674–693 (1989)

10. Vaishali, D., Ramesh, R., Christaline, J.A.: Autoregressive modelling: application of mitosis detection in brain cancer histopathology. Int. J. Biomed. Eng. Technol. **20**(2), 179–194 (2016)

11. Chu, A., Sehgal, C.M., Greenleaf, J.F.: Use of gray value distribution of run lengths for texture analysis. Pattern Recogn. Lett. **11**, 415–420 (1990)

12. Woods, J.W.: Two-dimensional discrete Markovian fields. IEEE Trans. Inf. Theory **IT-40**, 232–240 (1982)

13. Unser, M., Eden, M.: Multiresolution feature extraction and selection for texture segmentation. IEEE Trans. Pattern Anal. Mach. Intell. **11**, 717–728 (1989)

14. Daubechies, I.: The wavelet transform, time-frequency localization and signal analysis. IEEE Trans. Inf. Theory **36**, 961–1005 (1990)

15. Waheed, S., Moffitt, R.A., Chaudry, Q., Young, A.N., Wang, M.D.: Computer aided histopathological classification of cancer subtypes. In: 7th International IEEE Conference Bioinformatics and Bioengineering, pp. 503–508 (2007)

16. Dundar, M.M., et al.: Computerized classification of intraductal breast lesions using histopathological images. IEEE Trans. Biomed. Eng. **58**(7), 1977–1984 (2011)

17. Zhang, J., Wang, D., Tran, Q.N.: A wavelet-based multiresolution statistical model for texture. IEEE Trans. Image Process. **7**(11), 1621–1627 (1998)

18. Krishnan, M.M.R., et al.: Automated classification of cells in the sub-epithelial connective tissue of oral sub-mucous fibrosis - an SVM based approach. J. Comput. Biol. Med. **39**, 1096–1104 (2009)

19. Gurcan, M.N., Pan, T., Shimada, H., Saltz, J.: Image analysis for neuroblastoma classification: segmentation of cell nuclei. In: Proceedings of the 28th IEEE EMBS Annual International Conference, New York City, USA, August 2006 (2006)

20. Chan, K., Lee, T.-W., Sample, P.A., Goldbaum, M.H., Weinreb, R.N., Sejnowski, T.J.: Comparison of machine learning and traditional classifiers in glaucoma diagnosis. IEEE Trans. Biomed. Eng. **49**(9), 963–974 (2002)

21. Kuncheva, L.I., et al.: Decision templates for multiple classifier fusion: an experimental comparison. Pattern Recogn. **34**(2), 299–314 (2001)

22. Doyle, S., Feldman, M., Tomaszewski, J., Madabhushi, A.: A boosted Bayesian multiresolution classifier for prostate cancer detection from digitized needle biopsies. IEEE Trans. Biomed. Eng. **59**(5), 1205–1218 (2010)

23. Kather, J.N., et al.: Multi-class texture analysis in colorectal cancer histology. Sci. Rep. **6**(27988), (2016)

24. Boucheron, L.E., Bi, Z., Harvey, N.R., Manjunath, B.S., Rimm, D.L.: Utility of multispectral imaging for nuclear classification of routine clinical histopathology imagery. BMC Cell Biol. **8**(Suppl 1), S8. https://doi.org/10.1186/1471-2121-8-s1-s8

25. Chan, A., Tuszynski, J.A.: Automatic prediction of tumour malignancy in breast cancer with fractal dimension. R. Soc. Open Sci. **3**(12), 160558 (2016). pmid: 28083100

26. Kahya, M.A., Al-Hayani, W., Algamal, Z.Y.: Classification of breast cancer histopathology images based on adaptive sparse support vector machine. J. Appl. Math. Bioinform. **7**(1), 49 (2017)

27. Bardou, D., Zhang, K., Ahmad, S.M.: Classification of breast cancer based on histology images using convolutional neural networks. IEEE Access **6**, 24680–24693 (2018)

Ontology Evolution Using Recoverable SQL Logs

Awais Yousaf[1][✉], Asad Masood Khattak[2][✉], and Kifayat Ullah Khan[1][✉]

[1] IKMA Lab, National University of Computer and Emerging Sciences
(FAST-NUCES), Islamabad, Pakistan
kifayat.alizai@nu.edu.pk
[2] College of Technological Innovation, Zayed University, Dubai, UAE
asad.khattak@zu.ac.ae

Abstract. Logs of SQL queries are useful for building the system design, upgrading, and checking which SQL queries are running on certain applications. These SQL queries provide us useful information and knowledge about the system operations. The existing works use SQL query logs to find patterns when the underlying data and database schema is not available. For this purpose, a knowledge-base in the form of an ontology is created which is then mined for knowledge extraction. In this paper, we have proposed an approach to create and evolve an ontology from logs of SQL queries. Furthermore, when these SQL queries are transformed into the ontology, they loose their original form/shape i.e., we do not have original SQL queries. Therefore, we have further proposed a strategy to recover these SQL queries in their original form. Experiments on real world datasets demonstrate the effectiveness of the proposed approach.

Keywords: SQL logs · Bloom filter · Ontology evolution · Online transaction processing · RecSQL · DFS

1 Introduction

Online transaction processing systems (OLTP) are useful to facilitate the day to day operations of the organization. One specific and useful feature obtained as a by-product is the logs of SQL queries. These SQL logs can be used for design evaluation, upgrading, maintenance among others. We can build the Dimensional Model from SQL logs.

The size of such logs is very big in large OLTP systems, so it is hard to detect get trends, patterns, the relationship among queries and so on. To solve this matter, we can build an ontology from these logs for aforementioned purposes. As we know that an ontology is information-rich data repository and enhances the semantics of data. Therefore, it can be used to enhance the semantic of source data, and integration of homogeneous schema. For example, consider two very basic queries "select id, name from student" and "select id, name, marks from student". We find a minor difference between them as the second

© Springer Nature Switzerland AG 2021
H. Hacid et al. (Eds.): ICSOC 2020 Workshops, LNCS 12632, pp. 509–517, 2021.
https://doi.org/10.1007/978-3-030-76352-7_46

query contains only one additional attribute. Such minor difference can easily be detected manually. However, when we have a massive logs of queries which contain some related queries as above, so there is a dire need to have a proper mechanism for knowledge extraction.

The existing uses SQL logs for various purposes like recommendation systems [11], clustering [8] for access patterns learning, workload analysis based on term frequency of projection [9], and also for building the ontology [1] using materialized views. So, we find from the literature that a number of people has performed research on SQL logs for building the ontology however, we could not find any work where someone have used the SQL logs for ontology evolution purpose.

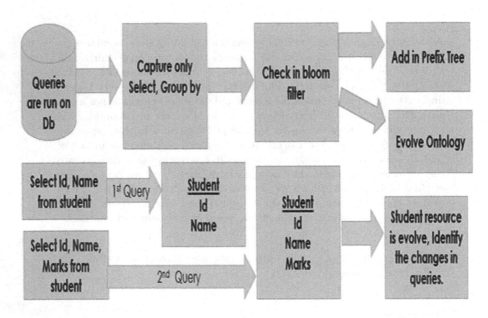

Fig. 1. Key step of the proposed approach.

Considering aforementioned utilization of SQL logs and research gap in the literature, we focus on ontology evolution. Figure 1 illustrate the major steps of our proposed approach where as soon as a query is received, it is first evaluated whether such query is previously processed or not. By process, we mean that if this is already incorporated in the ontology then it will not be processed further to evolve the ontology. On the other hand, it will be processed to be part of the ontology and hence the ontology gets evolved. Our change capture mechanism/approach is also useful in a sense that it allows us to accurately recover the original queries. One of the limitation of our work is that we only focus on simple select statements which may or may not have group by clause.

2   Related Works

In this section, we review some existing studies where we have divided them into 2 parts i.e. usage of query logs for various purposed and using logs for ontology building.

[6] analyze the query logs for assessing the structure of queries. These queries logs help in the recommendation systems and building the user profile. Agrawal et al. [2] rank the tuples of the SQL queries. They build rules for context and analyze the queries by using established rules. They identify the similarity of the query by cosine distance. Yang et al. [12] build the graph from logs to find the similar queries based on Jaccard coefficient. Aligon et al. [3] define their own similarity function for comparing the OLAP queries. They identify that important part of queries is the selection and join. They compute the similarity of join and group by, using the distance between attributes on different hierarchy levels. Compared to the existing works, we identify the changes in SQL logs instead of finding the similar queries.

In this aspect, Aadil [1] transform user needs/requirements into SQL queries and check their existence in database. They then create materialized views to build ontology. For building ontology, the materialized view is transformed into an OWL class, and the primary key attribute becomes property function. In this way, they avoid the problems like synonyms, equivalence, and identity. Whereas, their approach is quite comprehensive however, they do not study the task of ontology evolution. Rend et al. [10] build the ontology from source data and transform business requirements into the organization model. Then they perform dimensional modeling on the ontology and organizational model. However, the key drawback in work is that the changes happening in source data cannot be detected to evolve the ontology. Elfaki et al. [5] represent the relational tables by a knowledge graph which is then transformed into a knowledge base. For ontology evolution, Khattak et al. [7] consider 3 types of changes i.e., add, extend, and reduce. The changes are saved in the change history log in the semantic structure. By change history log, they reconcile the mapping to eliminate unreliable mappings and re-establish them for ontology evolution. However, they do not consider the nature of data as SQL queries.

3   The Proposed Methodology

In this section, we present our proposed approach. We refer to Fig. 2 to technically elaborate our methodology, where the SQL queries are used to evolve the ontology. Our approach is presented in following sub-sections.

3.1   Type of SQL Queries Handled by Our Approach

In this paper, we consider simple "select" statements which may or may not have "group by" clause. For example,

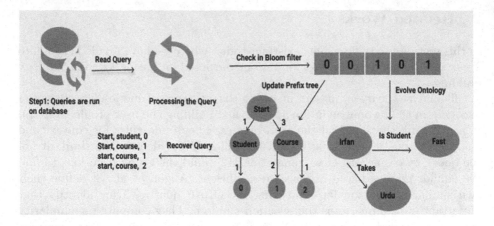

Fig. 2. We captured only those queries which have simple Select and Group By statements and check in Bloom Filter, then Add them in the prefix tree and Evolve the Ontology.

 (i) select id, name From student
 (ii) select id, title From course
(iii) select count(customerid), country from customers group by country

the generic form/syntax of these queries is

– Select ColNames From TableName
– Select ColNames From TableName Group By ColNames

3.2 Feature Engineering

In this sub-section, we perform featuring engineering for the queries to transform them into a format which is recognized by our approach. Our proposed feature engineering process initially converts a query into "lower case". We then remove the "select" and "from" keywords and add the table name as prefix followed by an underscore to attach it with the attribute name. The example is presented below.

```
{select course.id,course.title from course}
```

converted as

```
{course_id,course_title from course}.
```

3.3 Capturing Change Using Bloom Filter

Now we discuss how to capture the change in SQL logs for ontology evolution. For this purpose, we use bloom filter which helps us to identify whether certain element is already exists or not. If it does not exist, we add it into the ontology i.e. the ontology is evolved. On the other hand, if it is seen then we discard it. Along with the bloom filter, we build a prefix tree which is used to recover the transformed SQL queries.

Bloom filter [4] is a probabilistic and space-efficient data structure for searching. It checks whether the element is a member or not. It saves each seen value in the form of a bit vector and tells that whether the element is present or not in it.

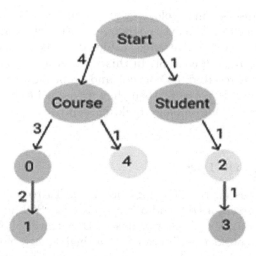

Fig. 3. A prefix tree obtained from given example SQL queries.

As soon as we engineer/transform the queries into our proposed form, we split it into separate terms and save each term in bloom filter and add into prefix tree. For instance, we consider following list of queries to demonstrate building of prefix tree.

- select course.courseid, course.title from course
- select courseid from course
- select courseid, title from course
- select id, name from student
- select year from course

and use following query as detailed/running example for clear understanding.

{Select course.id,Course.title from Course}

is converted as

{course_id,course_title from course}

After the transformation, we split the query into separate terms like this *course_id, course_title* and *course*. We then append the table name with each term as *course_id, course_name*. Each augmented term gets an index which is then added into bloom filter, which checks its existence. If this term is being seen for the first time, it gets added into the ontology and also added in prefix tree.

3.4 Ontology Evolution

In this section, we discuss our ontology evolution process. From the above example we got course_id, course_name, course_year, student_id and student_name. From these list of keywords, we make table name as a concept and column names as properties of each concept. In this way, an initial ontology is created. When a new query is received, its concept and properties are checked in bloom filter. If its properties do not exist then they are added either as new property of an existing concept. Our ontology building approach is based upon the idea of [1].

3.5 Recovering SQL Logs

The prefix tree holds all the SQL logs observed. To recover the query in its original form, we present a DFS and our proposed RecSQL approach. The DFS approach is based on the well known idea of DFS traversal. The issue of DFS approach is that it considers each branch as a single query, and hence unable to differentiate that a single branch may contain more than one query. For instance, for the above given queries certain are subset of others. This subset and superset queries are represented as a single branch. The proposed RecSQL approach is an extended/altered version of DFS traversal.

We present the steps of RecSQL using the Fig. 3. Prefix tree is taken as an input. Start at root node i.e. "Start" and check the edge having higher weight. We have edge weight 4 with the node "Course". Now we operate on this path for 4 times. On reaching the "Course", we further check the edge having higher weight. In this way, we reach the leaf node. Using the "Course", our first recovered items are "Course" and nodes "0", "1". Each traversal to leaf node and re-cursing back to "Course" node decrements the edge weights like the weight from "Course" to "0" is decreased to 2 from 3, and 2 to 1 for node "1". Going to "Course" again and decrementing the edge weights again, provide us "Course", "0", and "1". As the edge weight becomes 0, the edge is dropped. In this way, each branch is traversed equal to the edge weight number of times. Each recurse back provides a list of nodes which are infact the SQL queries which are 100% accurate. The Fig. 4 presents the running example to clearly explain our SQL recoverer.

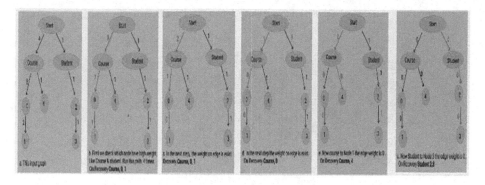

Fig. 4. This is illustration show the recovery of sql queries through our Proposed method RecSQL.

4   Experimental Evaluation

4.1   Experimental Setups

The algorithms were implemented in Python, and all the experiments were performed on a laptop with Core i3 with 2.0 GHz processor having 8 GB RAM.

4.2   Datasets

We used Kaggle dataset having 4985 SQL queries, containing only select statement without where clause and group by queries.

Evaluating the Query Recovery Time and Recovered Logs Accuracy.
The Fig. 5 compares the execution time of RecSQL method and DFS. We find that the proposed approach consumes more time. However, considering the accuracy of recovery is much lesser in DFS approach as shown in Fig. 6. As we see in Fig. 6 that in 1000 queries we recover the query 117 and 822 from DFS and proposed recover(RecSQL). In the from 5000 queries we recover the 749 queries from DFS and 4043 by RecSQL. We see that RecSQL recover more queries while DFS not. The DFS time taking is less but the issue is recovered is also less. This is because of DFS are not used edge weight. The DFS only take edge, not edge weight. Our proposed algorithm RecSQL recover the SQL logs by using the edge weight.

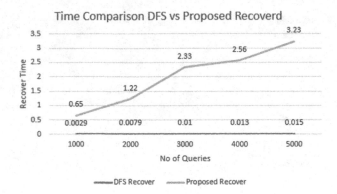

Fig. 5. Time comparison between DFS and proposed RecSQL on Kaggle dataset

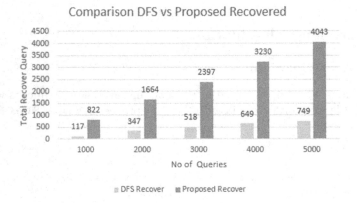

Fig. 6. Accuracy comparison between DFS and proposed RecSQL on Kaggle dataset

5   Conclusion

In this work, we proposed a simple yet robust and effective approach for ontology evolution based on SQL logs. The proposed approach recoveres the SQL queries with 100% accuracy using bloom filter and prefix tree. We recover the original SQL queires from ontology using the prefix tree. In this work, we only considered "select" and "group-by" queries. Whereas, we aim to consider the queries involving "join" and "where" clauses.

Acknowledgment. This research work was supported by Zayed University Cluster Research Fund 18038.

References

1. Aadil, B., Wakrime, A.A., Kzaz, L., Sekkaki, A.: Automating data warehouse design using ontology. In: 2016 International Conference on Electrical and Information Technologies (ICEIT), pp. 42–48. IEEE (2016)

2. Agrawal, R., Rantzau, R., Terzi, E.: Context-sensitive ranking. In: Proceedings of the 2006 ACM SIGMOD International Conference on Management of Data, pp. 383–394. ACM (2006)
3. Aligon, J., Boulil, K., Marcel, P., Peralta, V.: A holistic approach to OLAP sessions composition: the falseto experience. In: Proceedings of the 17th International Workshop on Data Warehousing and OLAP, pp. 37–46. ACM (2014)
4. Bloom, B.H.: Space/time trade-offs in hash coding with allowable errors. Commun. ACM **13**(7), 422–426 (1970)
5. Elfaki, A., Aljaedi, A., Duan, Y.: Mapping ERD to knowledge graph. In: 2019 IEEE World Congress on Services (SERVICES), vol. 2642, pp. 110–114. IEEE (2019)
6. Kamra, A., Terzi, E., Bertino, E.: Detecting anomalous access patterns in relational databases. VLDB J. **17**(5), 1063–1077 (2008)
7. Khattak, A.M., Pervez, Z., Khan, W.A., Khan, A.M., Latif, K., Lee, S.: Mapping evolution of dynamic web ontologies. Inf. Sci. **303**, 101–119 (2015)
8. Kul, G., et al.: Ettu: Analyzing query intents in corporate databases. In: Proceedings of the 25th International Conference Companion on World Wide Web, pp. 463–466. International World Wide Web Conferences Steering Committee (2016)
9. Makiyama, V.H., Raddick, J., Santos, R.D.: Text mining applied to SQL queries: a case study for the SDSS SkyServer. In: SIMBig, pp. 66–72 (2015)
10. Ren, S., Wang, T., Lu, X.: Dimensional modeling of medical data warehouse based on ontology. In: 2018 IEEE 3rd International Conference on Big Data Analysis (ICBDA), pp. 144–149. IEEE (2018)
11. Stefanidis, K., Drosou, M., Pitoura, E.: You may also like' results in relational databases. In: Proceedings International Workshop on Personalized Access, Profile Management and Context Awareness: Databases, Lyon. Citeseer (2009)
12. Yang, X., Procopiuc, C.M., Srivastava, D.: Recommending join queries via query log analysis. In: 2009 IEEE 25th International Conference on Data Engineering, pp. 964–975. IEEE (2009)

Artificial Intelligence in the IoT Security Services (AI-IOTS 2020)

IInternational Workshop on Artificial Intelligence in the IoT Security Services (AI-IOTS 2020)

The first International Workshop on Artificial Intelligence in the IoT Security Services (AI-IOTS 2020) was collocated with the 18th International Conference on Service Oriented Computing (ICSOC 2020), and held on December 14, 2020, on a virtual platform.

Artificial Intelligence (AI) is one of the disciplines of computer science that emerges in securing Internet of Things (IoT) services. AI techniques including machine learning, deep learning, and reinforcement learning approaches have been applied to emerging IoT applications in various fields such as smart cities, smart homes, smart grids, health care, smart transportation, smart farming, etc.

According to Gartner, the total number of IoT connected devices will reach 75.44 billion units worldwide by 2025. These devices generate an enormous amount of data, and the handling and analysis of this data are the current requirements of any application. This trend poses several challenges in building efficient and reliable IoT systems. Due to the advancement in AI technology, the connected devices are getting smaller and smarter but the provisioning of secured services and the constraints to be satisfied are increasingly complex and challenging. The AI plays a crucial role to manage huge data flows and storage in the IoT network. As IoT gains its full potential, AI will be at the forefront to promote the potential of IoT.

The focused themes of AI-TOTS 2020 were services in the Internet of Things (IoT) and security, privacy, and trust for services.

The workshop had a good response from the researchers of the IOT community and four papers were accepted after two to three rounds of reviews and revisions. The authors of the accepted papers were from Bulgaria, India, West Africa, and the USA. The papers were well presented and demos/results were included by every author, which received good feedback. The session chairs were Dr. P. Arun Raj Kumar, NIT Calicut, India, Dr. A. R. Vasudevan, NIT Calicut, India, and Dr. N.G. Bhuvaneswari Amma, IIIT Una, India.

The keynote talk on "Data replication and caching issues in large IoT infrastructures" was delivered by Prof. K. Ravindran, from the Department of Computer Science, City University of New York, USA. The talk focused on how data and software can be replicated in a secure IoT system, and the engineering and operational constraints therein.

Acknowledgement

We wish to thank all authors for their contributions, the Program Committee members for their detailed technical inputs and the ICSOC 2020 organizers for the successful conduct of AI-IOTS 2020.

Organization

Workshop Chairs

S. Selvakumar   IIIT Una, India
R. Kanchana   SSN College of Engineering, India

Program Committee

V. Jagadeesh Kumar   IIT Madras, India
Zhiyuan Chen   University of Maryland, Baltimore County, USA
Prakash Ranganathan   University of North Dakota, USA
Alex Bordei   Lentiq, UK
K. Ravindran   City University of New York, USA
G. Manimaran   Iowa State University, USA
B. Prabakaran   University of Texas at Dallas, USA
Srikant Srinivasan   IIT Mandi, India
Vrijendra Singh   IIIT Alahabad, India
B. S. Saini   NIT Jalandher, India
K. Chandrasekaran   NIT Karnataka, India
R. Leela Velusamy   NIT Tiruchirappalli, India
Arun Raj Kumar P.   NIT Calicut, India
Arun Adiththan   General Motors, Michigan, USA
Dr.Vasudevan, A. R.   NIT Calicut, India
 Wanling Gao   Institute of Computing Technology, Chinese Academy
                                   of Sciences, China

A Novel Automated System for Hospital Acquired Infection Monitoring and Prevention

Samyak Shrimali[✉]

Jesuit High School, Portland, OR, USA

Abstract. According to the World Health Organization (WHO), 1.7 million people suffer from hospital acquired infections each year in the United States alone, which accounts for 99,000 deaths. The most prominent reason for spreading these infections is poor hand hygiene compliance in hospitals. This paper proposes an automated system which can monitor and enforce proper hand hygiene compliance in hospitals as stipulated by WHO. The proposed system is a multi-module system based on microcontroller and multiple sensors that track hand hygiene compliance throughout a hospital, sends real-time compliance alerts to staff for immediate corrective actions, and provides automated compliance report generation for the hospital staff. This system is based on four modules, one module is worn by staff, it provides staff's unique ID to other modules and receives real time hand hygiene compliance alerts. The other three modules detect staff and use unique algorithms to do detailed hand hygiene compliance checks at patient beds, sinks, and alcohol dispensers. A custom software was developed to control all modules and upload compliance data to the central server. This system makes hospital hand hygiene compliance monitoring and tracking fully automated, real-time, and scalable. Once deployed it has the potential to significantly reduce the rate of infections and save many lives. With minor changes to the algorithms this system can find applications in other areas such as restaurants, shops, and households for hand hygiene monitoring.

Keywords: Hand hygiene compliance (HHC) · Radio frequency identification (RFID) · Force sensitive resistor (FSR) · Hospital-acquired infections (HAI) · World health organization (WHO) · Infrared (IR) · Printed circuit board (PCB) · Light emitting diode (LED) · Identification (ID) · Centers for disease control and prevention (CDC)

1 Introduction

Hospitals are meant to be a treatment facility to help the sick recover, but today rates of hospital-acquired infections have increased significantly making hospitals a threatening place to visit. In the United States about 1.7 million people suffer from hospital-acquired infections each year, which accounts for about 99,000 deaths, this number is significantly worse in developing and underdeveloped countries [1]. Hospital-acquired infections are infections that are acquired in a hospital by a visitor, staff, or patient during their visit or stay at the hospital [2]. These infections are also known as nosocomial infections [14]

© Springer Nature Switzerland AG 2021
H. Hacid et al. (Eds.): ICSOC 2020 Workshops, LNCS 12632, pp. 523–533, 2021.
https://doi.org/10.1007/978-3-030-76352-7_47

or HAI [3]. Of every 100 hospitalized patients, 7 in developed countries, 10 in developing/underdeveloped countries will acquire a hospital-acquired infection [4]. The most prominent reason for the spreading of these infections is poor hand hygiene compliance in hospitals. WHO stipulated strict hand hygiene guidelines to be followed in hospitals to reduce the rates of hospital-acquired infections. WHO guidelines suggest when hand cleaning is required and how to do proper hand washes/alcohol rubs [5, 6].

2   Related Work

There have been several researches that are targeted to tackle the problem of hospital-acquired infections [8]. Most hospitals still do manual and random compliance checks. They typically put poster-based hand hygiene guidelines throughout the hospital, and trust that the staff will follow them. A few hospitals use technological solutions like video monitoring, robot usage, or electronic monitoring but the solutions currently available are limited in scope, not effective, or too expensive. A few examples of the existing solutions are - Xenex LightStike robotic system, which focuses on disinfecting hospital surroundings using UV rays rather than staff hand-hygiene compliance monitoring [9]. There are a few RFID based systems that monitor compliance, like the nGageTM [10] and some electronic monitoring solutions like from Halyard [11], Debmed [12], and Biovigil [16] but their compliance checks are limited to alcohol dispenser usage at entry and exit of the room, not covering hand hygiene checks at the patient bed or sink area. All available systems have the capability to reduce rates of infections but so far, all existing systems are only focused on some parts of the problem, therefore achieve moderate results [15]. They also do not ensure whether all steps stipulated in WHO guidelines for hand cleaning are followed. In some cases, their compliance monitoring is done only at the hospital level, not at the individual staff level. Today, hospital employed systems are not very effective in tracking and monitoring hand hygiene compliance required by WHO and CDC [13]. Figure 1 below shows proper handwash and hand rub guidelines recommended by World Health Organization.

Fig. 1. Handwash and alcohol rub guidelines stipulated by WHO [7]

3   System Overview and Description

This paper presents a novel multi-module system based on a microcontroller and various sensors that tracks hand hygiene compliance throughout a hospital, sends real time alerts to staff for compliance status, keeps compliance records on a server, and enables automated report generation. The aim of this project was to design a complete solution that is automatic, real-time, cost-effective, efficient, and scalable. The plan was to develop a system consisting of four modules. One module is to be worn by staff to provides staff's unique ID to other modules and receives real-time compliance alerts. The other three modules are used to detect staff presence and track their hand hygiene compliance at patient beds, sinks, and alcohol dispensers. Data gathered by all these modules is automatically sent to a central server for report generation and compliance tracking. The system monitors and tracks whether all actions are followed for keeping proper hand hygiene, as depicted by WHO, including the required execution time of each cleaning step. The proper hand cleaning duration recommended by WHO for alcohol rub is 20–30 s and handwash is 40–60 s [7]. Here, Fig. 2 illustrates how this system will be used in hospitals. As soon as staff approaches any patient bed, they get an alert on wristband for hand hygiene compliance. Once the system detects staff's completion of alcohol rub, it provides compliance pass indication to staff and allows staff to attend patient. After performing duty, when the staff leaves the patient bed area, system again reminds staff to conduct either alcohol rub or hand wash before attending any other patients. The individual alcohol rub and sink stations, placed throughout the hospital, perform proper hand cleaning checks and alert staff.

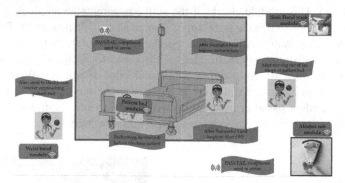

Fig. 2. Illustration of system usage in hospitals

All modules of the system were first designed and implemented on a breadboard and used for developing software algorithms and testing correct functionality. Once all modules were tested on breadboards, they were implemented on printed circuit boards (PCBs) and packaged in custom 3D printed enclosures to secure components and make a finished product.

The functional operation of each of these modules are detailed below:

- **Wristband Module:** This module is worn by staff on their wrist. This module has a microcontroller (Wi-Fi enabled), RFID tag, vibration motor, and LED. The RFID tag provides each staff member with a unique ID. It is detected by other modules placed at various compliance checkpoints across the hospital to identify staff. The other modules use this RFID tag to upload the staff's compliance status to the central server. The server processes this data and sends messages to the wristband module when hand wash or alcohol rub is needed or when hand hygiene compliance is violated. The wristband module decodes this message and alerts staff through vibration and LED indications (Fig. 3).

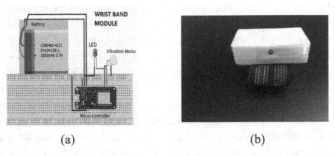

(a)   (b)

Fig. 3. (a) Wristband module circuit diagram (b) Wristband module

- **Alcohol Rub Module:** This module is attached to alcohol dispensers throughout the hospital. It consists of an RFID reader, microcontroller, IR sensor, and force-sensitive resistor (FSR). This module checks if the staff has completed an alcohol rub as per guidelines of WHO. When the staff comes near an alcohol dispenser, the RFID reader on this module reads the staff's unique tag, the IR sensor detects the presence of the staff's hand beneath the automatic alcohol dispenser for usage (or FSR will detect pushing of the dispenser in a non-automatic alcohol dispenser). When the above checks are satisfied a successful event is reported to the server else a fail event. The server then sends a pass/fail alert to the wristband module. Figure 4 and 5 shows alcohol rub module circuit diagram, prototype, and flowchart.

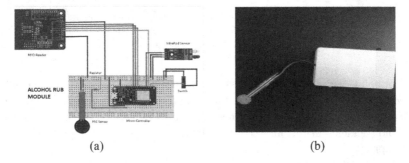

(a)   (b)

Fig. 4. (a) Alcohol rub module circuit diagram (b) Alcohol rub module

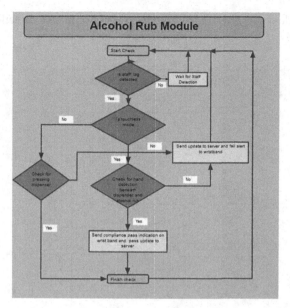

Fig. 5. Flow chart of the algorithm implemented for alcohol rub module

- **Sink Hand Wash Module:** This module is attached to the sinks throughout the hospital. It consists of an RFID reader, microcontroller, two IR sensors, and a water detection sensor. This module does a detailed check for proper hand washes conducted by staff, as recommended in the WHO guidelines. When a person comes near the sink for handwashing, the RFID reader reads the staff's unique tag and starts the compliance check. The IR sensor detects the presence of staff's hand, the water detection sensor detects the water flow for an initial rinse, the IR sensor on the soap dispenser checks for soap usage, and the IR sensor on water tap checks for the final hand wash step. When all the above checks are satisfied, a successful event is reported to the server else a fail event, the server then sends an immediate pass/fail alert to the wristband module. Figure 6 shows sink hand wash module circuit diagram and prototype.

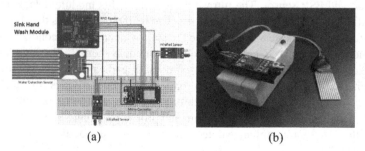

Fig. 6. (a) Sink hand wash module circuit diagram (b) Sink hand wash module

- Figure 7 provides overview of algorithm used for compliance detection in sink hand wash module.

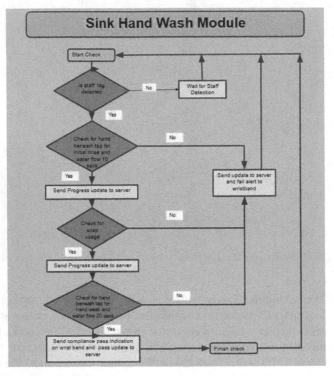

Fig. 7. Flow chart of the algorithm implemented for Sink hand wash module

- **Patient Bed Module:** WHO recommends that a health-care staff should do an alcohol rub when they enter a patient bed area and when they leave a patient bed area. The patient bed module tracks the compliance of this guideline. This module is attached to each patient's bed. It consists of an RFID reader, microcontroller, ultrasonic sensor, IR sensor, and FSR sensor. The ultrasonic sensor on this module detects the staff's entry within the 1-m range of a patient bed, The RFID reader reads the staff's unique tag and reports it to the server to provide the staff with an alert to conduct an alcohol rub, which will be placed near the bed. This module then checks for alcohol rub completion, updates compliance to the server, and sends a compliance alert to staff. It also checks when the staff leaves the patient area and provides the staff with an alert to conduct another alcohol rub, before attending any other patients. Figure 8 shows patient bed module circuit diagram and prototype. Figure 9 provides overview of algorithm used for compliance detection in patient bed module.

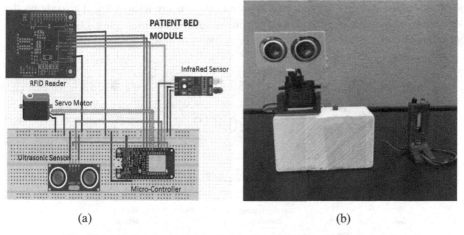

(a)   (b)

Fig. 8. (a) Patient bed module circuit diagram (b) Patient bed module

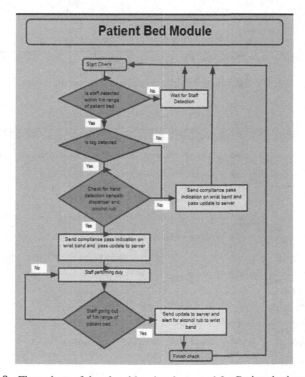

Fig. 9. Flow chart of the algorithm implemented for Patient bed module

4 Results

In this research a system was developed to monitor hand hygiene compliance in hospitals. Figure 10 shows the automatic compliance report generated with an arbitrary staff name and the compliance pass/fail reported to server from different modules. It also shows the detailed hand hygiene compliance report on the central server. Figure 11 captures each compliance check (staff names and ID, check pass/fail status, location of the check).

Staff Name	Staff ID	Compliance Success	Complaince Fail
Tom	F93F3959	3	4
John	9964159	12	1
Marry	F07F4FD3	4	1
Jeff	F0D74729	7	2
Harry	8EA38BB	0	0
unknown	5987EE6E	1	0

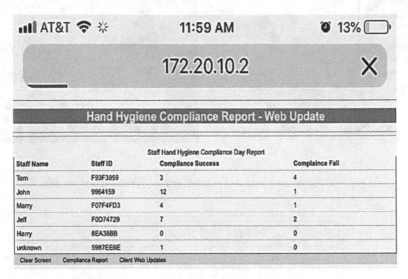

Fig. 10. Staff hand hygiene compliance summary report

Hand Hygiene Compliance Report - Web Update

For Tom (TagId - F93F3959): Person detected near Patient Bed module and Tag detected at Patient Bed module in due time 20 secs: Server sending alert to person's wrist band module for alcohol rub

For Tom (TagId - F93F3959): Alcohol Rub successful at Patient Bed module:server sending alert_off to person's wrist band module

For Marry (TagId - F07F4FD3): Person detected near Patient Bed module and Tag detected at Patient Bed module in due time 20 secs: Server sending alert to person's wrist band module for alcohol rub

For Marry (TagId - F07F4FD3): Alcohol Rub successful

For (TagId -): Warning:Alert:Health Care Person leaving patient bed area after procedure.Staff need to do alcohol rub before leaving chamber or continue with other work. Server sending alert to person's wrist band module

For Tom (TagId - F93F3959): Person detected near Patient Bed module and Tag detected at Patient Bed module in due time 20 secs: Server sending alert to person's wrist band module for alcohol rub

For Tom (TagId - F93F3959): Warning: Person detected near Patient Bed module but Hand not detected at Patient Bed module in due time 20 secs: check failed, send alert to person's wrist band module

For Harry (TagId - 8EA38BB): Person detected near Patient Bed module and Tag detected at Patient Bed module in due time 20 secs: Server sending alert to person's wrist band module for alcohol rub

For Harry (TagId - 8EA38BB): Alcohol Rub successful at Patient Bed module:server sending alert_off to person's wr

For John (TagId - 9964159): Person detected near Patient Bed module and Tag detected at Patient Bed module in due time 20 secs: Server sending alert to person's wrist band module for alcohol rub

For John (TagId - 9964159): Warning: Person detected near Patient Bed module but Hand not detected at Patient Bed module in due time 20 secs: check failed, send alert to person's wrist band module

For (TagId -): Warning:check_fail: Person detected near Patient Bed module but Tag not detected at Patient Bed module in due time 20 secs

For Marry (TagId - F07F4FD3): Person detected near Patient Bed module and Tag detected at Patient Bed module in due time 20 secs: Server sending alert to person's wrist band module for alcohol rub

For Marry (TagId - F07F4FD3): Alcohol Rub successful at Patient Bed module:server sending alert_off to person's wrist band module

For (TagId -): Warning:check_fail: Person detected near Patient Bed module but Tag not detected at Patient Bed module in due time 20 secs

Clear Screen Compliance Report Client Web Updates

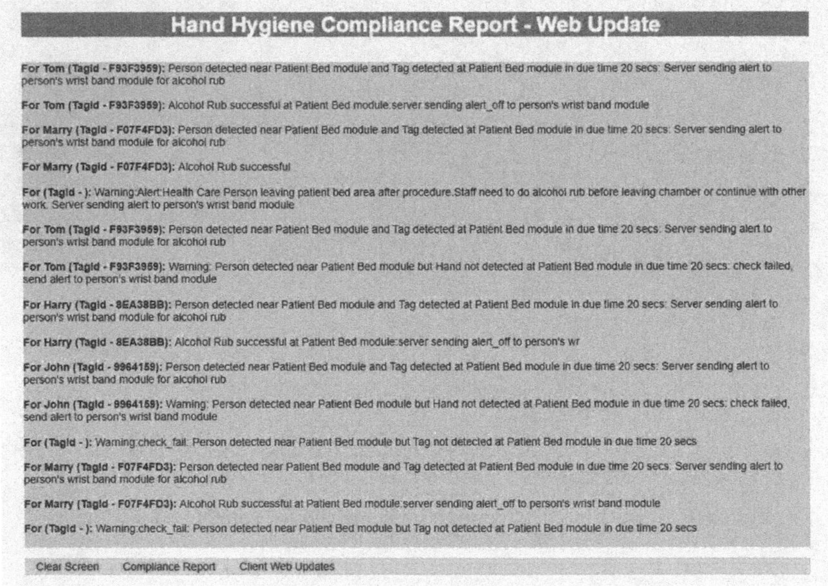

Fig. 11. Realtime staff hand hygiene compliance action report on server

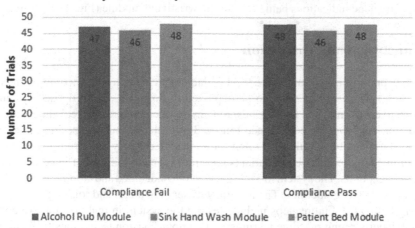

Fig. 12. Testing results graph (Realtime compliance messages reported to the central server, 50 trials conducted)

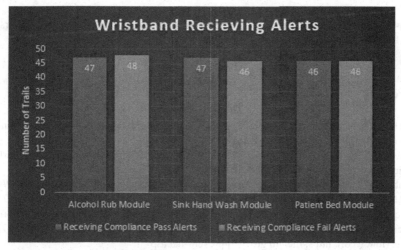

Fig. 13. Wristband module testing results (Realtime compliance alerts reported to staff, 50 trials conducted)

Multi-level testing and data analysis was done for the system. First, each sensor was tested independently to ensure proper operation, and then each sensor was calibrated for the correct required range. Next, each module was tested for their independent functionalities by stipulating hospital-like events for all possible scenarios. Lastly, all the modules were connected, and complete system level data was collected to see if they all work cohesively. The system collects and uploads hand hygiene compliance data followed by health care workers on a central server. The health care workers receive real time alerts. The whole system was tested 50 times for compliance reporting to the server and hand hygiene indications being sent to the wristband module (Fig. 12 and Fig. 13).

5   Conclusion and Discussion

This paper presents a novel automated and scalable system for hospital acquired infection monitoring and prevention that can be deployed in hospitals. The proposed system is designed to easily integrate into the hospital environment, operate in real-time and send updates for hand hygiene compliance followed by staff without hampering the normal medical staff activity. Result shows that the system accurately and promptly alerts the staff before and after contacting patients and tracks the compliance of hand hygiene at various checkpoints. The system module checks follow detailed and complete procedures stipulated in WHO guidelines. The accuracy of sending alerts and logging compliance is a very high 94%. Future steps include testing the system in real hospitals to collect data and incorporating machine learning and computer vision techniques to accurately and efficiently track specific hand cleaning motions as per the guidelines of WHO. With slight modifications, this system can be used in the food industry, schools, and even homes to track and monitor hand hygiene.

Acknowledgments. The author would like to acknowledge and thank teachers, mentors, and parents for their support during the research.

References

1. WHO - The burden of healthcare-associated infection worldwide. https://www.who.int/gpsc/country_work/burden_hcai/en/. Accessed 23 Apr 2013
2. Davis, Jr. C.P.: Definition of Nosocomial. https://www.medicinenet.com/script/main/art.asp?articlekey=4590. Accessed 12 Dec 2018
3. Stubblefield, H.: Hospital-acquired infection: definition and patient education. https://www.healthline.com/health/hospital-acquired-nosocomial-infections. Accessed 6 June 2017
4. Haque, M., Sartelli, M., McKimm, J., Abu Bakar, M.: Health care-associated infections - an overview. Infect. Drug Resist. **11**, 2321–2333 (2018). https://doi.org/10.2147/IDR.S177247
5. Overview. (n.d.). https://health.gov/hcq/prevent-hai.asp
6. Posters & leaflets. https://www.who.int/gpsc/5may/resources/posters/en/. Accessed 8 June 2011
7. How to handwash and hand rub. https://www.who.int/gpsc/tools/GPSC-HandRub-Wash.pdf
8. Clean hands protect against infection. https://www.who.int/gpsc/clean_hands_protection/en/. Accessed 8 June 2011
9. UV Disinfection with Pulsed Xenon to Combat HAIs. (n.d.). Retrieved from https://www.xenex.com/
10. Yarbrough, R., et al.: Efficacy of nGageTM by Proventix, an electronic hand hygiene surveillance and Feedback monitoring device, against healthcare associated infections. http://www.proventix.com
11. Hand hygiene compliance monitoring systems. (n.d.). https://www.halyardhealth.com/solutions/infectionprevention/compliance-monitoring.aspx

12. Debmed - Hand Hygiene Monitoring > Healthcare Hand Hygiene Compliance. https://www. debmed.com. Accessed 7 June 2019
13. Healthcare-associated infections. https://www.cdc.gov/hai. Accessed 4 Mar 2016
14. Inweregbu, K., Pittard, A., Dave, J.: Nosocomial infections. Oxford J. 5(1), 14–17 (2005). https://academic.oup.com/bjaed/article/5/1/14/339870
15. Shhedi, Z.A., Moldoveanu, A., Moldoveanu, F.: Traditional & ICT solutions for preventing the hospital acquired infection, pp. 867–873 (2015)
16. McCalla, S., Reilly, M., Thomas, R., McSpedon-Rai, D., McMahon, L.A., Palumbo, M.: An automated hand hygiene compliance system is associated with decreased rates of health care-associated infections. Am. J. Infect. Control 46(12) (2018)

A Novel Approach for Detecting IoT Botnet Using Balanced Network Traffic Attributes

M. Shobana(✉)🆔 and Sugumaran Poonkuzhali🆔

Rajalakshmi Engineering College, Chennai, TamilNadu, India
poonkuzhali.s@rajalakshmi.edu.in

Abstract. Over the evolution of internet technology give rise to the intelligence among tiny objects so called IoT devices. At the same time, this scenario increases the intrusion of malwares into the IoT devices e.g. Mirai, bashlite. Researchers have proposed many framework by addressing this issue. But the framework of those proposed work which are tested using Real time traffic of IoT devices is very fewer. In this work, the class imbalance problem has been identified in the BoT-IoT dataset. This problem is overcome by the random over sampling technique. Then this resultant dataset is further classified into normal and attack traffic using three machine learning classifier such as Support Vector Machine, Naive Bayes, and Decision Tree (j48) and deep learning technique such as deep neural network. The performance of the security model is evaluated using quality metrics like Precision, Recall, F-measure, Response time and ROC to identify the best classifier which is apt to detect malware in IoT devices.

Keywords: IoT Botnet · Sampling · IoT security · IoT malware

1 Introduction

In the era of recent technologies, IoT (Internet of Things) is playing a huge role in digital world. In 2016 the overall used IoT devices is almost 9 million and by in the year of 2025 the count of IoT device will be increased up to trillion annually [3]. Since the tremendous increase in short span of time leads IoT vendors to concentrate less on security for IoT devices rather that they concentrate well on developing the IoT devices across various fields such as medical, defense, Industry etc. Attackers makes use of this situation to pollute the smartness of the IoT device. On the last decades, there are many IoT devices were affected by malwares at very high speed. Out of these malwares, most of them are botnets such as brickerbot, Mirai, hajime, Leet etc., [20] and few of them are worms and trojan such as Hijame and Darlloz. Those IoT Botnets are specially designed to launch flood of DDoS attack from the IoT devices to many of the connected network and it is a great risk to stop those attack [4]. Some incidents have experienced PDoS (Permanent Denial of Service) attack, it is one

© Springer Nature Switzerland AG 2021
H. Hacid et al. (Eds.): ICSOC 2020 Workshops, LNCS 12632, pp. 534–548, 2021.
https://doi.org/10.1007/978-3-030-76352-7_48

of the worst scenario which makes the IoT device to shut down permanently and it cannot be recovered to earlier stage. Additionally, these malwares capable of launching many kinds of attacks like keylogging, Information Theft, DoS, OS Fingerprinting and Service scan. The traditional ways to mitigate those kind of malwares are signature detection, honeypot detection and behaviour or anomaly based detection. But these kind of approach are not comfortable to the IoT devices because of its resource constraint nature. Some of the researchers have proposed their work to defend against DDoS attack in IoT platform using SDN and this approach faces some hurdles in implementation stage. Recently many of the researchers have contribute their work towards enhancement of security for IoT devices using machine learning [2] and deep learning architectures such as LSTM, RNN, FNN, SNN, ANN etc. In this regard, the big challenge is getting real time network traffic of the IoT device. The main contribution of this work is summarized below

1. Class imbalance problem is identified in the dataset of IoT's network traffic.
2. Both oversampling and undersampling technique has been carried out to balance the IoT dataset.
3. Machine learning techniques such as SVM, Decision tree (j48) and Naïve Bayes as well as deep learning techniques were used to classify the normal and attack traffic generated by the IoT devices.

2   Related Work

The architecture of IoT has four layers viz perceptual layer, network layer, Middleware layer and application layer. The most reliable way of providing security to IoT device is to deploy the model at network layer. In this section, the security framework for IoT devices in network layer designed using artificial intelligence techniques such as machine learning and deep learning techniques has been discussed here briefly. To do this, an efficient real time dataset is required to test the designed model. Due to the non-availability of real time IoT network traffic, most of the below listed existing work is liable on traditional dataset like NSL-KDD, KDD CUP 99 and UNSW-15. Recently new real time IoT traffic is introduced such as N-BaIoT (2017) and BoT IoT (2018) for further realistic research in this field.

2.1   Using NSL-KDD Dataset

Pajouh et al. [19] implemented an intrusion detection system which consists of two layer feature reduction followed by two layer classification process. For feature reduction process component analysis and linear discriminate analysis technique was used. For classification, Naïve Bayes and Certainty Factor version of K-Nearest Neighbor was deployed. This work concentrate only on U2R (user to root) and R2L (Remote to local) attacks.

2.2 Using UNSW-15 Dataset

Nour Moustafa et al. [17] proposed a framework to detect IoT botnet with adaboost ensemble with three major classifier namely Decision Tree (DT), Naive Bayes (NB) and Artificial Neural Network (ANN). Author utilized both NIMS and UNSW-15 dataset to test the model. Nickolaos et al. [10] created a model using Decision Tree C4.5, Association Rule Mining, Artificial Neural Network and Naive Bayes to detect IoT Botnets using UNSW dataset and for feature selection information gain has been applied". Timčenko et al. [21] developed a model for mitigating IoT malware using three classifiers such as SVM, bagging and boosting algorithm".

2.3 Using RedIRIS

Lopez-Martin et al. [14] proposed an IoT network traffic classification model by combining both CNN and RNN. This work is especially for monitoring and managing the network activity of IoT devices through its traffic itself.

2.4 Using N-BaIoT Dataset

Meidan et al. [16] proposed a system for mitigating the infected IoT devices by its network traffic using Deep autoencoders. Authors trained his model using normal IoT traffic and the model shows its difference by mean square error while the arrival abnormal traffic to that particular device. Chawathe et al. [6] proposed a model for detecting IoT botnets using monitoring process of network activity. Author used simple classifiers such as ZeroR, oneR, JRip, J48, PART and Random Forest for attribute reduction. Nomm et al. [18] presented a common unsupervised model for all sorts of IoT device. Author have concentrate more on dimensionality reduction of the dataset. The approaches used for feature reduction are Hopkins statistics, Entropy and variance based feature reduction methods. Then the classifiers used for the classification process are SVM and Isolation forest.

2.5 Using BoT-IoT Dataset

Ibitoye et al. [9] presented an intrusion detection system for IoT network using two deep architecture namely feed forward neural network (FNN) and Self normalizing neural network (SNN). In this work they have proved that SNN outperforms FNN with the help of quality measures.

2.6 Self-generated Network Traffic

Domb et al. [7] suggested a detection framework for the intruders over the IoT environment. This approach deals with the analysis of traffic data of sensors using random forest. Luo et al. [15] proposed IoT Botnet mitigating technique in Wireless sensor network. Since WSN is stands as a backbone for IoT platform.

His model is deployed in two part i.e. one module in IoT cloud and another one in sensor platform. The malware detecting engine is designed using Autoencoders neural network. In the case of IoT security, traditional datasets may not be suitable because it does not hold any recent attacks happens in IoT platform to train and test the proposed model whereas N-BaIoT and BoTIoT dataset holds the different types of real IoT attacks. At the end of this survey, it can conclude that only few work is tested on these two real time dataset.

3   Motivation of the Work

The BoT-IoT dataset comprises of maximum attack instance and minimum normal instance, this case can increase the accuracy towards detecting the traffic of malware attack rather than detecting the normal traffic as normal. There is a chance of detecting normal instance as abnormal and it will leads to the IoT device in an idle state without receiving any authorized command from its server. Moreover minority class will not be detected and it does not reflect to the accuracy of the classifier. This kind of problem is said to be class imbalance problem [13].

4   Methods and Materials

4.1   Dataset Description

The BoT-IoT dataset [11] generated under the IoT environment of Cyber range lab of UNSW Canberra Cyber. The network traffic dataset is broadly classified into two types such as normal and attack. The attack instances comprises of four types of attack namely DDoS, DoS, Theft and Keylogging. This type of attack again has been divided into subcategory such as DDoS TCP, DDoS UDP, DDoS HTTP, DoS HTTP, DoS UDP, DoS TCP, OS Fingerprint, service scan, Data exfiltration and Key logging. Overall dataset has 9543 normal instances and 73360900 abnormal instances.

4.2   Correlation Based Feature Selection

Correlation based method is used to measure the linear dependence among the set of existing feature set. This method allocates values between -1 to $+1$ for each features. This value indicates the significance and type of correlation among feature set. The features having low correlation is considered to be more significance.

4.3   Support Vector Machine

SVM is a most commonly used learning algorithm for classification. This algorithm plots its n-features in n-dimensional space. Then the algorithm will create few hyper plane which cuts the data points into two classes. Among the few hyper plane one of the best plane is selected on the basis of classification.

4.4   Decision Tree (j48)

Decision tree is a branch like structure algorithm. Here each attribute acts as a node and the relationship between the nodes acts as branches. Each branches are divided based on some decision or condition. This algorithm works well for the classification task. In this case, this algorithm is utilized here to classify malicious clusters from the normal clusters.

4.5   Naive Bayes

Naive Bayes algorithm works based on the Bayes theorem. It states that any two features classified is independent of each other. This classifier is well suited for high dimensional dataset. This algorithm performs the classification based on its previous occurrence and it is termed as prior probability.

4.6   Deep Neural Network

In this work, a lightweight deep learning architecture, namely the deep neural network (DNN), is used to allow the deep intrusion detection system to capture the IoT botnet as it evolves over the network. Since the IoT network is said to be very resource constrained, the architecture of the deep neural network is designed with only three hidden layers, and the number of iteration/epoch is 3. A neural network stands as a basement for deep learning techniques because it works like the neurons of the human brain. In this regard, the deep neural network is one of the versions of the neural network that always operate on one input layer, one output layer and one more hidden layer. Each layer comprises of mathematical functions to process input data. The values of the hyperparameters are chosen after performing several experiments, and they are explored in the Table 1

Table 1. Hyperparameters values for deep neural network

Hyperparameters	Values
Epochs	3
Hidden layers	3
Batch size	300
Activation function	Hidden layer-relu function output layer-sigmoid function
Neurons	6 neuron-input layer 1024 -1st hidden layer 768-2nd hidden layer 512-3rd hidden layer 1-output layer
Optimizer	Adam
Metrics	Accuracy
Loss function	Binary cross entropy

5   System Architecture and Overview

Figure 1 shows the phases involved in the architecture of the proposed framework. In this section, the operations performed on each phase is explained in detailed manner.

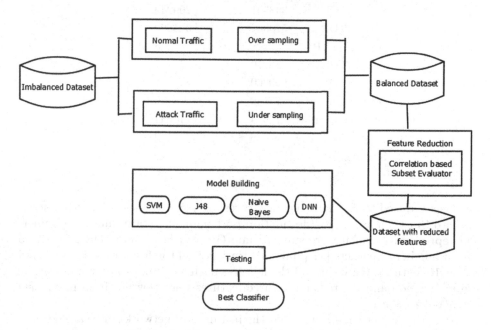

Fig. 1. Architecture of the proposed model

5.1   Sampling Phase

In this phase, class imbalance problem in the dataset has been handled without disturbing the significance of the dataset. Sampling technique is classified into two ways such as oversampling and undersampling. Oversampling technique is used to increase number of instance of the minority class in the dataset. Undersampling technique is quite opposite to oversampling whereas it decreases the instance of the majority class. In this case, both sampling technique done in parallel manner, the given attack instance is decreased as well as the normal instance is increased at reasonable rate. After the total normal instances is increased upto 999722 as well as malware instances is decreased to 100288. The detailed count of sampled instances is described in Table 2.

5.2   Feature Reduction Phase

Initially this dataset has 45 network attributes as features. Since the deployment of intrusion detection in network layer requires less space it is necessary to implement feature reduction technique. By using the correlation based attribute evaluator along with greedy hill climbing search method, the number of attributes

Table 2. Hyperparameters values for deep learning model

Class		Number of instances	
		Before sampling	After sampling
DDoS	HTTP	989	600
	TCP	348751	209249
	UDP	576884	345592
DoS	HTTP	8	3
	TCP	10	7
	UDP	650260	389968
Service scan		73168	44176
OS fingerprinting		17914	10641
Data exfiltration		6	3
Keylogging		73	49

has been reduced to 6. The reduced features are pkseqid (Row identifier), stime (Record Start Time), dport (destination port number), state_ number (Numerical representation of feature state), ltime (Record last time), TnP_ PerProto (Total number of packets per protocol). Here not all the features were extracted out of 10 features. Only three of the features such as stime, state number, ltime are in common and the remaining three features were new not from previously extracted 10 features.

After the reduction of feature set, the instances of network traffic are normalized by Min-Max normalization technique to ease the process of classification.

$$N' = \frac{N - min}{max - min} (max' - min') + min' \tag{1}$$

Using Eq. (1) min-max normalization has been calculated for all values in the dataset to scale in the range of [0 1]. Whereas max, min are considered to be initial value and the variables max', min' are considered as newly calculated instance.

5.3   Classification Phase

In this stage, three most familiar machine learning classifiers (Support vector Machine, Decision Tree (j48) and Naive Bayes) and one deep learning network (Deep neural network) for intrusion detection system was used. In order to test the efficiency of these algorithm, it is used in three different ways, they are given below:

1. All the instance is broadly classified into normal/Attack (Binary Classification)
2. All the instance were classified into DDoS, DoS, Theft, Reconssaince, and Normal (Mulitclass Classification-category wise)

3. All the instance were classified into DDoS TCP, DDoS UDP, DDoS HTTP, DoS HTTP, DoS UDP, DoS TCP, OS Fingerprint, service scan, Data exfiltration, Key logging and Normal (Mulitclass Classification-subcategory wise) After these training and testing phase, among these three classifier one of the best classifier will be chosen based on its performance on IoT network data.

6   Results and Discussion

The proposed model was implemented on a Core i3 Laptop with 2.30 GHz CPU and 4 GB RAM using Rstudio version 3.5.1 software environment. In the original dataset, the whole generated traffic is split into five bins for the case of easy accessibility. Among the five bins 2 million of records has been chosen in such a way that records contains all varieties of attack instances then it is combines into single network traffic dataset. On that combined dataset the sampling technique is done using ROSE (Random over Sampling Examples) package in R. In this package both undersampling and oversampling has been done at the same time. After sampling 6 features is extracted from the balanced dataset for classification.

6.1   Evaluation of Classifiers

The performance of three classifiers has been calculated using the following metrics.

True Positive (TP) denotes that attack instances is correctly predicted as attack.

True Negative (TN) denotes that normal instances is correctly predicted as normal.

False Positive (FP) denotes that normal behavior is wrongly detected as attack.

False Negative (FN) denotes that attack behavior is wrongly detected as normal.

Accuracy. The accuracy of the classifier can be defined as the total number of correctly predicted instance in the given dataset is shown in Eq. (2)

$$Accuracy = \frac{TP + TN}{TP + TN + FP + FN} \tag{2}$$

Precision. The Precision can be termed as the ability of the classifier should correctly label the malicious traffic as attack is shown in Eq. (3)

$$Precision = \frac{TP}{TP + FP} \tag{3}$$

Recall or Detection Rate. Recall or Detection rate can be defined as the number of correctly detected attack traffic connections is shown in Eq. (4)

$$Recall = \frac{TP}{TP + FN} \tag{4}$$

F-Measure. F-measure can be defined as the weighted harmonic mean of precision and recall is shown in Eq. (5)

$$F - measure = 2 * \frac{precision * recall}{precision + recall} \tag{5}$$

Response Time. The total time taken by the classifier to build the model accurately.

Error Rate. The error rate of the classifier is the ratio between total numbers of incorrectly classified instance to the total number of instances is shown in Eq. (6)

$$Error\,rate = \frac{FP + FN}{TP + TN + FP + FN} \tag{6}$$

False Alarm Rate. False alarm rate can be defined as the number of normal traffic is incorrectly detected as malicious traffic is shown in Eq. (7)

$$False\,alarm\,rate = \frac{FP}{TN + FP} \tag{7}$$

Table 3 illustrates the quality metrics of the classifiers for three kinds of classification. From this analysis it is proven that Decision tree (j48) outperforms than the other two classifiers such as SVM and Naïve Bayes for three types of classification in terms of its accuracy, Response time and false alarm rate. Naive Bayes and SVM stands in same position in their performance. While using full features also high performance is achieved in terms of accuracy but when comparing false alarm rate and response time is very high, it shows that minority class (normal) is not detected by the classifier due to the class imbalance problem in the original dataset. By using 10 features only decision tree alone is capable to perform well but in the case of 6 reduced features all the three classifiers able to perform well. The response time taken by SVM to build model is quite higher than the other two classifiers. It is also shown that the false alarm rate for 10 features is lies between 0 to 0.006 whereas the reduced features produce the false alarm rate from 0 to 0.001. So it is very easy to conclude that 6 effective feature capable to perform well for the three classifiers at low false alarm rate.

The values taken from Fig. 2 used to calculate the error rate for this work. The value of error rate is comparatively low for this work when compared to the previous work [13]. This improvement is achieved by the sampling technique to overcome the problem of class imbalance. This enhances the higher detection rate of minority class (i.e.) normal traffic instances given in the BoTIot dataset. The existing of class imbalance problem tends to detect the normal traffic as malicious one, this will be the serious in IoT kind of environment and it is overcome in this work using sampling techniques. Since the feature extraction phase was done after the sampling process, it extracted only minimum number of features.

Table 3. Performance evaluation for different sets of features

Two class classification

Method	6 features			10 features			Full features		
	Accuracy	Response time	False alarm rate	Accuracy	Response time	False alarm rate	Accuracy	Response time	False alarm rate
SVM	99.62	6325.45	0.002	99.62	3631.41	0.001	0.987	10345.55	0.56
NAÏVE BAYES	99.78	51.14	0.004	99.90	3.23	0.001	0.988	12345.78	0.54
DECISION TREE	100	69.9	0	100	14.56	0	0.990	13547.89	0.75
Category wise classification									
SVM	99.87	3791.23	0	99.92	9550.71	0	0.988	15768.66	0.60
NAÏVE BAYES	99.66	29.96	0	99.90	6.33	0.001	0.999	15887.4	0.75
DECISION TREE	100	38.02	0.006	99.99	64.47	0	0.999	16789.55	0.56
Subcategory wise classification									
SVM	99.85 4460.60	0	99.764	7373.01	0.001	0.998	14767.89	0.56	
NAÏVE BAYES	99.66	5.85	0.006	99.9003	12.12	0.001	0.998	14899.78	0.71
DECISION TREE	100	183.88	0	100	48.74	0	0.990	15789.99	0.89

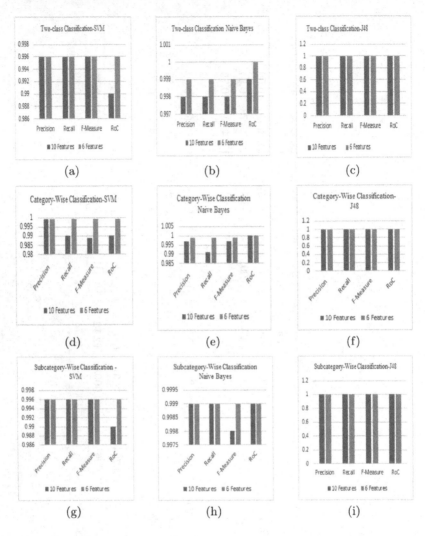

Fig. 2. Performance analysis of three classifiers for 10 features vs 6 features

Table 4. Performance evaluation of deep neural network

Type of classification	Accuracy	Precision	Recall	F1 measure
Binary classification	99.99	0.998	0.997	99.9
Category classification	99.99	0.999	0.999	0.999
Sub category classification	99.88	0.998	0.998	0.998

Table 4 describes about the performance of the deep neural network in terms of accuracy, precision, recall and F1-measure. By exploring these values it shows that due to sampling traditional machine learning techniques becomes overfit model, so by deploying deep learning technique this problem has been avoided. Table 5 describes about the additional metrics used for evaluating the multiclass classification.

Table 5. Detailed evaluation of deep neural network for multiclass classification

Class	TPR	FPR	FNR	FDR	TNR
Normal	1.000	0.999	0.0001	0.0001	0.00926
DDoS	1.000	0.999	0.0001	0.0001	0.0001
DoS	0.999	1.000	0.0001	0.0001	0.0001
Theft	0.992	1.000	0.0001	0.0001	0.0001
Reconssaince	0.999	1.000	0.0001	0.0001	0.0001

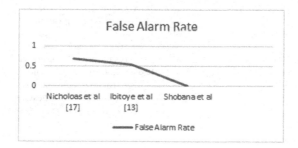

Fig. 3. Comparison of false alarm rate of various methods using BoTIoT

In Table 6 compares the accuracy of the proposed work with existing work and it proves that the accuracy value of this work is predominately high and the error rate is comparatively low when compares to the existing solution. This proves the efficiency of the designed solution is reliable for the IoT environment. In Fig. 3, the false alarm rate of existing methods using imbalanced dataset is compared with this proposed work designed using balanced dataset. It shows the importance of sampling technique by producing low false alarm rate than the existing model.

Table 6. Comparative summary of proposed work with existing work

	Author and Year	Methodology	Accuracy (%)
Existing work	Moustafa et al. [17] and 2018	Decision Tree	95.32
		Naïve Bayes	91.17
		Artificial Neural Network	92.61
		Ensemble (Adaboost)	99.54
	Koroniotis et al. [10] and 2017	ARM	86.45
		Decision Tree	93.23
		Naïve Bayes	72.73
		Artificial Neural Network	63.97
	Ibitoye et al. [9] and 2019	Feedforward Neural Network	95.10
	Domb et al. [21] and 2017	Random Forest	95
	Nomm et al. [18] and 2018	Isolation Forest	86.18
		Support Vector Machine	95.65
	Hodo et al. [8] and 2016	Artificial Neural Network	99.4
	Pajouh et al. [19] and 2016	k-nearest neighbor algorithm with Naive Bayes	84.82
	Alghuried [1] and 2017	Decision Tree	97
	Bezerra et al. [5] and 2018	One class support vector machine	96.65 to 99.33
	Kumar et al. [12] and 2019	Random Forest	88.8
		k-nearest neighbor algorithm	94.44
		Gaussian Naïve Bayes	77.78
Proposed work	Shobana et al.	SVM (Binary Classification)	99.62
		SVM (Categorywise Classification)	99.92
		SVM (Sub Categorywise Classification)	99.76
		Naive bayes (Binary Classification)	99.90
		Naive bayes (Category Classification)	99.90
		Naive bayes (SubCategory Classification)	99.90
		Decision Tree (Binary Classification)	100
		Decision Tree (Category Classification)	99.99
		Decision Tree (Subcategory Classification)	100

7   Conclusion

In this work, the class imbalance problem has been identified in the real time dataset of IoT device namely Bot IoT dataset. Furthermore this problem has been rectified using sampling techniques in an efficient manner. The ability of the balanced dataset has been tested by using three major machine learning classifiers. The major role of this sampling technique is to acquire the minority class (normal instances) as well as majority class (attack instances) at equivalent success rate. Out of these three classifiers decision tree algorithm achieved 100% accuracy rate in the balanced dataset and it is higher than the accuracy of the existing solution and it is seems to be overfit. In this regard, deep neural network has been deployed and it is well performed than the machine learning classifiers. Moreover the performance of this algorithm is measured using precision, Recall, F-measure and Response time. In future this work is further improved using deep learning architecture with some advanced sampling technique.

References

1. Alghuried, A.: A model for anomalies detection in Internet of Things (IoT) using inverse weight clustering and decision tree (2017)
2. Andročec, D., Vrček, N.: Machine learning for the Internet of Things security: a systematic. In: 13th International Conference on Software Technologies, vol. 4120, p. 97060 (2018). https://doi.org/10.5220/00068
3. Angrishi, K.: Turning Internet of Things (IoT) into internet of vulnerabilities (IoV): IoT botnets. arXiv preprint arXiv:1702.03681 (2017)
4. Bertino, E., Islam, N.: Botnets and Internet of Things security. Computer **50**(2), 76–79 (2017)
5. Bezerra, V.H., da Costa, V.G.T., Junior, S.B., Miani, R.S., Zarpelao, B.B.: One-class classification to detect botnets in IoT devices. In: Anais Principais do XVIII Simpósio Brasileiro em Segurança da Informação e de Sistemas Computacionais, pp. 43–56. SBC (2018)
6. Chawathe, S.S.: Monitoring IoT networks for botnet activity. In: 2018 IEEE 17th International Symposium on Network Computing and Applications (NCA), pp. 1–8. IEEE (2018)
7. Domb, M., Bonchek-Dokow, E., Leshem, G.: Lightweight adaptive random-forest for IoT rule generation and execution. J. Inf. Secur. Appl. **34**, 218–224 (2017)
8. Hodo, E., et al.: Threat analysis of IoT networks using artificial neural network intrusion detection system. In: 2016 International Symposium on Networks, Computers and Communications (ISNCC), pp. 1–6. IEEE (2016)
9. Ibitoye, O., Shafiq, O., Matrawy, A.: Analyzing adversarial attacks against deep learning for intrusion detection in IoT networks. In: 2019 IEEE Global Communications Conference (GLOBECOM), pp. 1–6. IEEE (2019)
10. Koroniotis, N., Moustafa, N., Sitnikova, E., Slay, J.: Towards developing network forensic mechanism for botnet activities in the IoT based on machine learning techniques. In: Hu, J., Khalil, I., Tari, Z., Wen, S. (eds.) MONAMI 2017. LNICST, vol. 235, pp. 30–44. Springer, Cham (2018). https://doi.org/10.1007/978-3-319-90775-8_3
11. Koroniotis, N., Moustafa, N., Sitnikova, E., Turnbull, B.: Towards the development of realistic botnet dataset in the Internet of Things for network forensic analytics: Bot-IoT dataset. Future Gener. Comput. Syst. **100**, 779–796 (2019)
12. Kumar, A., Lim, T.J.: Edima: Early detection of IoT malware network activity using machine learning techniques. In: 2019 IEEE 5th World Forum on Internet of Things (WF-IoT), pp. 289–294. IEEE (2019)
13. Longadge, R., Dongre, S.: Class imbalance problem in data mining review. arXiv preprint arXiv:1305.1707 (2013)
14. Lopez-Martin, M., Carro, B., Sanchez-Esguevillas, A., Lloret, J.: Network traffic classifier with convolutional and recurrent neural networks for Internet of Things. IEEE Access **5**, 18042–18050 (2017)
15. Luo, T., Nagarajan, S.G.: Distributed anomaly detection using autoencoder neural networks in WSN for IoT. In: 2018 IEEE International Conference on Communications (ICC), pp. 1–6. IEEE (2018)
16. Meidan, Y., et al.: N-BaIoT-network-based detection of IoT botnet attacks using deep autoencoders. IEEE Pervasive Comput. **17**(3), 12–22 (2018)
17. Moustafa, N., Turnbull, B., Choo, K.K.R.: An ensemble intrusion detection technique based on proposed statistical flow features for protecting network traffic of Internet of Things. IEEE Internet Things J. **6**(3), 4815–4830 (2018)

18. Nõmm, S., Bahşi, H.: Unsupervised anomaly based botnet detection in IoT networks. In: 2018 17th IEEE International Conference on Machine Learning and Applications (ICMLA), pp. 1048–1053. IEEE (2018)
19. Pajouh, H.H., Javidan, R., Khayami, R., Ali, D., Choo, K.K.R.: A two-layer dimension reduction and two-tier classification model for anomaly-based intrusion detection in IoT backbone networks. IEEE Trans. Emerg. Topics Comput. **7**, 314–323 (2016)
20. Shobana, M., Rathi, S.: IoT malware: an analysis of IoT device hijacking (2018)
21. Timčenko, V., Gajin, S.: Machine learning based network anomaly detection for IoT environments. In: ICIST-2018 Conference (2018)

KMeans Kernel-Learning Based AI-IoT Framework for Plant Leaf Disease Detection

Youssouph Gueye$^{(\boxtimes)}$ and Maïssa Mbaye$^{(\boxtimes)}$ 🄳

LANI (Laboratoire D'Analyse Numérique et Informatique),
Université Gaston Berger de Saint-Louis, Saint-Louis, Senegal
{gueye.youssouph1,maissa.mbaye}@ugb.edu.sn

Abstract. Development of IoT based solutions in agriculture is changing the sector with Smart Agriculture. Plant Leaf Disease Detection (PLDD) using ICT is one of the most active and challenging research areas because of its potential in the food security topic. Some of current solutions based on AI/Machine learning techniques (E.g. KNN, CNN) are very efficient. However, deploying them in the context of Africa will be challenging knowing that computation resources, connectivity to data centers, and electrical power supply won't be guaranteed. In this paper we propose an AI-IoT Framework based on KMeans Kernel Learning to build Artificial Intelligence services on Core Network and deploy it to Edge AI-IoT Network. AI-Service Segment selects leaves images that have representative characteristics of diseased leaves (Kernel-Images), uses KMeans machine learning algorithm to build clusters of Kernel-Images so that diseased regions are contained cluster. We call the resulting models KMeans Kernel Models. Main outcome of our proposal is designing a low-computation and economic Edge AI-AoT Framework as efficient as sophisticated methods. We have evaluated that our system is efficient and provides a very good result with a rate of 96% accuracy with a low number of training images. Our proposed framework reduces the need for large training datasets to be efficient (in comparison to KNN/SVM and CNN) and learned models are embeddable in IoT devices near the plants.

Keywords: AI-IoT · KMeans Kernel Learning · Edge-AI · Machine learning · Plant Leaf Disease Detection

1 Introduction

Plant diseases are one of the major sources of yield and quality declining for agriculture in the world including Africa. Considering this, early diagnosis of plant diseases is an important task against starvation. Traditional approach for detecting diseased plants consists in the naked eye observation by experts [7]. This method of visual assessment, being a subjective task, can lead to errors due to bias (psychological and cognitive phenomena) and optical illusions.

© Springer Nature Switzerland AG 2021
H. Hacid et al. (Eds.): ICSOC 2020 Workshops, LNCS 12632, pp. 549–563, 2021.
https://doi.org/10.1007/978-3-030-76352-7_49

On the other hand, laboratory tests such as molecular, immunological, or recommended approaches on the culture of pathogens are often time consuming and not responding in a timely manner.

Plant diseases most often take the form of an alteration of the plant which modifies or interrupts its vital functions such as photosynthesis, transpiration, pollination, fertilization, germination, etc. Disease manifestations are usually seen on the leaves, fruits and stems of the plant. These diseases are caused by pathogenic organisms (fungi, bacteria and viruses) or by unfavorable factors resulting in harmful changes in the shape, development or the whole of the plant thus causing partial damage or death of the plant [7].

Majority of existing digital methods so far are based on digital image processing. Most methods are based on the plant leaves aspects and color transformation due to the disease. Advantage of leaves-based techniques is that leaves are almost observable while flowers and fruits depend on the period of the season.

Artificial Intelligence (AI) techniques are intensively used for leaf plant diseases such as K-Nearest Neighbor (KNN); Artificial Neural Networks (ANN), Probabilistic Neural Network (PNN), Genetic Algorithm, Support Vector Machine, Linear Regression Analysis (LDA) Fuzzy logic and Deep Learning which is largely used nowadays [7].

However, these AI techniques classifiers in practice will face challenges in African agriculture's context. For instance, most of african computation resources, if they exist, are in town and are poorly connected to rural areas. So, having AI Solution that needs CPU, ubiquitous connectivity and large datasets wouldn't be successfully deployed. Operational challenges will be connectivity that doesn't exist in farms (at most GSM network is the most widespread) and power energy supply is not guaranteed. However, since smartphones are very widespread lastly, a lightweight and decentralized solution will me more adapted to this context.

In this paper we propose an IoT based Edge-AI framework that is based on KMeans Kernel method to provide an Core-AI service for Plant Leaf Diseases Detection. That framework has Edge AI-IoT Network Segment and Core AI Service Segment. These two segments are connected to share data and machine learning models. Our goal is to design lightweight AI that can be deployed in IoT devices to detect plant leaf diseases. The Machine Learning K-Means Kernel Learning is used for clustering methods on Kernel Images. And then these models are used for disease detection on plant leaf images. Main outcomes of our proposal are a low-computation and economic Edge AI framework that is as efficient as sophisticated methods such as CNN, SVM, KNN... This framework reduces the need for large training datasets to be efficient and the learned model is embeddable in IoT devices near the plants in rural area.

The paper is organized as follows: Section 2 presents related works on the AI-based Plant Disease detection researchs while Sect. 3 gives a general overview of plant diseases detection methods. In Sect. 4 we present the general architecture of the frameworks, the KMean Kernel Models, Kernel Image Selection algorithm and evaluation. Finally we conclude in Sect. 5.

2   Related Works

In last decades, several techniques has been developed for Plant Leaf Diseases Detection (PLDD). For Artificial Intelligence (AI)/Machine Learning based techniques, the most intensively used are classifiers such as K-Nearest Neighbors (KNN), Radial Basis Function (RBF), Artificial Neural Networks (ANN), Probabilistic Neural Networks (PNN), and the Back-Propagation Network (BPN), Support Vector Machines (SVM), KMeans Clustering, and SGDM [7].

Deep Learning related techniques (ANN, CNN, DNN, ...) show excellent results in the topic of plant leaf disease detection and took the attention of researchers [8] during the last years. Authors in [2] applied GLCM (Gray-Level Co-Occurrence Matrix) for image characteristics extraction (Homogeneity, Correlation, Entropy, ...) of tomato disease, then used CNN for convolutional Neural Network for other authors to classify the leaf into diseased leaf and healthy leaf. Comparisons of different methods in [2] gave the following accuracy results: Artificial Neural Network (ANN) 92.94%, (Deep Neural Network) AlexNet 95.75% and CNN 99.25% based on the Plant Village database. Works in [12] uses ANN to classify Banana Leaf and Fruit Diseases to detect them at their early stage. They collected 60 images and used 25 for detection and the rest for training. They presented visual results and mean, standard deviation, but have not given explicit accuracy parameters. Proposal in [1], is about a tomato crop disease detection system using a CNN based model. The CNN architecture has 3 convolutions and max pooling layers with varying number of filters in each layer. The training Dataset contains more than 10000 images and the average testing accuracy of the model is 91.2%.

Support Vector Machine classifier (SVM) is a supervised learning model formally defined by separating hyperplanes used also to identify the diseases on the leaves [6,11]. The authors in [10] have focused on implementing image analysis and classification techniques for detection of leaf diseases. They use K-means clustering to detect diseased areas; feature extraction by Gray-Level Co-Occurrence Matrix (GLCM) and finally Support Vector Machine for classification. Islam et al. in [6] an approach that integrates image processing and machine learning to allow diagnosing diseases from leaf images. This automated method classifies diseases on potato plants from a publicly available plant image database 'Plant Village'. Their segmentation approach and utilization of support vector machines demonstrate disease classification over 300 images with an accuracy of 95%.

The third category of classifier for plant leaf disease detection is K-Nearest Neighbors(KNN) based solutions. Authors in [13] propose to replace the SVM classifier with KNN classification. They extracted the characteristics resulting from the transformation of the RGB image to HSV to classify the latter with the KNN technique. With the 250 images selected, including 45 for training and 105 for testing, the authors were able to detect 4 types of disease: Early Leaf spot, Late Leaf spot, Rust and Bud Necrosis. Preliminary results seem promising with this technique but accuracy of their proposal has not been fully evaluated. In [5] the authors use the GLCM for the extraction of mean, Standard Deviation, energy, contrast, homogeneity and correlation characteristics. Then the KNN for the classification of 200 leaf images from five disease classes and 37 independent leaf images for classifier performance testing. With this method the authors obtained an efficiency of 96.76%.

K-means clustering is an unsupervised algorithm used to form different clusters of datasets so that similar data are grouped into K clusters identified by K Means. This technique is used by several works more specifically for segmentation [4]. Proposition in [9] uses KMeans for segmentation of images into clusters that isolate diseased part of the leaf image. P. Badar et. al. has used an approach of segmentation using KMeans Clustering [3] on various features of Potato leaf image samples such as color, texture, area, etc. and applied Back Propagation Neural Network algorithm for identifying and classifying the disease in the leaf image in which they obtained a classification accuracy of 92% [3].

However, all of these classifiers of AI techniques in practice may face two major challenges in the context of African agriculture:

– Most rural areas of Africa do not have computer equipment powerful enough in terms of CPU and energy to implement the resources necessary for the proper functioning of these classifiers. And funding for the acquisition of its equipment remains inaccessible.
– African rural areas also face difficulties in accessing the 3G or higher mobile network for transmitting and receiving data. However, smartphones remain widespread.

Our proposal aims at detecting diseased plants leaves with very low resource consumption, with a very fast and reliable execution time. This Framework is based on Kernel KMeans method to provide an edge AI service for Leaf Plant Disease detection. There exist other works using similar approaches however they all need large training datasets and more computation than our proposition. Works that use KMeans do Machine Learning models for each plant leaf image but we use Kernel instead to have mode general models than can be used as general Models.

3   Background Concepts for Plant Disease Detection

There are several techniques for detecting diseased plants based on the color or texture of the leaf. In a diseased plant leaf, the pixels can be grouped into two groups: the pixels of the infected part and the pixels of the healthy part to finally extract the information concerning the diseased pixels. This type of process is called Region based. Edge based methods like gradient, log, canny, sobel, laplacian, robert, leaf border based are also widely used to identify discontinuities that exist on the leaf. Threshold based which, as its name suggests, is based on threshold values obtained from the histogram of these edges of the original image. Another much more widely used method in disease detection is the clustering algorithm called K-means which allows the image to be segmented and classified into multiple clusters. It will be used in this article for the detection of the disease in plants.

4   Smart IoT Framework for Leaf Plant Disease Detection

4.1   General Architecture of the Framework

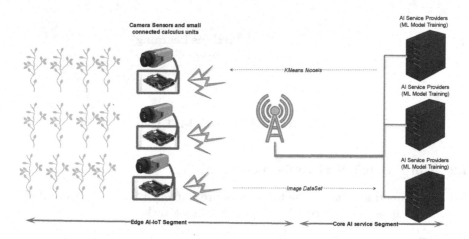

Fig. 1. General architecture for edge AI-IoT network

The general framework of our proposal is described in Fig. 1. The framework is composed of two segments: Edge AI-IoT Segment and Core AI Service Segment. These two segments are connected to share data and models.

Edge AI-IoT Segment: Edge AI-IoT Segment is the part of the IoT where the plant detection Machine learning models are deployed in AI-IoT-Nodes. These nodes are composed of at least a camera sensor and connected small computation unit like raspberry pi. Plant leaves images are taken and analyzed by AI-IoT nodes to detect leaf disease at its early stage (Fig. 1). The result of plant leaf disease detection is sent to the core AI-Core segment where farmers and agronomics can exploit the service.

Core AI Service Segment: This segment is in the cloud and hosts machine learning tools and image datasets for training them to detect plant leaf disease. Nodes in this segment are responsible for training machine learning KMeans Kernel models using kernel Images and deploy them to the AI-IoT Nodes. Operation of this framework is as follows: AI-Service Segment select kernel images that have core characteristics of diseased leaf (Kernel-Images), use KMeans machine learning algorithm to build clusters of Kernel-Images so that diseased regions are one cluster. We call the resulting models KMeans Kernel-Model. These models are deployed to the AI-IoT Nodes that use them to analyze the stream of pictures that comes from the camera sensor (Fig. 2).

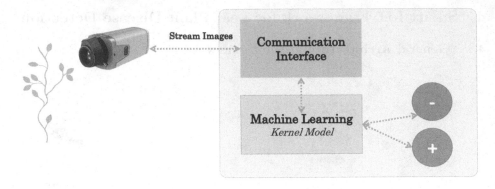

Fig. 2. Architecture of an AI-IoT node

4.2 Core AI-IoT Service Segment

The main functionalities of the framework are implemented in the Core AI-IoT Service Segment. The process is composed of three tasks:

– Select Kernel-Images
– Build K-Means Kernel Models
– Deploy the service to the Edge AI-IoT Segment

All these functionalities are built on KMeans clustering because it has low computation constraints and is efficient for feature extraction. K-Means Clustering is used here to separate the image into four clusters among which one cluster contains most of the disease region. Figure 3 illustrates the output of KMeans clustering in a diseased plant leaf. It shows well how we can detect regions of interest. using square root distance of pixels RGB or HSV color spaces. Something also interesting is that this clustering can work with tone variation of colors.

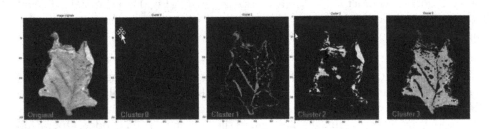

Fig. 3. Example of K-means clustering output on a diseased plant leaf image. Cluster 2 contains the biggest diseased region

4.3   KMeans Kernel Learning

The principle of KMeans Kernel Learning consists in creating KMeans models trained with selected images (Kernel Images). The clusters resulting from these Kernel Images are called Kernel Clusters and are then labeled diseased zones or healthy zones. Figure 4 shows how it works. The framework uses Kernel Image I_{k0} which is supposed to have representative features of a diseased plant leaf. This Kernel Image is used to build a KMean Kernel Model φ_{k0} and Kernel Clusters $\omega_{i,k0}$; i $\in$ [0,4]. Each cluster can be labelled *healthy* or *diseased*. In our context we orient KMeans algorithm so the cluster that contains most of diseased region is always named $\omega_{2,k0}$. KMean Kernel Models are just classifiers based on KMean that have been trained with data $\mathbb{R}^3$ composed by Kernel Image pixels components. We limited the number of clusters to 4 because we observed that the number of empty clusters increases when $K \geq 4$.

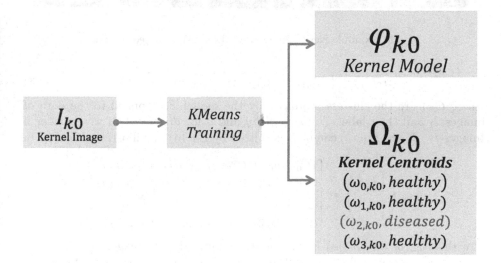

Fig. 4. Principle of KMeans Kernel models

More formally, considering I_{k0} a Kernel Image chosen we have the Eq. (1):

$$KMeans(I_{k0}) = \{\varphi_{k0}, \Omega_{k0}\} \tag{1}$$

Where φ_{k0} is the KMeans Kernel Model clustering model based on the I_{k0} kernel image and Ω_{k0} is the set of cluster centroids and their labels as a *healthy* or *disease* regions of the plant leaf. Ω_{k0} is defined by Eq. (2):

$$\Omega_{k0} = \{(\omega_{i,k0}, label)/i \in [0-3], label \in \{health, diseased\}\} \tag{2}$$

Where $\omega_{i,k0}$ is the centroid of the cluster number i (related to I_{k0}) and the label indicates if the cluster formed from this centroid belongs to a *diseased* region or *healthy*. We make the assumption that by taking the one cluster that contains most significant disease region we can make decision about the health of the plant leave. So only one cluster is labelled diseased and we always refer to it by $\omega_{2,k0}$. The Fig. 5 illustrates a Kernel Image and clusters built based on KMean Kernel Machine Learning. At this point KMeans Kernel Model is ready to be used to detect Plant Leaf Disease from any other image. For instance, considering Image I_1, the clustering based on $\{\varphi_{k0}, \Omega_{k0}\}$ is defined by Eq. (3):

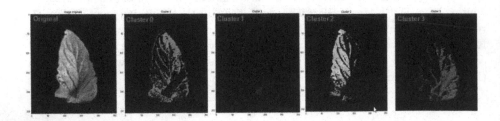

Fig. 5. Example Kernel Image and its Clusters based on φ_{k0}

$$\varphi_{k0}(I_1) = \{(C_{0,1}, h), (C_{1,1}, h), (C_{2,1}, d), (C_{3,1}, h)\} \tag{3}$$

where $C_{x,1}$ are the clusters produced by the model φ_{k0} applied to the data of images I_1 and their labels (h for *healthy* and d for *diseased*). If we consider n Images $I_1...I_n$ the Eq. (3) can be generalized by Eq. (4) and illustrated by Fig. 6.

$$\varphi_{k0}\left(\begin{bmatrix} I_1 \\ \cdots \\ I_n \end{bmatrix}\right) = \begin{bmatrix} (C_{0,1}, h_{0,1}) & (C_{1,1}, h_{1,1}) & \cdots & (C_{3,1}, h_{3,1}) \\ (C_{0,2}, h_{0,2}) & (C_{1,2}, h_{1,2}) & \cdots & (C_{3,2}, h_{3,2}) \\ \vdots & \vdots & \ddots & \vdots \\ (C_{0,n}, h_{0,n}) & (C_{1,n}, h_{1,n}) & \cdots & (C_{3,n}, h_{3,n}) \end{bmatrix} \tag{4}$$

In this Fig. 6, the trained Kernel model is applied to n Images. The result is clusters where cluster 2 ($C_{2,3}$) is the one that is more representative of diseased region of the plant leaf. It is a kind of projection of Kernel Clusters to the Images $I_1 \ldots I_n$. We also use a score to measure the quantity of clustering. It is a metric of how big the disease is and is defined in Eq. (4):

$$Score(\varphi_{k0}(I)) = \frac{\| C_2 \|}{\| I \|} * 100 \tag{5}$$

Where $\|C_i\|$ is the number of points pixels classed in this cluster. If the plant is healthy $\|C_2\| = 0$. Figure 7 is an illustration of $\varphi_{k0}([I_1 I_2])$ where I_1 and I_2 are respectively diseased leaf and a healthy leaf. So, the cluster 2 of I_2 is empty for the second plant because this cluster is supposed to contain diseased regions. This result shows that our plant disease detection algorithm is working properly.

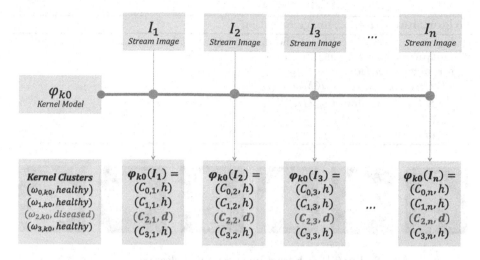

Fig. 6. Kernel Model application to a set of plant leaf images

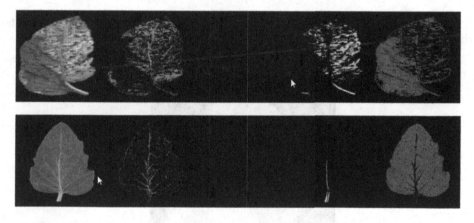

Fig. 7. Example of Kernel Clusters by the I_{k0} model. First is diseased leaf and the second is healthy

4.4   Kernel Image Selection Process

One parameter of our framework is the Kernel Image. It should be representative enough in terms of the features of diseased plants. It is very important to choose the kernel image or reference image because it has a big impact on efficiency of the algorithm. That's why we defined an algorithm to find the most accurate Kernel Image. The algorithm is quite simple, it computes the accuracy of all candidates images and picks most accurate (Algorithm 1). We executed the algorithm in the tomato bacterial spot dataset of the Plant village Dataset. The expected result is to find out the best Kernel Images. Figure 8 illustrated what the best kernel Image looks like. The highest accuracy Kernel Image was the

Algorithm 1: Kernel Image Selection

Input   : I set of healthy H and diseased D plant leaves Images
Output: I_{k0} Best Kernel Image
foreach I_{ki} *in* I **do**
  $\{\varphi_{ki}, \omega_{0i}, ...\omega_{3i}\} \leftarrow KMeans(I_{ki});$
  $S[I_{ki}] \leftarrow \text{Accuracy}(\varphi_{ki}(I_{ki}))|_D^H$;
end
return $\arg\max_I(S[I_{ki}])$

Fig. 8. Top 6 Best Kernel Images Using Selection algorithm

rightmost image (Fig. 9) and its results on healthy plant leaf images. Knowing this we have done larger performance evaluation.

Fig. 9. Best Kernel image output by Algorithm on Plant Village Dataset Tomato

Figure 10 represents the results of respectively $\varphi_{k0}([I_1, I_2])$ representing clustering of healthy Images and $\varphi_{k0}([I_3, I_4])$ representing diseased ones. We can see that cluster 2 ($C_{x,2}$) is almost empty for healthy leaves and capture diseased region for diseased plant leaves.

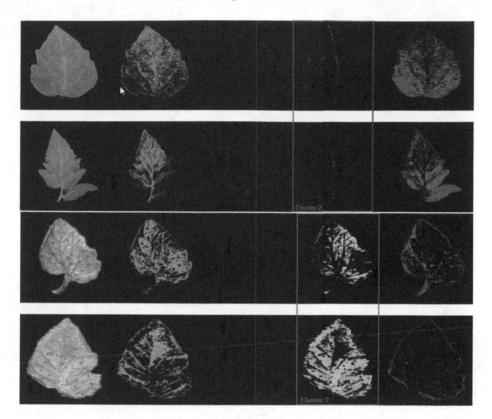

Fig. 10. Clustering of health and diseased images based on Best Kernel Image

4.5   Performance Evaluation

To evaluate the quality of our KMeans Kernel-Learning we used tomato images from plant Village DataSet (www.plantvillage.org) and we set aside two classes: healthy tomato leaves and diseased tomato leaves. Plant Village is a database of images accessible to the public. It is composed of 54,306 images of diseased plant leaves and healthy plant leaves from 14 crop species. All images in the dataset were captured under controlled conditions against a uniform background, resized to a dimension of 256 × 256 pixels.

Main results we have are reported in the confusion matrix (Fig. 11).

- Horizontally, out of the 50 healthy images (i.e.: 48 + 2), 48 were correctly classified by our KMeans Kernel Learning classification system as such and 2 were estimated as diseased (i.e.: 2 false-negatives),
- Horizontally, out of the 30 diseased images (i.e.: 1 + 29), 1 was estimated by the automatic Kernel KMeans classification system as healthy (i.e.: 1 false-positive) and 29 were estimated as diseased;
- Vertically, out of the 49 sheets (i.e.: 48 + 1) estimated by the automatic KMeans classification system as sound, 1 is in fact sound;

– Vertically, out of the 31 leaves (i.e.: 2 + 29) estimated by the automatic KMeans classification system as diseased, 2 are in fact sound;
– Diagonally (from top left, to bottom right), out of the initial 80 sheets, 77 (48 + 29) were estimated correctly by the automatic k-means classification system.

| | | Predicted Class (By the Kernel KMeans Classifier) | |
		Healthy leaf	Sick Diseased
Actual class	Healthy leaf	48 (True Positive)	2 (False Negative)
	Diseased leaf	1 (False positive)	29 (True Negative)

Fig. 11. Confusion matrix of the proposed algorithm

– Accuracy: We can define the accuracy as being the number of really diseased plants leaf found compared to the total number of potentially diseased plants, based on our reference plants.

$$\text{accuracy} = \frac{TP + TN}{TP + TN + FP + FN} = 96,25\%$$

– Precision: Precision is the ratio of correctly predicted positive observations to the total predicted positive observations. The question this metric answer is that of all the plants labeled as healthy, how many are really healthy? The high precision is related to the low rate of false positives.

$$\text{precision} = \frac{TP}{TP + FP} = 97,96\%$$

– The recall The recall is defined by the number of really diseased plants found with regard to the total number of diseased plants in the database.

$$\text{recall} = \frac{TP}{TP + FN} = 62,34\%$$

– F1-score: The F1 score is the weighted average of the precision and recall. Therefore, this score takes into account both false positives and false negatives.

$$\text{F1-Score} = \frac{2 * (Recall * Precision)}{Recall + Precision} = 76,19\%$$

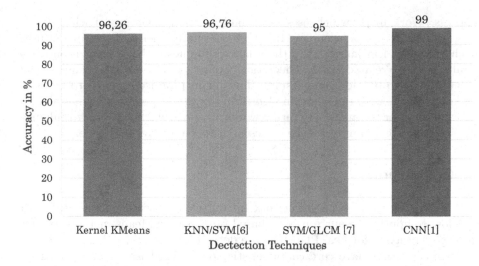

Fig. 12. Accuracy: Kernel KMeans based detection vs other techniques

Detection Methods	Size of training DataSet	Accuracy (in %)
Kernel KMeans	80	96,26%
KNN/SVM[6]	200	96,76%
SVM/GLCM [7]	237	95%
CNN[1]	10000	99%

Fig. 13. Accuracy and Training Dataset Size: Kernel KMeans based detection vs other techniques

We also compared our classification method with existing propositions for plant leaf disease detection that are [1,5,6]. We compared our accuracy with three other related works. These three other works respectively use the KNN associated with SVM [5], SVM with GLCM [6] and Convolutional Neural Network (CNN) [1].

Figure 12 shows the result of the comparison in term of accuracy and it shows that the CNN hase the best result while the Kernel Kmeans has a result which is comparable to other methods with 96% accuracy. The other element of comparison is given in Fig. 13 where we compare the size of the dataset of each method in comparison with the accuracy that is obtained. We can see that the KNN/SVM method uses 120 more images for training than the Kernel KMeans with an accuracy gain of only +0.5%. The SVM/GLCM uses 237 images for training and achieves an accuracy of less than 1.26% compared to the Kernel

KMeans. Work in [1] which uses CNN had to train their model with 10,000 images to achieve 99% accuracy. In summary our Kernel KMeans model uses much less plant leaf images than the other techniques to have a precision comparable to these related works. Theoretically our model needs only a single good Kernel Image to provide good results. However in practice we need to find out the kernel image with the selection algorithm to achieve precision comparable to other methods or techniques. In future works one possible direction is to enhance this selection algorithm to reduce de research space for Kernel Image.

5   Conclusion

In this paper we propose a framework for detecting diseased plants leaf with very low computing resource needed based on Kernel KMeans method to offer an Edge AI service for Leaf Plant Disease detection. AI-Service Segment selects kernel images that have core characteristics of diseased leaf (Kernel Images), use K Means machine learning algorithm to build clusters of Kernel-Images so that diseased regions are one cluster. We call the resulting models KMeans Kernel-Model. These models are deployed to the AI-IoT Nodes that use them to analyze the stream of pictures that comes from the camera sensor. When we evaluate our proposal, we note an accuracy of 96% of our system. The proposed method efficiently detects and classifies plant disease and could be applied to the detection and classification of several types of plants. We also compare it to other works and it give quite good results while needing less computation and images in the dataset. Future work in this study will focus on improving our algorithm by using Light Weighted KNN or Random Forest.

References

1. Agarwal, M., Singh, A., Arjaria, S., Sinha, A., Gupta, S.: "Tomato Leaf Disease Detection using Convolutional Neural Network" Procedia Computer Science, pp. 293–301, Elsevier (2020)
2. Ashok, S., Kishore, G., Rajesh, V., Suchitra, S., Sophia, S.G.G., Pavithra, B.: tomato leaf disease detection using deep learning techniques. In: 2020 5th IEEE, International Conference on Communication and Electronics Systems (ICCES), pp. 979–983. Coimbatore, India (2020)
3. Athanikar, M.G., Badar, M.P.: Potato leaf diseases detection and classification system. Int. J. Comput. Sci. Mobile Comput. 5(2), 76–88 (2016)
4. Bin Abdul Wahab, A.H., Zahari, R., Lim, T.H.: Detecting diseases in chilli plants using K-means segmented support vector machine. In: 3rd IEEE, International Conference on Imaging, Signal Processing and Communication (ICISPC), pp. 57–61. Singapore (2019)
5. Hossain, E., Hossain, M.F., Rahaman, M.A.: A color and texture based approach for the detection and classification of plant leaf disease using KNN classifier. In: International Conference on Electrical, Computer and Communication Engineering (ECCE) (2019)

6. Islam, M., Dinh, A., Wahid, K., Bhowmik, P.: Detection of potato diseases using image segmentation and multiclass support vector machine. In: 30th IEEE Canadian Conference on Electrical and Computer Engineering (CCECE). IEEE (2017)
7. Gavhale, K.R., Gawande, U.: An overview of the research on plant leaves disease detection using image processing techniques. IOSR J. Comput. Eng. (IOSR-JCE) **16**(1), 10–16 (2014)
8. Chobe, P.K.S.: Leaf disease detection using deep learning algorithm. Int. J. Eng. Adv. Technol. **29**(06), 3599–3605 (2020)
9. Mugithe, P.K., Mudunuri, R.V., Rajasekar, B., Karthikeyan, S.: Image processing technique for automatic detection of plant diseases and alerting system in agricultural farms. In: International Conference on Communication and Signal Processing (ICCSP), pp. 1603–1607. IEEE, Chennai, India (2020)
10. Prakash, R.M., Saraswathy, G.P., Ramalakshmi, G., Mangaleswari, K.H., Kaviya, T.: Detection of leaf diseases and classification using digital image processing. In: International Conference on Innovations in Information, Embedded and Communication Systems (ICIIECS), pp. 1–4, IEEE, Coimbatore (2017)
11. Reddy, J., Karthik, V., Remya, S.: Analysis of classification algorithms for plant leaf disease detection. In: 2019 IEEE International Conference on Electrical, Computer and Communication Technologies (ICECCT). https://doi.org/10.1109/ICECCT.2019.8869090
12. Saranya, N., Pavithra, L., Kanthimathi, N., Ragavi, B., Sandhiyadevi, P.: Detection of banana leaf and fruit diseases using neural networks. In: 2020 Second International Conference on Inventive Research in Computing Applications (ICIRCA), pp. 493–499. IEEE, Coimbatore, India (2020)
13. Vaishnnave, M.P., Devi, S.K.: Detection and Classification of Groundnut Leaf Diseases using KNN classifier. In: Proceeding of international Conference on Systems (2019)

System for Monitoring and Control of Vehicle's Carbon Emissions Using Embedded Hardwares and Cloud Applications

Tsvetan Tsokov$^{(\boxtimes)}$ and Hristo Kostadinov

Institute of Mathematics and Informatics,
Bulgarian Academy of Sciences, Sofia, Bulgaria
`hristo@math.bas.bg`

Abstract. Today, the electronic devices such as sensors and actuators, forming the Internet of Things (IoT) are presented naturally in the people's day to day life. Billions of devices are sensing and acting upon the physical world and exchange information. All major industries like transportation, manufacturing, healthcare, agriculture, etc. adopt IoT solutions. Not only people and industry are affected in a positive way by IoT, but also the nature and environment. The IoT is recognized as a key lever in the urge to save the climate. It has a major potential in reducing air carbon emissions and pollution. Taking into account the promising sectors of IoT application, this paper proposes a solution for monitoring and control of carbon emissions from vehicles. It consists of hardware device that ingests data related to vehicles' carbon emissions and cloud based services for data storage, analysis and representation. It controls the carbon emissions via notifications and vehicle's power restrictions.

Keywords: Clustering · Internet of things · Reduction of carbon emissions · Sensor data processing

1 Introduction

Nowadays, the smart physical devices are part of the everyday people life and generate huge amount of data, which drives applications and services on top of them, establishing the Internet of Things (IoT) ecosystem [1]. Applications are developed to ingest, store and analyze large amounts of data generated by the IoT devices in all economic sectors such as transportation, agriculture, health and education. According to CISCO Internet Business Solutions Group (CISCO IBSG), the total number of interconnected devices by 2015 was 25 billions and it's expected to be 50 billion by 2020 [2]. For the same time period it is expected 5.8 billion IoT endpoints and global economic revenue from endpoint electronics to total 389 billion US dollars.

One of the most impacted sectors by the IoT is the automotive industry. Recently, tens of millions of cars are said to be connected to the Internet and their number is expected to become hundreds of millions in the near future.

© Springer Nature Switzerland AG 2021
H. Hacid et al. (Eds.): ICSOC 2020 Workshops, LNCS 12632, pp. 564–577, 2021.
https://doi.org/10.1007/978-3-030-76352-7_50

According to a global industry analysis, the number of connected vehicles will become 125 million by 2022. In order to enable such a massive amount of connected vehicles to communicate with other IoT endpoints in real time, new network bridges and protocols are developed [3]. At the same time, mobile network technology is recognized to have considerable potential to enable carbon emissions reduction across a wide range of sectors. Currently, 70% of the carbon savings come from the use of machine-to-machine (M2M) technologies, according to Global e-Sustainability Initiative. The survey data shows that 68% of smartphone users are willing to adopt behaviours that could result in even more future reductions to personal carbon emissions. According to another study the power consumption of the radio access networks (RAN) and the production of mobile devices are major contributors in the carbon emissions [4]. There it is concluded that the usage of green technologies to reduce the power consumption offer big potential for reduction of the emissions. IoT is pointed as a key lever to reduce the carbon emissions.

The current hardware and software solutions for tracking vehicles give evidence for the efforts of using IoT technologies in automotive industry. Geotab provides a service for monitoring and analysis of vehicles using integrated hardware module that sends data to private cloud. The hardware module is connected to the onboard diagnostic bus of the vehicle and collects data about fuel consumption, traveled distance and other parameters. The data analysis, which is made in the cloud allows identification of vehicles with not optimal fuel consumption. Unfortunately, Geotab solution does not detect increased rate of carbon emissions and provide control over vehicle's parameters. Madgetech is a data logger, which provides functionality for regular monitoring of carbon dioxide levels (Data Loggers). It measures the carbon emissions by exhaust gas sensors in vehicles and sends data to private cloud. The measured data is visualized by mobile application, but analysis and control of the carbon emissions are not supported. In addition to these existing solutions, there are mandatory emissions inspection and certification procedures in many countries. But these procedures usually are done only one or several times per year, which is a small frequency and can not guarantee that the vehicles are working with optimal emissions for big period of time.

Inspired by the low-carbon roadmap of European union and the grate potential provided by the IoT technologies for reducing the carbon emissions, in this paper we propose a solution for real-time monitoring and detection of rising levels of carbon emissions from vehicles, called EcoLogic. The proposed solution includes hardware module, which collects sensor data related to vehicle's carbon emissions such as air pressure, air temperature and fuel mixture. The data is transferred to a cloud-based applications, where it's stored and analysed. The results from the analysis are used to control the carbon emissions through driver notifications and vehicle's power limitations. The source code of the main software components of the solution is publicly available and can be accessed and downloaded from the following locations: https://github.com/ttsokov/vehicle-monitor-controller-hw-controller, https://github.com/ttsokov/vehicle-monitor-controller-hw-proxy, https://github.com/ttsokov/vehicle-monitor-controller-backend. The repositories contain information on how to setup the projects and deploy them on hardware device and server or in the cloud. The source code and hardware modules are with open-source licenses, because the purpose of the solution is to be freely reviewed, used and extended by everyone who is interested to track and reduce emissions at global scale. Additionally the system can be integrated with the countries tax system supporting drivers with small emissions footprint to pay smaller taxes.

The rest of the paper is organized as follows. Section 2 presents the architecture of EcoLogic with its components, interconnections and protocols. Section 3 describes its components in implementation details and why the concrete technologies are used. Section 4 shows a case study that validates the feasibility of the proposed solution by comparing the results of the algorithm from two datasets: test dataset and real dataset. Finally, Sect. 5 concludes our research by giving the pros and cons of the system and points out directions for future work.

2   System Design

This section presents the architecture of EcoLogic system with its components, communication paths and protocols.

The EcoLogic system is composed of two big parts: hardware modules and cloud-native microservice applications. The hardware modules are located in vehicles and the microservice applications are deployed on a cloud platform. The architecture of the system is shown in Fig. 1. The history of physical computing technology showed us that each physical computing system contains at least one hardware device, but nowadays the Internet of Things paradigm, which extends the physical computing technology, is showing us that except hardware devices each IoT system should use also Cloud or Fog platform, which enables massive scale, reliability and efficiency. It becomes a standard for IoT systems to incorporate Cloud and Fog computing platforms in its architectures [5] as main components that give a lot of benefits like enormous computing power for artificial intelligence algorithms, including machine learning and data mining, high availability, disaster recovery, big storage and near real-time speeds.

The hardware module of the system has sensors and measures several physical parameters. It can also extract parameters from the onboard diagnostic system of the vehicle. The data is sent to cloud-native applications in the cloud platform. The measured physical parameters are:

- *Air/fuel ratio*, which is measured by lambda sonde sensor, which is located into the exhaust system of the vehicle.
- *Absolute pressure of the air* that is consumed by the engine, which is measured by sensor located in the air intake manifold of the vehicle.
- *Temperature of the air* that is consumed by the engine, which is measured by sensor located in the air intake manifold of the vehicle.

The cloud-native applications are implemented as microservices, which are designed in a platform agnostic way in order to have the possibility for deployment on different cloud platforms. The cloud applications store data in a relational database, which is represented by backing service from the cloud platform. They process the incoming data, store it into the database and analyse it.

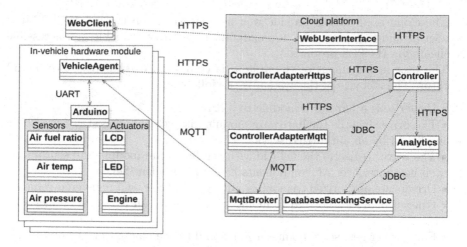

Fig. 1. EcoLogic general architecture

The hardware modules communicate with the cloud with wireless network via HTTPS or MQTT protocols. The following physical parameters are calculated on top of the incoming sensor data:

- Mass of the consumed air by the engine;
- Mass of the consumed fuel by the engine;
- Mass of the carbon dioxide emissions, exposed into the atmosphere.

The database contains all measured and calculated physical parameters. A cloud-native Analytics application executes an anomaly detection on the streamed data and outlines vehicles that have not optimal amount of carbon dioxide emissions or system failures. Clustering analysis is made in order to detect anomalies. The hardware module is notified when some vehicle is detected by the system as an anomaly, with suboptimal amount of emissions. Then the hardware actuator is started automatically to reduce the amount of emissions. This comprises a feedback control loop. In this way the system monitors and controls the amount of carbon dioxide emissions in the atmosphere in real time. The hardware modules are equipped with three actuators:

- *Liquid crystal display (LCD)*, which visualize the measured and calculated physical parameters to the driver.

- *Light-emitting diode (LED)*, which indicates to the driver that the amount of carbon dioxide emissions is not optimal or there is a system failure (not optimal parameters).

- *Actuator*, which controls the amount of injected fuel in the engine and regulates the amount of emissions.

Currently, only the display and LED actuator are implemented in the Eco-Logic. The purpose of the LED actuator is to notify the driver to intentionally reduce the acceleration and change the driving behaviour, which leads to reduction of the amount of burned fuel and emissions.

The cloud applications provide web user interface, on the base of HTML5, JavaScript and CSS resources. It's endpoint is publicly available and accessible by clients via HTTPS protocol.

The user management of the system is composed of two roles: driver and operator. The lifecycle of the system is the following:

- Dealer sells a hardware module to a driver.
- The driver installs the hardware module into vehicle. This step can also be accomplished by an authorized service.
- The driver registers the vehicle with the hardware module and sensors via the web user interface. All components have unique identifiers.
- Drivers are authorized to monitor and control their own registered vehicles.
- Operators are authorized to monitor and control all registered vehicles by regions.
- Each driver gets score points proportional to the amount of carbon dioxide emissions exposed in the atmosphere from their vehicles.
- The score points can be integrated with tax systems of countries and city halls. In this way the taxes can be reduced proportionally to the score points. Drivers can also participate in greenhouse gas trading schemes with their score points.

Important aspect of the design of the system is it's security. Security is major problem in all IoT applications, because of the big number of heterogeneous devices, the data, which is distributed across many locations and the need for high-speed communication. New protocols and schemes that incorporate symmetric and asymmetric cryptography are developed [6] in order to resolve these problems. The current system uses TLS v1.2 (TLS protocol) with it's built-in asymmetric and symmetric cryptography for the network communication between components.

3 System Implementation Elements

This section outlines the main components of EcoLogic system. First, the hardware module with it's components and the algorithm for calculation of carbon emissions are presented. Later, the cloud applications are described and the algorithm for data analysis.

3.1 Embedded Hardware System

The embedded hardware system contains two embedded subsystems: Arduino Uno and Raspberry Pi B+. The Arduino operates on the low-level hardware sensors and actuators in the vehicles, while the Raspberry Pi works on higher level. It aggregates the data from the low-level hardware and communicates with the cloud platform in bidirectional way.

Arduino Embedded Hardware System. The Arduino Uno embedded system has the following capabilities:

- Measure physical parameters of internal combustion engine.
- Visualize the measured parameters on 4×16 liquid crystal display.
- Control of actuator (light emitting diode).
- Communication with Raspberry Pi embedded system.

The physical parameters are measured in two ways: by sensors or extracted from the onboard diagnostic system (OBD2), which is provided by the electronic control module of the vehicle. If the onboard diagnostic interface exposes the necessary parameters in the Application Programming Interface (API), no additional sensors will be installed in the vehicle. If the onboard diagnostic interface API does not expose the necessary parameters, additional sensors, which measure these parameters, will be integrated. There is also a possibility to reuse the already existing default sensors of the vehicle. In this way the hardware module is flexible and can be installed on huge amount of vehicles.

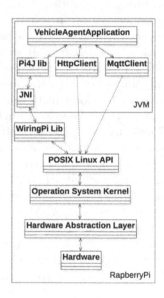

Fig. 2. Raspberry Pi architecture

The Arduino embedded system has deployed application in its local storage and executes it. The application is implemented on the C++ programming language. It is publicly available and can be downloaded from the following location: https://github.com/ttsokov/vehicle-monitor-controller-hw-controller (vehicle- monitor-controller- hw-controller source code). The repository from the link contains information for the build, deployment and execution on local Arduino-based hardware module. The application consumes Wiring library, which is part of the Arduino SDK (Software Development Kit). The Wiring library executes on the concrete Arduino microcontroller via drivers. Currently, the Arduino Uno hardware is used, which has Microchip ATmega328 microcontroller based on RISC architecture. This hardware module executes a critical calculations from physical sensors in real time. This is the reason why it's a separate module with own independent computational resources and with software application written in fast low-level programming language like C++.

Raspberry Pi Embedded Hardware System. The Raspberry Pi B+ embedded system delegates the communication between the Arduino embedded system and the applications in the cloud platform. Its architecture is shown in Fig. 2.

The communication between the Raspberry Pi B+ embedded system and the Arduino embedded system is via serial UART (Universal Asynchronous Receiver/Transmitter) protocol with a custom application messaging protocol on top of it. The Raspberry Pi B+ embedded system is connected to the cloud platform via 4G broadband cellular network. The measured physical parameters by the Arduino embedded system are ingested into the Raspberry Pi B+ embedded system, which caches them in a local cache storage for further processing and sends them to the cloud platform. It sends the data to the cloud platform via HTTPS or MQTT protocol. The Adapter cloud-application is the facade, which is the only endpoint seen by the Raspberry Pi B+ embedded system. When it sends the data to the cloud platform, it receives response that contain information about the state of the vehicle, including the amount of the carbon dioxide emissions. There are two possible states: optimal (eco) and not optimal (not eco). The Raspberry Pi module notifies the Arduino embedded system once the emissions are not optimal. Then the Arduino embedded system activates the low-level hardware actuator in order to reduce the quantity of emissions. The main component inside the Raspberry Pi embedded hardware is the System on a Chip (SoC), which is using ARM architecture and Linux based operation system, currently Raspbian. It executes the VehicleAgent application, which is implemented in the Java programming language and it runs on top of a Java Virtual Machine. The VehicleAgent application is publicly available and can be downloaded from the following location: https://github.com/ttsokov/vehicle-monitor-controller-hw-proxy (vehicle-monitor-controller-hw-proxy source code). The repository from the link contains information on how to build the application locally, how to deploy it on any Raspberry Pi-based hardware module and how to execute it.

The VehicleAgent application depends on the Pi4J and WiringPi libraries, which support the serial communication between the two embedded systems. An USB WiFi adapter is used for the 802.11n communication between the Raspberry Pi embedded hardware and a hotspot, which provides the 4G broadband cellular network to the cloud platform. The adapter is connected to one of the USB ports of the Raspberry Pi. The communication with the cloud platform can be achieved by several application layer protocols: HTTPS or MQTT.

The decision which protocol will be used is taken according to the supported protocol by the cloud platform. The modular architecture of the application enables easy extension with other application layer protocols. Because the communication network to the cloud platform is constrained it is expected to exist glitches and that's why the application caches the latest data in the local cache storage. After successful reestablishment of the connection with the cloud platform, the application retries to sent the locally cached data. The application can be configured by configuration file (config.xml), which is located into the local file system storage. It contains unique identifiers for the vehicle, sensors and type of the communication with the cloud platform (HTTPS, MQTT, etc.). The driver or operator, who registers the vehicle into the system, should populate the configuration file and save it.

This hardware module serves as a communication proxy and need to have capabilities to work with a lot of communication networks and protocols. This is the reason why it's a separate hardware module, which can work with many hardware adapters or antennas supporting different communication networks. The software application is written in high-level programming language like Java, because it needs to work with many communication protocols and serialize data in different formats.

3.2   Cloud Software Applications

The EcoLogic is composed of several cloud- native microservice applications, namely Controller application, Adapters applications, Web user interface and Analytics application. Each application is described in the following subsections.

Controller Application. The Controller application is the main cloud application, which manages all vehicles with their hardware modules and sensors, make calculations, persists data into database, executes artificial intelligence algorithms on the data and provides HTTP REST API. It is implemented using the Java Enterprise Edition programming language using JPA and Apache CXF services framework. The Controller application is publicly available and can be downloaded from the following location: https://github.com/ttsokov/vehicle-monitor-controller-backend (vehicle-monitor-controller- backend source code). The repository from the link contains information on how to build the application locally, how to deploy it on a web container in any cloud platform or local server and how to execute it. The application has the following functionality:

• Compose a data model, which has the following relations: users, which have vehicles, which have sensors, which have measurements of physical parameters.
• Orchestrates the lifecycle of all users, vehicles, sensors and measurements.
• Persists the data into a relational database, which is provided as a backing service exposed by the cloud platform. The application is database-agnostic, which means that it can work with any relational database, which provides Java connectivity. If it does not provide Java connectivity, an adapter can be used. The database is composed of four tables: User, Vehicle, Sensor and Measurement.

• Calculates the total mass of the carbon dioxide emissions disposed into the atmosphere by vehicles.

• Manages the state of each vehicle: optimal (eco) state and not optimal (not eco) state. This is a simple state machine with two states.

• Calls the API of an Analytics application, which executes clustering analysis and anomaly detection. The purpose is to find vehicles, which have not optimal amount of carbon dioxide emissions per region.

• Provides HTTP REST API which is used by the Adapter applications and web user interface.

Adapter Application. The data coming from the vehicle hardware modules is intercepted and adapted by the Adapter cloud-native applications. Then it is routed to the Controller application. The Adapter applications are implemented using the Java programming language. Currently there are two types of Adapter applications that handle HTTPS and MQTT network protocols: ControllerAdapterHttps and ControllerAdapterMqtt, respectively. Controller-AdapterMqtt application communicates with MQTT broker. It is broker independent, so: Mosquitto, HiveMQ, Mosca or other type of MQTT broker can be used. The MQTT broker and the Adapter applications have publicly available URL endpoints, which are called by the vehicle hardware modules. The traffic is routed by the cloud platform to the concrete application depending on the application layer protocol that is used. The traffic is routed to the Controller-AdapterHttps application if HTTPS protocol is used. The traffic is routed to the MQTT broker if MQTT protocol is used. This routing behaviour of the cloud platform is based on TCP routing mechanism. TCP routing enables cloud platforms to support applications, which communicate with different non-HTTP protocols. TCP routing is used by one of the industry standard cloud platforms, called Cloud Foundry (Cloud Foundry TCP Routing).

The flow of the MQTT traffic is the following:

1. On creation of new vehicle, the Controller application registers two topics with names vehicles/{id}/sensors/{id}/measurements and vehicles/{id}/state. The ControllerAdapterMqtt application subscribes for that topics.
2. The concrete hardware module also subscribes to both topics and start to publish new measurements to vehicles/{id}/sensors/{id}/measurements topic.
3. On receiving of new message with measurement by the ControllerAddapter-Mqtt application, it adapts the measurement and sends it to the Controller application via the HTTP protocol.
4. The ControllerAddapterMqtt application receives the state of the concrete vehicle and publishes it to the vehicles/{id}/state topic.
5. The concrete hardware module is subscribed to the state topic and receives a response with the state of the vehicle, whether it is in optimal state or not. In this way vehicle is notified.

This flow represents how ControllerAddapterMqtt application adapts the data from MQTT to HTTP in both directions.

Web User Interface. A web server serves static HTML5, JavaScript and CSS resources, which are assembling the web user interface. The web resources do not contain any back-end logic, but only front-end code, which assembles a responsive and user-friendly interface. This functionality for serving of static web resources is lightweight and it's supported by the most cloud platform providers. The concrete web user interface in the EcoLogic is based on the open-source JavaScript front-end web application framework OpenUI5. It makes many simultaneous AJAX (Asynchronous JavaScript and XML) requests to the HTTP REST API endpoint exposed by the Controller application. Public access is provided and the drivers and operators of the system are using it. The web user interface and the Controller application have different origins, because they have different domain names. The problem of the same-origin policy (OpenUI5), which postulates that one web application can access web resources from the same origin or only permitted web resources from another origin is resolved by most of the cloud platforms, usually by tokens.

The web user interface is based on the model-view-controller architecture and supports the following capabilities:

- Handles the whole lifecycle of the users, vehicles, sensors and measurements: supports create, read, update, and delete operations.
- Display all historic and live measured parameters in real-time.
- Static manual control of the state of all vehicles by appropriate user interface controls, which leads to control of the vehicle's actuator. This is achieved with manually configured threshold.
- Automatic dynamic control of the state of the vehicles, which is based on the value of emissions or parameters that are not optimal per region.
- Visualization of all vehicles with not optimal emissions or parameters - anomalies (outliers).

Analytics Application. The Analytics application executes a clustering analysis algorithm on the stored data. The algorithm takes two parameters as input: engine capacity of the vehicles and total mass of carbon dioxide emissions exposed in the atmosphere. In this way it partitions vehicles that have similar engine capacity and emissions amount in clusters and finds the vehicles which are anomalies. The analytics application uses K-Means algorithm for clustering analysis. The number of engine capacities of the registered vehicles represent the number of clusters (K) in the algorithm. The application is connected to the backing service and consumes the stored relational data.

4 Algorithm Results

This section proves the relevancy of the EcoLogic algorithm by presenting a case study with two separate datasets: test dataset and real dataset. The goal of the case study is to test the capability of the algorithm to find outliers in the dataset, which represent vehicles' with not optimal carbon dioxide emissions in the context of a region with many vehicles. The experiment is carried out with one real vehicle with installed hardware module. The cloud platform and services that are configured for the case study are as follows:

- All described cloud-native applications are deployed into SAP Cloud platform.
- The data is stored in a HANA relational database, provided as a backing service from the cloud platform [7].
- The Analytics application is represented by a SAP cloud platform predictive service, which provides an algorithm for clustering analysis used for anomaly detection.

4.1   Algorithm Results on a Dataset with Known Anomalies

First for validation of the correctness of the anomaly detection algorithm a test dataset, which contains known anomalies is used. The algorithm is executed on the test dataset and the returned result is compared with the known result. An official test dataset is used for that purpose (Mugglestone 2014). It has data for customers with the following parameters: id, name, lifespend, newspend, income and loyalty. The test dataset contains 152 rows. The returned results from execution of the clustering analysis with anomaly detection on the test dataset are shown in Fig. 3.

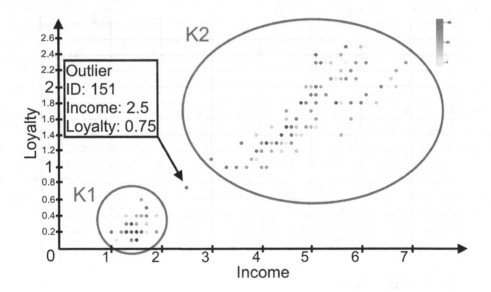

Fig. 3. Clusters of dataset with known anomalies

The x-axis contains the income of the customers. The y-axis contains the loyalty of the customers. The algorithm successfully make two clusters (K1 and K2) and detects one anomaly. The clusters represent two kinds of customers:

customers with low income and low loyalty and customers with high income and high loyalty. The anomaly, marked in Fig. 3 as Outlier, represents a data point, which is located too far from the centers of the clusters K1 and K2, according to other data points. The customer, which is detected as an outlier has ID: 151, income: 2.5 and loyalty: 0.75. The result obtained from the algorithm is the same as the known result from the test dataset, which proves the correctness of the algorithm for this dataset.

4.2   Algorithm Results on a Real Dataset

The hardware module of the EcoLogic is installed into a real vehicle with internal combustion engine with a capacity of 1800 cubic centimetres, working on petrol fuel. In order to acquire bigger dataset, the collected real data is expanded with proportional simulated data. The final operative dataset contains data for vehicles with many engine capacities and carbon emission amounts. The x-axis contains the engine capacity measured in cubic centimetres (cc). The y-axis contains the mass of the carbon dioxide emissions, measured in milligrams (mg). The outcome clusters obtained after execution of the clustering algorithm are shown in Fig. 4. The data points, which have unique IDs, corresponds to the vehicles.

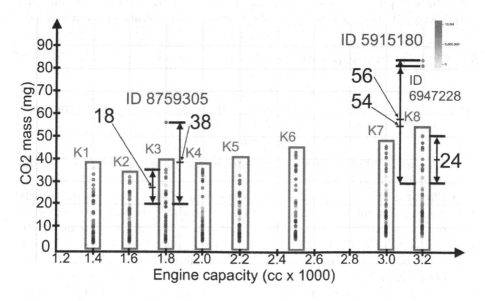

Fig. 4. Clusters of real dataset

The clustering algorithm should place vehicles, which have equal engine capacity and different amount of emissions preferably in the same cluster. In this way, the vehicles with not optimal emissions will not be placed in any cluster and should be detected as outliers. The algorithm returns 8 clusters (K1-K8) for the current dataset, which represent vehicles grouped in 8 different engine capacities – 1400, 1600, 1800, 2000, 2200, 2500, 3000 and 3200 cubic centimetres respectively. They have different carbon dioxide emissions values, which vary between 2.70 and 50.44 mg. Vehicle with ID 8759305 has engine capacity of 1800 cubic centimetres and emissions of 56.07 mg. The nearest cluster center located to vehicle with ID 8759305 is those of cluster K3, with distance of 38 mg. The maximal internal cluster distance in cluster K3 is 18 mg. Vehicle with ID 6947228 has engine capacity of 3200 cubic centimetres and emissions of 81.42 mg. Vehicle with ID 5915180 has engine capacity of 3200 cubic centimetres and emissions of 83.86 mg. The nearest cluster center located to vehicles with ID 6947228 and 5915180 is those of cluster K8, with distance of 54 and 56 mg respectively. The maximal internal cluster distance in cluster K8 is 24 mg. The vehicles with IDs 8759305, 6947228 and 5915180 are detected as outliers, because the distance from them to their nearest clusters is bigger than the internal cluster distance. These vehicles don't have optimal amount of carbon dioxide emissions in the context of the rest of the vehicles, which are located into clusters K1-K8. Important notice for the outlier with ID 8759305 is that this is the real vehicle with parameter values measured during the idle period on cold engine. The outliers with IDs 6947228 and 5915180 have simulated parameters on the base of the real vehicle's parameters on cold engine. The result obtained from the algorithm is feasible and proves it's correctness on the real dataset. The hardware modules of the detected anomalies are successfully notified and actuators are activated.

5   Conclusion and Future Work

In this paper we presented an IoT platform, called EcoLogic, for real-time monitoring, analysis and control of carbon dioxide emissions exposed into the atmosphere by vehicles with internal combustion engines.

The platform has the following advantages:

• High availability and velocity during work with big amounts of data due to cloud-native architecture.

• Platform-agnostic – possibility to work with different vehicles and clouds. The hardware modules are flexible and easy extendable to work with variety of sensors and/or collect data from the default onboard diagnostic bus or sensors of the vehicles. The implemented cloud-native applications are microservices, which are not vendor locked and can work on different clouds.

• Fully completed and validated solution for monitoring and control of carbon dioxide emissions from vehicles, which is ready to contribute in the fight against climate change and reduction of the carbon dioxide emissions in the atmosphere.

The following routes for further improvements are identified and will be addressed in the latter stages of the project:

• The engine capacity and carbon dioxide emission parameters are not adequate to appropriately regulate the amount of fuel injected. The solution could be extended to use also other parameters like vehicle weight and performance requirements for better control.

- The measured parameters on cold and hot (with normal working temperature) engine are not deviate between each other. The measurements on a cold engine leads to detection of the outliers. The solution could be optimized to use two separate data sets: data, which is measured on normally working engine and data, which is measured on cold engine.

- Different algorithm for anomaly detection or classification can be used. It can be a supervised machine learning algorithm, which takes the data measured on cold engine as a training dataset, which defines not optimal amount of carbon dioxide emissions.

- Implementation and usage of more lightweight application network protocols such as CoAP, DDS and AMQP.

- Implementation of predictive maintenance algorithm, which makes notifications for potential future failures in vehicles, based on the current and historical data.

- Integration of the project with other third party systems such as emissions trading systems, transportation tax institutions and smart cities. For example, the more eco-friendly drivers could pay smaller taxes in tax systems of countries and city halls. The amount of carbon dioxide emissions could determine the traffic routing by the traffic lights in concrete regions of smart cities.

Acknowledgements. This work was partially supported by the National Science Fund of Bulgaria under Grant KP-06-N32/2-2019.

References

1. Xia, F., Yang, L., Wang, L., Vinel, A.: Internet of things. Int. J. Commun. Syst. **25**(9), 1101–1102 (2012)
2. Evans, D.: The Internet of Things. How the Next Evolution of the Internet Is Changing Everything, Cisco IBSG (2011)
3. Carignani, M., Ferrini, S., Petracca, M., Falcitelli, M., Pagano, P.: A Prototype Bridge Between Automotive and the IoT. IEEE World Forum on Internet of Things, Milan, Italy (2015)
4. Fehske, A., Fettweis, G., Malmodin, J., Biczok, G.: The global footprint of mobile communications: The ecological and economic perspective. IEEE Commun. Mag. **49**(8), 55–62 (2011)
5. Munir, A., Kansakar, P., Khan, S.: IFCIoT: Integrated Fog Cloud IoT: a novel architectural paradigm for the future Internet of Things. IEEE Consum. Electron. Mag. **6**(3), 74–82 (2017)
6. Henriques, M., Vernekar, N.: Using symmetric and asymmetric cryptography to secure communication between devices in IoT. In: International Conference on IoT and Application, India (2017)
7. Färber, F., May, N., Lehrer, W., Grosse, P., Rause, H., Dees, J.: The SAP HANA database - an architecture overview. IEEE Data Eng. Bull. **35**(1), 28–33 (2012)

Cyber Forensics and Threat Investigations Challenges in Emerging Infrastructures (CFTIC 2020)

CFTIC 2020: 1st International Workshop on Cyber Forensics and Threat Investigations Challenges in Emerging Infrastructures

The International Workshop on Cyber Forensics and Threat Investigations Challenges in Emerging Infrastructures (CFTIC) is a newly-established forum for innovative ideas from research and practice in all areas of cybersecurity, digital forensics, incident response, and threat investigations. The first meeting took place on December 14, 2020, online from Dubai, United Arab Emirates.

The increasing proliferation of global cyber-attacks, at a frightening rate, has been facilitated by the connectivity growth using wired and wireless communication technologies between various heterogeneous systems. These attacks dangerously aim at a broad array of computing systems varying from data centers and personal machines to mobile devices and industrial control systems. The deficiency of existing cyber-defenses and strategies may hinder the ability of security practitioners in applying suitable and timely reactions against these threats. Additionally, the inadequacies in existing digital investigation techniques and tools pose many challenges related to determining the identity of the threat actors, discovering the root vulnerability to impede any subsequent exploitation of a related event, and identifying the threat actors' motivations for the development of efficient cybersecurity defense strategies.

Cyber forensics and threat investigations has rapidly emerged as a new field to provide the key elements for ensuring that the next generation of emerging infrastructures, such as service-oriented architecture and cloud environments, are secure, reliable, and trustworthy, where all future internet services and applications will be hosted and running. In this regard, a rethinking in the current digital investigations' practices, methods, and tools is required in order to adapt to this new context. The use of traditional methods may lead to the loss of valuable data hosted on various types of remote or uncommon cyber-infrastructures and cause a struggle to find relevant evidential data; this impact will continue to expand in the forthcoming years.

The main aim of this workshop is to bring researchers and practitioners together in a multi-disciplinary forum to disseminate current research and to provide insight for the discussion of the major research challenges and achievements in all areas of cyber-security, digital forensics, incident response, and threat investigations. The first edition of the workshop (CFTIC 2020) accepted only five research papers (an acceptance rate of 40%) focusing on important and timely research problems. The accepted papers reported current research issues and advances in cyber forensics and threat investigations. Papers were evaluated by two experts in the domain: members of the workshop Program Committee. CFTIC 2020 presented an attractive workshop program including two keynote presentations by leading international experts together with five technical presentations and discussions.

The first keynote was presented by **Christian Berg**, the CEO of **Paliscope** and CFO of **Safer Society Group**, **NetClean**, and **Griffeye**, and gave industry insights into the applications of AI and Natural Language Processing (NLP) for future investigations. The keynote highlighted the latest NLP and deep fakes technologies used to create texts, fake images, and videos, which are hard for a human to distinguish as fake. The challenges in introducing these new technologies in the workflow of an investigation were presented but the opportunities of using NLP and AI are massive and are part of the future. Online demos were presented showing how fake images and videos can be produced with a single click of a button.

The second keynote was presented by **Yuri Gubanov**, the CEO and founder of **Belkasoft**, and gave industry insights into the hot topic of iOS Forensics. The keynote highlighted the latest security measures in iPhones, iPads, and other iOS devices and how Apple's efforts translate to difficulties for investigators. Recent advances, such as checkm8 and checkra1n jailbreak, were also presented, which allow researchers to acquire the so-called "full file system", which typically contains more data than a regular iTunes backup. Online demos were presented showing iOS device acquisition with checkm8, data extraction from a locked iPhone, and Belkasoft Evidence Center's advanced capabilities for mobile forensics.

In the course of the workshop, the technical presentations featured important research contributions on botnet, threat hunting, activity recognition in smart homes, cybersecurity education, and peer-to-peer threat investigations. The workshop is technically supported by the Association of Cyber Forensics and Threat Investigators (www.acfti.org) and the Industrial Cybersecurity Center (www.cci-es.org).

CFTIC 2020 gathered about 52 attendees on Discord. The discussions following the presentations and the closing showed a big interest in the novel and emerging research fields of cyber forensics and threat investigations. The participants had ample opportunity for professional exchange and networking, so that the 1st edition of the event can be regarded as a complete success.

Acknowledgements

We would like to thank the authors for their submissions, the Program Committee for their reviewing work, and the organizers of the ICSOC 2020 conference for their support which made this workshop possible.

Organization

Workshop Program Chairs

John William Walker   Nottingham Trent University, UK
Ahmed Elmesiry   University of South Wales, UK

Workshop Coordinator

Mamoun Qasem   University of South Wales, UK

Technical Program Committee

Rossana M. de Castro   Federal University of Ceará, Brazil
  Andrade
Ali Jwaid   De Montfort University, UK
Karima Boudaoud   University of Nice Sophia Antipolis, France
Dmitri Botvich   Waterford Institute of Technology, Ireland
Tawfik Al-Hadhrami   Nottingham Trent University, UK
Mohamed Aborizka   Arab Academy for Science, Technology and Maritime
                               Transport, Egypt

An Information Retrieval-Based Approach to Activity Recognition in Smart Homes

Brendon J. Woodford[1(✉)] and Ahmad Ghandour[2]

[1] Department of Information Science, University of Otago, Dunedin, New Zealand
`brendon.woodford@otago.ac.nz`
[2] College of Business, Al Ain University, Abu Dhabi, United Arab Emirates
`ahmad.ghandour@aau.ac.ae`

Abstract. One of the principal challenges in developing robust Machine Learning (ML) classification algorithms for Human Activity Recognition (HAR) from real-time smart home sensor data is how to account for variations in 1) the activity sequence length, 2) the contribution each sensor has to an activity, and 3) the amount of activity class imbalance. Such changes generate observations that do not conform to expected patterns potentially reducing the efficacy of classification models. Moreover the architecture of prior solutions have been quite complex which have resulted in large training times for these approaches to achieve acceptable classification accuracy. In this paper we address these three issues by 1) proposing a data structure representing the duration and frequency information of each sensor for an activity, 2) transforming this data structure into an Information Retrieval (IR)-based representation, and finally 3) compare and contrast the utility of this IR-based representation using four different supervised classifiers. Our proposed framework in combination with a state-of-the-art ensemble learner results in more accurate and scalable ML classification models that are better suited toward off-line HAR in a smart home setting.

Keywords: Human activity recognition · Smart homes · Machine learning

1 Introduction

Developing robust Data Mining (DM) methods for Human Activity Recognition (HAR) from real-time smart-home or smart-phone sensor data has been reported by the likes of [4,16,21] to be a challenging task. One of the main problems with correctly classifying peoples' actions is caused by observations which do not conform to an expected pattern. This is because human beings going about their daily lives are unlikely to execute activities in exactly the same way so models that attempt to classify such activities based on sensor activations need to adapt accordingly [22,27].

State-of-the-art methods have been influenced by work conducted in data stream clustering [23], concept drift adaptation [11], and activity recognition

© Springer Nature Switzerland AG 2021
H. Hacid et al. (Eds.): ICSOC 2020 Workshops, LNCS 12632, pp. 583–595, 2021.
https://doi.org/10.1007/978-3-030-76352-7_51

with labelled and unlabelled data [25] but there are still issues with determining the most appropriate representation of the data streams themselves which may enhance the performance of these approaches as van Kasteren [24] had recognized over a decade ago. There have been a number of different activity-based representations vary from widow-based techniques [21] to more knowledge driven ontology-based methods [4]. Even through there have been substantial advances in what representations are most appropriate for effective activity recognition [19], we argue that this problem has not been completely solved and that representation requires additional research.

To do so involves investigating the larger picture of intra and inter-sequencing of activities i.e. not all sensor sequences which take place have the same ordering, take the same amount of time, or belong to the same activity. Although, at face value, these issues could be regarded as confounding factors in assigning an activity class to a sequence of sensor readings, but we feel they may offer valuable information on how such data contribute to identifying it. Finally, there is also the question of class imbalance which may change over short or long periods of time [1] and the interleaving of activities when there's more than one occupant in a house [12].

Nevertheless, the fundamental problem to solve for HAR in smart homes is to correctly associate a sequence of sensor readings with an activity label. Knowing how many individual sensor readings should be somehow segmented or clustered to classify an activity is the subject of many recent works [11]. Recently, there have a been a number of authors who have concentrated their efforts applying on-line segmentation of sensors readings using fixed or dynamic windows combined with ensemble-based learning [10,22]. These state-of-the-art approaches have attempted to address the aforementioned challenges but to various degrees of success.

As the basis of our work we focus on how the frequency of sensor activations contribute to activity recognition. Although Jurek et al. [15] applied this method in their Cluster-Based Ensemble-Classifier (CBCE) framework, our approach is to consider the frequency information quite differently by transforming it into a Term Frequency-Inverse Document Frequency (TF-IDF) representation inspired by [13].

Our contributions are that we extend both these ideas into a new framework well suited to learning compact representations of data streams is employed creating a model with less complexity than prior solutions but still achieves comparable accuracy. As opposed to [15] we also incorporate discriminating between up to two residents' interleaving activities and adopt more sensors than those reported in [13] in an attempt to improve classifier performance.

The rest of the paper is organized as follows. In Sect. 2, we briefly review relevant prior work in techniques for classification of smart home sensor data in smart homes. Section 3 briefly describes our proposed methodology. The details of the methodology and experiments applying our framework are described in Sect. 4. The results and discussion of the experiments are presented in Sect. 5. Finally, in Sect. 6 we revisit our key objectives in the context of the results we obtained leading to areas of improvement in future work.

2   Background

The specific motivation for our work has been driven by the increased attention to smart homes in healthcare [1]. Amiribesheli et al. [2] surveyed the state of these technologies, which when combined, form the infrastructure to model and recognize activities to support the elderly in their independent living especially for those individuals who have chronic conditions (e.g. dementia). Moreover, detecting these activities may also help to identify anomalous behaviors to aid such individuals to maintain their independence.

To identify activities requires that sensor activations somehow be combined to form an instance of an activity. Because the duration of the activity is not known in advance work conducted by [4] investigated how a sliding fixed window coupled with a knowledge-based approach could be used to segment sensor readings into activities. [21] extended this approach proposing a dynamically sized window that could shrink and expand to better accommodate the correct number of sensors readings contributing to an instance of an activity. Similarly, [19] adopted this same approach but also past contextual information e.g. that a 'Enter_Home' activity would normally happen after a 'Leave_Home' activity. However there are weaknesses in this prior work such as the undue complexity of the windowing algorithms and the inability to handle significant class imbalance.

As opposed to the conventional techniques for analyzing this type of data on-line which are largely based on processing the readings at the individual sensor-level, [14] describes an off-line method employed to learn multiple models for each activity using a novel context-aware hierarchical clustering algorithm which retains information about the sequences of activities that a residents' regular behavior is composed of.

[15] performed similar work where the frequency of sensor activations for each activity are computed and fed into a clustering-based ensemble learning model. Although this method addresses the class imbalance issue, it loses the sequence in which the sensors were activated. In addition, the two representations of sensor activations used were based on simple frequency-based information; either how many times a sensor activated during an activity or whether or not that a sensor was active during an activity.

However, we argue that such frequency-based information could be of use if it provided more insight into which sensor(s) contributed more to classifying an instance of an activity. Furthermore, by doing so such a method could still produce prototypical configurations of various activities and these prototypes can then be analysed clustered, compared, and classified.

3   Methodology

Our proposed framework consists of three distinct parts:

1. A data pre-processing step which constructs sensor activation sequences into prototype activities.

2. Transformation of these prototype activities into a TF-IDF representation.
3. Use the TF-IDF representation as input to four different supervised learning algorithms.

4   Framework Implementation

4.1   Data Preprocessing

As mentioned beforehand, in this article we adopted a different approach compared with previous work. Instead of generating variable-length windows of sensor sequences as proposed by [21], we create a vector representing the duration and frequency information of each sensor for an activity similar to the work of [14,15]. In this way each vector can be investigated and compared with each other as they have the same length. This "chunking" or meta-representation also lends itself to a simpler and more appropriate representation for subsequent TF-IDF transformation. An additional advantage of this approach is that the representation can be built up independent of the number of residents living in the home.

To elaborate, each instance of the original data set is a six-dimensional vector where each instance consists of a mandatory date, time, sensor number, a corresponding sensor status, and an optional activity type and activity status. Algorithm 1 presents a version of the method used to transform these individual sensor readings into a generic representation suitable for any supervised or unsupervised learning algorithm. Algorithm 1 assumes that for each identified activity there is a corresponding 'end' activity status for each 'begin' activity status.

All generated activity prototypes are represented using six features, $\{f_1, f_2, \ldots, f_6\}$. f_1 is the date and start time of the activity, f_2 is the elapsed time in seconds for an activity, f_3 is the sensor that activated at the start of the activity, f_4 is the sensor that activated at the end of the activity, f_5 is a vector containing the frequency information for each sensor that activated during the occurrence of the activity, and finally f_6 is the activity type itself.

For example, presented in Fig. 1 is a section from the Aruba data set [6]. After Algorithm 1 is run over this section, a single activity vector is constructed as presented in Table 1.

With a significantly reduced number of prototypes compared with the number of instances in the original data set as shown in Table 3, training of a suitable classifier should take less time as there are fewer examples to learn. This is an advantage of the representation for these prototypes but the real strength of it is this representation affords a wider range of data analysis techniques to be applied since we are now not restricted to only on-line data stream classification methods.

4.2 Generating the TF-IDF Representation

Our assumption is that some sensors would be more important than others to aid activity recognition process. In order to determine these critical sensors, we

---

ALGORITHM 1: Activity Feature Selection

Input:

S: A data set of sensor readings.

activityTypes: A vector of activity type labels.

sensorTypes: A vector of sensor labels.

Output:

$SEFMatrix$: A set of activity prototypes each labelled with a corresponding activity type.

1 $resStorage \leftarrow$ **Size**(*activityTypes*);
2 $FVMatrix \leftarrow$ **Size**(*sensorTypes*);
3 $actInd \leftarrow 1$;
4 **for** $i \leftarrow 1$ **to Size** (S) **do**
5       $SR \leftarrow$ instance of data set S_i;
6   Identify activity type, AT from SR;
7   Determine activity status AS from AT;
8   **if** AS status is `'begin'` **then**
9   Set $resInd$ to the index in *activityTypes* where AT was found;
10   Set $resStorage[resInd].startTime$ to the start timestamp of the activity;
11   Set $resStorage[resInd].startSensor$ to the activated sensor in SR;
12          $resStorage[resInd].activityTypes \leftarrow AT$;
13   **else if** AS status is `'end'` **then**
14   Set $resInd$ to the index in *activityTypes* where AT was found;
15   Set $resStorage[resInd].endSensor$ to the activated sensor in SR;
16   Set *activityDuration* to the number of seconds elapsed between the start and end times of the activity;
17          $SEFMatrix[actInd, 1] = resStorage[resIndex].startTime$;
18          $SEFMatrix[actInd, 2] = activityDuration$;
19          $SEFMatrix[actInd, 3] = startSensor$;
20          $SEFMatrix[actInd, 4] = endSensor$;
21          $SEFMatrix[actInd, 5] = resStorage[resInd].FVMatrix$;
22          $SEFMatrix[actInd, 6] = AT$;
23          $resStorage[resIndex] = activityInstance$;
24          $actInd += 1$;
25   **else if** $resIndex > 0$ **then**
26   Set *theIndex* to the location in *sensorTypes* where the activated sensor in SR was found;
27          $resStorage[resInd].FVMatrix(theIndex) += 1$;
28   **end**
29 **end**

```
. . .
2010-11-04        05:40:51.303739 M004 ON Bed_to_Toilet begin
2010-11-04        05:40:52.342105 M005 OFF
2010-11-04        05:40:57.176409 M007 OFF
2010-11-04        05:40:57.941486 M004 OFF
2010-11-04        05:43:24.021475 M004 ON
2010-11-04        05:43:26.273181 M004 OFF
2010-11-04        05:43:26.345503 M007 ON
2010-11-04        05:43:26.793102 M004 ON
2010-11-04        05:43:27.195347 M007 OFF
2010-11-04        05:43:27.787437 M007 ON
2010-11-04        05:43:29.711796 M005 ON
2010-11-04        05:43:30.279021 M004 OFF Bed_to_Toilet end
. . .
```

Fig. 1. Extract of sensor data from the Aruba data set [6].

Table 1. Breakdown of an activity prototype

Feature	Description	Value
f_1	Start timestamp	'2010-11-04 05:40:51.303739'
f_2	Activity duration	158.9753
f_3	First sensor	'M004'
f_4	Last sensor	'M004'
f_5	Sensor activations	0,0,0,6,2,0,4,0,0,0,0,0,0
		0,0,0,0,0,0,0,0,0,0,0,0,0
		0,0,0,0,0,0,0,0,0,0,0,0,0
		0,0,0,0,0,0,0,0,0,0,0,0,0
f_6	Activity label	'Bed_to_Toilet'

adopt the TF-IDF measure from the field of Information Retrieval (IR) [18] to evaluate sensor importance.

Consider a set of terms $T = \{t_1, t_2, \ldots, t_m\}$ and a set of documents $D = \{d_1, d_2, \ldots, d_n\}$. The common use of TF-IDF is to evaluate how important a term, $t \in T$, is to a document, $d \in D$. More formally, TF-IDF is defined in Eq. 1 as

$$\text{TF-IDF}(t, d) = \text{TF}(t, d) \times \text{IDF}(d, t, D) \tag{1}$$

$\text{TF}(t, d) = \text{times}(t, d)/(\text{times}(t_1, d) + \text{times}(t_2, d) + \ldots + \text{times}(t_m, d))$ where $\text{times}(t, d)$ is how many times the term, t, appears in the document d. $\text{IDF}(t, d, D)$ $= \log(|D|/(1 + |\{d|d \in D \text{ and } \text{times}(t, d) > 0|\}))$.

We can adapt this equation to determine how important a sensor is to an activity instance. So, instead of a set of terms, T, we have a set of sensors, $S = \{s_1, s_2, \ldots, s_m\}$. Similarly instead of the set of documents, D, we have a set of activity instances $AI = \{ai_1, ai_2, \ldots, ai_n\}$. TF-IDF$(s, ai)$ is therefore defined as TF-IDF(s, ai) = TF(s, ai) × IDF(ai, s, AI). By applying TF-IDF in this way, sensors that occur often in a specific activity instance but not in all activity instances will have higher weights assigned to them. To this end we can transform $SEFMatrix$ by Eq. 1 into a TF-IDF representation.

Finally, as the range of values produced by TF-IDF sometimes could be outside the boundaries of acceptable values for input to a learner, the TF-IDF values are both normalized using both Eq. 2 and Eq. 3 as defined by [13]. Performing these transformations on the original (Raw) $SEFMatrix$ produces three additional data sets 1) TF-IDF, 2) LogSig(TF-IDF), and 3) TanSig(TF-IDF). Our assumption in having the original $SEFMatrix$ and its three aforementioned variants of each smart home data set is that the Raw representation would provide a baseline performance for the classifier it was trained and tested on. A similar classifier would then exhibit improved performance when trained and tested on each of the three variants. By demonstrating this improvement in performance would help to justify the utility of these IR-based representations of smart home sensor data.

$$\text{LogSig}(\text{TF-IDF}(s, ai)) = \frac{1}{1 + e^{(\text{TF-IDF}(s,ai))}} \tag{2}$$

$$\text{TanSig}(\text{TF-IDF}(s, ai)) = \frac{e^{\text{TF-IDF}(s,ai)} - e^{-(\text{TF-IDF}(s,ai))}}{e^{\text{TF-IDF}(s,ai)} + e^{-(\text{TF-IDF}(s,ai))}} \tag{3}$$

To evaluate our method prototypes from two smart home data sets, Aruba [6] and Kyoto [7], were used for all experiments reported in this work. A feature of Algorithm 1 described in Sect. 4 is that it has been designed to discriminate between interleaving activities of up to two residents which is the case with the Kyoto data set. The number and type of sensors for each smart home is summarized in Table 2. As can be seen, the configuration of the sensors in the two smart homes were varied but this was a feature of these smart homes as we wanted to test our framework on as many diverse smart home data sets as possible.

Table 2. Set-up of each smart home and number of sensors.

Data set	Sensor type				
	Motion	Door	Temperature	Pressure	Other
Aruba	31	3	5	N/A	3
Kyoto	51	15	5	1	16

Table 3. Breakdown for the number of instances of the two data sets.

Data set	Individual sensor readings	Prototypes created
Aruba	1719559	6477
Kyoto	2804813	3744

Table 4. Breakdown of the number of prototypes for each activity class.

Aruba		Kyoto	
Activity class	No. Instances	Activity class	No. Instances
Meal preparation	1606	R1 Bathing	31
Relax	2919	R1 Bed toilet transition	24
Eating	257	R1 Eating	26
Work	171	R1 Enter home	85
Sleeping	401	R1 Housekeeping	1
Wash dishes	65	R1 Leave home	147
Bed to toilet	157	R1 Meal preparation	131
Enter home	431	R1 Personal hygiene	550
Leave home	431	R1 Sleep	274
Housekeeping	33	R1 Sleeping not in bed	4
Resperate	6	R1 Wandering in room	15
		R1 Watch TV	110
		R1 Work	504
		R2 Bathing	55
		R2 Bed toilet transition	6
		R2 Eating	71
		R2 Enter home	52
		R2 Leave home	68
		R2 Meal preparation	184
		R2 Personal hygiene	534
		R2 Sleep	305
		R2 Sleeping not in bed	2
		R2 Wandering in room	5
		R2 Watch TV	117
		R2 Work	432
Totals	6477		3743

Table 3 and Table 4 show the breakdown of the data sets had been processed by Algorithm 1. Not only were there differences in the number of activity classes, but there was large variability in the number of instances representing each class (ranging from 1 to 2919).

Four different supervised learning methods were used to compare and contrast the utility of the four different representations of the sensor data. A Support Vector Machine (*SVM*) [8], decision tree (*DT*) [26], ensemble of decision trees using bagging (*Ensemble*) [9], and the recent and popular extreme gradient boosting (*XGBoost*) framework [5]. All experiments were conducted using Python's scikit-learn and *XGBoost* packages on a laptop using a dual-core Intel i5-2540M processor with 8GB memory running Ubuntu 20.04. The use of the

Python hyperopt[1] package was used to fine-tune the learning parameters for all algorithms. With the number of activity classes present in the data sets and the obvious presence of class imbalance, created many challenges for supervised machine learning techniques to accurately classify them. For this reason, the results of five-fold cross validation was used to evaluate the performance of the supervised learning models. Performance measures used were weighted Precision, Recall, and F1-Scores as there were more than two classes in each of the data sets. The results are summarized in Tables 5 and 6.

5 Results and Discussion

In terms of the suitability of our proposed framework to processing the smart home data sets, it did generate quite different representations for training and testing each of the four classifiers. We also expected that a classifier built using

Table 5. Classifier performance of the four Aruba data set variants.

Classifier	Train			Test		
	Precision	Recall	F1-Score	Precision	Recall	F1-Score
Raw						
SVM	0.9340	0.9376	0.9331	0.9255	0.9298	0.9241
DT	**0.9850**	**0.9847**	**0.9848**	0.9201	0.9244	0.9217
Ensemble	0.9822	0.9822	0.9820	0.9128	0.9252	0.9187
XGBoost	0.9787	0.9786	0.9785	**0.9406**	**0.9452**	**0.9428**
TF-IDF						
SVM	0.9381	0.9359	0.9307	0.9340	0.9406	0.9368
DT	**0.9848**	**0.9846**	**0.9846**	0.9272	0.9190	0.9226
Ensemble	0.9822	0.9822	0.9817	0.9289	0.9352	0.9315
XGBoost	0.9739	0.9737	0.9737	**0.9527**	**0.9522**	**0.9487**
LogSig(TF-IDF)						
SVM	0.9395	0.9386	0.9344	0.9263	0.9313	0.9271
DT	**0.9828**	**0.9826**	**0.9827**	0.9272	0.9236	0.9248
Ensemble	0.9808	0.9807	0.9803	0.9355	0.9355	0.9379
XGBoost	0.9789	0.9788	0.9787	**0.9396**	**0.9506**	**0.9449**
TanSig(TF-IDF)						
SVM	0.9330	0.9375	0.9332	0.9249	0.9313	0.9256
DT	0.9816	0.9815	0.9815	0.9203	0.9174	0.9182
Ensemble	**0.9826**	**0.9824**	**0.9821**	0.9235	0.9313	0.9268
XGBoost	0.9818	0.9817	0.9816	**0.9355**	**0.9452**	**0.9399**

---

[1] http://hyperopt.github.io/hyperopt/. Last accessed 30th September, 2020.

the original *SEFMatrix* would perform comparatively poorly. To this end, the results reported for the variants of the Aruba data set in Table 5 suggest that there is no one classifier that achieves the best training performance. All classifiers exhibited good performance when assessed using the three performance measures. The best training performance was split between the *DT* and *Ensemble* classifiers but the testing performance was consistently better when the *XGBoost* classifier was adopted. Finally, all classifiers benefited from the TF-IDF representation of *SEFMatrix* but the performance of all four classifiers degraded a little when trained and tested using the LogSig(TF-IDF) and TanSig(TF-IDF) data sets.

Table 6. Classifier performance of the four Kyoto data set variants.

Classifier	Train			Test		
	Precision	Recall	F1-Score	Precision	Recall	F1-Score
Raw						
SVM	0.7269	0.7287	0.7110	0.6014	0.6101	0.5925
DT	0.9327	0.9318	0.9320	0.6240	0.6128	0.6131
Ensemble	0.9372	0.9369	0.9364	0.6693	0.6742	0.6634
XGBoost	**0.9937**	**0.9947**	**0.9942**	**0.7569**	**0.7583**	**0.7490**
TF-IDF						
SVM	0.7292	0.7274	0.7120	0.6195	0.6395	0.6215
DT	0.9248	0.9245	0.9244	0.6412	0.6355	0.6348
Ensemble	0.9299	0.9292	0.9287	0.6786	0.6849	0.6766
XGBoost	**0.9973**	**0.9977**	**0.9975**	**0.7566**	**0.7664**	**0.7589**
LogSig(TF-IDF)						
SVM	0.6980	0.7050	0.6877	0.6534	0.6689	0.6526
DT	0.9168	0.9151	0.9151	0.6361	0.6382	0.6352
Ensemble	0.9149	0.9131	0.9129	0.6759	0.6903	0.6772
XGBoost	**0.9710**	**0.9719**	**0.9706**	**0.7085**	**0.7130**	**0.7023**
TanSig(TF-IDF)						
SVM	0.6858	0.6973	0.6798	0.6277	0.6489	0.6310
DT	0.9185	0.9161	0.9163	0.6266	0.6182	0.6191
Ensemble	0.9153	0.9135	0.9132	0.6922	0.7063	0.6935
XGBoost	**0.9849**	**0.9846**	**0.9839**	**0.7356**	**0.7330**	**0.7277**

Table 6 suggests that there is more variability of classifier performance when trained and tested on the four variants of the Kyoto data set. *SVM* performance was poor across all data set variants when compared with the other three classifiers. Furthermore, although the *DT*, *Ensemble*, and *XGBoost* classifiers exhibited good training performance, there was a marked decrease in testing performance for these three classifiers possibly due to the number of classes (Kyoto (25) vs. Aruba (11)). In general, however, the *XGBoost* classifier exhibited consistently better performance when trained and tested on the TF-IDF representation of the *SEFMatrix* and potentially be adopted subsequent to judicious hyper-parameter tuning to improve the testing performance.

6 Conclusion

In this paper we have proposed a new framework for investigating smart home sensor data with a view to discovering how sensor activation frequency and duration relates to the various activities carried out by the residents of smart homes. We employ a new algorithm as described in Algorithm 1 to generate prototypical activities representative of movements of residents going about their daily lives. Unlike current state-of-the-art methods whose efforts are channelled into on-line activity recognition, we were motivated by more off-line analyses which would help to assist in identifying latent information contained in these sensor activations which improve the recognition of the activity. Moreover, we comprehensively assessed the overall performance of this representation and its variants by using four popular supervised classifiers including the more recent *XGBoost* classifier. Our overall conclusion is that there are benefits in adopting the proposed IR-based representation for off-line analysis of smart home sensor data.

Future work will be in the areas of extending the framework to better handle uncertainties in the smart home environments themselves specifically those activity recognition techniques which have been identified by [17] applying concepts for an on-line rule-learning Type-2 fuzzy classifier originally proposed by [3] and also investigating how more knowledge-based algorithms can be adopted based on the work of [20]. Finally, one other direction would be to consider how the work of [25] who developed a method to recognize and model activities in smart homes using both labelled and unlabelled data can be incorporated into our framework.

References

1. Abidine, M.B., Fergani, L., Fergani, B., Fleury, A.: Improving human activity recognition in smart homes. Int. J. E-Health Med. Commun. **6**(3), 19–37 (2015). https://doi.org/10.4018/IJEHMC.2015070102
2. Amiribesheli, M., Benmansour, A., Bouchachia, A.: A review of smart homes in healthcare. J. Amb. Intell. Hum. Comput. **6**(4), 495–517 (2015). https://doi.org/10.1007/s12652-015-0270-2
3. Bouchachia, A., Vanret, C.: GT2FC: an online growing interval type-2 self-learning fuzzy classifier. IEEE Trans. Fuzzy Syst. **22**(4), 999–1018 (2014)
4. Chen, L., Nugent, C.D., Wang, H.: A knowledge-driven approach to activity recognition in smart homes. IEEE Trans. Knowl. Data Eng. **24**(6), 961–974 (2012)

5. Chen, T., Guestrin, C.: XGBoost: a scalable tree boosting system. In: Proceedings of 22nd ACM SIGKDD International Conference on Knowledge Discovery and Data Mining, KDD 2016, pp. 785–794. ACM, New York (2016)
6. Cook, D.J.: Learning setting-generalized activity models for smart spaces. IEEE Intel. Syst. **27**(1), 32–38 (2012)
7. Cook, D.J., Schmitter-Edgecombe, M.: Assessing the quality of activities in a smart environment. Method Inf. Med. **48**(5), 480–485 (2009)
8. Cortes, C., Vapnik, V.: Support-vector networks. Mach. Learn. **20**(3), 273–297 (1995)
9. Dietterich, T.G.: An experimental comparison of three methods for constructing ensembles of decision trees: bagging, boosting, and randomization. Mach. Learn. **40**(2), 139–157 (2000)
10. Duarte, J., Gama, J., Bifet, A.: Adaptive model rules from high-speed data streams. ACM Trans. Knowl. Discov. Data **10**(3), 30:1–30:22 (2016)
11. Gama, J., Žliobaitė, I., Bifet, A., Pechenizkiy, M., Bouchachia, A.: A survey on concept drift adaptation. ACM Comput. Surv. **46**(4), 44:1–44:37 (2014)
12. Gu, T., Wu, Z., Tao, X., Pung, H.K., Lu, J.: epSICAR: an emerging patterns based approach to sequential, interleaved and concurrent activity recognition. In: Proceedings of 7th IEEE International Conference on Pervasive Computing and Communications, pp. 1–9. IEEE (2009)
13. Guo, J., Mu, Y., Xiong, M., Liu, Y., Gu, J.: Activity feature solving based on TF-IDF for activity recognition in smart homes. Complexity **37**, 1–10 (2019)
14. Hoque, E., Dickerson, R.F., Preum, S.M., Hanson, M., Barth, A., Stankovic, J.A.: Holmes: a comprehensive anomaly detection system for daily in-home activities. In: 2015 International Conference on Distributed Computing in Sensor Systems, pp. 40–51. IEEE Press, June 2015
15. Jurek, A., Nugent, C., Bi, Y., Wu, S.: Clustering-Based Ensemble Learning for Activity Recognition in Smart Homes. Sensors **14**, 12285–12304 (2014)
16. Kim, E., Helal, S., Cook, D.: Human activity recognition and pattern discovery. IEEE Perv. Comput. **9**(1), 48–53 (2010)
17. Kim, E., Helal, S., Nugent, C., Beattie, M.: Analyzing activity recognition uncertainties in smart home environments. ACM Trans. Intell. Syst. Technol. **6**(4), 52:1–52:28 (2015). https://doi.org/10.1145/2651445
18. Kondylidis, N., Tzelepi, M., Tefas, A.: Exploiting TF-IDF in deep convolutional neural networks for content based image retrieval. Multimed. Tools Appl. **77**(23), 30729–30748 (2018)
19. Krishnan, N.C., Cook, D.J.: Activity recognition on streaming sensor data. Perv. Mob. Comput. **10**(Part B), 138–154 (2014)
20. Lühr, S., Lazarescu, M.: Incremental clustering of dynamic data streams using connectivity-based representative points. IEEE Trans. Knowl. Data Eng. **68**, 1–27 (2009)
21. Okeyo, G., Chen, L., Wang, H., Sterritt, R.: Dynamic sensor data segmentation for real-time knowledge-driven activity recognition. Perv. Mob. Comput. **10**(Part B), 155–172 (2014)
22. Sagha, H., Bayati, H., Millán, J.D.R., Chavarriaga, R.: On-line anomaly detection and resilience in classifier ensembles. Patt. Recogn. Lett. **34**(15), 1916–1927 (2013)
23. Silva, J.A., Faria, E.R., Barros, R.C., Hruschka, E.R., Carvalho, A.C.D., Gama, J.: Data stream clustering: a survey. ACM Comput. Surv. **46**(1), 13:1–13:31 (2013)
24. Van Kasteren, T., Noulas, A., Englebienne, G., Kröse, B.: Accurate activity recognition in a home setting. In: Proceedings of 10th International Conference on Ubiquitous Computing, pp. 1–9. ACM (2008)

25. Wen, J., Zhong, M.: Activity discovering and modelling with labelled and unlabelled data in smart environments. Expert Syst. Appl. **42**(14), 5800–5810 (2015)
26. Wu, X., et al.: Top 10 algorithms in data mining. Knowl. Inf. Syst. **14**(1), 1–37 (2007)
27. Zhu, C., Sheng, W., Liu, M.: Wearable sensor-based behavioral anomaly detection in smart assisted living systems. IEEE Trans. Autom. Sci. Eng. **12**(4), 1225–1234 (2015)

Botnet Sizes: When Maths Meet Myths

Elisa Chiapponi[1([⊠])], Marc Dacier[1], Massimiliano Todisco[1], Onur Catakoglu[2], and Olivier Thonnard[2]

[1] Eurecom, Biot, France
{elisa.chiapponi,marc.dacier,massimiliano.todisco}@eurecom.fr
[2] Amadeus IT Group, Biot, France
{onur.catakoglu,olivier.thonnard}@amadeus.com

Abstract. This paper proposes a method and empirical pieces of evidence to investigate the claim commonly made that proxy services used by web scraping bots have millions of residential IPs at their disposal. Using a real-world setup, we have had access to the logs of close to 20 heavily targeted websites and have carried out an experiment over a two months period. Based on the gathered empirical pieces of evidence, we propose mathematical models that indicate that the amount of IPs is likely 2 to 3 orders of magnitude smaller than the one claimed. This finding suggests that an IP reputation-based blocking strategy could be effective, contrary to what operators of these websites think today.

1 Introduction

This work has been realised in close collaboration with a major IT provider for the airline industry which hosts several dozens of airline websites. These sites are protected by one of the leading commercial anti-bots services, placed in front of them. This service checks the origin and the fingerprints associated with each request against a large number of "signatures"[1].

Bots have been a plague for the Internet for more than 20 years. Early warnings date back to the 2000s with the early DDoS attacks against major websites [3]. Since then, they have continuously evolved from relatively rudimentary pieces of software to very sophisticated components such as the numerous "all in one sneaker bots" (e.g., aiobot.com) that automate the buying process of luxury goods in high demands. To increase their resilience, the bots take advantage of proxy services publicly available on the web, for a fee. Thanks to these services, the bots use temporarily IP addresses that are owned and used by legit users. There are, supposedly, tens of millions of such IPs made available to bots. Would the targeted websites decide to block each IP which is considered to behave like a bot, they would quickly deny access to millions of IPs, some of them belonging to potential customers. Clearly, an IP blocking solution does not appear to be a viable approach due to the, supposedly, sheer volume of IPs, available all over the world.

---

[1] This is a simplistic explanation. We refer the interested reader to [23] for more information on such existing commercial offerings.

© Springer Nature Switzerland AG 2021
H. Hacid et al. (Eds.): ICSOC 2020 Workshops, LNCS 12632, pp. 596–611, 2021.
https://doi.org/10.1007/978-3-030-76352-7_52

In this paper, we use empirical evidence to investigate the conjecture that such IP blocking strategy will always fail. We reach the conclusion that the situation might not be as bleak as it might seem.

In order to present our findings, the paper is structured as follows. In Sect. 2, we outline the problem faced and our contributions. Section 3 presents the state of the art on web scraping bots prevention. Section 4 describes the experimental setup and the data it produced. Section 5 briefly describes the raw results obtained over a period of 56 days. Section 6 assesses the credibility associated with the belief that these botnets have millions of IP addresses at their disposal. Mathematical analysis confronted with the empirical pieces of evidence leads us to adjudicate against that belief. In Sect. 7, we gather additional information about the IPs observed in order to consolidate the ideas developed in Sect. 6. In Sect. 8, we discuss the lessons learned thanks to our experiment and analysis. A conclusion as well as thoughts for future work are offered in Sect. 9.

2 Problem Definition and Contributions

A 2019 Imperva report [6] describes how the airlines industry is heavily impacted by large armies of bots. In 2017, according to that report, the proportion of bad bots traffic to airline websites was 43.9%. Almost all these bots are used to gather free information from the airlines' sites about flights and ticket prices. It is commonly agreed that the actors behind these bots activities are unauthorized business intelligence companies, online travel agencies and data aggregators. Indeed, a large part of their business relies on web scraping and using bots instead of having a paying agreement with the targeted websites is much more profitable for them. They harness information, increasing dramatically the amount of requests to be served by airlines websites. Responding to these requests, due to the price ticketing process, is an expensive task which well-behaving organisations normally pay for. The bots aim at getting the same service *for free*. By doing so, they misuse the service provided by airlines companies to individual users.

An arms race exists between bot makers and anti-bot providers. The bot detection relies on a number of different fingerprinting techniques to recognize malicious agents [23]. As soon as a family of bots is identified and blocked, their bot masters replace them with new ones. Blocking all the IP addresses of identified bots is usually not seen as a viable option because it is well known that the real IP addresses of the bots remain hidden behind a large amount of proxy IP addresses provided by professional services. These services claim to offer to their customers millions of residential IP addresses, leaving any IP blocking solution doomed to potentially block a large amount of legit customers.

Quoting one of these websites [12], we see that they offer to their customers to "use [their] rotating residential proxies comprised of real user devices, making them undetectable when used correctly". The owners of these real devices, also called exit nodes, "[...] agreed to route [...] traffic through their hosts in exchange for free service" [12]. A quick search on the Internet returns more than a dozen similar proxy service offerings. We prefer not to offer them some additional advertisement by listing them all here. Suffice it to say that, for instance, both [12,15] claim to offer more than 70 millions of IP addresses whereas [20] supposedly has more than 10 millions IPs. Others have similar claims.

The benefits of hiding behind this very large pool of IP addresses is threefold for the web scraping actors: first, linking a scraping campaign to any known organisation is impossible, thus no attribution and legal recourse; second, the impressive number of frequently changing IP addresses used renders any IP blocking strategy impractical; third, they can run these campaigns with a very limited amount of powerful machines on their back end without the need of any vast and highly distributed infrastructure.

In [2], we describe an experiment designed to analyse the behavior of these bots. That experiment did confirm the existence of these advanced persistent bots (APBs) and the proxies they were relying on. It also raised questions regarding the real amount of IPs put at the disposal of the bots. In this work, we carry out an in-depth investigation of that question. By doing so, we provide the following contributions:

- We provide additional empirical pieces of evidence of the existence of very stealthy APBs and confirm the usage of proxy servers by these bots
- Using two distinct approaches, we show that i) IP addresses provided to the bots are not randomly assigned and that ii) the pool of IPs they are taken from is *two* to *three* orders of magnitude smaller than what is announced by the proxy websites.
- We explain how the idea of IP-blocking could be rejuvenated to defeat such sophisticated bots.

3   State of the Art

Botnets, collections of hosts controlled by a bot, have been used for the years for nefarious activities, such as scraping web pages of different industries [5]. Applying IP reputation to mitigate the threats of web scraping is not a new idea [4]. Moreover, this technique has already been largely used against spam bots [10].

However, as shown in the Imperva Report 2020 [5], recent years have witnessed the rise of traffic produced by Advanced Persistent Bots (APBs). These bots produce few requests per IP staying below the rate limits and protecting their reputation. They rely on professional proxy services that make large numbers of IP addresses available for these activities [19]. These services claim to have access to tens of millions of residential IPs and to be able to rotate them among the different requests of each client. For these reasons, the report [5] asserts that IP blacklisting has become "wholly ineffective". Doubtlessly, millions of different IP cannot be blacklisted all together and e-commerce websites cannot risk to block requests coming from real customers.

In 2019, Mi et al. [13] proposed the first comprehensive study of Residential IP Proxy as a Service. Even if their methodology has been partially criticized for the fingerprinting process of the devices [16,17], they created a successful infiltration framework that enabled them to study residential proxy services from the inside. They collected 6.18 milions of IPs, of which 95,22% are believed to be residential. Among their findings, it is peculiar to see a discrepancy between the number of IPs claimed by Luminati [12] (30 millions) and the ones collected by them for the same provider (4 millions using 16 million probings). The authors provide no clear explanation for this gap. Furthermore it is noteworthy to mention the discovery of two providers using the same pool of IPs, while another one built its network on top of Luminati [12]. Our paper also aims at better understanding the residential IP proxies ecosystem by providing a different view point.

Nowadays, web site owners usually take advantage of third party anti-bot services to perform bot management. These commercial solutions analyse the incoming requests to the websites. As described in [23], multiple parameters are collected from the environment in which the request is generated, thanks to fingerprinting. This set of parameters can be used to identify the same actor who launches different requests, potentially from different IP addresses. If a signature is recognized as coming from a bot, the corresponding traffic can be blocked or other mitigation actions can be put in place.

Azad et al. [1], propose an empirical analysis of some anti bot services. Unfortunately, their findings indicate that these solutions are mostly efficient against basic bots but not against the truly sophisticated ones. Indeed, an arms race is taking place between anti bot services trying to fingerprint and bots trying to circumvent the detection. This has led the actors behind the bots to perform only small amounts of requests per IP, with the goal of remaining undetected.

4   Experimental Setup

In [2], we have described a honeypot-based experimental setup designed to analyze the behavior of web scraping bots. We offer in this Section a brief recap of that experiment as it is the source of the data we will be analysing in the rest of the paper. The interested reader is referred to [2] for more detailed information.

This experiment was run in close collaboration with a major IT provider for airlines websites. This party handles the calculation of the fares and the booking process for multiple airlines. The airlines companies pay the IT provider an amount proportional to the transactions served. Naturally, an excess of bot traffic dramatically increases the volume of transactions and thus the infrastructure costs for both the airlines and the IT provider. In recent years, bots started to perform an intensive price scraping activity towards airline's websites, producing up to 90% of the requests on some domains [6].

To mitigate this phenomenon, the IT provider with which we collaborate, is using a commercial bot detection service provider. A box is put in front of the provider's booking domains and it detects bots thanks to browser fingerprinting and machine learning. Every request is studied and a signature is assigned to it. If the signature matches the one of a bot, an action is taken such as blocking, serving a CAPTCHA [24] or a JavaScript challenge. However, sophisticated bots, dubbed APBs for Advanced Persistent Bots [5], can overcome these countermeasures and/or change their parameters to avoid detection [14].

This solution works but has a major drawback, which is to provide feedback to the bots when they are identified. They use this information to morph as soon as they detect that they have been unmasked. By doing so, they defeat the mitigation process provided by the anti-bot solution. To overcome this problem, we have decided to create a new action associated with a signature match: requests coming from identified bots would now be redirected to a real-looking, yet fake, web page. This web page, which can be seen as an application layer honeypot, serves two distinct objectives: i) reduce the workload of the production servers, ii) study the behaviour of the bots.

This honeypot is able to produce responses that are, syntactically, indistinguishable from the real ones. However, semantically, they differ because we use cached values or, sometimes, modified values for the tickets. Cached values dramatically reduce the cost of computing the responses. Modified values enable us to analyse to what extent the bots are capable of detecting erroneous information provided to them.

We have designed and implemented such a platform in collaboration with the IT provider and a specific airline company. We chose a company whose traffic was highly impacted by bots. At the time of our experiment, that company was receiving, on average, 1 million requests per day, of which 40% were detected as bot traffic by the anti bot solution. Unfortunately, the anti-bot solution is not capable of blocking all bots. Each signature is associated with a confidence level indicating the uncertainty whether the request comes, or not, from an ill behaving bot. Depending on that value, that IP will be blocked, challenged (e.g. with a CAPTCHA) or simply put under scrutiny (e.g. to be blocked later if it sends a suspiciously large number of requests). We focused on that last category and found, for that airline, a signature that was matched every day, always in the same small time window of 40 min by, roughly, the same amount of IPs. Last but not least, almost none of these IPs ever booked a ticket. All these elements gave us great confidence that that signature, while not blocked by the anti bot solution, was reliably identifying members of a specific botnet. We have then configured the anti-bot solution to redirect all requests matching that signature to our honeypot.

For this publication to be self-contained, we offer in the next Section a synthetic presentation of the raw results obtained and some statistical results.

5   Experimental Results

The experiment ran for 56 days, between 7th January and 2nd March 2020. We have had no match for our signature after that date. We believe the reason has to do with the business needs of the actor behind these bots. Indeed, that date coincides with the beginning of the worldwide pandemic. Furthermore, the airline, subject of our experiment, is the main one for a country whose government issued its first major travel restriction on the 2nd of March, practically shutting down airline travel to and from that country. Without any customer interested in buying tickets to/from that country, there was no incentive for the

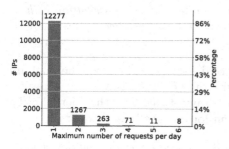

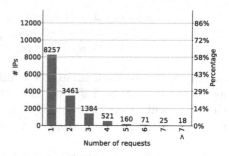

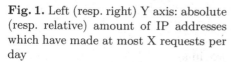

Fig. 1. Left (resp. right) Y axis: absolute (resp. relative) amount of IP addresses which have made at most X requests per day

Fig. 2. Left (resp. right) Y axis: absolute (resp. relative) amount of IP addresses which have made at most a grand total of X requests during the whole period of the experiment.

malicious actor to keep collecting ticket prices from that company. This most likely explains the disappearance of these bots.

Over the duration of the experiment, the honeypot has received 22,991 requests. The daily average amount was 410 with a standard deviation of 33 queries. All requests arrived at the same time of the day. The signatures were only seen during a small time window of 38.18 min, on average. The amount and the timing of the requests were in line with those of that bot signature before the beginning of the experiment.

The 22,991 requests were issued by 13,897 unique IPs. Figure 1 shows that most of the IPs (97% of the total) made at most two requests per day, with the vast majority (88%) making only one request per day. Figure 2 shows the total amount of requests made per distinct IP over the whole experiment. Here, we see that 8,257 IPs have sent only one request. This value is to be compared with 12,277 of Fig. 1. It highlights the fact that a large amount of IPs have shown up on at least two different days, issuing a single request every time. This is confirmed by Fig. 3 where we see that almost 30% of the IPs have been seen on at least two different days.

That number is surprisingly high. Indeed, at this stage, we have to remind the reader that these IPs are proxy IPs and that the actual client machine sending a request is hidden behind. The proxy service offers a pool of addresses to be given to these clients. Let us call P the size of that pool. Figure 3 shows how many times a given address has been picked over a period of 56 days. The fact that there are 2,801 that have been used twice over that period is inconsistent with the assumption that the addresses would be randomly picked out of a very large pool of millions of IPs. Indeed, to calculate the probability that a given IP got picked twice over this period comes down to resolving the classical birthday paradox which can be generalized as follows:

Given n random integers drawn from a discrete uniform distribution with range [1,d], what is the probability p(n; d) that at least two numbers are the same? (d = 365 gives the usual birthday problem.) [22]

In our case, n is equal to 56, the number of days where IPs from the pool are assigned to clients and d is equal to the size of the pool P. We want to assess the probability that the same IP would be drawn twice over that period of 56 days. We can rephrase the birthday problem for our needs as follows:

Given 56 random integers drawn from a discrete uniform distribution with range [1,P], what is the probability p(56; P) that at least two numbers are the same?

The formula $1 - (\frac{P-1}{P})^{\frac{56(55-1)}{2}}$ gives an approximate result:

- If $P = 10000000$ then $p(56, 10M) \approx 0.000154$
- If $P = 1000000$ then $p(56, 1M) \approx 0.001538$
- If $P = 100000$ then $p(56, 100K) \approx 0.015282$

Clearly, considering that we have seen more than 30% of the IPs drawn at least twice, either P is significantly lower than the number announced by the proxy services, or the assignment of IPs is not randomly done, or both.

Regarding the total amount of IPs, we saw only 13,897 different ones. Every day the number of distinct IPs, on average 371 (shown in yellow in Fig. 4), was similar to the number of requests, on average 410. Thus, it is clear that most IPs send a single request and reappear some time later. In the same figure, the green columns represent the cumulative number of unique IPs observed in our honeypot since the beginning of the experiment. The figure shows that the daily increment decreases over time, suggesting that it will eventually reach a maximum.

To better characterize and understand the threats ecosystem we are facing, we try to find a mathematical model that approximates as closely as possible the assignment of IPs made by the proxy provider. We use that model to derive the most likely size of P. This is done in the next Section.

6   Modeling Results

6.1   Introduction

We propose two distinct modeling approaches to assess the most likely size of the pool of IPs P put at the disposal of the stealthy APBs we have observed. Both models deliver a value which is below 70K, i.e. three orders of magnitude less than the 70M IPs supposedly provided by [12].

In the first approach (subsection 6.2), we look at the IPs assigned every day by the proxy to the bots. We model this as a drawing process made in a pool of size P and we try to find the best probability distribution function that would

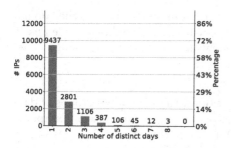

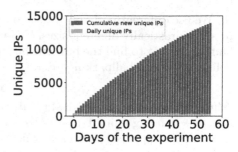

Fig. 3. Left (resp. right) Y axis: absolute (resp. relative) amount of IP addresses which were seen in X distinct days during the whole experiment.

Fig. 4. Cumulative curve of the new unique IPs in comparison with the daily unique IPs.

produce similar results to the ones we have witnessed. From there, we derive the value of P.

In the second approach (Subsect. 6.3), we look for a fitting curve to approximate the one shown in Fig. 4 and, by extrapolating it, see what maximum value it would reach, and when.

6.2   IP Assignment as a Drawing Process

General Principle. Figure 3 tells us how many IPs have been assigned to a bot only once, or twice, or three times ... over the duration of the experiment. We model this assignment process by a daily probabilistic drawing process without replacement. We arbitrarily define a pool size $\mathcal{P}$. On a given day, we draw from our pool, without replacement, a number of values equal to the amount of distinct IPs seen that day. We do this every day, keeping track of which value got drawn several times during this exercise. We use these accumulated results to produce a histogram similar to Fig. 3.

We use the Wasserstein[2] distance to assess the similarity between this histogram and the one from Fig. 3, making the reasonable assumption that the values of the produced histogram are derived from the real one by small and non-uniform perturbations.

We have no reason to believe that the drawing is done every day instead of every 2 days or 3 days or more. We thus repeat the process with other window sizes s (2 to 10), but we proceed with a drawing with replacement. We impose an additional constraint though. A given value cannot be drawn more than s times, i.e., once per day. Once a value has been drawn s times, it is not replaced in the pool anymore.

---

[2] This distance is known as the earth mover's distance, since it can be seen as the minimum amount of "work" required to transform one histogram into another, where "work" is measured as the amount of distribution weight that must be moved, multiplied by the distance it has to be moved [11].

We use different probability distribution functions to ensure that they do not produce drastically different sizes. Various other functions could have been used. Our goal is not to find the best one but to show that several "good enough" ones deliver the same ballpark figure for P.

Algorithm Used. We studied the distribution of the IPs for subgroups of days of size s ranging from 1 to 10. To group the days, we have used juxtaposed windows (as opposed to sliding windows) to ensure that our final histogram contained the same amount of values as the one in Fig. 3. We chose juxtaposed windows to not count twice the IPs of a singular day and reproduce thus a coherent replica of the observed data.

We have run simulations with different population sizes P. We have incremented P by 10,000, starting with the initial value of 10,000 up to 100,000. Moreover, we have tested values from 100,000 to 200,000 with an increment of 20,000. For a given time window, we have produced as many histograms as distinct population sizes. Each histogram is obtained thanks to 100 simulations. Each simulation produces its own histogram and we compute the Wasserstein distance between this histogram and the empirical one. An average Wasserstein distance value is then obtained from these 100 simulations. The lowest value of this average distance corresponds to the size P which best represents the observed data. For each window size, for each population size, for each simulation, we have plotted the distances obtained using a boxplot representation. This algorithm has been applied using three distinct probability distribution functions, as explained here below.

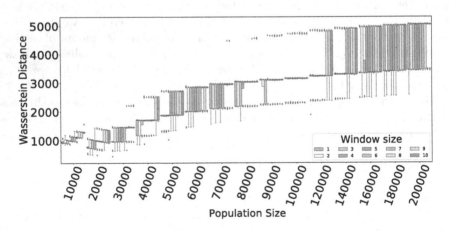

Fig. 5. Uniform distribution: for each population size on the x axis, a group of boxplots displays the Wasserstein Distances (y axis) obtained in the 100 experiment for that population size. Each color represents the window size used for the simulation.

Uniform Distribution. The simplest model is the one where all IPs, every day, have the same probability of being assigned to a bot. To model this, we use a uniform distribution as the probability distribution function in our drawing process. Figure 5 shows for each window size (colored legend) and for each population size (X-axis), the boxplots of Wasserstein Distances (Y-axis) obtained in all the experiments. We clearly see that for $\mathcal{P}$ bigger than 30K, the bigger its value, the more different is the obtained histogram with respect to Fig. 3. The best distances are obtained for the low value of $\mathcal{P}$ of 20K, for all time window sizes. The Wasserstein distance is quite high though, around 1,000 and we have looked for other distributions with the hope of obtaining smaller distances.

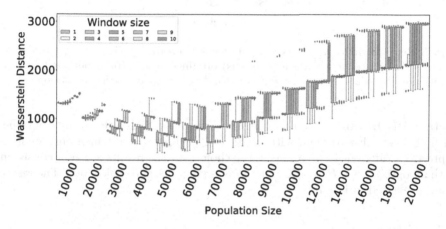

Fig. 6. Gaussian distribution: for each population size on the x axis, a group of boxplots displays the Wasserstein Distances (y axis) obtained in the 100 experiment for that population size. Each color represents the window size used for the simulation.

Gaussian Distribution (aka normal). It is reasonable to imagine the existence of a bias in the IP assignment process that would lead some IPs to be more frequently used whereas others would be rarely picked. This could be due, for instance, to the simple fact that some residential IPs might be more frequently available (online) than others. Another reason could be that proxies, to ensure a better quality of service, assign preferably IPs "close" to their customers. Our goal is not to identify the causes of these biases but, simply, to assume that they could exist and, thus, model this possibility. To do so, we have run our algorithm with a Gaussian distribution. For the sake of concision, the results presented here correspond to the parameters $mu=0.5$ and $sigma=0.1$. Other choices lead to the same lessons learned and this combination offers the best distances. We offer in Fig. 6 a similar representation as in Fig. 5. This model seems to be a better approximation since the best Wasserstein distance is now half of the one obtained for the uniform distribution. As expected, the size $\mathcal{P}$ does grow since a number of IPs are now very rarely chosen. Its value, around 60K, is still three orders of magnitude below the claimed 70M.

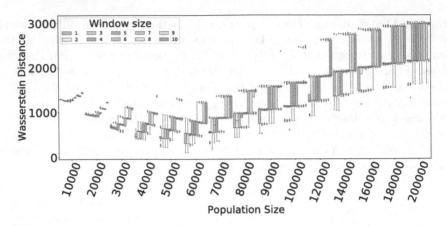

Fig. 7. Beta distribution: for each population size on the x axis, a group of boxplots displays the Wasserstein Distances (y axis) obtained in the 100 experiment for that population size. Each color represents the window size used for the simulation.

Beta Distribution. Last but not least, we present also the results obtained with the Beta distribution, with $alpha = 1$ and $beta = 5$; which enables us to represent a different form of bias in the choice but the results are very consistent with an optimal size P of 60K and a Wasserstein distance below 500. The results are represented in Fig. 7.

6.3   Fitting Curve

As explained before, a distinct approach to get an informed estimate of the size P consists in starting from the values observed in Fig. 4, in finding a fitting function and in extrapolating its values.

To do so, after having looked at the data at our disposal, we have observed that, roughly speaking, the amount of new IPs (i.e., never seen so far) observed on a daily basis was decreasing linearly over time. We were thus hoping to be able to find a good fitting function [18], thanks to an exponentially decaying one. We found out by means of simulations that the best fit was achieved with the following function:

$$a * (1 - e^{-(x-b)/c}) \tag{1}$$

The parameters that provide the best fit are:

$$a = 2.77313369e + 04$$
$$b = -4.77879543e - 01$$
$$c = 8.04885708e + 01$$

The fitted curve is represented in Fig. 8. To assess their similarities, we calculate the Pearson correlation factor [21] and obtain the value 1.000 which indicates a total positive linear correlation, confirming the adequacy of our fitting function which is visible by the quasi superposition of both curves in Fig. 8

We can now use that fitting function to extrapolate the total amount of distinct IPs we would have seen, had we been able to run the experiment for 3 years. Figure 9 shows how the curve reaches a plateau after a bit more than a year. Thus, according to this distinct approach, the bots we have observed only have a couple of tens of thousands of IPs at their disposal, a value which is consistent with the ones found with the first approach.

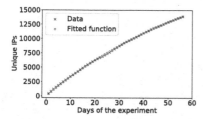

Fig. 8. Projection of the real data on the fitting curve values

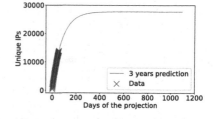

Fig. 9. 3 years prediction of the number of different IPs that would have been seen in the honeypot

7   Complementary Results

In this Section, we present additional pieces of evidence to those already provided in [2], which confirm that the IPs we have analysed are, indeed, quite likely provided by proxy services.

These IPs are supposed to be residential IPs; i.e., they belong to legit users who could, possibly, be interested in buying tickets. To verify this, we have looked for the presence of these IPs in the logs of 17 other airlines. We found out that during the experiment, five bookings have been realised by 5 of our IPs. In Table 1 we indicate when the booking was done vs. when the same IP was seen in our honeypot logs. As expected, the dates differ greatly. Moreover, none of these requests had the bot signature associated with them. They look perfectly legit. This confirms two things i) some of these IPs are likely used by legit users, ii) the risk of blocking legit customers when blocking identified proxy IPs remains extremely small.

On the other hand, the simplest way to implement a proxy is to open some ports and have a proxy server listening behind it. This should thus be detectable by the various actors who scan the Internet continuously, looking for threats and, or, vulnerabilities. We have used two such systems to see if they had identified our IPs as behaving like proxies. First, we have used IPInfo.io [7] which provides a boolean value for each IP in the categories "VPN", "Proxy", "Hosting". According to the provider of that service, VPNs are usually encrypted traffic endpoints so typically, if there is a VPN operating on an IP address, there will

Table 1. Timestamp of the bookings and the honeypot requests made by the same IPs.

Booking time	Request time
2020-01-17	2020-02-01
	2020-02-05
	2020-02-14
2020-02-26	2020-01-10
	2020-01-23
2020-02-29	2020-02-01
2020-02-06	2020-02-23
2020-02-07	2020-01-24
	2020-02-02
	2020-02-19

Table 2. Distribution of the fraud score of IPQualityScore

Score (S)	% of IPs	# of IPs
$S < 75$	28%	3958
$S \in [75, 85]$	46%	6371
$S \geqslant 85$	26%	3568

be either encrypted traffic or ports open which will obviously show a VPN is being used. Proxies are usually just a "HTTP forwarding" service and redirect traffic to somewhere else (internal domains, other servers, etc.) [8]. "Hosting" category specifies if the IP belongs to hosting providers.

Table 3 shows that a couple of IPs have been categorized as involved in suspicious activities but not as many as expected. However, the results obtained with IPQualityScore [9] are much more aligned with our expectations. As explained in their documentation, this service tells if an IP has been used in "automatic fraudulent behavior" in the "Bot status" category, while indicating a positive value of "Recent Abuse" if the IP has been involved in a "recently verified abuse across their network". The abuse behavior includes charging back, compromised devices, fake app installation. Moreover, the "VPN" category indicates server or data center IPs which allow tunneling. Finally, the "Proxy"[3] category, identifies a device being infected by malware, a user selling bandwidth from their connection or other types of proxy like SOCKS, Elite, Anonymous, Tor, etc. With this service, we can notice that the number of IPs involved in malicious activity is much higher in comparison to the first one. Furthermore, this service provides a general fraud score for the IP: this value ranges form 0 to 100, indicating a suspicious activity when higher than 75 and an high risk when greater than 85. Table 2 tells that around 72% of the IPs show a suspicious behavior, of which 28% are classified as high risk. This is quite consistent with the idea that malicious actors are hiding behind them, ruining the reputation of these IPS.

To dig deeper into the analysis of the malicious behaviors associated with these IPs, we looked for their presence in anti-spam DNS blocklists. Using the Python library Pydnsbl we checked multiple blocklists and we found out that 76% of the IPs were blocked at least in one of them at the time of our analysis

---

[3] A "VPN" is automatically a "Proxy" according to their definitions.

Table 3. IPInfo.io classification of the IPs

Type	Number of Ips	Percentage
VPN	180	0.013
Proxy	59	0.004
Hosting	1733	0.125

Table 4. IPQualityScore classification of the IPs (*From the total number of positive matches, 10213, we subtracted the number of positive values of VPN)

Type	Number of Ips	Percentage
VPN	9138	0.658
Proxy*	1075	0.077
Recent abuse	3878	0.279
Bot status	2780	0.200

(July 2020). Hence, we had the confirmation that these IPs were doing malicious activity also outside of our environment.

8   Discussion

The whole point of our experiment was to obtain, over a long period of time, a meaningful set of IPs that we could confidently say were behaving as they were members of the very same botnet. The very strong correlation in their activity patterns, detailed in [2], is as close as a ground truth one could hope for. The anti-bot detection solution identifies many more IPs as behaving like bots but our experience in looking at the logs gives us no assurance that IPs flagged with a given signature belong to the same botnet. Indeed, the goal of each signature is to fingerprint *"a bot"*, not *"the bot from botnet X, Y or Z"*. The analysis we have carried out in this paper required a *clean* dataset in order to be able to derive meaningful conclusions. We are very well aware though that, compared to all the bots that are out there, our dataset is relatively small and we do not pretend that our conclusions can, or should, be extended to all botnets that are in activity. Our results do only apply to the botnet we have studied. Having said so, all elements at our disposal, explained in the previous pages, indicate that this botnet is a perfect example of so called APBs, Advanced Persistent Bots, and is thus representative of the many others that are scraping websites. Therefore, we have good reasons to believe that our results could probably be generalized to many other botnets, without having, at the moment the data to support this claim.

If true, this would mean that large websites victims of web scraping bots would see the same IPs coming back regularly and that the grand total of IPs they would have to watch for would remain manageable (in the order of tens of thousands instead of tens of millions). An IP blocking strategy could thus be rejuvenated: seeding their sets of IPs with the ones clearly identified as behaving as bots, that strategy could enable them to catch the most evasive bots when they show up with a known bad IP. Redirecting these IPs to a fake web site instead of blocking them would also enable them to keep watching their behavior and, possibly, redirect them to the real web site if their requests are not consistent with those of known bots (i.e., in the case of a false positive).

The results presented in this paper have helped us in convincing our partner, the major IT provider, to move forward into building such an environment and the work is under way. We felt it was important to share already now our preliminary results with the community not only in order to let other benefit from the gained insights but also, possibly, to obtain feedback on important elements we could have missed. We do hope our contributions will participate in diminishing the negative impact created by these bots on the global Internet ecosystem.

9   Conclusion

In this paper, we have studied in detail a specific web scraping botnet that is representative of the plague most airline websites are suffering from.

Thanks to two distinct mathematical models, we have shown that the total amount of IPs at the disposal of this botnet was most likely in the low tens of thousands. We have also given pieces of evidence that these IPs were provided by proxy services, thought to be able to provide tens of millions of IPs to their customers. If our finding applies, as we think it does, to other botnets then an IP-blocking strategy could be applied, contrary to the common belief. We encourage others to carry out similar experiments to confirm, or deny, our findings while we are in the process of testing our conjecture in a new large scale experiment.

References

1. Amin Azad, B., Starov, O., Laperdrix, P., Nikiforakis, N.: Web runner 2049: evaluating third-party anti-bot services. In: 17th Conference on Detection of Intrusions and Malware & Vulnerability Assessment (DIMVA 2020), Lisboa, Portugal (2020)
2. Chiapponi, E., Catakoglu, O., Thonnard, O., Dacier, M.: HoPLA: a honeypot platform to lure attackers. In: C&ESAR 2020, Computer & Electronics Security Applications Rendez-vous, Deceptive Security Conference, Part of European Cyber Week, Rennes, France (2020). http://www.eurecom.fr/publication/6366
3. Dietrich, S., Long, N., Dittrich, D.: Analyzing distributed denial of service tools: the shaft case. In: Proceedings of the 14th USENIX Conference on System Administration, pp. 329–339. New Orleans, Louisiana, USA (2000)
4. Haque, A., Singh, S.: Anti-scraping application development. In: 2015 International Conference on Advances in Computing, Communications andInformatics (ICACCI), pp. 869–874, Kochi, India (2015)
5. Imperva: Imperva bad bot report (2020). https://www.imperva.com/resources/resource-library/reports/2020-bad-bot-report/
6. Imperva: how bots affect airlines (2019). https://www.imperva.com/resources/reports/How-Bots-Affect-Airlines-.pdf
7. Comprehensive IP address data, IP geolocation API and database - IPinfo.io. https://ipinfo.io/
8. IpInfo.io: personal communication, August 2020
9. Fraud prevention—detect fraud—fraud protection—prevent fraud with IPQS. https://www.ipqualityscore.com/

10. Jung, J., Sit, E.: An empirical study of spam traffic and the use of DNS blacklists. In: Proceedings of the 4th ACM SIGCOMM Conference on Internet Measurement, IMC 2004, p. 370–375. Association for Computing Machinery, Taormina, Sicily, Italy (2004)

11. Levina, E., Bickel, P.: The earth mover's distance is the mallows distance: some insights from statistics. In: Proceedings Eighth IEEE International Conference on Computer Vision, ICCV 2001, vol. 2, pp. 251–256, Vancouver, Canada (2001)

12. World's leader in web data collection and proxy for businesses— luminati.io. https://luminati.io/

13. Mi, X., et al.: Resident evil: understanding residential IP proxy as a dark service. In: 2019 IEEE Symposium on Security and Privacy (SP), pp. 1185–1201. San Francisco (2019). https://doi.org/10.1109/SP.2019.00011, ISSN: 2375-1207

14. Motoyama, M., Levchenko, K., Kanich, C., Mccoy, D., Voelker, G., Savage, S.: Re: CAPTCHAs–understanding CAPTCHA-solving services in an economic context. In: Proceedings of the 19th USENIX Security Symposium, pp. 435–462. Washington, DC (2010)

15. Oxylabs: gather data at scale with an innovative proxy service—oxylabs. https://oxylabs.io/

16. Samarasinghe, N., Mannan, M.: Another look at TLS ecosystems in networked devices vs. web servers. Comput. Secur. **80**, 1–13 (2019). https://doi.org/10.1016/j.cose.2018.09.001

17. Samarasinghe, N., Mannan, M.: Towards a global perspective on web tracking. Comput. Secur. **87**, 101569 (2019). https://doi.org/10.1016/j.cose.2019.101569

18. Scipy.optimize curve fit function. https://docs.scipy.org/doc/scipy/reference/generated/scipy.optimize.curve_fit.html

19. Scraper api. https://www.scraperapi.com/blog/the-10-best-rotating-proxy-services-for-web-scraping/

20. The best residential proxy network with 40M+ IPs. https://smartproxy.com/

21. Stigler, S.M.: Francis Galton's account of the invention of correlation. Stati. Sci. **4**(2), 73–79 (1989). http://www.jstor.org/stable/2245329

22. Suzuki, K., Tonien, D., Kurosawa, K., Toyota, K.: Birthday paradox for multi-collisions. In: Rhee, M.S., Lee, B. (eds.) ICISC 2006. LNCS, vol. 4296, pp. 29–40. Springer, Heidelberg (2006). https://doi.org/10.1007/11927587_5

23. Vastel, A., Rudametkin, W., Rouvoy, R., Blanc, X.: FP-Crawlers: studying the resilience of browser fingerprinting to block crawlers. In: Starov, O., Kapravelos, A., Nikiforakis, N. (eds.) MADWeb 2020 - NDSS Workshop on Measurements, Attacks, and Defenses for the Web. San Diego, United States, February 2020. https://doi.org/10.14722/ndss.2020.23xxx

24. von Ahn, L., Blum, M., Hopper, N.J., Langford, J.: CAPTCHA: using hard AI problems for security. In: Biham, E. (ed.) EUROCRYPT 2003. LNCS, vol. 2656, pp. 294–311. Springer, Heidelberg (2003). https://doi.org/10.1007/3-540-39200-9_18

Cyber Security Education and Future Provision

Gaynor Davies[1], Mamoun Qasem[2(✉)], and Ahmed M. Elmisery[2]

[1] School of Computing, Engineering, Science and Computing, Ystrad Mynach Campus, Twyn Rd, Ystrad Mynach, Hengoed CF82 7XR, UK
Gaynor.Davies@cymoedd.ac.uk
[2] Faculty of Computing, Engineering and Science, University of South Wales, Treforest, Pontypridd CF37 1DL, Wales, UK
{Mamoun.qasem,ahmed.elmesiry}@southwales.ac.uk

Abstract. Cybersecurity education is a crucial element to provide a workforce for the future to have an awareness together with the skills and knowledge enabling them to adapt and diversify in the field. Cybercrime is covered by the need for distinct aspects of security and measures of control on subsequent systems and devices. As the need for Cybersecurity specialists has increased in recent years and during the exceptional circumstances of the Covid-19 pandemic, the provision of education in secondary, post-16 and higher education sectors needs to be met. Utilisation of strategies and innovations to meet industry and educational expectations is key for future provision. Drive for different strategies and innovations in place by governments and organisations throughout the world, as provision in education is not balanced enough to cope with the increasing demand for a cybersecurity workforce. This study will inform recommendations for effective provision of future cybersecurity education.

Keywords: Cybersecurity · Education · Provision · Future

1 Introduction

Cyber security is prevalent with the inherent use of digital technology and devices connected to the internet. Therefore, increasing the need for individuals and organisations to have an awareness and knowledge for the prevention of cyber threats. This identifies the need for individuals to at least have an awareness of cyber security and the preventative measures that can be adopted to prepare them for threats against their cyber environment. The National Cybersecurity Strategy 2016–2020 [16], refers to cyber security as the protection of information systems (hardware, software, and associated infrastructure), the data held on them, and the services they provide, from unauthorised access, harm, or misuse. This includes harm caused intentionally by the operator of the system, or accidentally, because of failure to follow security procedures.

Crimes of this nature are anonymous with no geographic boundary, therefore allowing crimes to be difficult to pursue and punish the perpetrator. Crime figures for England and Wales pertaining to cybercrime estimated that the number of fraud incidents increased by 15% to 3,863,000. The subcategory "bank and credit account fraud"

© Springer Nature Switzerland AG 2021
H. Hacid et al. (Eds.): ICSOC 2020 Workshops, LNCS 12632, pp. 612–626, 2021.
https://doi.org/10.1007/978-3-030-76352-7_53

accounted for much of the volume increase in total fraud, rising by 17% from 2.3 to 2.7 million offences, with a rise of computer misuse offences in April and May [34].

There are many different methods used to commit fraud, however the banks are developing new technology to combat fraud with the use of behavioural biometric tools that will allow them to stay one step ahead of a fraudster. This is only one area of cybersecurity that is needed in today's world especially with the Covid-19 pandemic that has changed the lives of everyone within organisations, the general population, schools, colleges, and universities.

The infrastructure of organisations and susceptible systems and the increased need to use the internet to be connected to thousands of individuals working from home can be severely lacking in terms of connectivity and the sustainable bandwidth needed. High numbers of cyber related threats are occurring, and the education of current and future generations is paramount to address the daily occurrence of the possible threats accordingly without escalation or recurrence. Escalation of these types of crime are intrinsic with the need for education at all levels for future prevention on a personal and professional level.

The emphasis now is to educate school children to become the next generation fighting the war against cybercrime and its prevention as cyber security is becoming a crucial element in curricula at all education levels. The foundational knowledge upon which the field of cyber security is being developed is fragmented and, as a result, it can be difficult for both students and educators to map coherent paths of progression through the subject [30]. The current trends for the use of a computer for personal, professional use, provide a gateway for criminal intent by individuals who have the ability to capture 'digital details' that are stored in more places than people realise, thereby making users delusional when it comes to privacy and the way that they use and communicate using a digital device [36]. People inadvertently click on links or open emails expecting something that will help them. Many of these are related to the NHS, HMRC, Bank payments, holiday scams and online shopping. This has been exponential during the Covid-19 pandemic, especially since countries across the world were put on lockdown [2]. In a recent publication [34], it was found that remote mobile banking fraud cases were up by 132% from the previous year.

This indicates that the threats to individuals and security does not keep people free from danger as there is always an adversary where individuals intentionally or unintentionally create a cybersecurity situation. Within organisations there is the need to keep information private, thereby addressing confidentiality and with the continual growth in the economic sector through legal protection and technical skills that enable encryption, and access control measures. Integrity is the confidence in the organisation that data held is not accessible without authorisation. Availability offers the user access to the system. Although there are threats and vulnerabilities in security, the differences are that a vulnerability is a weakness in the security of the system, yet the weakness does not constitute a danger if no one wants to exploit that weakness. Someone who is interested in exploiting a vulnerability constitutes a threat. A single vulnerability may have multiple threats. Therefore, a person who wants to attack cybersecurity creates a cyberthreat.

1.1 Problem Statement

The increase in cybercrime has led to gaps in the cybersecurity workforce. The gap in the workforce has been escalated by exceptional circumstances especially within the education sector. This has resulted in differences in the provision of cybersecurity education and the curriculum design across education sectors. The growing need for an efficient cybersecurity workforce brings about a need for change within the education system. This will address expectations from an industry perspective as well as engaging learners in the field of cybersecurity. The divide in the workforce has evolved into specialist areas. Diversity to enable a broader knowledge area of cybersecurity is needed for progression to further and higher education. Future provision in education will inherently, benefit from learners who have the background knowledge and skills in cybersecurity to meet expectations and provide scope for the challenges of the future especially under exceptional circumstances.

1.2 Scope of the Research

Different organisations and education establishments around the world provide cybersecurity education as either part of the curriculum or as additional learning, apart from higher education where courses are specific to different fields that include cybersecurity. However, despite government initiatives and incentives there is still a shortage in geographical locations where the skills needed for the future of cybersecurity are not adequate or included as part of mainstream curriculum in schools and colleges. This project aims to identify gaps in the field of cybersecurity education and the provision that is currently available in Wales compared to England.

The difference between educating learners to be aware not just at GCSE, BTEC or degree level is varied dependent on the provision of a curriculum that incorporates cybersecurity across secondary, further, and higher education. This is also determined by resources and the knowledge and competence of staff required to deliver the curriculum. The study will determine current and future provision and the implications for the future workforce and to inform the design of cybersecurity education from secondary to postgraduate including continuing professional development.

The threats to society and businesses across the world are on the rise especially during uncertain times. The growing need to protect systems and devices from attack has increased the workload of cybersecurity professionals as the need to protect the infrastructure and connectivity of the use of the internet has grown. There is a need for educating more learners to pursue a career in the prevention of attacks. The problem of a resilient cybersecurity workforce has become a challenge that has to be addressed by the education system.

2 Related Works

There are individuals who do not understand the concept of cyber security. Education is vital to encourage awareness and a desire to encompass cyber security as part of daily life especially for those individuals and organisations where a device/system is connected to the internet using wired or wireless technology. There is also the culture of cloud computing, IoT and smart technologies within the home and business' of today. Current practices relating to cyber security and educational resources for development of curriculum in academic institutions at secondary, further, and higher education differ according to education establishment, delivery and level of qualification. Collaborative bodies who encourage the use of organisational involvement with education from an industry perspective, therefore, allowing room for growth in the realm of cyber security.

Future development in schools and colleges to aid individuals and organisations to educate themselves and their workforce through continual professional development both in house and at academic levels. Future collaborative research through government and other initiatives in the field of cyber security. Is this enough for the ever increasing trend in cybercrime? Everyone who connects a device or system to the internet is vulnerable to the possibility of becoming a victim of cybercrime. Repercussions of any form of attack can be devastating for individuals and organisations, whether they are commercial or provide a service. The number of cyber-attacks against individuals and organisations has increased exponentially during 2020. The Covid-19 pandemic has resulted in a number of scams and many attempts to infiltrate systems and devices.

Governments across the world are looking at solving cyber-attacks which can greatly affect the infrastructure of a physical or virtual system. This creates a challenge within the world economy as problems currently encountered need to be addressed with effective solutions to reduce the negative impact. A proactive stance on cyber defense has been made by international governments. In 2014, the Australian government established the ACSC [1], now part of the ASD, in order to help make the country the safest place to connect online by improving cyber resilience and providing advice and information for individuals, their families and business'. In 2016, the UK Government published the National Cyber Security Strategy. The vision of the strategy for 2021 is that the UK is secure and resilient to cyber threats, prosperous and confident in the digital world [3]. This is also evident on international levels with the USA investing in the National Initiative for Cybersecurity Education [28], led by NIST promotes a framework that increase the number of people that currently work or wish to work in cybersecurity with the knowledge, skills, and abilities to perform the tasks required [27]. The NICE Cybersecurity Workforce Framework is a combination of government, academia and the private sector to address the current and future education of the challenges faced by the cybersecurity workforce to develop standards and best practices [28]. These strategic government strategies and initiatives recognised throughout the world are have addressed the need for cybersecurity to be an integral part of current and future curricular for a number of years.

Cyber threats ignore boundaries or borders of organisations or locations as they do not exist in the virtual world. NATO [23], is an alliance consisting of 30 independent member countries who engage with partner countries and organisations internationally to enhance security. Initiatives such as the Cyber Defence Pledge [5], adopted in 2016, are central to enhancing cyber resilience. An Active Cyber Defence initiative was launched in the UK in 2017 by the UK government [24], designed to protect the majority of people in the UK from the majority of the harm caused by the majority of attacks, the majority of the time. As a result, UK-hosted phishing attacks fell by about 20% in the 18 months prior to February 2018, even as global volume itself rose by nearly 50% [20]. The CCCS in Canada work to protect and defend their country's' cyber assets in the private and public sector [4]. New Zealand has CertNZ who also work other government agencies and organisations to help better understand and stay resilient to cyber security threats [6]. An advisory committee been set up by collaborative research of five nations including the UK, USA, Australia, Canada and New Zealand. Their purpose is to highlight best practices for incident investigations [9]. The collaboration of different nations identifies the worldwide strategies and initiatives to combat cybercrime and provide resilience in the form of cyber defence is effective and proven as a deterrent giving people and organisations confidence for the future.

2.1 Challenges

There are different governments within the UK, where devolution has created a national Parliament in Scotland, a Welsh Parliament called Senedd Cymru and a national Assembly in Northern Ireland. This process transfers varying levels of power from the UK Parliament to the UK's nations, yet still keeping authority over the devolved institutions in the UK Parliament itself [35]. The education minister has the responsibility to. to direct funding for education in Wales [17]. Reductions in funding are a key factor as budget deficits are an increased stress that education sectors are facing, resulting in the struggle to meet the demands for tuition and resources especially where the lifespan of technology is short. Although resource spending is through departments within Further and Higher Education it is difficult to keep up to date with technology and provide suitable resources within the classroom. This greatly affects recruitment as outdated resources lead to insufficient scope in practical skills. To find students who can succeed in education at higher levels, it is imperative that the institution can provide specialist knowledge and skills for prospective cyber security graduates. The demographics for different education sectors, defines what is available.

There is a need to increase the innovations to attract learners and continue to keep them up to date by using innovative practices and technologies. This enables future development of the curriculum with content and the learning environment using up to date resources suitable for preparing students to become part of the evolving cybersecurity workforce [18]. The underlying issue within education is it is student centric where a curriculum is needed that is up to date with technology and includes a focus on cybersecurity in schools and colleges in both England and Wales. Covering the basics begins at school leads to interest, leading to wanting to go to university. NCSC highlighted in their annual Cyber security survey that 40% of universities had Cyber Essentials certification and almost a third of colleges have followed suit [11]. Therefore, the gap is still evident of the divide in provision across all sectors as a point of further research.

Cyber security threats in HE and FE are on the increase with data breaches, ransomware and malware and patch management as the top 5 listed. Within FE, the third most mentioned threat is ransomware/malware followed by DDoS attacks, malicious attacks from inside and data breaches [8].

Cyber security training within all education sectors for staff and students should be compulsory as computers are used every day. Findings from the Cyber Security Posture survey [8], shows that compulsory staff training is more common than student training for HE and FE. This highlights the need for awareness within organisations and the need for training both staff and learners alike. However, the amount of compulsory training provided over a 3 year period indicates that staff in FE have the same amount of training as in the previous year (55%). Whereas in HE there is an increase to 81%. Student information security awareness training on the other hand has decreased by 7% from the previous year for HE. This highlights that there is not enough compulsory training awareness for staff or students within FE. The key to provide such training so that it is accessible to both staff and students is to ensure that funding is available. Due to the impact that Covid-19 has had on education at all levels, there is currently government funding available for schools, colleges and universities in England [13], and Wales, [7]. Subsequent research has identified a gap in Wales who only have a small percentage of the workforce and the focus of UK analysis is predominantly in England where employees are leaving or moving across the border from Wales for higher paid jobs. Factors that impede a successful cybersecurity workforce falls on the educational system. There is required research in this field to provide both qualitative and quantitative analysis of views from a professional perspective to explore the true challenges faced by education establishments.

With the growing sophistication of cyberthreats and the need to improve on the current strategies for education with informed collaboration with government, academia, and industry to work together as an advocation for advanced cyber security competencies. The continuing trend in cybercrime is on the increase and it is apparent that education provision is struggling to keep pace with workforce requirements. Education and training institutions in the United States have so far found it difficult to keep pace with the growing need for cyber talent [10]. As there are many innovations and incentives for younger learners to engage in cybersecurity as a career path, there is still a shortage of learners progressing onto further and higher education courses. The level of qualification required is dependent on the job role being advertised. A recent government study on the cyber security skills of the UK labour market [15], identifies the minimum education requirements for both core and cyber-enabled postings from September 2016 to August 2019.

Even though there are a number of innovations and incentives for younger learners to engage in cybersecurity as a career path, there is still a shortage of learners progressing onto further and higher education courses. Efforts were not focused on the complete set of initiatives as these vary from country to country, however there is a significant difference in the UK, especially between England and Wales, [15]. A study of cyber security initiatives for schools in South Africa and the UK in 2017 [19], focuses on the awareness initiatives for school learners whether present or absent and the reasons for these. School is compulsory up until the age of 16 in both England and Wales where children are online both at home and at school.

As there are initiatives there is also investigations into education strategies for pursuing the need for appropriate education provision in this sector. A study by Chris Ross, looks at how schools and colleges are working to prevent students from online threats. More than half of respondents who attended the Bett show in London in 2017 claimed that their organisation had invested in cyber security training for staff (56%), whilst training for pupils was being offered in 42% of organisations [31]. In the USA, the government recognises the importance for of cybersecurity and has made efforts to promote cybersecurity programs and education, including the NICE initiative led by NIST and the CAE program led by the NSA and DHS [38].

Competitions increase learner interest and knowledge promoting teamwork and cooperation by making students work in groups on real life scenarios with questionable responses to navigate and protect defences. Therefore, cultivating a mindset for a changing environment. This changing environment is challenging for industry as expectations in the field of cybersecurity not only depend on technological skills, team working, problem solving, oral and written communication skills are vital to being able to work in this challenging environment, [10]. Cyber UK brings together with government, academia, and industry to develop global expertise in a crucial industry to keep people safe. The development for the next generation saw the introduction of CyberFirst [25], that covers a range of opportunities including girls' competitions and courses for 11–17 year olds helping to introduce the younger generation to the world of cybersecurity and a bursary/degree apprenticeship. This is an excellent opportunity that allows a choice for a career path in cybersecurity after 6th form or college.

There are also online resources aimed at younger learners that are available through NCA as part of the CSC UK for free access to CyberLand, a range of interactive modules designed for teaching the fundamentals of cybersecurity [25]. The Education network has a number of online resources form key stage 1 to 4 which include links to ICT, STEM modules for use within the classroom. Some of the resources are freely available and others need a log in as they are only licensed to schools [26]. Further incentives launched by the NCSC include recognition and support for educators and the development of the Cybersecurity Body of Knowledge [30]. The variety of resources and incentives available to academia could provide to be invaluable for use in directed learning as the only resource needed is access to a computer or other device.

2.2 Provision

Most schoolchildren today have grown up using digital devices and are familiar with IT/ICT. In secondary education, learners use applications to problem solve, present information, and communicate. However, this is limited to the requirements of the curriculum and the capabilities of their teacher in terms of confidence and competence in the field of computing. GCSE Computer Science encompasses a broad knowledge with skills in more technical aspects such as coding, security, and networks. Different awarding bodies also have elements of cybersecurity built in, however, it is a small part of the qualification that is theory based with no practical element of combatting cybersecurity.

As a facilitator in education, it has become apparent that there is a need for a case study on the provision of cybersecurity education across all levels of academia. This includes a detailed analysis of secondary, further, and higher education provision and the need for the competence in the delivery of skills required for the practise of cybersecurity prevention. The delivery of the specialised education falls on universities to deliver recognised degree courses in this area. However, the current provision is varied across the UK, with a number of higher education providers being sector leaders in this context. The current provision in schools and colleges does not provide enough scope of the in-depth knowledge and skills needed for the next generation of cybersecurity specialists. This was addressed in 2014 from a different perspective by Mason and Pike who stated that the education system was ill prepared to meet the challenge of producing an adequate number of cybersecurity professionals, but programs that use competitions and learning environments to teach depth and knowledge are filling this void. However, this research is based on the number of hours of practice and preparation needed to learn from this type of cybersecurity education. Additional objectives for curriculum needs related to Cybersecurity were the result of a workshop sponsored by NSF to address the need for and to build a cybersecurity workforce to meet the needs of industry and government [21].

Training and staff development in the field of cybersecurity is not compulsory within schools in England and Wales, however research has found that the workforce within education establishments do not always teach their specialist area. An Estyn report highlighted that schools need guidance to integrate cybersecurity into the curriculum with clear indicators to build opportunities. Staff training was also identified to build confidence and competence in order to teach the complexities of cybersecurity whilst also giving them time to implement initiatives.

Schools currently follow a set curriculum which is set by the Government and are restricted due to the knowledge and expertise required to deliver cybersecurity as a specialist area. Initiatives are needed to involve leaners to engage in realistic scenarios to promote cybersecurity awareness. This form of development can lead to the prospect of a career in cybersecurity.

The NCSC [25], also conducted research within schools where of 430 schools across the UK who took part, it was found that 92% of respondents would welcome more cybersecurity training for teachers and staff. NCSC is now developing cybersecurity training packages for schools. At the time of writing, the author was unable find such resources, however information cards are available for download. The NCSC also engage with colleges helping to improve their cyber resilience.

There are varied factors affecting the provision of education within schools and colleges. Analysis of data conducted on teaching workforce data for secondary education in England [14] and Wales [12]. There are disproportionate numbers of teaching staff over 50 was 17% in England and 25% in Wales with the highest teaching age from 30 to 39 was 33% in England compare to 30% in Wales. This indicates that there is an ageing population in the teaching workforce for colleges in both England and Wales. A contributory factor in the provision of specialism in areas such as cybersecurity. There is also a decline in the number going into the teaching profession. As teachers approach retirement, they tend to opt for part-time leaving the gap for 'new blood'. However, the ramifications of replacing these staff are vast as teachers in specialist areas are scarce.

Current provision within colleges of Further Education within the UK is varied depending on resources and staff who are conversant in this area. The availability of courses that can be tailored for delivery to include cybersecurity is dependent on the knowledge and skills of the lecturer(s). Staff development is helpful as it can provide the knowledge and understanding through higher education courses that are taught locally at higher education institutions. This can be at a considerable cost, however in different areas where there is funding available with no extra cost to the individual. At present there are 168 colleges in England with only 14 in Wales. An ICT review conducted in 2018 by Qualifications Wales [29], identified the need for a change in the curriculum for both secondary schools and colleges. The following table is a snapshot of the current qualifications offered in Wales by BCS, City & Guilds, OCR, Pearson and WJEC.

Higher level courses are available across the UK, with sector leaders working with organisations and government to provide in depth knowledge and understanding of the scope of cybersecurity together with associated modules relating to current and future trends in this area. Findings from What Uni.com [37], found 9 universities offering 36 BSc Computer Science courses in England that featured cybersecurity and 14 courses were found in Wales. A further search found 9 universities offering 11 post graduate courses including cybersecurity in England. The offering for wales was 9 universities offering 122 courses. However, subsequent searches revealed different results which does not give a true picture of courses available. NCSC highlighted that their annual Cyber security survey that 40% of universities had Cyber Essentials certification and almost a third of colleges have followed suit [11]. There are, however, universities that do not deliver cybersecurity as a specialist course, only courses with aspects of cybersecurity included.

As the skills gap has widened with the barriers to hiring a competent workforce in cybersecurity, it has become apparent that there is a need for roles need to be filled with candidates who have the relevant experience and capabilities. These requirements are normally something that has come with experience in the field even though there are specialist fields. Applications for posts in cybersecurity pose a daunting adversary as the position requires skills in distinct aspects, thereby putting them on the spot prior to an interview even taking place. This leads to industry requirements for specialism of multifaceted origins. At the other end of the scale is the learner who has an interest in cybersecurity and does not know which path they should take through their education to get into their chosen career.

The future is uncertain with the use of the internet and users who lack the skills needed even though free resources are available to individuals and organisations. The use of these resources is not fully implemented or understood. However, the innovative and connectedness of the internet as a resource is valuable as an asset to the education of today, the future of industry and the underlying growth of the economy both in the UK and worldwide. The industry expects a workforce that has at least a basic knowledge and skills of cybersecurity to progress in the world of work. This is a diligent factor in an industrial context as a potential employer needs to find a person who can do the job that they are expected to do.

Learner expectations are varied in that they do not know what to expect from the world of work as they have limited experience. Higher education provides suitable work experience/placement and prepares an individual with valuable experience in preparation for their future after university. Although there are challenges throughout the world in the cybersecurity workforce, it is evident that governments are recognising the need for provision in education, especially within secondary and further education. Current provision is fragmented, however the innovations and resources that have and will be available should start to address the problem. Staff development in key areas of cybersecurity fundamentals needs to be timely and factored into each establishment to provide the knowledge and confidence for delivery of a specialist curriculum. Universities already have a wide range of course available in this field. The advancement of cybersecurity skills and knowledge at secondary and further education will push the drive towards industry expectations to provide the workforce of the future.

2.3 Workforce Gap

As cybercrime trends continue to escalate, it is apparent that education provision is struggling to keep pace with workforce requirements. This is evident in the USA where education and training institutions have so far found it difficult to keep pace with the growing need for cyber talent [10]. The credibility and qualifications are skills and knowledge gained from previous education or experience.

The level of qualification required is dependent on the job role being advertised. The minimum education requirements are specified on job applications for core and cyber enabled postings. Without relevant qualifications in key areas is a barrier to employment in this field. A person applying for a job may have the skills and knowledge and not the qualifications which may be favourable as experience is invaluable as well as an advantage. There are many skills requested in job applications for roles in a cybersecurity workforce. The specifics of these skills are dependent on the job application. For someone to be proficient in a number of these skills, they would have had to have vast experience in previous work in cybersecurity.

The need for implement these skills was highlighted in the findings report for the cybersecurity skills in the labour market for 2020 [15]. A third of vacancies were hard to fill as technical skills and knowledge was a contributory factor. Clarity of job types was also identified as cyber security or cyber enable jobs as there are skills gaps and shortages still affecting a large number of organisations [15].

The jobs market is quite diverse especially with the geographical divide between England and Wales. Wales represents 1.4% of the UK workforce for core cyber roles where the region is known.

The workforce gap has identified the need for successful applicants to be employed in cybersecurity. Qualifications and core skills are needed for different job roles that are dependent on ability and competency. However, communication, problem solving and team working skills are a desirable asset. Job postings in separate locations have major differences in salary which is also dependent on experience.

2.4 Future Provision

The future developments for collaborative initiatives from government, industry, academia, and professional bodies on the future of cyber security and protection giving industry and academia qualified personnel. Establishments should make use of CyBOK [30], as a guide to organisations for staff development and within education establishments to aid learning and to map to the curriculum to provide the basis of different knowledge areas.

The workforce gap widened because of exceptional circumstances The closure of schools, colleges and universities affected teaching and learning on a vast scale. The loss of information from formal assessments resulted in many learners having 'predicted grades where assessments were adapted, postponed, or cancelled until learners could be assessed in a 'safe environment'.

Education provision across all sectors around the globe changed overnight to the virtual classroom. Future provision of education for the next academic year will be a combination of blended learning with a set number of hours within a classroom environment and other parts of the curriculum delivered remotely.

Disadvantages of these methods are that it is debatable as to who completes the work behind the screen in this scenario, which can lead to cheating and/or plagiarism where it is difficult to prove the authenticity of the individual as being the person who did the work. Organisations throughout the world have provided links to education resources. One example is NIST who said that during this unusual time in our lives, many of us find we want to improve our knowledge, skills or even prepare for new career opportunities [32]. The National Centre for Computing Education also offers professional development courses to support the teaching of computing from Key Stages 1–4 [33].

2.5 Integrity and Ethics

Learners are responsible for their own education, resulting in acceptable morals and social values in terms of the challenges of digital technology in their education to be able to expand beyond the traditional confines of the classroom environment to the convenience of remote learning. Although ethics serve a myriad of purposes, within an educational context, they preserve the integrity of the educator, the reputation of the individual and that of the academic institution.

With the advent of the provision of education becoming virtual in these uncertain times, there are underlying issues that could lead to young people being/becoming bullied, exploited, or even radicalised. The global pandemic affected provision with the virtual learning environment changing people's lives with the seclusion of lockdown, uncertainty of grades and completion of work.

3 Findings

The survey respondents were sent communication via email due to Covid 19 and social distancing. Individual emails were sent to schools, college and universities throughout England and Wales. Results were dependent on the provision of cybersecurity and not the sector.

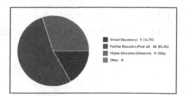

Fig. 1. Education sector

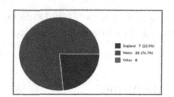

Fig. 2. Location

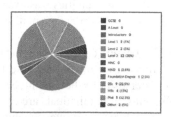

Fig. 3. Qualifications offered

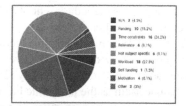

Fig. 4. Barriers to provision

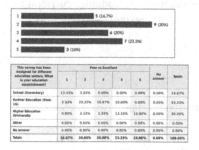

Fig. 5. Education provision rating

The survey itself consisted of 13 questions relating to the provision of cybersecurity currently delivered within the education sector at schools, colleges and universities in the UK.

Figure 1 identifies the different educational establishments for different sectors in both England and Wales. Figure 2 Specific locations were requested for either England or Wales. The majority of the respondents were from Wales (76.7%) with only 23.3% from England. Figure 3 represents the different types of qualifications offered. Figure 4 identifies barriers relevant to the improvement of the provision within their sector. Figure 5 displays overall ratings for their establishment ranging from poor to excellent for cyber security education provision within their sector.

4 Recommendations

Everyone needs to at least have an awareness of cyber security. It is imperative that the education of the current and future generation for cyber security prevention against the ongoing threats that occur daily can be addressed accordingly without escalation or recurrence A change of direction is needed to include benchmarks for success to ensure that curriculum content includes cybersecurity as a separate concept within STEM.

- The range of qualifications available should be streamlined to ensure content is up to date and relevant to the real world
- Awarding bodies share resources to enable a balanced curriculum suitable to the needs of industry expectations.

- Investment by schools and colleges in appropriate staff development
- Promote STEM as a core subject with engagement with professionals and organisations.

5   Conclusion

The number of establishments who scored high for excellent provision was non-existent for schools and poor for colleges, however the provision for universities was excellent. The exception was secondary schools who had next to no provision in the small sample from either England or Wales.

The Covid-19 pandemic has inadvertently affected education opportunities and influenced cybersecurity and the effectiveness of education, government and industry in different ways. Education, especially recruitment for courses in further and higher education is uncertain as the pandemic has meant that parents and individuals are cautious to pursue education opportunities The effectiveness of education is questionable with the uncertainties pertaining from the unknown.

The future of education is important and the curriculum design should consider the recommendations to aid industry to fill the gap in the workforce by updating curriculum to include cybersecurity as a core subject area as it is an increasing cause for concern especially during these uncertain times.

References

1. Australian Cyber Security Centre: About The ACSC (2020). https://www.cyber.gov.au/. Accessed 24 July 2020
2. Action Fraud: UK Finance Reveals Ten Covid-19 Scams the Public Should Be on High Alert For | Action Fraud (2020). https://www.actionfraud.police.uk/news/uk-finance-reveals-ten-covid-19-scams-the-public-should-be-on-high-alert-for. Accessed 30 July 2020
3. Assets.publishing.service.gov.uk. National Cyber Security Strategy 2016–2021 (2020). https://assets.publishing.service.gov.uk/government/uploads/system/uploads/attachment_d ata/file/567242/national_cyber_security_strategy_2016.pdf. Accessed 18 June 2020
4. CCCS: Canadian Centre for Cyber Security (2020). https://www.cyber.gc.ca/en/. Accessed 12 Aug 2020
5. Ccdcoe.org: Cyber Defence Pledge (2016). https://ccdcoe.org/uploads/2018/11/NATO-160 708-CyberDefencePledge.pdf. Accessed 12 on Aug. 2020
6. CERT NZ: About| CERT NZ (2020). https://www.cert.govt.nz/about/. Accessed 12 Aug 2020
7. Chapman, J.: New Publication: A Quick Guide to Post-16 Education Funding. IN BRIEF (2018). https://seneddresearch.blog/2018/04/05/new-publication-a-quick-guide-to-post-16-education-funding/. Accessed 7 Aug 2020
8. Chapman, J., Francis, J.: Cyber Security Posture Survey Results 2019. Repository.jisc.ac.uk (2019). https://www.jisc.ac.uk/reports/cyber-security-posture-survey-results-2019. Accessed on7 July 2020
9. CISA: Technical Approaches to Uncovering and Remediating Malicious Activity | CISA (2020). https://us-cert.cisa.gov/ncas/alerts/aa20-245a. Accessed 1 Sept 2020
10. Crumpler, W.: The Cybersecurity Workforce Gap. Csis.org (2019). https://www.csis.org/ana lysis/cybersecurity-workforce-gap. Accessed 29 July 2020

11. Daisy, S.: Universities and Colleges Take Action on Cyber Security. Ncsc.gov.uk (2019). https://www.ncsc.gov.uk/blog-post/universities-and-colleges-take-action-on-cyber-security. Accessed 4 July 2020
12. Ewc.wales: Statistics Digest 2019 (2020). https://www.ewc.wales/site/index.php/en/statistics-and-research/education-workforce-statistics.html. Accessed 5 Aug 2020
13. FE News: How Has the 2020 Budget Impacted The FE Sector? £1.8Bn For#FE Colleges Plus £3Bn National Skills Fund. FE News (2020). https://www.fenews.co.uk/fevoices/43647-sector-response-to-the-2020-budget. Accessed 7 Aug 2020
14. GOV.UK: Explore-education-statistics.service.gov.uk. 2020. School Workforce in England, Reporting Year 2019 (2020). https://explore-education-statistics.service.gov.uk/find-statistics/school-workforce-in-england
15. GOV.UK. 2020. Cyber Security Skills in The UK Labour Market 2020 (2020). https://www.gov.uk/government/publications/cyber-security-skills-in-the-uk-labour-market-2020/cyber-security-skills-in-the-uk-labour-market-2020#current-skills-and-skills-gaps. Accessed 29 July 2020
16. GOV.UK.: National Cyber Security Strategy 2016 To 2021 (2017). https://www.gov.uk/government/publications/national-cyber-security-strategy-2016-to-2021. Accessed 20 July 2020
17. GOV.WALES: Kirsty Williams MS: Minister for Education| GOV.WALES (2020). https://gov.wales/kirsty-williams-ms. Accessed 5 Aug 2020
18. Grajek, S.: Top 10 IT Issues, 2020: The Drive to Digital Transformation Begins. EDUCAUSE Review (2020). https://er.educause.edu/articles/2020/1/top-10-it-issues-2020-the-drive-to-digital-transformation-begins#issue1. Accessed 17 June 2020
19. Kritzinger, E., Bada, M., Nurse, J.: (PDF) A Study into the Cybersecurity Awareness Initiatives for School Learners in South Africa and the UK. ResearchGate (2017). https://www.researchgate.net/publication/314543784_A_Study_into_the_Cybersecurity_Awareness_Initiatives_for_School_Learners_in_South_Africa_and_the_UK. Accessed 25 July 2020
20. Levy, I.: Active Cyber Defence – One Year On. Ncsc.gov.uk (2018). https://www.ncsc.gov.uk/information/active-cyber-defence—one-year-on. Accessed 26 Aug 2020
21. McGettrick, A., Cassel, L., Dark, M., Hawthorne, E., Impagliazzo, J.: Toward effective cyber-security education. IEEE J. Mag. IEEExplore.ieee.org (2014). https://ieeexplore.ieee.org/abstract/document/6682988. Accessed 25 June 2020
22. Ofqual: The Register of Regulated Qualifications: Home Page. Register.ofqual.gov.uk (2020). https://register.ofqual.gov.uk/. Accessed 7 Aug 2020
23. NATO: Cyber Defence (2020). https://www.nato.int/cps/en/natohq/topics_78170.htm. Accessed 24 July 2020
24. Ncsc.gov.uk: Active Cyber Defence (ACD) (2020). https://www.ncsc.gov.uk/section/products-services/active-cyber-defence. Accessed 12 Aug 2020
25. Ncsc.gov.uk: Cyberfirst Schools (2020). https://www.ncsc.gov.uk/cyberfirst/cyberfirst-schools. Accessed 22 July 2020
26. NEN.gov.uk, n.d. Resources "NEN". Nen.gov.uk. https://www.nen.gov.uk/resource/. Accessed 7 Aug 2020
27. NIST: National Initiative for Cybersecurity Education (NICE) (2017). https://www.nist.gov/itl/applied-cybersecurity/nice. Accessed 18 June 2020
28. Nvlpubs.nist.gov: National Initiative for Cybersecurity Education (NICE) Cybersecurity Workforce Framework (2020). https://nvlpubs.nist.gov/nistpubs/SpecialPublications/NIST.SP.800-181.pdf. Accessed 30 July 2020
29. Qualifications Wales: DELIVERING DIGITAL Sector Review of Qualifications and the Qualifications System in Information and Communication Technology (2018). https://qualifications.wales/media/4044/itc-review-2018-e.pdf. Accessed 5 Aug 2020

30. Rashid, A., et al.: Scoping the cyber security body of knowledge. IEEE J. Mag. IEE-Explore.ieee.org (2018). https://ieeexplore.ieee.org/document/8395134. Accessed 11 May 2020
31. Rogers, C., n.d. Education Sector Increases Focus on Cyber Security - Education Technology. [online] Education Technology (2018). https://edtechnology.co.uk/comments/education/. Accessed 26 June 2020
32. Smith, J.: NIST Providing Online Cybersecurity Training Resources – Meritalk. Meritalk.com (2020). https://www.meritalk.com/articles/nist-providing-online-cybersecurity-training-resources/. Accessed 7 Aug 2020
33. Teach Computing: Courses - Teach Computing (2020). https://teachcomputing.org/courses. Accessed 7 Aug 2020
34. UK Finance: FRAUD - THE FACTS 2020 The Definitive Overview of Payment Industry Fraud (2020). https://www.ukfinance.org.uk/system/files/Fraud-The-Facts-2020-FINAL-ONLINE-11-June.pdf. Accessed 29 July 2020
35. UK Parliament: Parliament's Authority (2020). https://www.parliament.uk/about/how/role/sovereignty/. Accessed 5 Aug 2020
36. Volonino, L., Anzaldua, R.: Computer Forensics for Dummies, pp. 33, 36, 307. Wiley, Hoboken, NJ (2008)
37. Whatuni: Compare the Best University Degrees Courses UK| Whatuni. Whatuni.com (2020). https://www.whatuni.com/. Accessed 12 Aug 2020
38. Wu, X., Tian, S.: Student-Centered Learning in Cybersecurity in Summer Semester - IEEE Conference Publication. IEEEplore.ieee.org (2015). https://ieeexplore.ieee.org/abstract/document/7344154. Accessed 22 July 2020

Efficient Threat Hunting Methodology for Analyzing Malicious Binaries in Windows Platform

Ahmed M. Elmisery[1], Mirela Sertovic[2], and Mamoun Qasem[1(✉)]

[1] Faculty of Computing, Engineering and Science, University of South Wales, Pontypridd, UK
Mamoun.qasem@southwales.ac.uk
[2] Threat Defense Unit, Concept Tech Int. Ltd, Belfast, UK

Abstract. The rising cyber threat puts organizations and ordinary users at risk of data breaches. In many cases, Early detection can hinder the occurrence of these incidents or even prevent a full compromise of all internal systems. The existing security controls such as firewalls and intrusion prevention systems are constantly blocking numerous intrusions attempts that happen on a daily basis. However, new situations may arise where these security controls are not sufficient to provide full protection. There is a necessity to establish a threat hunting methodology that can assist investigators and members of the incident response team to analyse malicious binaries quickly and efficiently. The methodology proposed in this research is able to distinguish malicious binaries from benign binaries using a quick and efficient way. The proposed methodology consists of static and dynamic hunting techniques. Using these hunting techniques, the proposed methodology is not only capable of identifying a range of signature-based anomalies but also to pinpoint behavioural anomalies that arise in the operating system when malicious binaries are triggered. Static hunting can describe any extracted artifacts as malicious depending on a set of pre-defined patterns of malicious software. Dynamic hunting can assist investigators in finding behavioural anomalies. This work focuses on applying the proposed threat hunting methodology on samples of malicious binaries, which can be found in common malware repositories and presenting the results.

Keywords: Malicious binaries · Malware · Threat hunting · Digital investigations

1 Introduction

Nowadays, cybersecurity can be considered one of the most crucial topics in the computer science discipline [1]. Enterprises are obligated by law to protect sensitive information. They usually rely on the available security controls such as firewalls, intrusion prevention, and application whitelisting systems. However, some attacks still occur in these organizations even when their systems are secured with these controls. The effect of a single successful attack may be more severe in the organizational and/or personal context. If a malicious software infiltrates the network, it may cause real damage with even

© Springer Nature Switzerland AG 2021
H. Hacid et al. (Eds.): ICSOC 2020 Workshops, LNCS 12632, pp. 627–641, 2021.
https://doi.org/10.1007/978-3-030-76352-7_54

less possibility of attributing the adversary behind that. In enterprise environments, there is a high probability of being targeted with sophisticated attacks. Hence, relying solely on existing security controls is not sufficient. The current system mechanisms rely on the use of predefined characteristics or behavioural analysis of previous/known attacks [2]. Using signature-based detection, only malicious software that is already known and vulnerable can be detected. The behavioural-based detection can help to mitigate so-called zero-day attacks, which are newly created malicious software utilizing zero-day vulnerabilities with no threat information defining it. Only malicious software that is already identified and analysed can be detected using signature-based identification. Behavioural-based identification may help to alleviate the impact of so-called zero-day attacks and any newly evolved malicious software that exploits these vulnerabilities, even without prior threat information describing it. However, given the number of successful attacks and their impact, there is a need for quick and efficient methodologies to analyse malicious files and attachments.

Digital investigations are scientifically proven methodologies that lean towards examining and analyzing digital artifacts. The success of each methodology relies heavily on the competence of the investigators in manually analyzing large amounts of digital data to identify appropriate artifacts. This requires enormous analytical resources given the volume of data collected. The proliferation of cyber-crimes cases, Increases the demands and needs for new methodologies and paradigms to cope with such a critical situation. One of the emerging themes of digital investigations is threat hunting. Threat hunting can simply be defined as a preemptive methodology to cyber-defense that encompasses continuous and reiterated searches across internal systems for complex threats and possible vulnerabilities that evade current security controls. Threat hunting differs from the existing digital investigation methodologies, which utilize a passive approach, which detect cyber threats when pre-configured rules deployed within security solutions are violated. Threat hunting can immensely support the creation of a robust digital perimeter, which constantly tracks digital artifacts that could lead to the early detection of unusual threats or destructive activities on the local systems. This, in turn, can have a profound impact on protecting businesses and people.

In the past, threat actors utilized a wide range of malvertising campaigns linked to malevolent exploit kits to target selected victims. This vector resulted in a large number of attacks but less return in success. In recent years, threat actors have begun to utilized targeted attacks for specific high-value industries. These targets are tempting because they store in their servers a gigantic sensitive data and/or their digital operations are critical to the national infrastructure. Example for such common targets are governmental contractors, healthcare providers, and financial institutions. Targeted attacks against other enterprises and individuals are continuing at a less frequent rate [3]. The hunting for malicious activities relies on a combination of different utilities and techniques to gain different views of events occurring in the system. This helps the investigators to analyse and locate evidences regarding the potential maliciousness of a certain process or a file exists in the environment. The realization of the threat hunting relies heavily on the anticipated methodology that the investigators use to conduct a quick and efficient analysis of security threats on large amounts of unlinked data to determine appropriate evidence of misconduct, which in turn will require significant time given the amount of data involved in the hunt. The main aim of this research is to present an efficient threat hunting methodology that includes a combination of different utilities and techniques to facilitate the immediate detection of malicious binaries against the windows platform. The suggested methodology will have a significant impact on alleviating the need for performing a lengthy binary reverse engineering. Since only high confidence evidence of misconduct related to the malicious binaries is provided to the investigators. The presented methodology will help to resolve the uncertainty of evidence acquisition, as early copies of any malicious binaries or processes are analysed and compared together from multiple internal and external sources. This is considered to be beneficial in detecting malicious software propagation in the internal systems. Finally, the logs obtained from several security controls can collectively support the insights extracted from the proposed methodology, which in turn, makes the extracted evidence much stronger as well as ensures the validity of a threat to the internal systems.

In this research, a threat hunting methodology is suggested to explicitly extract significant information related to any binaries under investigation using multiple utilities that implement various techniques. This methodology has been designed to be carried on windows systems. The presented methodology will help investigators to easily follow a proactive approach for threat hunting tasks. Any suspicious binaries from different windows systems are collected and analysed using our proposed methodology to restrain any probable risks of data breaches or systems outages. During the analysis phases of the threat hunting process, any beneficial patterns can easily be found, which can be preserved to facilitate the classification process related to emerging threats. This paper has been organized as follows, In Sect. 2, relevant works were summarized. In Sect. 3 the proposed threat hunting methodology envisioned in windows systems is outlined along with the study of selected utilities and techniques. Discussions on static and dynamic hunting tools and techniques were presented in Sects. 4 and 5 respectively. Finally, the conclusions and future directions were given in Sect. 6.

2 Related Works

Presently, digital investigations are potent technical approaches used by nearly all large companies. The digital investigations' processes are usually conducted only after a violation has occurred. Recently, the focus of this domain is shifting from reactive approaches towards more proactive ones, where defense strategies would identify risks and flaws in the internal system to deter violations from emerging. An Earlier detection of any malicious actions or possible weaknesses, a greater chance of limiting or preventing any losses that could arise. This strategy can be defined as threat hunting, and has grown rapidly as a hot trend in the context of cyber-forensics. However, there is a noticeable lack of research studies on this modern approach. Each cyber defense provider appears to support its own interpretation of threat hunting in order to distinguish its own offering as a threat hunting system. This in essence, tends to increase the uncertainty regarding its definition. There are several common meanings that describe this concept established on the basis of its interpretation within the cyber-security context. For example, threat hunting can be described as the operation of finding adversaries in local systems before successfully carrying out attacks [4]. Sqrrl describes threat hunting as a systematic and iterative scanning across networks and databases to identify potential threats that can ultimately bypass current security controls [5]. For the context of this study, threat hunting can be briefly described as preventive practices that finds evidence of misconduct in local systems. The research in [6], a framework was proposed to models multi-stage attacks in a manner that defines the attack methods and the predicted consensuses of each attacks. The basic aim of their study is to simulate misconduct using the cyber kill-chain and intelligence driven defense patterns. Their suggested approach was implemented on Apache Hadoop. In [7], the authors introduced a method that uses text mining techniques to systematically compare the collected evidence of security-related incidents with a database of attack patterns. The goal of their work is to reduce the hunting time and improve the efficiency of attack detection. In [8] an approach was presented to incorporates the identification of systemic abnormalities from communication networks with the psychological profiling of users. Systematic anomaly recognition utilizes graph analysis and learning techniques to recognize systemic abnormalities in diverse knowledge networks, while psychological profiling dynamically blends human psychological characteristics with their behavioural trends. The authors continue to classify threats by connecting and rating the varied findings of systematic anomaly recognition and psychological profiling of the systems' users. The authors in [9], a component of threat identification has been incorporated in their research, which depends on the recognition of irregular variations in the weight of the edges over time. Wavelet decomposition technique was used to separate the transient behaviour from the stationary behaviour in these edges. The research in [10] Introduced the use of data mining techniques with visualization techniques to address various threats. In their study, two new simulation methods have been suggested to visualize threats. The authors in [11], a botnet discovery methodology has been designed, which uses cluster analysis to classify the correlation trends of C&C communication and behaviour flows. The methodology proposed in this work begins with sniffing the network traffic, then conducting two kinds of parallel investigations, one investigation is conducted to identify a group of hosts with identical traffic patterns, while the other examines packet payloads to identify anomalous behaviours. The behaviours are later pooled together to detect a swarm of hosts with similar malignant activity. The cross-correlation method is used to combine the findings of prior analyses into coherent classes of hostile hosts that may constitute a Botnet. In [12] introduced a threat hunting approach to implicitly elicit the relevant threat reputation groups from the multiple feeds of event logs. The proposed approach has been designed to be carried at the smart homeowner side, it also attains security, privacy for event logs which can help smart homeowners to easily adopt a proactive approach for threat hunting in a privacy-preserving manner. Two-stage concealment protocols, which was previously presented in [13–17], used for masking the event logs of smart homeowners when being released for threat hunting. A detailed survey in [18], presented a set of detection techniques for cryptographic ransomware. However, the authors have not centered their research on producing a practical methodology that links these detection algorithms to tools that can assist the investigators in the threat hunting process. Finally, the research work in [19] employed static and dynamic analysis to study WannaCry ransomware. The goal of this research is to uncover crucial information regarding encryption functionality, dynamic link libraries, and Windows API used by WannaCry. This can assist in implementing effective mitigation mechanisms to detect WannaCry and other strains of ransomware that have a similar profile. Most of the related work in this field has the same direction. The main aim of this work is to propose a general threat hunting methodology that is not tailored to a particular strain. Our proposed methodology provides the building blocks and popular methods that can assist the investigator in identifying evidence of misconduct related to the malicious binaries quickly and efficiently.

3   Proposed Threat Hunting Methodology

Malicious binaries serve the disruptive aims of the threat actors, and these aims range broadly from each payload to the next, from ransomware attacks on small to medium-sized businesses to large state-funded operations. Malicious payloads can also be distributed in a range of manners, such as manipulating browser vulnerabilities, insecure networking services, and social engineering. Cyber defense providers typically supply signatures to their security controls that try to align signatures against data located on the user's device. In order to build these signatures, malignant artifacts must first be detected and then assessed for patterns of misconduct, using two separate examination methods, static analysis, and dynamic analysis [20].

The static analysis mainly relies on decompilers where a reverse engineering process is conducted to convert the malicious binary into assembly or C code. Subsequently, the generated code can be rigorously examined to gain deeper insights into these patterns and artifacts that can be considered malicious. The malicious binary will never be executed during the various static analysis tools. Although static analysis alone generates useful insights regarding various functions of the binary under investigation, it can be used as an entry point for dynamic analysis, for example, it can be used to identify main functions that need further investigation using dynamic analysis tools [21]. Analyzing malicious binaries using static analysis is a complicated process that requires certain skill sets. Malicious software programmers use concealing strategies that can make it incredibly difficult for static analysis tools to read generated code and assess the control flow of its functionality, as well as resolve API calls at the run-time to make it harder to identify the main ones that can attribute any binary as malicious.

Dynamic analysis is conducted by executing the malicious binary in a restrained environment. It offers useful information about the binary under investigation, as the conducted examination can intercept API calls, and evaluate the memory contents assigned to the binary. Other valuable insights can also be gained when examining the requested URLs, files generated/updated during/after the execution, and registry keys inserted/edited during/after the execution. Currently, there are numerous research efforts underway to integrate artificial Intelligence with conventional dynamic analysis tools. This is achieved by using sandbox environments to construct detailed activity reports that could be used to detect harmful payloads that are not commonly identified with signature-based monitoring [22]. Performing dynamic analysis has multiple benefits, it especially aids in detecting any concealment process supported in the malicious binary, as it is not susceptible to evasion techniques. However, it does have a significant downside, as it requires the malignant binary to run on the system, and this is an essential step to ensuring that the malicious payload is fully implemented and the occurrence of all that all modifications into the system or changes to the file system are present. Due to that, it is highly recommended to carry out the dynamic analysis in a protected and isolated environment, such as containers or virtual systems, which can always be reverted to its original state. The workflow of the proposed methodology is presented in Fig. 1 depicted below. Although there are numerous tools that can be allocated to each stage of the proposed methodology, we have decided to select a set of tools, that have been found efficient and easy to handle in most of the scenarios encountered. The presented methodology combines a set of static and dynamic hunting tools to help investigators to easily follow a proactive approach for analyzing any suspicious binaries from different Windows systems. Beneficial patterns can easily be preserved, which can be utilized to facilitate the future classification of emerging threats.

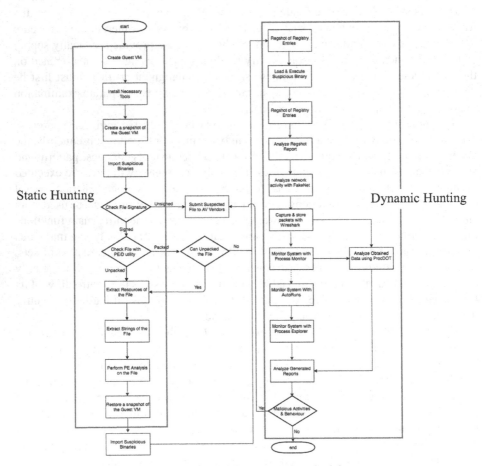

Fig. 1. Workflow of the proposed methodology

3.1 Selected Static Hunting Tools

A selected set of static hunting tools was employed to retrieve unique information from the binaries under investigation. Although there are multiple tools that can be utilized in the different stages of static hunting, only efficient tools and utilities that operate in a similar manner have been selected, and which can also extract useful data to complement other data generated by other selected tools as well. PE header extractors were used to collect header information of the portable executable samples, these tools are also capable of recognizing signatures of packers used to obscure the binary sample. The PEiD utility was used to identify the used packers, and compilers. The latest version of PEiD is equipped with detection algorithms that are capable of detecting more than 600 unique signatures of popular packers and compliers in binaries [23]. PE Tools is a lightweight executable control utility. It appeared for the first time in 2002 and contains a small set of tools such as process viewer and manager, HEX editor, EXE rebuilder and so on [24]. The PEview utility was utilized to retrieve a list of imported Windows APIs required to implement the binary. This will provide insights regarding the various functions inside the binary. for example, it can reveal if some functions within the binary were employed to manipulate the system's registry and file system. PEview can be used to inspect Windows 32-bit and 64-bit binaries to retrieve a list of imports required by its functions. It also presents the essential dependencies required to implement the binary as well as its debugging information and paths [25].

3.2 Selected Dynamic Hunting Tools

A selected set of dynamic hunting tools was employed to audit the behaviours of various processes and modules operating in the system environment. The process monitoring utility was utilized to check the behaviour of the current operating system. It can also track the behaviours of every main and child processes running in the system. It is also attainable for the process monitoring utility to document in real-time, file system manipulations, as well as changes to the registry. Process monitoring is a very useful utility in dynamic hunting. it would be beneficial to implement it in an isolated environment prior to executing the malicious binary, as it can be used to record system changes made by the malicious binary such as the filesystem or registry manipulation [26]. Process Explorer is another utility that has been utilized in dynamic hunting. It displays the currently active processes along with their profiles, called API and loaded DLLs. It is a valuable tool for detecting the type of APIs and DLLs are assigned to the binary under investigation to implement particular tasks [27]. ProcDOT utility [28] is used to take the output of Process Monitor and Wireshark, then visualize how a specific process is executed in the form of a flowchart. This present and interpret the collected pieces of data in order. This aids the investigator to know what has happed and in which order. Registry monitoring is an open-source utility that can be utilized to record a snapshot of the registry entries to be evaluated at a future stage with a second snapshot to highlight the modifications made to the registry entries due to certain events. It can be used to detect which updates in the registry entries have materialized when the malicious binary implemented [29]. FakeNet is a network emulation utility that has been used to mimic the communication infrastructure for tracking DNS requests and outbound IP connections. It is a valuable tool in assessing the network activity of binaries under investigation. It was employed in dynamic hunting to monitor and track DNS requests, detect any dropper URIs, and classify malignant external IP connections to C&C services [30]. Wireshark is a network packet sniffing that has been employed to record the IP traffic of various modules. It has a broad variety of functions, such as live logging and off-line examination. It is a valuable tool for recording traffic information from binaries under investigation for future evaluation or for use with other tools as well [31].

3.3   Selected Dataset for Threat Hunting

It is important to assess the effectiveness of the proposed methodology in detecting malignant artifacts of misconduct in any binary file under investigation, these artifacts can be preserved later to facilitate the classification process related to emerging threats. It is imperative to make use of a dataset made up of different malware families that are known to have many variants. Since there are no common data sets for malware families, VirusTotal [32] was employed to gather our own sample set. Few malwares and their reports can be obtained using the VirusTotal API. Hence, a selected sample set was gathered and analysed for the suitability of the inclusion on a weekly basis for the period of 3 months. We believe that this is a fair amount of time to gather a sample of malicious software that reflects the different Windows malware families that are known to have many variants. The following malware families were selected for our sample set, which includes OnlineGames, Bifrose, Delf, Lmir, Zbot, Zlob, Banload, Vundo, Yimfoca, Farfli, Banker, Vapsup, Swizzor, PcClient, and NSAnti. Some of these malware families have mechanisms to evade specific analysis techniques, which have been determined to be an advantage for testing the accuracy of the proposed methodology in real-world situations.

4   Discussion on Static Hunting Tools

The analysis platform comprised an open-source hosted hypervisor for x86 virtualization entitled Oracle VM VirtualBox with a virtual network in host-only network setting. Two virtual machines were created, one is a Windows 8.1 machine, and the other is Windows 7 machine. Windows 8.1 was selected to permit the proper execution of various analysis tools required for the proposed research. Moreover, minimal configurations in the selected tools will successfully modify them to report the results properly. The following programs will be employed for the path of the required analyses: sigcheck, 7zip, strings, PEview, ultimate packer for eXcutables, fakenet, regshot, autoruns, process monitor, and process explorer. The previously mentioned programs have been copied on specified virtual machine then installed. The sample files were installed in the pair of Windows machines. In addition, these Windows machines were fully backed up and isolated from the internet to assure their digital hygiene. Finally, the network was configured to connect to the internet through an intermediary VM running Parrot OS.

On one sample file, the signature can be extracted with a utility named 'sigcheck'. Extracting the file signature from an executable file is a quick and easy way of determining whether the file is malicious. If the file is not digitally signed, it is a clear indication that the executable has been modified. Running sigcheck against any benign file, will show legitimate information related to the digital signature of the file. However, running sigcheck against the sample file appearing to be a legitimate file indicates that the executable has not been digitally signed and is a strong indicator that the file may be malicious.

Malicious executables are usually packed by obfuscating their malicious payload, then combines the packed executable with deobfuscation code to create a self-extracting archive. When the obfuscated executable is executed, the deobfuscation code recreates the original code before executing it. Most obfuscation packers deobfuscate the original code into memory. As the obfuscated payload is not readable, packed executables are commonly used by malware to evade detection by security controls and to make it harder to be analysed. PEiD utility can be used to determine if obfuscation packer was used to obfuscate the executable, applying the PEiD utility on one of the sample files, obfuscation using *'ultimate packer for executables'* utility was detected. PEiD also shows other information such as the entry point and file offset of the original executable. The sample file can be deobfuscated using *'ultimate packer for excutables'* utility.

To extract the sample file, an archiving program such as 7zip was used to display the file structure of the executable. Many archiving programs have the ability to extract the resources of an executable file. Using this technique on legitimate executables displays its structure. However, applying the same technique against one of the sample files, displays its contents. Continuing to extract these executables to a separate folder, facilitates displaying their resources. The extracted resources can further be analysed.

'Strings' is command-line utility used to extract the ASCII representation of the hex values within the sample file. Only ASCII strings that are found to be 3 consecutive ASCII characters or longer are displayed. While this technique does extract a lot of false-positive information, it is a simple way of quickly extracting information from an executable. Running the strings command on the sample file extracted previously provides 3360 matches. Quickly looking at the output shows that sample file is likely to be some sort of keylogger as shown in Fig. 2. Examining the output in detailed, a set of IRC commands can be found within the extracted strings. It can be suggested that the keylogger will periodically log into an IRC server and upload some log files. Other

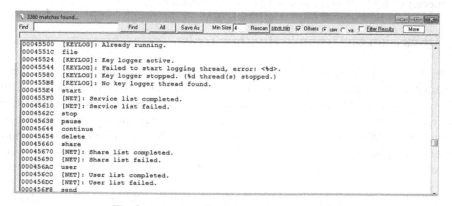

Fig. 2. Extracting the strings of the sample file

commands such as '*RegCreateKeyExA*' were spotted as well, which suggests that a set of registry keys are installed to provide persistence on the system PE analysis can be performed using PEview, that was used to extract the PE header information. Expanding the '.idata' section and selecting the import name table, a list of imports can be displayed as seen in Fig. 3.

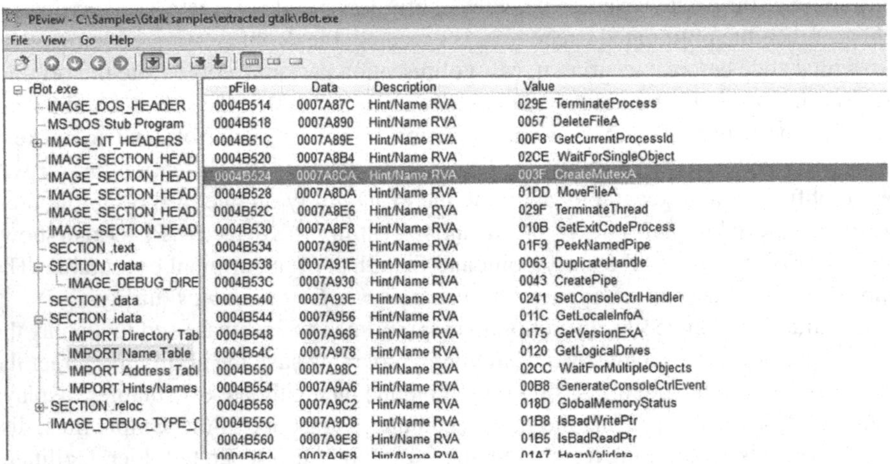

Fig. 3. List of imports of the sample file

There are several imports that can be found in the sample files that could be interesting for further analysis as. For example, the *CreateMutexA* import is commonly used by malware to set a value in a predetermined location on disk, in order to not infect the same target multiple times. It is important to note that legitimate programs may use *CreateMutexA* to ensure that only a single instance of the program is running on the system to prevent conflicts. Other imports such as *SetHandleCount* and *SetStdHandle* are used to create a handle to a process and redirect input, which can be used to redirect keystrokes from the user to a file. Additionally, the *ReadProcessMemory* import is used to interact with victim processes and extract information.

5   Discussion on Dynamic Hunting Tools

In the dynamic hunting step, the selected sample files were executed 10 times, clean snapshots of the VMs were acquired before each execution plan. After completion, all VMs are restored to the saved snapshots to obtain immaculate states in every attempt. The analysis platform was prepared as stated in the subsection, with two virtual machines, one is running Windows 7 and the other runs Windows 8.1. Both are running on a host only network, through an intermediary VM running Parrot OS. The Windows VMs were prepared with these IP addresses 172.16.22.10/70, while the Parrot VM was prepared with the IP address 172.16. 22.40.

To simulate a network, the FakeNet utility was used to allow the analyst to observe the network activity of the sample file within a safe environment. When FakeNet runs, it listens to multiple ports of multiple protocols. After running one of the sample files, an executable suspected to be a malware sample shows that a connection was attempted to IP address 58.65.232.68 on port 443 as shown in Fig. 4. Conducting a who-is lookup on the extracted IP address shows that it is an IP address originating from foreign network. An unsolicited outbound request to an external network suggests malicious activity and is a good indication that the executable is indeed a malicious binary.

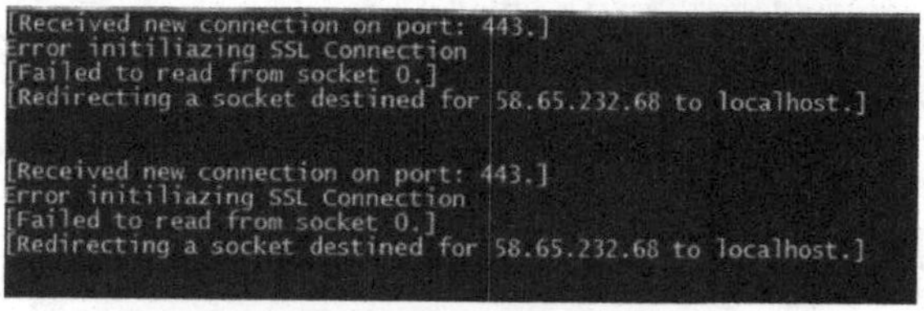

Fig. 4. FakeNet shows an unsolicited outbound request

To extract and monitor network traffic on a specific interface, the Wireshark utility is used within a safe environment. The utility has the ability to capture webpages and traffic and then store this data in a file that dumps all network elements that were transferred while this utility was running.

Regshot utility was employed to determine which registry values were added, altered or modified by the malicious binaries. All unnecessary services and applications running on the VM have been terminated before running of RegShot. A registry snapshot was taken from the registry before running the malicious binaries. After, the complete run of the malicious binary, a second registry snapshot was taken. The output of the second registry snapshot is compared with the output of the previously taken snapshot to determine which registry entries where modified, as well as folders and files within the operating system that have been added, altered or modified.

An example of a report of changes generated by Regshot is provided in Fig. 5, which shows that two registry keys were deleted, 1346 keys were created, 13 files have been modified and 246 folders have been added. Examining this report in detail, it is possible to detect where malicious files have been installed and which registry entries have been maliciously modified.

AutoRuns utility was used as start-up observer to monitor and list all processes that are set to run on the start-up programs. Snapshots of the current system configuration can be made and saved to a file then later compared to previous snapshots to determine changes. After the complete run of the sample file, the saved snapshot can be compared to the current start-up programs.

Process monitor utility is an advanced monitoring tool for windows, it is a vital utility to understand the changing behaviours of any running binaries. It excessively offers different views of the running processes by swapping between different filters that can depict multiple insights about the binary file in runtime. The process monitor can analyse and watch in real-time each running process, created files, accessed registries entries, etc. The utility is equipped with various filters to search for patterns of interest, it also has the ability to filter out all other processes that can interfere with the required results. Process monitor can be used to record all operations completed by any running

```
------------------------------------
Keys deleted: 2
------------------------------------
HKLM\SYSTEM\ControlSet001\services\PROCMON23\Enum
HKLM\SYSTEM\CurrentControlSet\services\PROCMON23\Enum

------------------------------------
Keys added: 1346
------------------------------------
HKLM\SOFTWARE\Microsoft\Windows\Currentversion\explorer\AutoplayHandlers\Handlers\VLCPlayCDAudioOnArrival
HKLM\SOFTWARE\Microsoft\Windows\Currentversion\explorer\AutoplayHandlers\Handlers\VLCPlayDVDAudioOnArrival
HKLM\SOFTWARE\Microsoft\Windows\Currentversion\explorer\AutoplayHandlers\Handlers\VLCPlayDVDMovieOnArrival
HKLM\SOFTWARE\Microsoft\Windows\Currentversion\explorer\AutoplayHandlers\Handlers\VLCPlayMusicFilesOnArrival
HKLM\SOFTWARE\Microsoft\Windows\Currentversion\explorer\AutoplayHandlers\Handlers\VLCPlaySVCDMovieOnArrival
HKLM\SOFTWARE\Microsoft\Windows\Currentversion\explorer\AutoplayHandlers\Handlers\VLCPlayVCDMovieOnArrival
HKLM\SOFTWARE\Microsoft\Windows\Currentversion\explorer\AutoplayHandlers\Handlers\VLCPlayvideoFilesOnArrival

------------------------------------
Files [attributes?] modified: 13
------------------------------------
C:\Users\scientwolf\AppData\Local\Microsoft\Windows\History\History.IE5\index.dat
C:\Users\scientwolf\AppData\Local\Microsoft\Windows\Temporary Internet Files\Content.IE5\index.dat
C:\Users\scientwolf\AppData\Local\Microsoft\Windows\UsrClass.dat
C:\Users\scientwolf\AppData\Local\Microsoft\Windows\UsrClass.dat.LOG1
C:\Users\scientwolf\AppData\Roaming\Microsoft\Windows\Cookies\index.dat
C:\Users\scientwolf\AppData\Roaming\Microsoft\Windows\Recent\AutomaticDestinations\1b4dd67f29cb1962.automaticDestinations-ms
C:\Users\scientwolf\NTUSER.DAT
C:\Users\scientwolf\ntuser.dat.LOG1
C:\Windows\AppCompat\Programs\RecentFileCache.bcf
C:\Windows\ServiceProfiles\LocalService\AppData\Local\Microsoft\Windows\History\History.IE5\index.dat
C:\Windows\ServiceProfiles\LocalService\AppData\Local\Microsoft\Windows\Temporary Internet Files\Content.IE5\index.dat
C:\Windows\ServiceProfiles\LocalService\AppData\Roaming\Microsoft\Windows\Cookies\index.dat
C:\Windows\System32\LogFiles\Scm\eaca24ff-236c-401d-a1e7-b3d5267b8a50

------------------------------------
Folders added: 246
------------------------------------
C:\Program Files (x86)\VideoLAN
C:\Program Files (x86)\VideoLAN\VLC
C:\Program Files (x86)\VideoLAN\VLC\locale
```

Fig. 5. Report of changes generated by Regshot

processes. By setting process monitor to capture data and then running the malicious binary sample, all the performed operations can be captured. If the malicious payload is hidden within a legitimate installer, filters could be applied to filter out any unneeded process operations. The 'process tree' view can be selected where the legitimate process is displayed along with the malicious payload.

Process Explorer is a powerful utility for managing windows processes. It can be used to present insights about all the processes running in the system. Every running process is displayed in a tree-like structure that shows the relations between parent and child processes.

Process Explorer is considered an advanced task manager with enhanced features. Some of these features are the hierarchical coloured view of processes which simplify the analysis, the ability to identify input/output operation of a specific process such as opening or locking specific file or folder and loading a DLL, the ability to terminate or suspend any specific process tree with all of its spawned processes, and smooth integration with VirusTotal database that permits submitting and comparing cryptographic hashes of all running executables against those stored on VirusTotal then displays the number of detections without needing to isolate each executable file and then upload it separately for review. The granularity of information offered by process explorer can help the analyst to trace DLL versioning and memory leaks problems for any executable file. An example of an output generated by process explorer is provided in Fig. 6, a process name 'newbos2.exe' (that is a new child process of 'explorer.exe' process) has 47 out of 55 detections. This sharp detection rate confidently indicates that this executable is malicious. The Properties window of the malicious binary can provide additional useful information to the investigator such as, the character strings in memory, the user under whom the process is being executed, the active network connections, active threads, and the location of the executable pertaining to the process on the desk. In order to reduce the interpretation time for the investigator, the aggregated traces of process monitor and the collected network traffic dump can be visualized by ProcDOT as seen in Fig. 7. This application can generate a call graph that represents behavior of the sample under investigation by combining the previously collected data.

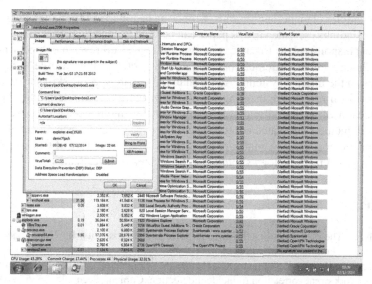

Fig. 6. An output generated by 'Process Explorer'

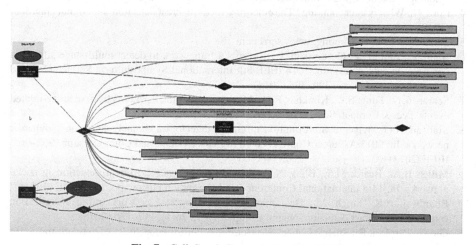

Fig. 7. Call Graph Generated by 'ProcDOT

6 Conclusions and Future Work

This research presents our efforts to propose a methodology to distinguish malicious binaries from benign binaries in a quick and efficient manner. The proposed methodology consists of static and dynamic hunting techniques. Using these two techniques, the proposed methodology is not only capable of identifying a range of signature-based anomalies but also to pinpoint behavioural anomalies that arise in the operating system when malicious binaries are triggered. The proposed threat hunting methodology was applied to samples of malicious binaries, which can be found in common malware repositories. The findings presented in this paper can be used to construct other methodologies and incident response plans for other emerging threats. Full reverse engineering of the malicious binary can be extremely beneficial. However, it is not feasible to do this for all binaries on the internal systems. Additionally, there are specific skill sets required to properly perform reverse engineering for a binary. More research is required on automatic detection of advanced malware using artificial intelligence techniques. Future research will include automated approaches to uncover various concealment algorithms and evasion methods used by malware developers. The integration of Cuckoo Sandbox with the proposed methodology should also be an area of investigation.

References

1. Dowdy, J.: The cyber-security threat to us growth and prosperity. In: Cyberspace: A New Domain for National Security (2012)
2. Friedberg, I., Skopik, F., Settanni, G., Fiedler, R.: Combating advanced persistent threats: from network event correlation to incident detection. Comput. Secur. **48**, 35–57 (2015)
3. Connolly, L.Y., Wall, D.S.: The rise of crypto-ransomware in a changing cybercrime landscape: taxonomising countermeasures. Comput. Secur. **87**, (2019)
4. Lord, N.: What is threat hunting? The emerging focus in threat detection. In: Digital Guardian (2018)
5. Sqrrl. Cyber Threat Hunting. www.sqrrl.com
6. Bhatt, P., Yano, E.T., Gustavsson, P.: Towards a framework to detect multi-stage advanced persistent threats attacks. In: 2014 IEEE 8th International Symposium on Service Oriented System Engineering, pp. 390–395 (2014)
7. Scarabeo, N., Fung, B.C., Khokhar, R.H.: Mining known attack patterns from security-related events. Peer J. Comput. Sci. **1**, (2015)
8. Mahyari, A.G., Aviyente, S.: A multi-scale energy detector for anomaly detection in dynamic networks. In: 2013 Asilomar Conference on Signals, Systems and Computers, pp. 962–965. IEEE (2013)
9. Miller, B.A., Beard, M.S., Bliss, N.T.: Eigenspace analysis for threat detection in social networks. In: 14th International Conference on Information Fusion, pp. 1–7. IEEE (2011)
10. Bhardwaj, A.K., Singh, M.: Data mining-based integrated network traffic visualization framework for threat detection. Neural Comput. Appl. **26**(1), 117–130 (2015)
11. Gu, G., Perdisci, R., Zhang, J., Lee, W.: Botminer: clustering analysis of network traffic for protocol-and structure-independent botnet detection (2008)
12. Elmisery, A.M., Sertovic, M.: Privacy preserving threat hunting in smart home environments. In: Anbar, M., Abdullah, N., Manickam, S. (eds.) Advances in Cyber Security (ACeS 2019) Communications in Computer and Information Science, vol. 1132, pp. 104–120. Springer, Singapore (2020). https://doi.org/10.1007/978-981-15-2693-0_8
13. Elmisery, A.M., Botvich, D.: Privacy aware recommender service using multi-agent middleware-an IPTV network scenario. Informatica **36**(1) (2012)
14. Elmisery, A.M., Rho, S., Botvich, D.: A fog based middleware for automated compliance with OECD privacy principles in internet of healthcare things. IEEE Access **4**, 8418–8441 (2016)
15. Elmisery, A.M., Rho, S., Botvich, D.: Collaborative privacy framework for minimizing privacy risks in an IPTV social recommender service. Multimedia Tools Appl. **75**(22), 14927–14957 (2016)

16. Elmisery, A.M., Botvich, D.: Enhanced middleware for collaborative privacy in IPTV recommender services. J. Converg. **2**(2), 10 (2011)
17. Elmisery, A.M., Doolin, K., Roussaki, I., Botvich, D.: Enhanced middleware for collaborative privacy in community based recommendations services. In: Yeo, S.S., Pan, Y., Lee, Y., Chang, H. (eds.) Computer Science and its Applications. Lecture Notes in Electrical Engineering, vol. 203, pp. 313–328. Springer, Dordrecht (2012)
18. Berrueta Irigoyen, E., Morató Osés, D., Lizarrondo, M., Izal Azcárate, M.: A survey on detection techniques for cryptographic ransomware. IEEE Access **7**, 144925–144944 (2019)
19. Akbanov, V.G., Vassilakis, I.D. Moscholios, Logothetis, M.D.: Static and dynamic analysis of WannaCry ransmware
20. Aman, W.: A framework for analysis and comparison of dynamic malware analysis tools. Int. J. Netw. Secur. Its Appl. **6**(5), 63–74 (2014). arXiv preprint arXiv:1410.2131
21. Wichmann, B.A., Canning, A., Clutterbuck, D., Winsborrow, L., Ward, N., Marsh, D.: Industrial perspective on static analysis. Softw. Eng. J. **10**(2), 69–75 (1995)
22. Firdausi, I., Erwin, A., Nugroho, A.S.: Analysis of machine learning techniques used in behavior-based malware detection. In: 2010 Second International Conference on Advances in Computing, Control, and Telecommunication Technologies, pp. 201–203. IEEE (2010)
23. Snaker (ed.): Softpedia (2008). https://www.softpedia.com/get/Programming/Packers-Crypters-Protectors/PEiD-updated.shtml
24. Petoolse (ed.): Github (2018). https://github.com/petoolse/petools
25. Miller, S. (ed.): Dependency walker (2015). http://www.dependencywalker.com
26. Microsoft (ed.): Process explorer (2019). https://docs.microsoft.com/en-us/sysinternals/downloads/process-explorer
27. Microsoft (ed.): Process monitor. https://docs.microsoft.com/en-us/sysinternals/downloads/procmon
28. Wojner, C. (ed.): ProcDOT, a new way of visual malware analysis. Austrian National CERT (2015). https://www.procdot.com/
29. Maddes, X. (ed.): Regshot download (2018). https://sourceforge.net/projects/regshot/
30. Hungenberg, T., Eckert, M. (ed.): INetSim: internet services simulation suite (2013)
31. Wireshark, F.: Wireshark-Go Deep, vol. 15. Retrieved Oct 2011
32. Sistemas, H. (ed.): VirusTotal (2004). https://www.virustotal.com/gui/

Peer-to-Peer Application Threat Investigation

Mohamed Mahdy[✉]

OWASP Cairo-Chapter, Cairo, Egypt
mohamed.mahdy@owasp.org

Abstract. Understanding the layers of an application leads to a better threat investigation outcome as well as helping with developing proper controls with optimized cost. Starting with the blockchain as the peer-to-peer application we will analyze the peer-to-peer networks and how they provide the underlay for blockchain. We'll have a look at the layers of the peer-to-peer application starting from the network layer, communication flows and communication ports through analyzing packet captures collected from both client side and network taps, up to the blockchain layer and client-side processes where we start to have a look at imported functions, memory and CPU utilization. We aim to have a structured approach for threat investigation for peer-to-peer applications.

Keywords: Blockchain · Cryptocurrency · Mechanism design · Peer-to-peer · Mining · Threat investigation · Cyber threats · API

1 Introduction

1.1 Blockchain and P2P Connected

Blockchain concept started to gain attention back in 2008 with the introduction of Bitcoin by S. Nakamoto in [1]. Blockchain uses peer-to-peer networks as an underlay and provides a universal data set that every peer can trust even though they might not know or trust each other. It provides a shared and trusted chain of blocks, where encrypted and immutable copies of this chain are shared with other network peers. It can be considered both a target and a medium for initiating sophisticated cyber-attacks.

1.2 Blockchain Investigation Challenges

The main concern around how blockchain affect endpoints pushes enterprises toward investing in endpoint solution as the main control to secure or investigate blockchain systems.

While endpoint solutions provide tremendous amount of information and provide the capability to intercept function calls at a very low level within endpoint memory, the cost associated with such deployments is very high specially with the Industrial Control systems (ICS) in the picture.

Enterprises either rely solely on the endpoint solutions or ignore the threat related to blockchain systems as the cost of securing it is very high.

© Springer Nature Switzerland AG 2021
H. Hacid et al. (Eds.): ICSOC 2020 Workshops, LNCS 12632, pp. 642–650, 2021.
https://doi.org/10.1007/978-3-030-76352-7_55

In this research we are going through the layers contributing to the blockchain system, to identify how different defenses and controls can be implemented at each layer.

This approach helps more with identifying at which kill chain stage the organization team can engage based on the existing controls and how they can make the best use of those controls to build a proper well-informed hypothesis.

2 Background

2.1 Blockchain Changed the Game

Blockchain solved challenges that faced previous implementations of digital payment systems (e.g., Digicash) see [2]. Some of these challenges are listed below.

- Sybil attack: This attack targets system reputation by creating large number of pseudo nodes to use them to gain a malicious influence, see [2]
- Double spending: Digital information can be used multiple times by moving or copying them. Unlike physical assets (e.g., bills, coins, etc.), they require expensive effort to be copied.

Blockchain made it possible to trust the output of the system without the need to trust any of its members and removed the need for third parties like banks and authorities, see [2].

2.2 Blockchain Main Components

- Distributed Ledger [3]: It can be described as a chain of blocks, where each block contains some transactions and a hash of the prior block, this guarantees the chain integrity back to the first block (genesis block).
- Proof-of-work: This enables distributed control and statistical protection for the blockchain system which makes it very expensive to cheat the system solving the double spending problem.

Blockchain makes use of the 3 aspects; P2P network, Game Theory and Cryptography as illustrated in Fig. 1.

(1) P2P Network (Peer-to-peer network): The concept of the decentralized applications in peer-to-peer networks is the underlay for applications like blockchain. P2P provides the communication channel on top of which blockchain runs the required protocols, and on top of the blockchain we have the relative application protocol (e.g., bitcoin protocol).
(2) Inverse game theory (mechanism design) [4]: It's related to economic theory, where in normal situations economists try to predict outcomes that are generated by certain processes. In mechanism design it's the opposite; we start with the outcomes and work backward to design a system and mechanisms to lead to those outcomes.
(3) Cryptography: It uses public key cryptography and hash functions to preserve system integrity.

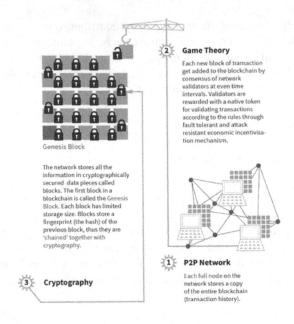

Genesis Block

The network stores all the information in cryptographically secured data pieces called blocks. The first block in a blockchain is called the Genesis Block. Each block has limited storage size. Blocks store a fingerprint (the hash) of the previous block, thus they are 'chained' together with cryptography.

2 **Game Theory**

Each new block of transaction get added to the blockchain by consensus of network validators at even time intervals. Validators are rewarded with a native token for validating transactions according to the rules through fault tolerant and attack resistant economic incentivisation mechanism.

1 **P2P Network**

Each full node on the network stores a copy of the entire blockchain (transaction history).

3 **Cryptography**

From the Book "**Token Economy**" by Shermin Voshmgir, 2019
Excerpts available on **https://blockchainhub.net**

Fig. 1. Blockchain components

2.3   Tokens and Blockchain

Tokens are not a new concept; it represents any form of economic value. Daily life examples of tokens are vouchers and bonus points.

In blockchain implementations, digital tokens are used to reward the peers contributing their resources to the network.

This is how the transaction in Fig. 2 flows.

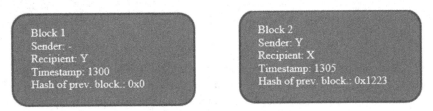

Block 1
Sender: -
Recipient: Y
Timestamp: 1300
Hash of prev. block.: 0x0

Block 2
Sender: Y
Recipient: X
Timestamp: 1305
Hash of prev. block.: 0x1223

Fig. 2. Blockchain flow example

1. Person X creates a transaction.
2. Transaction is sent to the peers.
3. The network determines the difficulty of validating the transaction.
4. Miners start their work to validate the block of transactions.

5. Each miner is rewarded with a token and the ledger is updated.
6. Ledger (The blockchain list) will contain Y:1, X:1, Z:3 with X, Y acting as participants, Z represents a node that was rewarded with tokens for the work done (proof-of-work).

3   Blockchain Traffic Analysis

In this section we will go through the traffic types within blockchain systems, examples used in this section can be found in [5]. Let's discuss the traffic types within the blockchain network as in Fig. 3.

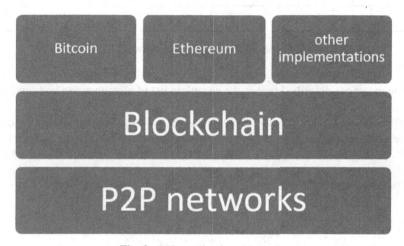

Fig. 3. P2P applications hierarchy

- P2P network messages similar to the below:

  - Connect messages: These messages are used to negotiate the version and send the block, the timestamp and addresses within the transaction scope.
  - Initial block download message: This is used to Sync and exchange the blocks between peers.
  - Relay message: This is used after inventory checks for the blocks and validating the transaction to confirm chain validity.

- API Calls: They are used for communication between network peers.
- Mining process: This will be running at peer side to calculate the hashes for tasks like transaction creation, validation, etc.

4   Blockchain Threat Matrix

After this quick summary on blockchain system history and traffic types. We'll discuss a more structured approach to have our network ready for threat investigation activities.

4.1 Network Layer

Traffic capture gives us a great amount of information by inspecting protocol level messages as in [6]. Wireshark has some specific blockchain related filters (e.g., bitcoin protocol).

Peers stat to connect by exchanging a set of control messages, starting with control messages as in Fig. 4.

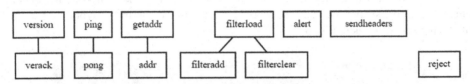

Fig. 4. P2P control messages

Some messages described below, the complete list of messages can be found at [7]:

1. Version message: This message contains the version to be used, block and the current time, this message provides information about the sender node to the recipient node.
   ```
   72110100 ......................... Protocol version:
   70002
   0100000000000000 ................... Services:
   NODE_NETWORK
   bc8f5e5400000000 .................. [Epoch time][unix
   epoch time]: 1415483324
   ```
2. Verack message: This message is sent from the recipient node to the sender node to acknowledge the version message sent previously, this message has no payload, similar to the below.
   ```
   f9beb4d9 ................... Start string: Mainnet
   76657261636b000000000000 ... Command name: verack + null
   padding
   00000000 .................. Byte count: 0
   5df6e0e2 .................. Checksum:
   SHA256(SHA256(<empty>))
   ```
3. Address message: Once above messages are successfully exchanged, the client sends the address message to get the remaining peers addresses.

```
fde803 ............................. Address count: 1000

d91f4854 ........................... [Epoch time][unix
epoch time]: 1414012889
0100000000000000 ................... Service bits: 01
([network][network] node)
00000000000000000000ffffc0000233 ... IP Address:
::ffff:192.0.2.51
208d ............................... Port: 8333


[...] ............................... (999 more addresses
omitted)
```

Once Version and address messages are completed, the peers start the data related communication as in Fig. 5.

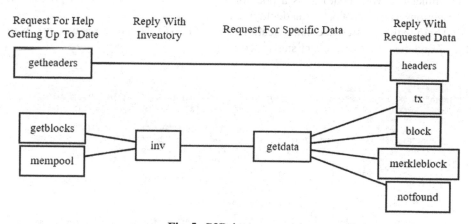

Fig. 5. P2P data messages

4.2 API Layer

In addition to network captures, inspecting REST API calls as in Fig. 6 in blockchain system allows for more information to be available during threat investigation activities.

As blockchain started to be used in private deployments (banking sector, government-related transactions, etc.), the use of API gateways between network peers shall give more visibility over the transactions.

API gateways can provide additional security to this type of traffic, via leveraging some ciphers that can't be supported by the network peers and can provide more availability in the network, more about API management in [8].

```
POST / HTTP/1.1
Host: 127.0.0.1
Connection: close
Authorization: Basic X19jb29raWVfXzo2OWE5OGQxMWU1YTMyMzg4MjNiM
Content-Length: 61

{"method":"getnewaddress","params":["sharkfest2018"],"id":1}
HTTP/1.1 200 OK
Content-Type: application/json
Date: Mon, 25 Jun 2018 04:21:12 GMT
Content-Length: 68
Connection: close

{"result":"3KWrA6cuQK2CgEPqQFBTvBuMLMKCr8HqEx","error":null,"i
```

Fig. 6. API capture

4.3 Mining

The miner (network node) joins a pool in a peer-to-peer network that's processing transactions as part of blockchain deployment.

There're three different communications stages, two of them over the network as in Fig. 7 and the third one is client-side processing.

```
{"method":"login","params":
{"login":"47oXJPtiqdjc3Dm7U8HWRt5jaCfVn1vD1iWptGcj7oEvQDMm2mwi3REXFWPhe7YcTrjXuiu2fw61tbjduxgEt3c2
7UjGedC","pass":"x","rigid":"sharkfest2018","agent":"xmr-stak/2.4.5/b3f79de/master/lin/cpu/aeon-
cryptonight-monero/20"},"id":1}
{"id":1,"jsonrpc":"2.0","error":null,"result":{"id":"6daf6419-c290-4e28-a49d-b486ff610d96","job":
{"blob":"0707dc8eb0d9053835eef85a228e398876a095ccd1fca8541170f381ebf2c5bc147444682ab23e00000000f28
f1abe51e0298e696f2d37d5deafd4a4299f684a13ff60dc78aa373ba36ede02","job_id":"rFD94RL/61pxfdf8G1cmigj
wMUrX","target":"711b0d00","id":"6daf6419-c290-4e28-a49d-b486ff610d96"},"status":"OK"}}
{"method":"submit","params":{"id":"6daf6419-c290-4e28-a49d-
b486ff610d96","job_id":"rFD94RL/61pxfdf8G1cmigjwMUrX","nonce":"d3120000","result":"381dad5f4194b92
d386d262e798d4749bd1c812e1c71a2692a405c3650260300"},"id":1}
{"id":1,"jsonrpc":"2.0","error":null,"result":{"status":"OK"}}
{"jsonrpc":"2.0","method":"job","params":
{"blob":"0707dc8eb0d9053835eef85a228e398876a095ccd1fca8541170f381ebf2c5bc147444682ab23e000000004db
f3706ea02c415daec09f1b79bade3214eaf4e99710b2055be2fab7b6d809702","job_id":"ZpDmTztAEu7xtXWa0O10QQB
DbFUo","target":"b6600b00","id":"6daf6419-c290-4e28-a49d-b486ff610d96"}}
{"jsonrpc":"2.0","method":"job","params":
{"blob":"0707a88fb0d905d55f3ef40bdec0efd83b3d5b9233fb38ccf9b32cec95f18d471f330ab201b20500000000a8f
3b2dd2205d06b978b1c569795f9fe3c3d6e872b3ea2b99925b86cd50379e705","job_id":"yL6suvdS0R8pbmQ02dMBXxA
IRGEb","target":"b6600b00","id":"6daf6419-c290-4e28-a49d-b486ff610d96"}}
{"jsonrpc":"2.0","method":"job","params":
{"blob":"0707a88fb0d905d55f3ef40bdec0efd83b3d5b9233fb38ccf9b32cec95f18d471f330ab201b205000000000dc
86e4bfabf4350d1855d972dd65f02fb31bd8007e28dc51148245999b7d5ad05","job_id":"hrGEvxG/+Tw18Z5vY6sJSfL
oK24t","target":"711b0d00","id":"6daf6419-c290-4e28-a49d-b486ff610d96"}}
```

Fig. 7. Mining capture

- Network node login to the pool of miners.
- Processing start at this new node to validate transactions or solve the shared hash, more about hash algorithms in [9].
- Once done the solution is shared over the network to the peers.

5 Conclusion

We broke down the blockchain application into network and application layers and client-side traces. Traffic flow artifacts – open ports, IP addresses, protocol level messages and application-level messages – were extracted using packet captures collected either from client node or network taps. Client-side artifacts related to mining processes and imported functions were extracted from investigating client machine, memory and CPU utilization and performance.

This structured approach shows the different layers for peer-to-peer application (e.g., blockchain) and how each layer provides a valuable input to securing or investigating such applications.

Enterprises don't need to invest deeply in sophisticated end point solutions to secure or investigate P2P applications, they can make a perfect use of analyzing existing network flows, API calls and client machines performance stats to build a well-informed hypothesis around their security posture with respect to P2P applications.

In Table 1, we summarized the communication types within the blockchain system, the collected traces and the suggested tools. This will allow for a more structured approach for blockchain applications threat investigation.

Table 1. Blockchain threat matrix

	P2P network	API calls	Mining
Network traces	Protocol level communication (IP: Port)	Unencrypted REST API calls	Protocol level communication (IP: Port)
Client-side traces	Listening ports	HTTP/HTTPS calls Listening ports	Mining processes Listening ports
Memory/Processes level traces	Network processes running	HTTP/HTTPS processes	Mining & Network processes running
Suggested tools' functions to be used	Network threat analytics SIEM solutions that collect network traces L4/L7 Protocol level inspection	API gateway for private deployments Proxy for external HTTP/HTTPS traffic HTTP/HTTPS level inspection	Network related traces as in P2P communication monitoring Process monitoring for mining activities

Table 1 can help a lot for the preparing for malicious activities as well as in having through visibility that supports future threat investigation activities.

References

1. Nakamoto, S.: Bitcoin: A Peer-to-Peer Electronic Cash System (2008)
2. Buford, J., Yu, H., Lua, E.: P2P Networking and Applications (2008)

3. Voshmgir, S.: Token Economy: How Blockchains and Smart Contracts Revolutionize the Economy, 2nd, BlockchainHub Berlin (2020)
4. Maskin, E.: Introduction to mechanism design and implementation. Trans. Corp. Rev. **11**(1), 1–6 (2019)
5. Traffic analysis of cryptocurrency and blockchain network. https://sharkfestus.wireshark.org/assets/presentations18/31.pdf
6. Bitcoin Protocol documentation. https://en.bitcoin.it/wiki/Protocol_documentation
7. P 2P Network. https://developer.bitcoin.org/reference/p2p_networking.html#addr
8. Brajesh De.: API Management: An Architect's Guide to Developing and Managing APIs for Your Organization, Apres (2017)
9. Bitcoin mining the hard way. http://www.righto.com/2014/02/bitcoin-mining-hard-way-algorithms.html

Author Index

Printed in the United States
by Baker & Taylor Publisher Services